Hydrogen Supply (

Design, Deployment and Opera

Hydrogen Supply Chain

Design, Deployment and Operation

Edited by

Catherine Azzaro-Pantel

ACADEMIC PRESS

An imprint of Elsevier

Academic Press is an imprint of Elsevier
125 London Wall, London EC2Y 5AS, United Kingdom
525 B Street, Suite 1650, San Diego, CA 92101, United States
50 Hampshire Street, 5th Floor, Cambridge, MA 02139, United States
The Boulevard, Langford Lane, Kidlington, Oxford OX5 1GB, United Kingdom

Notices

Knowledge and best practice in this field are constantly changing. As new research and experience broaden our understanding, changes in research methods, professional practices, or medical treatment may become necessary.

Practitioners and researchers must always rely on their own experience and knowledge in evaluating and using any information, methods, compounds, or experiments described herein. In using such information or methods they should be mindful of their own safety and the safety of others, including parties for whom they have a professional responsibility.

To the fullest extent of the law, neither the Publisher nor the authors, contributors, or editors, assume any liability for any injury and/or damage to persons or property as a matter of products liability, negligence or otherwise, or from any use or operation of any methods, products, instructions, or ideas contained in the material herein.

Library of Congress Cataloging-in-Publication Data

A catalog record for this book is available from the Library of Congress

British Library Cataloguing-in-Publication Data

A catalogue record for this book is available from the British Library

ISBN 978-0-12-811197-0

Publisher: Mara Conner
Acquisition Editor: Raquel Zanol
Editorial Project Manager: Mariana L. Kuhl
Production Project Manager: Anitha Sivaraj
Designer: Victoria Pearson

Typeset by SPi Global, India

Contents

Contributors

Catherine Azzaro-Pantel (3, 37, 349), Laboratoire de Génie Chimique, Université de Toulouse, CNRS, Toulouse, France

Lars Baetcke (167), Hamburg University of Technology (TUHH), Institute of Environmental Technology and Energy Economics (IUE), Hamburg, Germany

Mohamed Becherif (293), UBFC-UTBM, FCLab FR CNRS 3539, FEMTO-ST UMR CNRS 6174, Belfort Cedex, France

Olivier Bethoux (293), LGEP, University of Paris Sud 11, Gif sur Yvette Cedex, France

Cyril Bourasseau (271), LITEN, CEA, DRT, Grenoble Cedex 9, France

Katy Cabaret (293), UBFC-UTBM, FCLab FR CNRS 3539, FEMTO-ST/RECITS UMR CNRS 6174, Belfort Cedex, France

Camille Cany (271), I-tésé, CEA, DAS; Laboratoire Genie Industriel, CentraleSupélec, Université Paris-Saclay, Gif-sur-Yvette, France

Bela Csukas (461), Research Group on Process Network Engineering, Institute of Methodology, Kaposvar University, Kaposvar, Hungary

Sofía De-León Almaraz (3, 37, 349), Laboratoire de Génie Chimique, Université de Toulouse, CNRS, Toulouse, France

Chun Deng (423), State Key Laboratory of Heavy Oil Processing, College of Chemical Engineering, China University of Petroleum, Beijing, China

Michael Ehrenstein (389), Centre for Process Systems Engineering (CPSE), Imperial College London, London, United Kingdom

Xiao Feng (423), School of Chemical Engineering & Technology, Xi'an Jiaotong University, Xi'an, China

Didier Grouset (111, 207), Université de Toulouse, IMT Mines Albi, UMR CNRS 5302, Centre RAPSODEE, Campus Jarlard, Albi, France

Gonzalo Guillén-Gosálbez (389), Centre for Process Systems Engineering (CPSE), Imperial College London, London, United Kingdom

Benjamin Guinot (271), LITEN, CEA, DRT, Grenoble Cedex 9, France

Zaza Nadja Lee Hansen (521), Department of Management Engineering, Division of Management Sciences, Denmark's Technical University, Kongens Lyngby, Denmark

Wang Jingdai (405), State Key Laboratory of Chemical Engineering, Department of Chemical and Biological Engineering, Zhejiang University, Hangzhou, PR China

Martin Kaltschmitt (81, 167), Hamburg University of Technology (TUHH), Institute of Environmental Technology and Energy Economics (IUE), Hamburg, Germany

Yuki Kudoh (499), Research Institute of Science for Safety and Sustainability, National Institute of Advanced Industrial Science and Technology, Tsukuba, Japan

Alain Le Duigou (271), I-tésé, CEA, DAS, Université Paris Saclay, Gif-sur-Yvette, France

Jochen Linssen (309), Institute of Energy and Climate Research—Systems Analysis and Technology Evaluation (IEK-STE), Forschungszentrum Julich, Julich, Germany

Paul Lucchese (271), CEA and Capenergies, Domaine du Petit Arbois, Aix-en-Provence Cedex 4, France

Christine Mansilla (271), I-tésé, CEA, DAS, Université Paris Saclay, Gif-sur-Yvette, France

Frank Markert (521), Department of Civil Engineering, Section of Building Design, Denmark's Technical University, Kongens Lyngby, Denmark

Ange Nzihou (111), Université de Toulouse, IMT Mines Albi, UMR CNRS 5302, Centre RAPSODEE, Campus Jarlard, Albi, France

Jesus Ochoa Robles (3, 37, 349), Laboratoire de Génie Chimique, Université de Toulouse, CNRS, Toulouse, France

Akito Ozawa (499), Research Institute of Science for Safety and Sustainability, National Institute of Advanced Industrial Science and Technology, Tsukuba, Japan

Doan Pham Minh (111), Université de Toulouse, IMT Mines Albi, UMR CNRS 5302, Centre RAPSODEE, Campus Jarlard, Albi, France

Thanh Son Phan (111), Université de Toulouse, IMT Mines Albi, UMR CNRS 5302, Centre RAPSODEE, Campus Jarlard, Albi, France

Fabienne Picard (293), UBFC-UTBM, FCLab FR CNRS 3539, FEMTO-ST/RECITS UMR CNRS 6174, Belfort Cedex, France

Haithman Ramadan (293), UBFC-UTBM, FCLab FR CNRS 3539, FEMTO-ST UMR CNRS 6174, Belfort Cedex, France; Zagazig University, Faculty of Engineering, Electrical Power and Machines Dept., Zagazig, Egypt

Cyrille Ridart (111, 207), HERA—ALBHYON, Technopole Innoprod, rue Pierre Gilles de Gennes, Albi, France

Anne Rödl (81), Hamburg University of Technology (TUHH), Institute of Environmental Technology and Energy Economics (IUE), Hamburg, Germany

Tan Ji Siang (111), Faculty of Chemical and Natural Resources Engineering, University Malaysia Pahang, Lebuhraya Tun Razak, Gambang, Kuantan, Malaysia

Nicoals Simoncini (293), UBFC-UTBM, FCLab FR CNRS 3539, FEMTO-ST/RECITS UMR CNRS 6174, Belfort Cedex, France

Monika Varga (461), Research Group on Process Network Engineering, Institute of Methodology, Kaposvar University, Kaposvar, Hungary

Dai-Viet N. Vo (111), Faculty of Chemical and Natural Resources Engineering, University Malaysia Pahang, Lebuhraya Tun Razak, Gambang, Kuantan, Malaysia

Christina Wulf (81, 309), Institute of Energy and Climate Research—Systems Analysis and Technology Evaluation (IEK-STE), Forschungszentrum Julich, Julich, Germany

Yang Yongrong (405), State Key Laboratory of Chemical Engineering, Department of Chemical and Biological Engineering, Zhejiang University, Hangzhou, PR China

Petra Zapp (309), Institute of Energy and Climate Research—Systems Analysis and Technology Evaluation (IEK-STE), Forschungszentrum Julich, Julich, Germany

Liao Zuwei (405), State Key Laboratory of Chemical Engineering, Department of Chemical and Biological Engineering, Zhejiang University, Hangzhou, PR China

Preface

Hydrogen Supply Chains: Design, Deployment, and Operation

Hydrogen produced from renewable sources and used in fuel cells for both mobile and stationary applications constitutes a very promising energy carrier for the energy transition. The introduction to the market of new propulsion systems, such as the fuel cell electric vehicle (FCEV), can be one alternative to treat problems, such as energy security, urban air quality, and global warming potential. Yet, the cost of hydrogen is still considered prohibitive compared to the fossil fuels used in the transportation system, even if the development of some technologies associated with hydrogen has improved its competitiveness.

A key point in the development of hydrogen supply chains (HSC) is the demonstration of the feasibility of their infrastructure, while many technical, economic, and social obstacles must be overcome. Some strategic roadmaps have been regularly published about the energy potentialities of hydrogen at continental (for instance, European), national, and regional levels. Their main objective is to evaluate some industrial, technological, environmental, and social issues and to identify the main obstacles associated with the hydrogen economy. The literature review of recent dedicated scientific publications also agrees on the need to develop systemic studies in order to demonstrate the feasibility of the sector and to validate the techno-economic interest in the production of hydrogen produced from renewable sources.

A challenging and motivating issue is that the pathway that can be followed to develop a hydrogen economy is very flexible because many energy sources, production processes, and transportation and storage modes exist. Hydrogen can favor the connection of different energy sectors and energy transmission and distribution networks, thus enhancing the operational flexibility of future low-carbon energy systems.

The purpose of this book is to consider the interconnection of the different items contributing to the hydrogen supply chain (HSC) and to explore them from different angles, including techno-economic, environmental, and the safety and social aspects.

The design of an HSC for fuel use can be viewed as a framework considering different levels of scale (for instance national and regional scales), combining multiple energy sources with various production and storage technologies, while considering the transportation modes to link hydrogen demand to its supply involving multiple uses. Particular emphasis is devoted to the multiobjective formulation in which cost, environmental impact, and safety must be simultaneously taken into account at the early design stage.

It must be emphasized that the scope of supply chains is classically addressed in the industrial engineering field. Yet the objective here is to focus on the specific features of hydrogen supply chain management in the context of the integration of hydrogen energy systems in the overall energy system as an important and complex subject.

The ambition of the book is to provide guidelines that could be followed to develop a hydrogen supply chain, leading to a reference hydrogen energy system architecture that can be used for energy analysis, modeling, and integration for a subsequent HSC implementation, while taking into account all stakeholder perspectives and convincing policy makers. The efficiency of the integration is strongly related to the search for optimum pathways for hydrogen supply, with an advanced knowledge of the supply chain components for energy modeling purposes.

Our goal is to provide a concise overview of the methods and tools that can be used for the deployment, design, and operation of the HSC. This necessarily means that we cannot cover every possible hydrogen pathway, but we have tried to provide the basics that are needed by engineers, managers, engineering students (MSc), and researchers (PhD, for instance) in various fields, such as Process Systems Engineering, Energy Engineering, Industrial Engineering, and Sustainable Engineering. We have tried to capture both the common threads and the diversity, being aware that the originality of a hydrogen supply chain comes from the fact that there is no unique supply chain, so that it is difficult to embed in one generic formulation all the various possibilities that may be encountered, as the design of the supply chain may vary according to the target in view.

Our book is divided into two parts.

PART I—EXPLORING THE CHALLENGES AND SCALES OF HSC DESIGN, DEPLOYMENT, AND OPERATION

Chapter 1: Hydrogen as a Pillar of the Energy Transition

This introductory chapter is devoted to the exploration of the major roles that hydrogen is likely to play in the economy, with a specific focus on decarbonization. The concept of Power-to-Gas used in hydrogen supply chains is presented. The vision that is laid out is based on a systemic view of the potential of hydrogen in the energy system, in particular for mobility purposes.

Chapter 2: Hydrogen Supply Chain Design: Key Technological Components and Sustainable Assessment

The objective of this chapter is to present the concept of the hydrogen supply chain and its main activities, including multiple sources/multiple uses, production, storage, transportation and distribution, multiples stakeholders, multiperiod strategies in a context of uncertainty (for instance demand). The criteria to be taken into account in a sustainable development context are also highlighted.

The pillars of the HSC supply chain – production, storage, and distribution – are explored in the three following chapters, which will present hydrogen production processes at various scales from macroscopic to process scale, as well as state-of-the art reviews.

Chapter 3: Assessment of Selected Hydrogen Supply Chains—Factors Determining the Overall GHG Emissions

This chapter analyzes different hydrogen production options and their respective transport needs from an environmental viewpoint.

Chapter 4: Hydrogen Production From Biogas Reforming: An Overview on Steam Reforming, Dry Reforming, Dual Reforming, and Tri-reforming of Methane

At a more local and operational viewpoint, this chapter presents a case study of green hydrogen production from biogas as a renewable resource using a multistep process, including mainly biogas reforming, water-gas-shift reaction, and hydrogen separation. It is focused on different methane reforming processes: steam reforming, dry reforming, dual reforming, and tri-reforming. This case study is particularly interesting from a process engineering viewpoint because the methodological aspects are explored for each process, encompassing thermodynamic equilibrium, process at industrial scale or research laboratory development, kinetic models, and mechanistic study.

Hydrogen storage, along with distribution, is a key technology associated with the wide use of hydrogen. Both of these two group technologies seem to be the most limiting factors, currently and in the near future, to the deployment of hydrogen in the energy system. This is the core of the two following chapters.

Chapter 5: Hydrogen Storage for Mobile Application—Technologies and Their Assessment

The various storage technologies and concepts that have been developed to date are presented in this chapter: high-pressure storage tanks, storage as liquid hydrogen, and storage of the hydrogen gas in metal hydride. Other options, such as hydrogen storage for mobile applications, such as liquid organic hydrogen carriers (LOHC), activated carbon, Metal-Organic-Frameworks (MOFs), and others, are also discussed.

Chapter 6: Lowering Energy Spending Together With Compression, Storage, and Transportation Costs for Hydrogen Distribution in the Early Market

The optimization of cost and energy consumption for compression, transportation, and storage of hydrogen for vehicle refueling in the current hydrogen emerging market is addressed in this chapter by considering a recurrent issue in hydrogen supply chain development, that is, the location of a refueling station on a hydrogen production site and the case of a production unit supplying hydrogen to several distant refueling stations.

A HSC is globally viewed as demand-driven and hydrogen has many applications in the energy market, for example, direct use in an internal combustion engine, replacing the need for automotive gasoline, use in fuel cells to create electricity, use to power and heat buildings, alternative to natural gas in its application to heating and cooling homes, interest in hydrogen for the power-to-gas market, etc.

Chapter 7: Hydrogen Applications: Overview of the Key Economic Issues and Perspectives

The chapter explores the large range of potential applications of hydrogen from industry to the transport sector, currently and in the longer term. For each of these applications, the potential for economic competitiveness is discussed by highlighting the main drivers and variables.

Chapter 8: Social Aspects of H_2 Supply Chains

The deployment of a hydrogen economy is not without trials and controversies, doubts, and value oppositions. Using hydrogen as energy storage for renewables is particularly sound in an island context. The Corsican PV-Hydrogen MYRTE Platform serves here as a test bench for investigating the close links between technical aspects and political obligations, and the role of stakeholders.

Chapter 9: Power-to-Gas—Concepts, Demonstration and Prospects

In this vision, hydrogen from electrical energy via electrolysis is viewed as the first possible end-product of the so-called Power-to-Gas process chain or can be further converted to synthetic methane via methanation, a process requiring the feed-in of CO_2. The potentials, opportunities, and limitations of PtG are presented in this chapter.

PART II—EXPLORING METHODS AND TOOLS FOR HSC DESIGN, DEPLOYMENT, AND OPERATION

Methods and tools that can be selected to develop the methodology of the optimal design of a hydrogen supply chain are proposed, as well as their performance drivers and metrics (economic, environmental, societal). They are illustrated by significant case studies to show their large range of potential application.

Chapter 10: Methods and Tools for Hydrogen Supply Chain Design

This chapter presents the methods and tools classically used for hydrogen supply chain (HSC) design, generally involving optimization strategies. The most current trend is based on multiobjective formulations. Decision-aid methods to search for tradeoff solutions are often used. We also examine how the HSC design optimization framework can be linked with geographic information systems (GIS).

Chapter 11: Multiobjective Life Cycle Optimization of Hydrogen Supply Chains

The formulation of the design of hydrogen networks as a mixed-integer linear programming (MILP) problem is proposed in this chapter, including environmental objectives along with economic ones. From these optimal designs, stakeholders must select the one that best matches their preferences, considering the applicable legislation and technical, economic, and environmental constraints. The case study of the future supply chain (SC) for vehicle use in the UK is considered.

In addition to PtG supply chains, it must not be forgotten that hydrogen has been an increasingly important component of refining in the process industries, particularly in view of the increased demand for clean fuels. Refinery hydrogen networks typically interconnect many producers, consumers, and purification units with different pressures, purities, and operating objectives. This kind of gas network is the core of the two following chapters, dedicated to the formulation of the operational optimization model of a hydrogen pipeline network.

Chapter 12: Engineering Robust Strategy for Solving Optimization Problems of Refinery Hydrogen System

A robust engineering strategy that has been applied to hydrogen pipeline networks of a large-scale refinery has been developed for optimal scheduling of the hydrogen system to reduce energy cost and carbon emissions in refineries.

Chapter 13: Optimal Design of Refinery Hydrogen System With Purification Unit

In this chapter, both the pinch technique and mathematical programming approaches are introduced for the synthesis of a hydrogen network with a purification unit.

Chapter 14: Metamodeling of Hydrogen Supply Chains: A Programmable Structure Based Representation

Besides the well-developed mathematical programming–based optimization methods, the design and operation of these large-scale, long-term processes might also optionally require easily extensible, generic dynamic simulation. This chapter shows how the programmable structure of process models can be generated from the description of a process network (optionally geographically determined and multiscale) and from two general functional metaprototypes.

Chapter 15: Life Cycle Assessment of Hydrogen Supply Chain—A Case Study for Japanese Automotive Use

One of the supporting methodologies for environmental impact evaluation is life cycle assessment (LCA). The idea of integrating LCA into supply chains has received increased interest. A "Well-to-Wheel" (WtW) analysis in the Japanese context was conducted to understand the role of hydrogen in reducing greenhouse gas (GHG) emissions in the vehicle transport sector from a life cycle perspective.

Chapter 16: Risk Analysis of Complex Hydrogen Supply Chains

This chapter focuses on methods to assess safety risks in the future hydrogen-based infrastructure. The development of new large-scale infrastructure is viewed as a gradual procedure requiring different decision support tools, including cost-benefit assessments, sustainability assessments, optimization of supply chains, the best placement of buildings and process equipment in a growing market, and, last but not least, safety risk assessment and management.

Catherine Azzaro-Pantel
INP Toulouse, ENSIACET
1 August 2018

Part I

Exploring the Challenges and Scales of HSC Design, Deployment, and Operation

Chapter 1

Hydrogen as a Pillar of the Energy Transition

Jesus Ochoa Robles, Sofía De-León Almaraz and Catherine Azzaro-Pantel
Laboratoire de Génie Chimique, Université de Toulouse, CNRS, Toulouse, France

ACRONYMS

ADEME	Agency for Environment and Energy Management (Agence de l'environnement et de la maîtrise de l'énergie)
BEV	battery electric vehicle
CaFCP	California Fuel Cell Partnership
CCS	carbon capture and storage
CCU	carbon capture and utilization
CHP	combined heat/power
FC	fuel cell
FCEV	fuel cell electric vehicle
GHG	greenhouse gas
HHV	high heating value
HRS	hydrogen refueling station
ICE	internal combustion engine
IEA	International Energy Agency
LHV	low heating value
METI	Ministry of Economy, Trade, and Industry (Japan)
OPECTS	Parliamentary Office for the Evaluation of Scientific and Technological Options (Office Parlementaire d'Evaluation, des Choix Scientifiques et Technologiques)
PHEV	plug-in hybrid electric vehicle
PtG	Power-to-Gas
PtH	Power-to-Hydrogen
PtM	Power-to-Methane
SMR	steam methane reforming

Hydrogen Supply Chain. https://doi.org/10.1016/B978-0-12-811197-0.00001-4

SNG synthetic natural gas
TPES total primary energy supply
WtW Well-to-Wheel

1.1 INTRODUCTION

The global demand for energy production and environmental concerns are among the most significant issues in the 21st century (Dincer and Acar, 2015). One of the biggest challenges is to meet growing energy demand in an environmentally benign and sustainable manner, as highlighted in the Paris Climate Agreement, which aims to keep global average temperatures from rising by 2°C above preindustrial levels, and to pursue efforts to limit the temperature increase even further to 1.5°C (United Nations, 2015).

Fig. 1.1 shows the world fuel shares of total primary energy, electricity generation, and CO_2 emissions in 2015. According to the International Energy Agency (IEA, 2017a), in 2015, global total primary energy supply (TPES) was 13,649 Mtoe, electricity generation was 24.2 billion MWh, and final consumption was 9384 Mtoe. These numbers are expected to increase with continuing consumption and population increases. In 2050, according to (World Energy Council, 2013), global electricity generation is expected to increase to 53.6

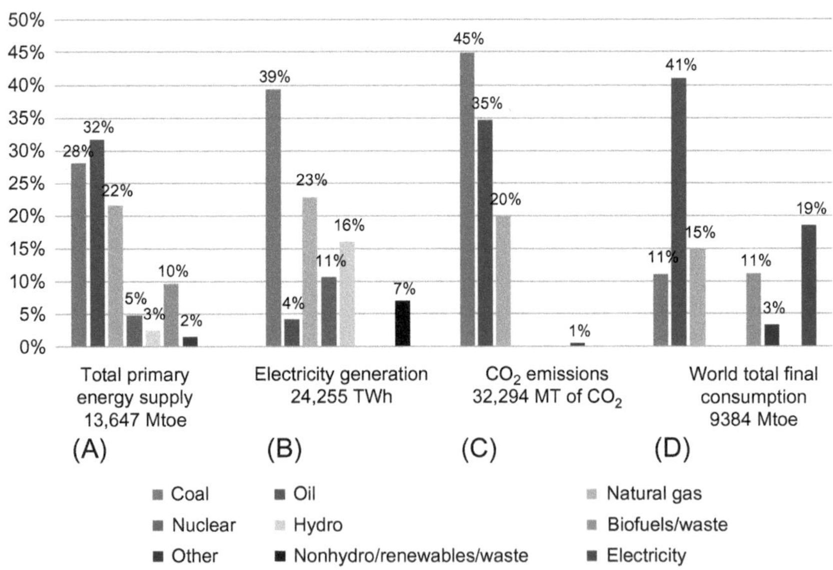

FIG. 1.1 World fuel shares of (A) total primary energy supply, (B) electricity generation, (C) CO_2 emissions in 2015, and (D) world total final consumption by fuel. *(Adapted from IEA, 2017a. International Energy Agency Technical Report. Key world energy statistics [WWW Document]. https://www.iea.org/publications/freepublications/publication/KeyWorld2017.pdf (Accessed 29 December 2017).)*

billion MWh (scenario Jazz) and to 47.9 billion MWh by 2050 (scenario Symphony). More than 80% of the global energy supply comes from fossil fuels (World Energy Council, 2013).

The massive utilization of fossil fuels causes economic and technical issues because they derive from resources that are finite and unequally distributed across the globe. This, in turn, may create a dependency of some countries on others, and thus generate tension. Another issue is that fossil fuel reserves are becoming less accessible as the easily accessible sources are consumed, so that an increase in the price of fossil fuels can be expected. In addition to the economic and technical issues, most of the emissions of human-caused (anthropogenic) greenhouse gases (GHG), mainly CO_2 emissions, come primarily from burning fossil fuels (coal, hydrocarbon gas liquids, natural gas, and petroleum) for energy use. Fig. 1.1 also shows that 99% of global GHG emissions were caused by fossil fuels. If emissions follow a commonly used business-as-usual scenario, there is a 93% chance that global warming will exceed 4°C by the end of this century, as recently highlighted in Brown and Caldeira (2017).

The development of clean energy solutions is, then, a key prerequisite to pave the way for the energy transition in which there is a switch from a system fueled primarily by nonrenewable, carbon-based energy sources to one fueled by clean, low-carbon energy sources. Four main levers are available to decarbonize the energy system: improving energy efficiency, promoting renewable energy sources, switching to low- or zero-carbon energy carriers, and developing carbon capture and storage (CCS) as well as utilization (CCU).

In that context, the concept of mitigating climate change by transitioning to an energy system with fewer greenhouse gas emissions, and more sustainable, even circular, consumption and production, is particularly attractive.

For that, hydrogen, as a near zero-emission energy carrier if produced from renewable energy sources, is viewed as an attractive candidate to overcome the challenges surrounding the energy transition. Some of the advantages have been listed in Dincer and Acar (2015): (i) high energy conversion efficiencies; (ii) production from water with no emission; (iii) abundance; (iv) different forms of storage (e.g., gaseous, liquid, or in together with metal hydrides); (v) long distance transportation; (vi) ease of conversion to other forms of energy; (vii) higher HHV (high heating value) and LHV (low heating value) than most of the conventional fossil fuels.

This introductory chapter is devoted to exploring the potential of hydrogen to fulfill these objectives and to presenting the barriers that must be overcome. The remainder of this chapter is organized as follows. Section 1.2 is dedicated to the major roles that hydrogen is likely to play in the economy, with a specific focus on decarbonization. The concepts of Power-to-Gas and hydrogen supply chains are then presented. The vision that is laid out is based on a systemic view of the potential of hydrogen in the energy system. Section 1.3 illustrates hydrogen supply chains for mobility purposes. Section 1.4 then presents the barriers that must be overcome and the roadmaps that have been reported for hydrogen deployment. They are often used by international organizations, businesses, and industry to address the challenges of hydrogen scaleup. This section also discusses transition plan scenarios. This chapter concludes that more precise analyses are necessary to design the hydrogen supply chain.

1.2 MAJOR ROLES OF H_2 IN THE ECONOMY

1.2.1 Decarbonization of Key Sectors of the Economy

Hydrogen is likely to play a major role in decarbonizing key sectors of the economy. Seven actions have been identified in Hydrogen Council (2017a), as illustrated in Fig. 1.2:

- *Favoring large-scale, efficient renewable energy integration.* Hydrogen offers valuable advantages, as it avoids CO_2 and particle emission if produced from renewables, can be deployed at large scale, and can be made available everywhere. Hydrogen can improve the efficiency and flexibility of an energy system in two ways: (1) the excess of electricity can be converted by electrolysis into hydrogen and be used in other sectors, such as transport, industry, and buildings; and (2) it can be used as a carbon-free storage medium as a way to store energy for long periods. Three modes of storage can be considered for hydrogen, namely, as a compressed gas, as a liquid, or as solid absorbing metals (Florida Solar Energy Center, n.d.).

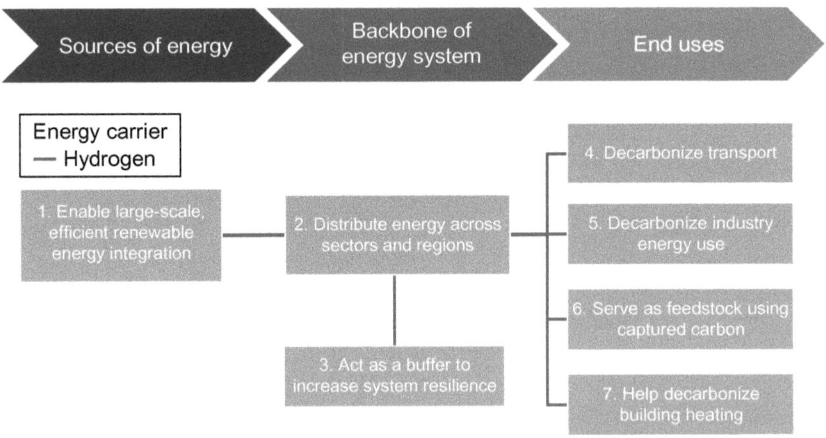

FIG. 1.2 Roles of hydrogen in decarbonizing major sectors of the economy. *(Adapted from Hydrogen Council, 2017a. How hydrogen empowers the energy transition [WWW Document]. Rep. January. http://hydrogeneurope.eu/wp-content/uploads/2017/01/20170109-HYDROGEN-COUNCIL-Vision-document-FINAL-HR.pdf (Accessed 29 December 2017).)*

- *Energy distribution across sectors and regions.* Some countries are not well positioned to generate energy with wind or solar power alone. For other countries, time may be needed to raise the necessary investment. As hydrogen has a high energy density and can be easily transported, it can (re)distribute energy effectively and flexibly. Hydrogen may be transported as a pressurized gas or as a cryogenic liquid. Gaseous hydrogen can be transported by highly pressurized pipelines or by tube trailers. Liquefied hydrogen can be transported in tankers (Dagdougui, 2012).
- *Acting as a buffer to increase system resilience.* Hydrogen can help align global energy storage with changing energy demand. Its physical characteristics make it well suited to serve as an energy buffer and strategic reserve. By 2030, 250–300 TWh of surplus renewable electricity are expected to be stored in the form of hydrogen for use in other segments (Hydrogen Council, 2017a).
- *Transport decarbonization.* Nowadays, battery electric vehicles (BEV) are already used to reduce CO_2 emissions. However, fully decarbonizing transport will require deployment of fuel cell electric vehicles (FCEVs), having the same performance as gasoline vehicles. FCEVs have several advantages: they actually zero emissions, have a good autonomy (500 km), and refuel quickly (from 3 to 5 min) (Ball and Weeda, 2015). Decarbonizing transport is particularly challenging because it represents a large share of total energy and more than 30% of hydrogen's total CO_2 abatement potential is expected in this sector (Hydrogen Council, 2017a).
- *Decarbonization of industry energy use.* Fossil fuels are the most used energy sources for industrial processes. Hydrogen can be an alternative when it is available as a byproduct of the chemical industry or when a specific industry needs an uninterruptable power supply. As hydrogen can be burnt in hydrogen burners or be used in fuel cells, it offers a zero-emission alternative for heating. Nowadays, hydrogen is only used in industry for low-grade heat applications (process heating and drying), but is expected to be used with fuel cells in the future for not only low-grade, but also high-grade heat needs (Hydrogen Council, 2017a).
- *Serve as feedstock using captured carbon.* Hydrogen could be used to convert captured carbon into usable chemicals, such as methanol, methane, formic acid, or urea. This technology is still in the research phase, and it is expected to be developed in the next 15 years. For example, in Iceland, geothermal CO_2 is used to generate electricity to produce hydrogen and ethanol, by two thermophilic bacteria (Koskinen et al., 2008). By 2030, 10–15 million tons of chemicals may be produced from such renewable feedstock (Hydrogen Council, 2017a).
- *Contribution to the decarbonization of building heating.* Heat generation in buildings and industry accounts for more than half of global final energy consumption and a third of global energy-related carbon dioxide (CO_2)

emissions (Dodds et al., 2015; IEA, 2014). Hydrogen technologies, such as fuel cell micro CHPs (combined heat/power units), serve as energy converters. Possible roles for hydrogen and fuel cell products include the substitution of hydrogen for natural gas in some processes, and the use of CHP technologies. For the purpose of illustration, about 190,000 buildings are already heated with hydrogen-based fuel cell micro CHPs, mainly in Japan.

1.2.2 Hydrogen Supply Chains and the Power-to-Gas (PtG)/Power-to-Hydrogen (PtH) Concept

The hydrogen supply chain is a concept in the life cycle perspective, consisting of several echelons, including selection of energy source, hydrogen production, hydrogen transportation, hydrogen refueling, and hydrogen utilization subsystems (see Fig. 1.3). Of course, there is not a unique hydrogen supply chain. Even if there is clear evidence for the use of renewable sources, as already highlighted from an environmental viewpoint, the switch to a 100% renewable scheme can only be gradual, in order to satisfy both economic and environmental concerns as well as to take into account the availability of the energy source.

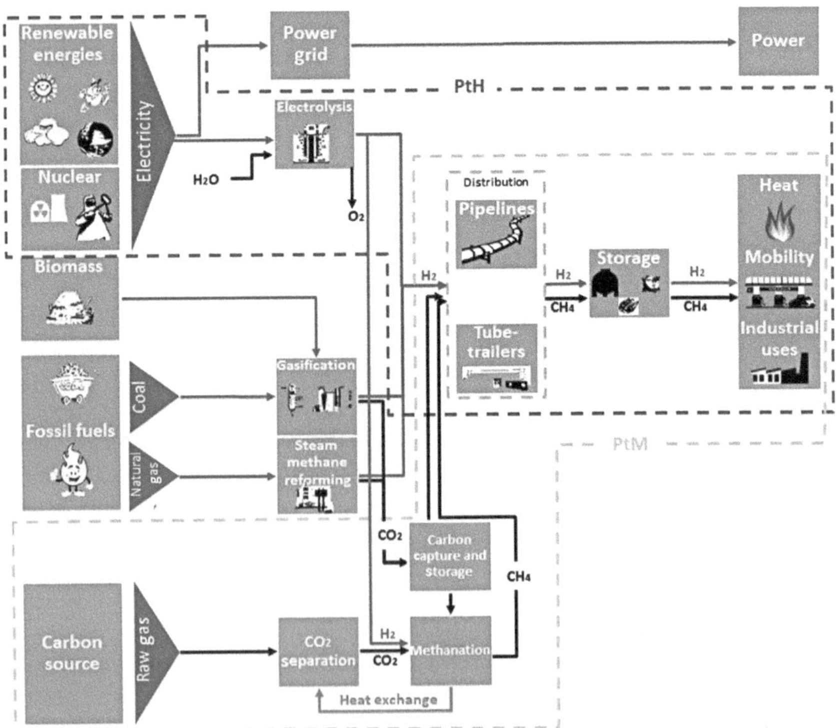

FIG. 1.3 Power-to-Gas supply chains.

Hydrogen can thus be produced using different energy sources (renewables or fossil fuels) and with different technologies (mainly steam methane reforming (SMR), electrolysis, and gasification) and distributed via pipelines or tube trailers. Fig. 1.3 also embeds other options that may be encountered.

The key roles of hydrogen in the future energy system emphasize the PtG concept, in particular the PtH one (brown dotted line in Fig. 1.3). PtG refers to the process in which electrical energy is converted into chemical energy via gas production. The main purpose is to store surplus electricity from fluctuating renewable sources by generating hydrogen (H_2) via water electrolysis, with optional methane (CH_4) synthesis from carbon dioxide (CO_2) and H_2 (methanation process, yellow dotted line). This "green hydrogen" produced by renewable resources without pollution allows for the storage, transportation, and reuse of the energy when needed.

The production of synthetic methane (synthetic natural gas, SNG) results in lower total efficiency but could be advantageous in terms of feeding the produced energy carrier into the gas distribution grid. In contrast to the case of pure H_2, the injection of SNG is not limited in amount. The SNG or H_2 can be used not only in electricity production, but also in other applications, such as mobility via fuel cells or natural gas vehicles (Fig. 1.3).

A Power-to-Gas supply chain, as shown in Fig. 1.3, is ultimately a network of integrated facilities, or nodes, that are interconnected and work together in a specific way. The network begins with primary energy sources and terminates with end uses. A supply chain is not unique, and one typical feature of a PtG supply chain, as a segment of a hydrogen supply chain, is the large number of configurations that can be encountered from energy sources, production, distribution, and storage to final uses.

The conversion to hydrogen and methane makes the transport of renewable energy outside the power grid possible, also allowing large-scale, long-term storage. The chemical energy carriers can also be converted to electricity and a multitude of other pathways are possible, resulting in different efficiencies of the total system.

As highlighted in Lehner et al. (2014), hydrogen is the first possible end product of the Power-to-Gas process chain. The efficiency of the conversion of methanation is reported to be 70%–85% in the case of the chemical path, and greater than 95% for the biological path (Grond et al., 2013). The main asset of SNG is its unrestricted compatibility with the natural gas grid. The so-called "repowering" of methane to electricity in combined cycle plants opens the possibility of producing electric power in areas far away from the renewable power sources, connected by an already existing gas grid. However, the efficiency of this option is the lowest of all possibilities (see Table 1.1).

Slightly better conversion efficiencies can be achieved by producing electricity from hydrogen. Gas turbines, fuel cells, or reverse fuels cells can be utilized for this purpose. The efficiency for PtG systems is increased with recovery of the released heat of the system, for example in district heating or

TABLE 1.1 Efficiencies for Different Power-to-Gas Process Chains (Sterner et al., 2011)

Path	Efficiency	Boundary Conditions
Electricity to gas		
Electricity → Hydrogen	54–72	Including compression to 200 bar (underground storage working pres.)
Electricity → Methane (SNG)	49–64	
Electricity → Hydrogen	57–73	Including compression to 80 bar (feed in gas grid for transportation)
Electricity → Methane (SNG)	50–64	
Electricity → Hydrogen	64–77	Without compression
Electricity → Methane (SNG)	51–65	
Electricity to gas to electricity		
Electricity → Hydrogen → Electricity	34–44	Conversion to electricity: 60%, compression to 80 bar
Electricity → Methane → Electricity	30–38	
Electricity to gas to combined heat and power (CHP)		
Electricity → Hydrogen → CHP	48–68	40% electricity and 45% heat, compression to 80 bar
Electricity → Methane → CHP	43–54	

in industrial plants nearby (Table 1.1). The pressure level has a significant influence on the global efficiency.

A thorough investigation of PtG systems is particularly interesting from a systemic viewpoint in the context of the energy transition and the different pathways should not be considered in isolation. Some recent investigations have highlighted that the PtG (Götz et al., 2016) might play an important role in the future energy system. However, technical and economic barriers must be solved and a critical aspect of the PtG process is the availability of CO_2 sources. Concerning methanation, biological and thermochemical methanation processes have potential for integration into the PtG process chain. Biological methanation is a simple process that tolerates gas impurities but induces slower reaction times and has higher power requirements, leading to a lower process efficiency than for thermochemical methanation. Thermochemical methanation is attractive for its high reaction rates and the high temperature level of thermochemical methanation results in more options for process integration, yielding more efficient processes. Due to the higher process temperature and the resulting higher reaction velocity, thermochemical methanation requires much lower reactor volumes for a certain feed gas flow than biochemical methanation. However, full CO_2-conversion in a single step thermochemical methanation reactor cannot be achieved due to thermodynamic equilibrium limitations (Götz et al., 2014).

In this chapter, specific attention is given to hydrogen supply chains for which many of the required technologies are already available today (Hydrogen Council, 2017a).

1.3 HYDROGEN SUPPLY CHAINS FOR MOBILITY PURPOSE

Hydrogen infrastructure and technologies are seen as an important part of the future energy mix, due to their advantages in terms of CO_2 reduction potentials in the transport sector, which has been shown to be one of the hardest to decarbonize (IEA, 2017a).

1.3.1 Environmental and Energy Benefit

Fig. 1.4A presents GHG emissions in the European Union (EU) and shows that energy use is the main source of GHG. Among the various sectors, the energy industry (29%) and transport (21%) are the most polluting. It can be observed that the decrease in GHG emissions is mainly due to significant declines in the energy industry (−7%) and residential-tertiary (−15%) sectors (Institute for Climate Economics and Ministère de l'Environnement, de l'Énergie et de la Mer, 2016).

As GHG emissions are declining in the industrial sector, the transport sector remains as one of the challenges to be tackled. Hydrogen thus represents an interesting fuel alternative in the transportation sector because this sector contributes approximately 836 Mt. CO_2 eq emissions in Europe and 14 Mt. CO_2 in France (Fig. 1.4B), and technologies related to FCVs are being developed rapidly (Institute for Climate Economics and Ministère de l'Environnement, de l'Énergie et de la Mer, 2016).

In France, as in most countries worldwide, there is a dependency on fossil fuels. Fig. 1.4B shows the contribution of GHG emissions by sector. The transport sector contributes significantly so that special attention must be paid to it. To more effectively reduce pollution, the transportation sector requires the development of both new vehicular technologies and new fuels (Cipriani et al., 2014). Hydrogen, which can be used in vehicles equipped with the

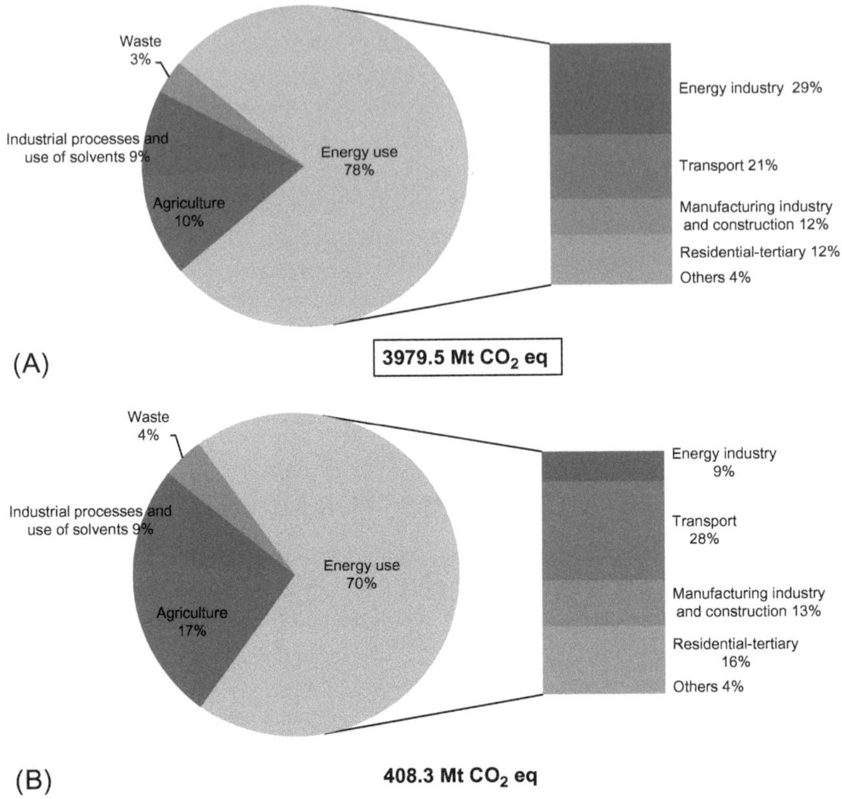

FIG. 1.4 Greenhouse gas emissions by sector in (A) Europe and in (B) France in 2014. *(Adapted from Institute for Climate Economics, Ministère de l'Environnement, de l'Énergie et de la Mer, 2016. Chiffres clés du climat France et Monde ÉDITION 2017.)*

technology for converting hydrogen into electricity, is thus particularly attractive because the carbon emissions of FCEVs are very low when the whole lifecycle is considered. Even if hydrogen is entirely produced from natural gas through steam methane reforming (SMR) without the use of carbon capture, FCEV emissions are 20%–30% lower than those of ICEs. In total, an FCEV powered by green or clean hydrogen in our example could achieve combined CO_2 emissions of 60–70 g per km (Fig. 1.5) (Hydrogen Council, 2017b).

Local air emissions, responsible for particulate matter, ozone, and acid rain, as well as noise, could be significantly reduced by the introduction of hydrogen fuel cell vehicles. Emissions of NOx, SO_2, and particulates can be reduced by 70%–80% compared to a case without hydrogen (Ball and Wietschel, 2009). Due to the growing number of megacities worldwide, the importance of improving urban air quality is of major importance. According to (Mobilité Hydrogène France, 2016), the societal cost savings are about 500 M€ over the 2015–30

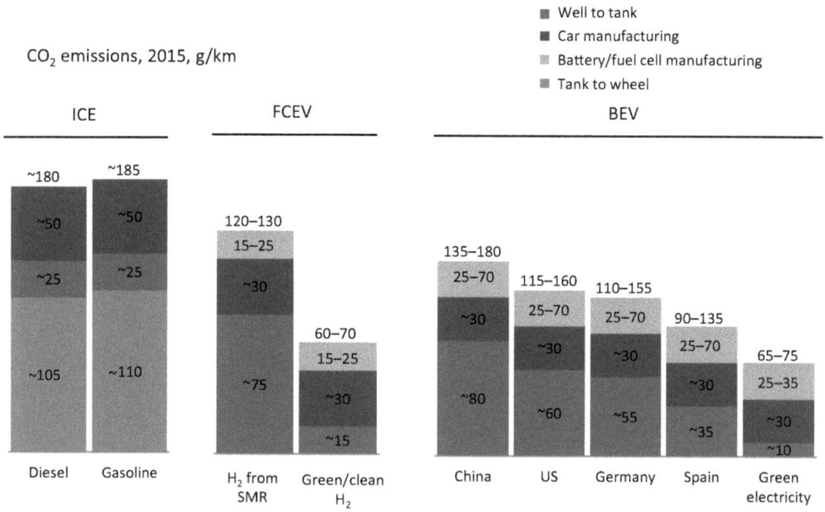

Assumption: compact car (C-segment) as reference vehicle (4.1 l/100 km diesel; 4.8 l/100 km gasoline; 35.6 kWh battery).120,000 km lifetime average grid emissions in China, Germany, Spain in 2015; EV manufacturing (excl. fuel cell and battery) 40% less energy-intensive than ICE manufacturing; 10 kg CO_2/kg H_2 from SMR; 0.76 kg H_2/100 km; 13 kWh/100 km

FIG. 1.5 CO_2 emissions for the ICE, FCEV, and BEV vehicles. *(Adapted from Hydrogen Council, 2017b. Hydrogen scaling up. A sustainable pathway for the global energy transition [WWW Document]. http://hydrogencouncil.com/wp-content/uploads/2017/11/Hydrogen-scaling-up-Hydrogen-Council.pdf (Accessed 4 January 2018).)*

period; in particular, the societal cost of the CO_2 emissions, noise, and pollutants evaluated for an ICE (internal combustion engine) vehicle, which came to 510 € per year, is reduced to 160 € for an FCEV.

These elements contribute to make hydrogen an accepted clean energy carrier worldwide, because it is source independent and has a very high energy content per mass compared to petroleum or the actual fuels (120 MJ/kg versus 46 MJ/kg, respectively) (Dutta, 2014). Also, liquid hydrogen possesses a very low density of 0.07 g/cm^3, one-tenth that of gasoline. These properties provide some advantages and disadvantages. On the one hand, an advantage is that hydrogen stores around twice the energy of gasoline/diesel, being energetically more efficient than gasoline. On the other hand, it requires a storage volume four times greater than gasoline (Sharma and Ghoshal, 2015).

(Ball and Weeda, 2015) developed a study describing the process and use of hydrogen in transportation. Currently, in order to achieve a deep decarbonization of road transport, three options exist:

- battery electric vehicles (BEV), using electricity as fuel;
- fuel cell electric vehicles (FCEVs), using hydrogen as fuel; and
- plug-in hybrid electric vehicles (PHEV), combining a battery system with a fuel cell system.

Nowadays, approximately 500,000 electric vehicles (PHEVs and BEVs) are used globally (mainly in the United States, Europe, and China), and in the coming years, most of these vehicles will come from China (Ball and Weeda, 2015).

The main advantage of this kind of vehicle is that the distribution infrastructure of the energy source or fuel (the electric energy) is available, and thus, automatically, the costs are reduced in comparison with alternatives that are in the development phase.

The market for hydrogen is expected to increase in the future. In France, the demand for 2030 is expected to reach 90,000 t of hydrogen, versus only 3,000 t in 2016 (Fig. 1.6). Obviously, electricity demand will also increase, reaching 3 TWh in 2030. In the transport sector, almost 800,000 FCEVs are expected to be found in 2030, requiring the installation of 600 hydrogen refueling stations (HRS) to satisfy the automotive demand. This situation contrasts with the current one, with only 23,000 FCV and 96 HRS to cover vehicle refueling needs.

1.3.2 Hydrogen FCEV

Hydrogen is one of the principal alternatives in the future of road transport (Ball and Weeda, 2015):

- FCEVs using hydrogen as fuel are real zero-emission vehicles, and in the last years, the efficiency of hydrogen has increased (50%–60%).
- As FCEVs are basically electric vehicles, they combine the benefits of electric driving (silence and smoothness) with the power of using hydrogen as a fuel (autonomy of 500 km and 3–5 min of refueling time). Their autonomy is two or three times an electric car (Bettayeb, 2017).
- Because of more favorable energy density characteristics compared to batteries, hydrogen and fuel cells are better suited to electrify a wide range of road vehicles, ranging from small cars to buses and light-duty trucks.
- Hydrogen can be produced from many different energy sources, from CO_2-free and renewable energy sources (wind, water, sun, etc.) to fossil fuels (natural gas, coal). Clearly, the use of renewable energy sources is the priority in order to make hydrogen a real and totally CO_2-free fuel.

Fig. 1.7 presents FCEV vehicles that are available in the market, with the emblematic Toyota Mirai. In the United States, a Toyota Mirai is available for sale at $57,500 USD and for lease at $349/month (IEA, 2017b).

Fuel cells (with a capacity of 10–30 L) are composed of an electrode sandwiched between two electrodes (anode and cathode), as seen in Fig. 1.8. The bipolar plates at the extreme sides of the cell help to distribute and collect the gases. Then, hydrogen flows through the labels to the anode, where the hydrogen molecules are separated into protons and electrons. The electrons follow a circuit until the cathode (they are electricity ready to be used). The oxygen gas, obtained from the air, flows to the cathode. After the electrons are used and

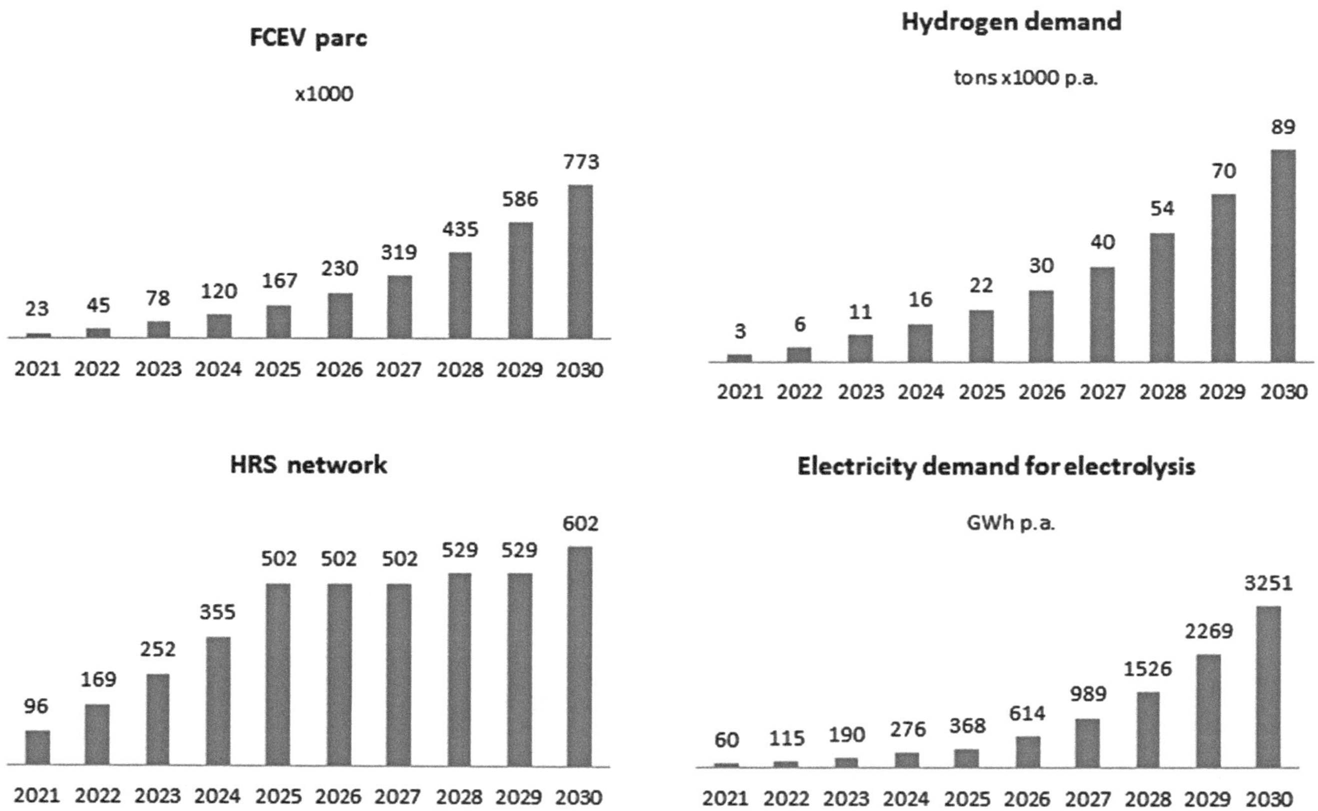

FIG. 1.6 FCEV market evolution in France. *(Adapted from Mobilité Hydrogène France, 2016. H2 MOBILITÉ FRANCE. Study for a Fuel Cell Electric Vehicle National Deployment Plan [WWW Document]. http://www.fch.europa.eu/sites/default/files/Smart%20Spec%20Fabio%20Ferrari%20%28ID%202436338%29%20%28ID%202497336%29.pdf (Accessed 31 December 2017).)*

FIG. 1.7 FCEV vehicles (IEA, 2017b).

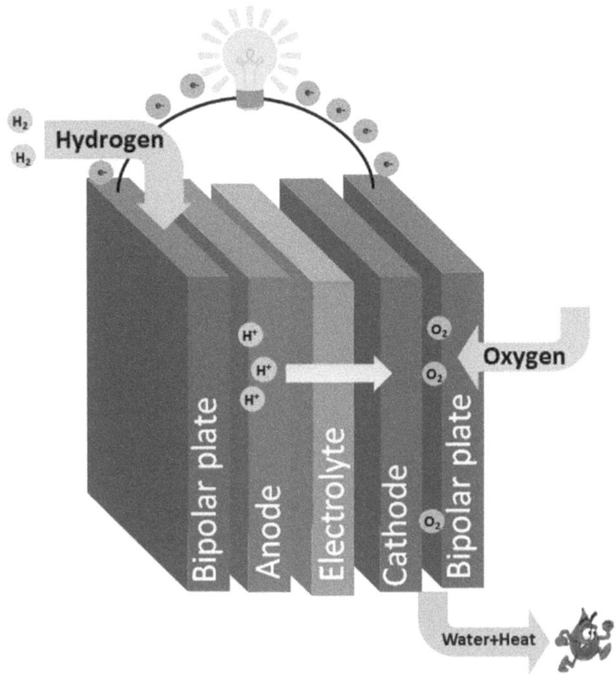

FIG. 1.8 A single fuel cell. *(Adapted from DOE Hydrogen Program, 2006. Hydrogen Fuel Cells.)*

return, they react with oxygen and the hydrogen protons at the cathode and form water that is released from the fuel cell (DOE Hydrogen Program, 2006).

1.3.3 Hydrogen Safety

Hydrogen is a flammable gas with a wide flammability range (4%–75% by volume) and relatively low ignition energy (0.02 mJ) (McCarty et al., 1981). It has a very low density and therefore must be stored at high pressure to achieve enough mass for practical use. The ease of ignition and high storage pressure of hydrogen are responsible for a large portion of the risk associated with hydrogen usage.

Hydrogen also has the ability to attack—and damage to the point of leakage—certain materials that are used for the construction of storage

TABLE 1.2 Autoignition Temperature of Comparable Fuels (Mazloomi and Gomes, 2012)

Fuel	Autoignition Temperature (°C)
Hydrogen	585
Methane	540
Propane	490
Butane	405
Methanol	358
Gasoline	246–280
Diesel	210

containers, piping, valves, etc., referred to as hydrogen embrittlement (Cramer and Covino, 2003).

To ensure the safe use of hydrogen, leakage must be prevented, because hydrogen is flammable and explosive, and any "confined" situation can be dangerous, requiring the use of appropriate safety devices[1] (fans, sensors, etc.). A good knowledge of these dangers and their consequences is needed in order to implement safe designs for systems using hydrogen. Besides, hydrogen is nontoxic and is very volatile.

The low density and high diffusion coefficient of hydrogen are safer than other fuels. Generally, wider ignition limits, lower energies, and lower temperatures for ignition make a fuel less safe, as they increase the beginning and extent of fire. One of the most important safety parameters is the autoignition temperature, that means, the temperature at which the material will ignite without any external ignition source. As seen in Table 1.2, hydrogen has the highest autoignition temperature, and this is a positive safety characteristic among various fuels.

A fuel is considered to be less safe if it possesses higher flame temperature, explosion energy, and flame emissivity, because its fire would be more damaging. When comparing hydrogen, gasoline, and methane, hydrogen turns out to be the safest fuel with a safety factor of 1, compared with 0.8 and 0.53 for methane and gasoline, respectively (Sharma and Ghoshal, 2015).

Hydrogen is nontoxic, yet extremely flammable. However, the flame temperature is almost the same as the others fuels, and its fire lasts 0.1–0.2 times that of a hydrocarbon consuming fire with the same volume, and the inhalation

1. It is difficult to detect without appropriate sensors because it is both colorless and odorless, and airborne flame is almost invisible.

of its smoke is absolutely harmless (Mazloomi and Gomes, 2012). Because of its nontoxicity, a hydrogen leak cannot cause environmental damage, and cannot be detected by simple smell.

In 1990, the International Standard Organization (ISO) established a technical committee to develop standards in the field of production, storage, transport, and various applications of hydrogen as, for example, the European Integrated Hydrogen Project (EIHP), which makes proposals for the regulation of FCEVs and hydrogen activities (Devillers et al., 2000). Another project concerned with safety issues on a technical level is the European Network of Excellence, HySafe (HySafe—Safety of Hydrogen as an Energy Carrier, 2007). Placing hydrogen at public fueling stations and using it in vehicles has created a need for new safety requirements. Hydrogen storage is regarded as one of the most critical issues that must be solved before a technically and economically viable hydrogen infrastructure can be implemented. In fact, without effective storage systems, a hydrogen economy will be difficult to achieve (Dagdougui, 2012).

1.4 DEPLOYMENT STRATEGIES OF HYDROGEN SUPPLY CHAIN

Many of the required technologies are already available today and a current challenge is to deploy hydrogen infrastructure and scale up manufacturing capacities so as to achieve competitive cost and mass market acceptance. The hydrogen contribution to the energy transition has accelerated over the last years, following the phases of precommercialization in all sectors. Several improvements along the entire value chain of hydrogen need yet to be made, mainly in the field of cost and performance (Fig. 1.9) (Hydrogen Council, 2017a).

1.4.1 Barriers to be Overcome for Hydrogen Supply Chain Deployment

A key point in the development of hydrogen supply chains is the demonstration of the feasibility of the infrastructure while economic and social obstacles must be overcome.

1.4.1.1 Economic Barriers

Economic obstacles include the cost of FCs and of hydrogen and the lack of cash flow and of a supply base during the first phase of deployment. The main institutional hurdles are difficulties of policy and regulatory frameworks for disruptive technologies moving from demonstration to large-scale deployment across the "valley of death" (Fig. 1.10). Societal barriers include insufficient coverage of FCs and hydrogen technologies (Cantuarias-Villessuzanne et al., 2016; Creti et al., 2015), such as the lack of recognition of hydrogen and its benefits in

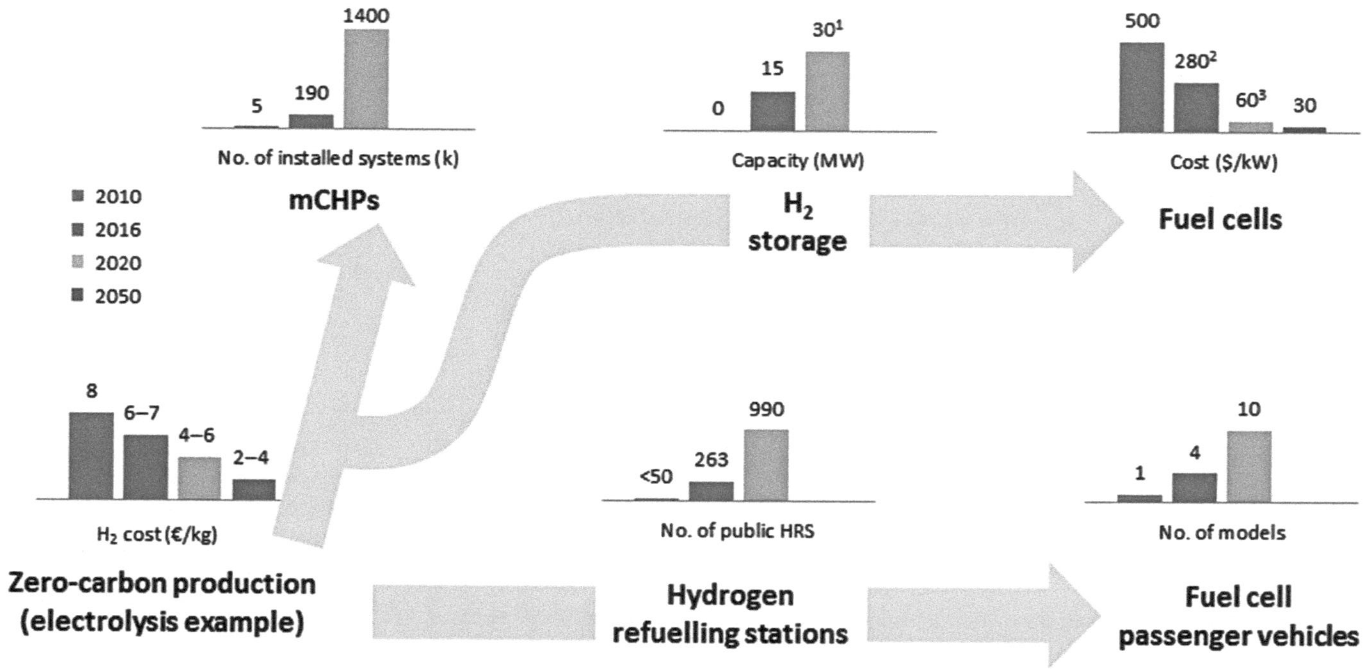

[1] Extrapolating the growth to 20 MW in 2017/2018 from outstanding projects.
[2] Assuming 20k units production per year.
[3] Assuming 100k units production per year in 2025.

FIG. 1.9 Examples of hydrogen technologies and the continuous improvements along the entire value chain. *(Adapted from Hydrogen Council, 2017a. How hydrogen empowers the energy transition [WWW Document]. Rep. January. http://hydrogeneurope.eu/wp-content/uploads/2017/01/20170109-HYDROGEN-COUNCIL-Vision-document-FINAL-HR.pdf (Accessed 29 December 2017).)*

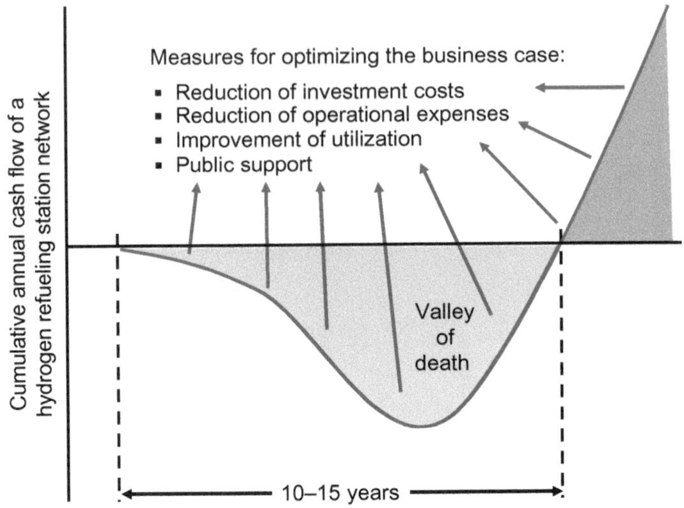

FIG. 1.10 Valley of death in hydrogen deployment from IEA (2015).

the energy transition, the absence of mechanisms to mitigate and share the long-term risks of the initial investments, a lack of coordinated action across stakeholders, a lack of fair economic treatment of a developing technology, and limited technology standards to drive economies of scale (Hydrogen Council, 2017a).

As highlighted in many investigations, the deployment of the hydrogen infrastructure is expected to be gradual. The main problem lies in its chicken-and-egg dilemma. Without a convenient hydrogen refueling infrastructure, there is a lack of interest in buying FCEVs, and the car manufacturers are not motivated to produce these vehicles. If there are no options for FCEVs to create a demand and market conditions, there is no interest in building a network of hydrogen refueling stations (HRS) (Ball et al., 2015). For that to occur, both sides of the network must be developed simultaneously and gradually. During the first decade of developing a hydrogen supply infrastructure for mobility, the costs are highest, exceeding 10 €/kg, mainly caused by the low rate of utilization of the HRS and the small number of FCEVs on the road. Over time, hydrogen costs are expected to fall. In the next 5 years, in the early commercial phase, the number of HRS and FCEVs is expected to increase, reducing the delivery cost to 7–9 €/kg, thus making hydrogen competitive with fossil fuels for the first time. With the arrival of the commercial phase, with a mature market, a well-developed HRS network and the current use of FCEVs, the hydrogen cost is expected to be in the range 4–5 €/kg. In fact, in 2030, 800,000 FCEVs (including large, mid-size and light commercial vehicles), 600 HRS and a demand of 90,000 t of hydrogen are expected to be reached in France, versus only 45 FCEVs, 96 HRS and a hydrogen demand of 6 t in 2022 (Mobilité Hydrogène France, 2016).

Risk mitigation hinges upon close collaboration among many stakeholders involved in the hydrogen supply chain, such as the different actors directly involved in the various echelons of the supply chain, power grid providers, car manufacturers, as well as local, regional, and national authorities.

1.4.1.2 Barriers Related to Social Acceptance and Safety

Effective public education will be necessary to achieve the widespread social acceptance of hydrogen technologies. Yet, as aforementioned, H_2 is commonly used in industry and is actually no more dangerous than other flammable fuels, such as gasoline and natural gas. Nevertheless, under specific conditions, hydrogen can behave dangerously. The burning or explosion of hydrogen causes most fatal accidents, so that the development of hydrogen infrastructure requires stringent safety considerations (Kim and Moon, 2008; Hake et al., 2006).

In the transition to a hydrogen economy, the public perception of safety is a critical issue. Although the public view on hydrogen is in general positive, an early large accident could change this perception quickly.

Convincing the consumer that FCEVs are safe will be one of the major tasks during the early market introduction phase. International hydrogen technology-related training programs are the core of the European HySafe project that will be further developed (IEA, 2017b).

The social perception of the use of hydrogen in FCEVs has been investigated (Bellaby et al., 2016). This research combines a quota sampling survey of 1003 adults across three disparate "travel-to-work areas" in England with representative focus groups. According to this study, "participants highlighted the benefits in hydrogen energy, not presenting any issues for its safety, discussing their risks, although the costs were considered a problem."

The study AIDHY/CEA (Le Duigou, 2010) conducted in 2010 was a decision support for the identification of societal changes brought by new hydrogen technologies. It analyzed the acceptability among users and experts. According to this presentation, the public was not worried by safety issues surrounding the use of hydrogen. Potentially 77% of people said they would be interested in this fuel, and strong acceptance appeared in highly urban areas.

The organization of public debates is also of tremendous importance (e.g., BEV vs. FCEV); one of the challenges is to demonstrate that they are part of a long-term vision. In France, the National Debate for the Energy Transition (Débat national sur la transition énergétique, 2013) allows for the discussion of some issues. How can France move toward energy efficiency and energy conservation? How to achieve the energy mix targets? Which renewable energies should France rely on? These issues have been discussed by civil society and experts. The Regional Innovation Agency MPI (Midi-Pyrénées Innovation) organized this debate on June 3, 2013 in Toulouse, treating the "hydrogen as fuel" topic based on the results obtained in a case study of the Midi-Pyrénées region (De León Almaraz, 2014).

Communication efforts will always be important in the transition phase, because each stakeholder has its own "language" and a large quantity of analysis about potential scenarios regarding the H_2 economy is mandatory.

1.4.2 Initiatives for Infrastructure Development and Roadmaps

Hydrogen supply chains have been the subject of various roadmaps and scenarios that have been developed at regional, national, and multinational levels. These roadmaps identify the steps needed to accelerate the implementation of radical technology changes, in order to help enable governments, industry, and financial partners to make the right choices (De León Almaraz, 2014). Their main objective is to evaluate some industrial, technological, environmental, and social issues and to identify the main obstacles associated with the hydrogen economy. They all agree that a key point in the development of the hydrogen supply chain is the demonstration of the feasibility of its infrastructure, though many technical, economic, and social obstacles must be overcome (Hydrogen Council, 2017b; IEA, 2017b).

This assessment agrees with the observation made from a literature review of recent scientific publications that emphasizes the need to develop systemic studies in order to demonstrate the feasibility of the sector and to validate the technical and economic interest in the production and recovery of hydrogen produced from renewable sources. The use of hydrogen for transport applications has generated the most interest and, subsequently, has been studied more extensively than stationary applications. Very few of the national roadmaps and hydrogen scenarios reviewed here consider both of these applications.

A factual and well-documented report entitled "A portfolio of power-trains for Europe: a fact-based analysis" (McKinsey & Company, 2010) provided a comparison of four different powertrains—BEVs, FCEVs, PHEVs and ICEs—on economics, sustainability, and performance across the entire supply chain (Wheel to Wheel, WtW) between 2010 and 2050, based on confidential and proprietary industry data. Various scenarios with different potential hydrogen demand behaviors (three scenarios: 5%, 25%, and 50% in 2050) were built. The study was conducted at a continental level. This study considered the interconnection with many stakeholders of the HSC, including vehicle producers (BMW AG, Daimler AG, Ford, General Motors LLC, Honda R&D, Hyundai Motor Company, Kia Motors Corporation, Nissan, Renault, and Toyota Motor Corporation) and industrial gas companies (Linde and Air Liquide). From 2010 to 2020, all cost and performance projections are based on proprietary industry data and on projected learning and annual improvement rates after 2020.

Some significant initiatives undertaken at the worldwide, European, and national levels are discussed below, although the list is not exhaustive.

1.4.2.1 Worldwide Level

The Technology Roadmap "Hydrogen and Fuel Cells," developed by the IEA, 2015 has as its purpose to lay out the potential of hydrogen and its limitations in different energy sectors, aiming to provide extensive information about the function and cost of hydrogen technologies, to identify the sectors in which hydrogen can offer the maximum added value, and also the actions required to deploy hydrogen technologies (IEA, 2015).

According to the different scenarios that have been evaluated, the adoption of renewable hydrogen, as opposed to fossil-derived hydrogen (with or without CCS), strongly depends on its economic competitiveness. The relationships between natural gas price, electricity price, annual full-load hours, carbon price, and the resulting cost of hydrogen have been studied in depth. Even under optimistic assumptions with regard to the techno-economic parameters of the electrolyzer, electrolytic hydrogen remains more expensive than hydrogen from natural gas reforming, unless very low-cost renewable electricity is available and carbon or natural gas prices are high.

1.4.2.2 European Level

In the *HyWays project* (2007–08), over 50 member-state (MS) workshops were conducted with key stakeholders. The HyWays project combines technology databases and socio-techno-economic analyses to evaluate selected stakeholder scenarios for future sustainable hydrogen energy systems. In this project, market scenarios for hydrogen end-use applications were also developed. Each country outlined its own preferences (Finland, France, Germany, Greece, Italy, the Netherlands, Norway, Poland, Spain, and the United Kingdom). The HyWays project differs from other road mapping exercises because it integrates stakeholder preferences, obtained from multiple member state workshops, with extensive modeling in an iterative way. The stakeholder validation process, which takes into account country specific conditions, is a key element of the road mapping process. In Europe, the prospect of the hydrogen economy plays a major role, especially because of the aggregation of many countries that have various specific institutions, opportunities, conditions, and territorial and socio-economic barriers (Dagdougui, 2012). Another well-known roadmap is *H_2 Mobility* (2010), which outlines a plan to introduce the use of FCEVs in Europe, starting in Germany and the United Kingdom (Williamson, 2010) with others following in 2013. In this program, the main car manufacturers and gas producers are involved.

Hydrogen Mobility Europe 2 (H2ME 2) is a six-year project that runs to the end of June 2022, and which brings together 37 partners from eight European countries. It includes the deployment and operation of 1230 FCEVs and the addition of 20 new hydrogen refueling stations (HRS), and it will test the ability of electrolyzers to simultaneously feed hydrogen stations and help balance the electrical grid (H2ME 2 launched in Europe to grow hydrogen fueling infrastructure network and vehicle fleet, 2016).

The HYRREG project, founded by the Program of Cooperation of Southwest Europe (SUDOE) has the objective of developing a platform to generate hydrogen-related programs and a roadmap between the countries of the southeast of Europe, namely, France, Portugal, and Spain. The HYRREG roadmap, mainly based on a qualitative analysis, is focused on stakeholder preferences and region-specific conditions, highlighting the importance of hydrogen in the transport sector and its production from renewable sources.

1.4.2.3 Some National Levels

Iceland can be identified as the first hydrogen economy in the world. *The Icelandic New Energy* is a partnership that was established between the Icelandic government, Shell, Norsk Hydro, Ford, and the University of Reykjavik. It was a pioneering initiative to create the first hydrogen economy in the world by 2040 (Dunn, 2000). The project aims to promote energy independence by exploiting the vast renewable resources of the island, such as geothermal and hydropower (Dutta, 2014). In 2003, the country inaugurated the first hydrogen station in the world to supply the three hydrogen buses that ran through the ECTOS program until the end of 2006. In 2007, a new demonstration phase began with 13 SMART-H_2 model cars and boats (Bento, 2010).

With very few natural resources, Japan has been a pioneer in investment in hydrogen. The world's largest national hydrogen program (WE-NET) was established in 2003 in Japan with market penetration targets on FCVs and hydrogen energy. More recently, the Ministry of Economy, Trade, and Industry (METI) in Japan established in 2014 the Strategic Road Map for Hydrogen and Fuel Cells in order to formulate a Roadmap toward the implementation of a "hydrogen society" (Agency for Natural Resources and Energy, 2014). The METI proposed three phases to achieve this goal: first, the expansion of the scope of application for fuel cell technology, such as fuel cells for households and for vehicles, scheduled to begin in 2014; second, the development of a system for supplying the hydrogen using energy resources imported from other countries, while introducing hydrogen power generation with a time frame in the late 2020s; third, establishment of a carbon-dioxide-free hydrogen supply system using renewable and other energy around 2040.

In Germany, the National Innovation Program for Hydrogen and Fuel Cell Technology NIP (2006–16) has contributed significantly to fostering the market development of hydrogen technology. It was founded in 2006 as a joint effort of German policymakers, industry leaders, and the research community. Germany covers the length of the supply chain with 20 manufacturers and 12 distributors of fuel cells and components.

Various studies have been conducted to estimate the development of a hydrogen economy in the United States and its implications in terms of emissions as well as infrastructure requirements. The *US* Department of Energy

(DOE) has developed a hydrogen program in cooperation with industry, national laboratories, universities, and government agencies since 1980 (Bento, 2010; Patay, 2008).

One of the most famous H_2 plans can be found in California. Since 1999 demonstration projects have been managed by the *California Fuel Cell Partnership* (CaFCP). This partnership was established between Ballard Power Systems, Daimler Chrysler, Ford Motor Company, Shell Hydrogen, and Chevron (formerly ARCO), and the California agencies (California Air Resources Board and California Energy Commission). Its mission is to facilitate the commercialization of fuel cells for transportation. The CaFCP has supported the establishment of stations and had experience with more than 170 hydrogen vehicles. Stations are normally financed by the state government (Bento, 2010). More recently, two programs, the California Fuel Cell Partnership and the *California Energy Commission and Air Resources Board* are in effect. The former is a roadmap for fuel cell vehicles and hydrogen stations and the latter is a program of incentives for vehicles and fueling infrastructure from state agencies.

In France, HyFrance3 is a project to support the HyWays project. This project was launched to focus on the competitiveness of different steps of the hydrogen chain, from production to end use, with a time horizon of 2030 (Le Duigou et al., 2013). The roadmap describes and analyzes the French hydrogen industrial markets, including different scenarios. The infrastructure to distribute hydrogen in France for automotive applications up to 2050 is also studied, taking into account hydrogen storage to balance the supply and demand characteristics.

The "Office Parlementaire d'Evaluation des Choix Scientifiques et Technologiques" (OPECTS) in France has also presented a roadmap to implement a national hydrogen network, promoting the use of FCEV via tax elimination (AFHYPAC, 2012).

The French Association for Hydrogen and Fuel Cells (Association Française pour l'Hydrogène et les Piles à Combustible*: AFHYPAC*) started a French study on H_2 power and mobility in May 2013. "Mobilité Hydrogène France" is a consortium of private and public stakeholders formed by the French Association for Hydrogen and Fuel Cells (AFHYPAC). The consortium spans the gamut from energy companies to end customers.

A first analysis of market segments highlighted the role of clusters of captive fleets. Under the 2015 FCH-JU call, the deployment started in 2015 with 10 hydrogen stations and a first wave of fuel cell vehicles. The consortium members continue to develop the next phase of the rollout strategy for hydrogen mobility in France, to support the introduction of fuel cell passenger cars for fleets and private citizens.

A Hydrogen Territory initiative was launched in France in 2016, which aims to demonstrate, on the scale of a particular territory, the technical and economic feasibility as well as the environmental benefits of deploying hydrogen in energy networks or local energy applications. The Occitania region has been awarded the metaproject HyPort, which combines advanced innovation in hydrogen fuel cell applications for aeronautics, green hydrogen production, and H_2 mobility deployment for the Toulouse International Airport, Tarbes-Lourdes-Pyrénées regional airport, and the vast urban, rural, and tourist perimeters connected to them.

Table 1.3 summarizes some roadmaps. The advantage of these plan scenarios is the potential to use the information generated in other scenarios, and also the demonstration of the potential of hydrogen in automobile applications.

1.4.3 Implementation Steps

Transition plan scenarios can be taken as an important basis for more precise studies in which the different potential activities of the network can be measured and analyzed to design the hydrogen supply chain.

The different roadmaps and transition plan scenarios that already exist and that have been presented constitute necessary information and can be considered as the starting point to launch more detailed analyses. They highlight general targets as well as coordination and communication efforts from which valuable information is shared. The principal limitations of these macro studies are the difficulty in generating specific results on the location, size, and number of production, storage, or transport units, and also the lack of interconnection between the different objectives. Their main advantage is that they provide valuable information that can be used to implement scenarios of interest, thus demonstrating or not the potential of hydrogen infrastructure.

According to these roadmaps, hydrogen demand for road transport is assumed to be a major parameter and likely to grow in three phases according to Energy Trends 2030 scenario (Ozkan, 2009):

- *Infrastructure phase I*: the demonstration phase during which a few large-scale first user centers are situated across Europe.
- *Infrastructure phase II*: the early commercialization phase with 3–6 user centers per country, possibly including a network of transit roads connecting these centers.
- *Infrastructure phase III*: the full commercialization phase, which encompasses existing user centers as well as newly developed regions, and a dense, long-distance road network. Phase III is assumed to develop in three subphases.

These large-scale deployment initiatives must be supported by long-term policy frameworks in countries that are early adopters. These deployment initiatives should use current activities as a platform to scale their successes up to a national level (Hydrogen Council, 2017b).

A common assumption in all roadmaps is that hydrogen use will take off in densely populated urban areas at first and then gradually expand outward through

TABLE 1.3 Plan Scenarios of Various Roadmaps

Roadmap	Scale	Year	Description
IEA	Worldwide	2017	Evaluation of some industrial, technological, environmental, and social issues, and identification of the main obstacles associated with the hydrogen economy
Hydrogen Council			
IEA Hydrogen Technology Roadmap		2015	This roadmap details the steps governments, industry, and researchers need to take to foster and to track the deployment of hydrogen technology
The Technology Roadmap. Hydrogen and Fuel Cells			Description of the potential of hydrogen and its limitations in different energy sectors
H2ME	Europe	2016	Largest EU-funded project for hydrogen mobility and FCEV deployment
			Integrated in Mobility Europe project (now called H2ME 1)
			Involves Germany, Scandinavia, France, and the United Kingdom
			Plans for 300 FCEVs and 29 hydrogen refueling stations in 2020 and the operation of 1230 FCEVs and 20 more HRS in 2022
HYRREG	France, Portugal, and Spain	2013	Development of a platform to generate hydrogen-related programs and a roadmap between the countries of the southeast of Europe through the HyFrance3, HiPo, and EDEN projects
HyWays	Finland, France, Germany, Greece, Italy, Netherlands, Norway, Poland, Spain, and the United Kingdom	2008	Compilation of all pivotal technological and socioeconomic factors related to a future hydrogen infrastructure, providing a number of scenarios under different assumptions. Some market scenarios for hydrogen end-use applications are developed (Finland, France, Germany, Greece, Italy, the Netherlands, Norway, Poland, Spain, and the United Kingdom)

Continued

TABLE 1.3 Plan Scenarios of Various Roadmaps—cont'd

Roadmap	Scale	Year	Description
Australian National Hydrogen Study	Australia	2003	Assessment of the necessary steps to use hydrogen to meet its future energy requirements Three potential uses were addressed: road transport, portable electrical appliances, and generation. Three scenarios based on the demand (low, medium, and high) are described within the timeline 2020–50
Joint Research Centre Ispra of the European Communities and the Government of Quebec	Canada	2002	The idea of this project is to provide renewable primary energy from the already available hydroelectricity in Quebec for hydrogen production via electrolysis. Hydrogen will be shipped to Europe to be stored and used in different applications (Momirlan and Veziroglu, 2002)
H2 Mobilité France	France	2016	Creation of hydrogen clusters that could provide nationwide coverage by 2030
HyFrance3		2013	Description and assessment of the French hydrogen industrial markets, including different scenarios
OPECTS		2012	Proposes that hydrogen be tax free for a transition period, except for the hydrogen produced from fossil fuels
Roadmap on hydrogen energy (ADEME)		2011	Four scenarios for hydrogen energy in 2050, in which each one represents specific infrastructures and interactions with other energy industries

National Innovation Programme for Hydrogen and Fuel Cell Technology NIP (2006–16)	Germany	2017	Three objectives are involved: (1) Secure Germany's position as a technology leader in hydrogen and fuel cells; (2) Accelerate the development of the hydrogen and fuel cell markets; and (3) Strengthen the industry along the whole hydrogen and fuel cell value chain
Icelandic New Energy Ltd.	Iceland	1999	Development of feasibility studies to produce hydrogen locally from hydro and geothermal power
National Hydrogen Energy Roadmap	India	2014	Production of hydrogen through biological and biomass routes, using renewables sources (Dutta, 2014)
METI	Japan	2014	Establishment of a carbon-dioxide-free hydrogen supply system using renewable and other energy in different steps until 2040
California Fuel Cell Partnership	United States	2012	Implementation of FCEVs in California, concluding that the refueling stations must precede the vehicles, must have free access, and must be easy to use

so-called "hydrogen corridors." For the sake of illustration, the Hydrogen Corridor between Spain, France, and Andorra (H2PiyR Initiative) aims to develop a cross-border corridor of refueling stations for hydrogen vehicles connecting Spain (Catalonia and Aragon), France, and Andorra with central and northern Europe, where the deployment of infrastructure with this type of mobility is more advanced, through the installation of 10 hydro generators (ABC.es, 2017). The corridor will include the Spanish cities of Zaragoza, Huesca, Fraga and Tarragona, as well as Andorra (city) and Pamiers (France). They will join the already existing stations in Huesca and Zaragoza, and the French ones in Rodez and Albi (Afhypac - Newsletter Hynovations - #62, 2016). The project includes 16 FCVs (Fuel Cell Vehicles) (6 passenger cars, 8 commercial vehicles, and 2 buses) (L'hydrogène, de Rodez à Sarragosse, 2016).

Some studies assume that hydrogen can also be used for stationary applications between the commercialization and mass use phases.

The economic viability of hydrogen mobility from an infrastructure or retail operator's perspective can be helped with a fast rollout of FCEVs. The introduction of hydrogen-fueled captive fleets could further help hydrogen development (Ball et al., 2015). A captive fleet can be defined as a fleet of vehicles with predictable driving and refueling patterns that make regular visits to a parking lot or depot. A cluster is based on multiple fleets of customers in a defined area (Mobilité Hydrogène France, 2016). At least one HRS must be founded in each cluster. The development via clusters maximizes the utilization of the HSR installed. The next steps in the development of the hydrogen infrastructure consist of analyzing consumer behavior and developing a widespread network for passenger car drivers (Fig. 1.11).

Post offices in Audincourt (Doubs) and Périgny (Jura) in France have distributed letters since 2013 by small FCVs (Fig. 1.12), developed under the auspices of the project MobyPost (Bettayeb, 2017).

A total of 50 FCVs (in 2017) have been sold to different companies under the framework of the Hyway project. The end of phase 1 consists in building two hydrogen refueling stations (in Grenoble and Lyon). For the second phase, the objective is to produce "green" hydrogen, from renewable energy (solar, wind).

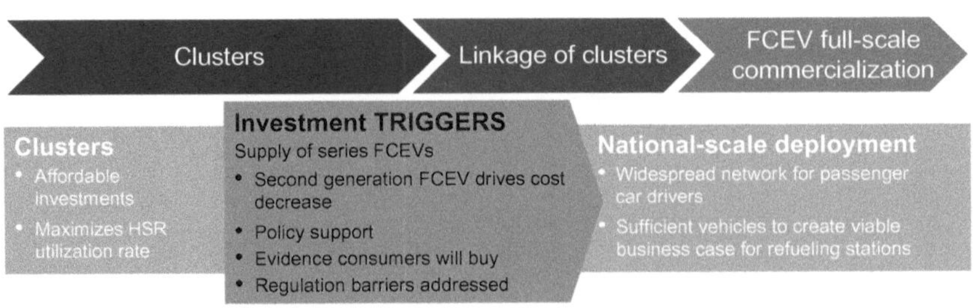

FIG. 1.11 Captive fleet approach: A 3 phase national roll-out (Mobilité Hydrogène France, 2016).

FIG. 1.12 The small FCVs used by the post offices.

The projects in Corsica with the Myrte platform (Mission renewable hydrogen for the integration to the electrical) and Jupiter 1000 in Fos-sur-Mer are perfect examples of this second stage (HyWay, 2017).

1.5 CONCLUSIONS

This chapter aims to shed some light on the challenges and opportunities that lie ahead for the development of hydrogen energy policies and for the deployment of hydrogen infrastructure. Hydrogen produced from renewable sources and used in fuel cells, both for mobile and stationary applications, constitutes a very promising energy carrier in the context of sustainable development. Hydrogen and fuel cell technologies have significant potential to improve energy security and mitigate the effects of climate change and other harmful environmental impacts, thus effecting a switch toward a clean, low-carbon energy system.

Yet hydrogen should not be seen as the unique solution to the world's energy problems and in particular not as the unique answer to the challenges that the transport sector has to cope with. The energy challenge will probably be solved by a much more diversified portfolio of fuels in the future.

Many energy sources, production processes, transportation modalities, storage modes, and end uses exist for hydrogen, so that the way that the hydrogen economy could be developed is very flexible. The concept of Power-to-Gas considered in isolation for hydrogen or extended to its coupling with methanation constitutes an attractive perspective.

Nowadays, the cost of hydrogen is considered prohibitive compared to the fossil fuels currently used in the transportation system, but the development of some technologies associated with hydrogen, coupled with high oil prices, has improved its competitiveness.

Some strategic roadmaps have been published about the potential of hydrogen at the continental, national, and regional levels. Their main objective has been to evaluate some of the industrial, technological, environmental, and social issues, and to identify the main obstacles in the development of the hydrogen economy. These roadmaps are particularly useful for accelerating the deployment of hydrogen technologies, with the support of policymakers, the private sector, and society; according to these studies, although many of the required technologies are already available today, the deployment of a hydrogen infrastructure constitutes a challenging task for the development of a hydrogen economy that is cost competitive and accepted by the mass market.

In this context, methodologies for the deployment and design of hydrogen supply chains need to be addressed, taking into account the variety of energy sources, production and storage technologies, transportation modes, and stakeholders, to link hydrogen demand to its supply with economic, environmental, and societal considerations.

REFERENCES

ABC.es, 2017. España tendrá 20 estaciones de servicio para hidrógeno antes de 2020 [WWW Document]. abc. http://www.abc.es/motor/reportajes/abci-espana-tendra-20-estaciones-servicio-para-hidrogeno-antes-2020-201703021822_noticia.html. (Accessed 5 September 2017).

AFHYPAC, 2012. Rapport d'activité de la France sur l'Hydrogène et les Piles à Combustible, période juin 2012 à novembre 2014 [WWW Document]. http://www.afhypac.org/documents/publications/rapports/AFHYPAC_RA_bd.pdf.

Afhypac - Newsletter Hynovations - #62, 2016. (WWW Document). http://www.afhypac.org/newsletter/62/actualites-internationales-81/. (Accessed 5 September 2017).

Agency for Natural Resources and Energy, 2014. Summary of the Strategic Road Map for Hydrogen and Fuel Cells. Ministry of Economy, Trade and Industry, Japan.

Ball, M., Weeda, M., 2015. The hydrogen economy—vision or reality? Int. J. Hydrog. Energy 40, 7903–7919. https://doi.org/10.1016/j.ijhydene.2015.04.032.

Ball, M., Wietschel, M., 2009. The future of hydrogen—opportunities and challenges. Int. J. Hydrog. Energy 34, 615–627. https://doi.org/10.1016/j.ijhydene.2008.11.014.

Ball, M., Basile, A., Veziroglu, T.N., 2015. Compendium of Hydrogen Energy: Hydrogen Use, Safety and the Hydrogen Economy. Woodhead Publishing, Oxford. https://doi.org/10.1016/B978-1-78242-364-5.09987-5.

Bellaby, P., Upham, P., Flynn, R., Ricci, M., 2016. Unfamiliar fuel: how the UK public views the infrastructure required to supply hydrogen for road transport. Int. J. Hydrog. Energy 41, 6534–6543. https://doi.org/10.1016/j.ijhydene.2016.03.024.

Bento, N., 2010. The Transition to a Hydrogen Economy: Infrastructure and Technological Change. Université de Grenoble, France.

Bettayeb, K., 2017. Les défis de la voiture à hydrogène. CNRS J. [WWW Document]. https://lejournal.cnrs.fr/articles/les-defis-de-la-voiture-a-hydrogene. (Accessed 5 September 2017).

Brown, P.T., Caldeira, K., 2017. Greater future global warming inferred from Earth's recent energy budget. Nature 552, 45. https://doi.org/10.1038/nature24672.

Cantuarias-Villessuzanne, C., Weinberger, B., Roses, L., Vignes, A., Brignon, J.-M., 2016. Social cost-benefit analysis of hydrogen mobility in Europe. Int. J. Hydrog. Energy 41, 19304–19311.

Cipriani, G., Di Dio, V., Genduso, F., La Cascia, D., Liga, R., Miceli, R., Ricco Galluzzo, G., 2014. Perspective on hydrogen energy carrier and its automotive applications. Int. J. Hydrog. Energy 39, 8482–8494. https://doi.org/10.1016/j.ijhydene.2014.03.174.

Cramer, S.D., Covino, B.S., 2003. Corrosion: Fundamentals, Testing and Protection. ASM International, Ohio, USA.

Creti, A., Kotelnikova, A., Meunier, G., Ponssard, J.-P., 2015. A cost benefit analysis of fuel cell electric vehicles, Research Report, Report 1: 1–41.

Dagdougui, H., 2012. Models, methods and approaches for the planning and design of the future hydrogen supply chain. Int. J. Hydrog. Energy 37, 5318–5327. Optimization approaches to hydrogen logistics. https://doi.org/10.1016/j.ijhydene.2011.08.041.

De León Almaraz, S., 2014. Multi-Objective Optimisation of a Hydrogen Supply Chain. INPT, Toulouse.

Débat national sur la transition énergétique, 2013. (WWW Document). http://www.transition-energetique.gouv.fr/liste-des-debats-en-france. (Accessed 25 November 2013).

Devillers, C., Pehr, K., Stoll, D., Wetzel, F., Duffield, J., Grubel, H., Zisler, S., Vandenborre, H., Driessens, T., Gonzalez, A., Vegas, G., Wurster, R., Kesten, M., Machel, M., Michel, F., Heurtaux, F., Adams, P., 2000. European integrated hydrogen project (EIHP). In: Proceedings of the Hydrogen Energy Progress Xiii.Vols. 1 and 2. pp. 139–149.

Dincer, I., Acar, C., 2015. Review and evaluation of hydrogen production methods for better sustainability. Int. J. Hydrog. Energy 40, 11094–11111. https://doi.org/10.1016/j.ijhydene.2014.12.035.

Dodds, P.E., Staffell, I., Hawkes, A.D., Li, F., Grünewald, P., McDowall, W., Ekins, P., 2015. Hydrogen and fuel cell technologies for heating: a review. Int. J. Hydrog. Energy 40, 2065–2083. https://doi.org/10.1016/j.ijhydene.2014.11.059.

DOE Hydrogen Program, 2006. Hydrogen Fuel Cells. U.S. Department of Energy Hydrogen Program. https://www.hydrogen.energy.gov/pdfs/doe_fuelcell_factsheet.pdf.

Dunn, S., 2000. The Hydrogen Experiment. World Watch Institute, Washington, DC.

Dutta, S., 2014. A review on production, storage of hydrogen and its utilization as an energy resource. J. Ind. Eng. Chem. 20, 1148–1156. https://doi.org/10.1016/j.jiec.2013.07.037.

Florida Solar Energy Center, n.d. Hydrogen Basics—Storage [WWW Document]. http://www.fsec.ucf.edu/en/consumer/hydrogen/basics/storage.htm (Accessed 3 October 2017).

Götz, M., Koch, A.M., Graf, F., 2014. State of the art and perspectives of CO2 methanation process concepts for power-to-gas applications.International Gas Union Research Conference, Copenhagen.

Götz, M., Lefebvre, J., Mörs, F., McDaniel Koch, A., Graf, F., Bajohr, S., Reimert, R., Kolb, T., 2016. Renewable power-to-gas: a technological and economic review. Renew. Energy 85, 1371–1390. https://doi.org/10.1016/j.renene.2015.07.066.

Grond, L., Schulze, P., Holstein, J., 2013. Systems Analyses Power to Gas: Deliverable 1: Technology Review. DNV KEMA Energy & Sustainability, Groningen.

H2ME 2 launched in Europe to grow hydrogen fueling infrastructure network and vehicle fleet, 2016. Fuel Cells Bull. 2016, 1. https://doi.org/10.1016/S1464-2859(16)30131-6.

Hake, J.-F., Linssen, J., Walbeck, M., 2006. Prospects for hydrogen in the German energy system. Energy Policy 34, 1271–1283. https://doi.org/10.1016/j.enpol.2005.12.015.

Hydrogen Council, 2017a. How Hydrogen Empowers the Energy Transition [WWW Document]. Rep. January. http://hydrogeneurope.eu/wp-content/uploads/2017/01/20170109-HYDROGEN-COUNCIL-Vision-document-FINAL-HR.pdf. (Accessed 29 December 2017).

Hydrogen Council, 2017b. Hydrogen Scaling Up. A Sustainable Pathway for the Global Energy Transition [WWW Document]. http://hydrogencouncil.com/wp-content/uploads/2017/11/Hydrogen-scaling-up-Hydrogen-Council.pdf. (Accessed 4 January 2018).

HySafe—Safety of Hydrogen as an Energy Carrier, 2007. (WWW Document). http://www.hysafe.org. (Accessed 8 February 2018).

HyWay, 2017. Tenerrdis [WWW Document]. https://www.tenerrdis.fr/fr/projets/hyway/. (Accessed 8 February 2018).

IEA, 2014. Heating Without Global Warming: Market Developments and Policy Considerations for Renewable Heat. International Energy Agency, Paris.

IEA, 2015. Technology Roadmap. Hydrogen and Fuel Cells [WWW Document]. https://www.iea.org/publications/freepublications/publication/TechnologyRoadmapHydrogenandFuelCells.pdf. (Accessed 31 December 2017).

IEA, 2017a. International Energy Agency Technical Report. Key World Energy Statistics [WWW Document]. https://www.iea.org/publications/freepublications/publication/KeyWorld2017.pdf. (Accessed 29 December 2017).

IEA, 2017b. Global Trends and Outlook for Hydrogen [WWW Document]. http://ieahydrogen.org/pdfs/Global-Outlook-and-Trends-for-Hydrogen_Dec2017_WEB.aspx.

Institute for Climate Economics, Ministère de l'Environnement, de l'Énergie et de la Mer, 2016. Chiffres clés du climat France et Monde ÉDITION 2017. .

Kim, J., Moon, I., 2008. Strategic design of hydrogen infrastructure considering cost and safety using multiobjective optimization. Int. J. Hydrog. Energy 33, 5887–5896. https://doi.org/10.1016/j.ijhydene.2008.07.028.

Koskinen, P.E.P., Beck, S.R., Örlygsson, J., Puhakka, J.A., 2008. Ethanol and hydrogen production by two thermophilic, anaerobic bacteria isolated from Icelandic geothermal areas. Biotechnol. Bioeng. 101, 679–690. https://doi.org/10.1002/bit.21942.

L'hydrogène, de Rodez à Sarragosse, 2016. Media12 [WWW Document]. http://www.media12.fr/lhydrogene-de-rodez-a-sarragosse/. (Accessed 5 September 2017).

Le Duigou, A., 2010. Témoignage sur le projet AIDHY. CEA, I-tésé, France.

Le Duigou, A., Quéméré, M.-M., Marion, P., Menanteau, P., Decarre, S., Sinegre, L., Nadau, L., Rastetter, A., Cuni, A., Mulard, P., Antoine, L., Alleau, T., 2013. Hydrogen pathways in France: results of the HyFrance3 project. Energy Policy 62, 1562–1569. https://doi.org/10.1016/j.enpol.2013.06.094.

Lehner, M., Tichler, R., Steinmüller, H., Koppe, M., 2014. Power-to-Gas: Technology and Business Models. Springer Briefs in Energy. https://doi.org/10.1007/978-3-319-03995-4_2.

Mazloomi, K., Gomes, C., 2012. Hydrogen as an energy carrier: prospects and challenges. Renew. Sust. Energ. Rev. 16, 3024–3033. https://doi.org/10.1016/j.rser.2012.02.028.

McCarty, R.D., Hord, J., Roder, H.M., 1981. Selected properties of hydrogen (engineering design data). Final report (No. PB-81-211492). National Engineering Lab. (NBS), Boulder, CO.

McKinsey & Company, 2010. A Portfolio of Power-Trains for Europe: A Fact-Based Analysis. The Role of Battery Electric Vehicles, Plug-in Hybrids and Fuel Cell Electric Vehicles [WWW Document]. 2010.

Mobilité Hydrogène France, 2016. H2 Mobilité France. Study for a Fuel Cell Electric Vehicle National Deployment Plan [WWW Document]. http://www.fch.europa.eu/sites/default/files/Smart%20Spec%20Fabio%20Ferrari%20%28ID%202436338%29%20%28ID%202497336%29.pdf. (Accessed 31 December 2017).

Momirlan, M., Veziroglu, T.N., 2002. Current status of hydrogen energy. Renew. Sust. Energ. Rev. 6, 141–179. https://doi.org/10.1016/S1364-0321(02)00004-7.

Ozkan, N., 2009, UKSHEC Working Paper No. 3. A Review of Hydrogen Demands in National Roadmaps. Policy Studies Institute.

Patay, E., 2008. Déploiement de la chaîne logistique de l'hydrogène pour le marché des carburants en 2050: Conception et développement d'un outil d'aide à la décision pour l'analyse de scénarios. Ecole Centrale Paris-Laboratoire Génie Industriel, Paris.
Sharma, S., Ghoshal, S.K., 2015. Hydrogen the future transportation fuel: from production to applications. Renew. Sust. Energ. Rev. 43, 1151–1158. https://doi.org/10.1016/j.rser.2014.11.093.
Sterner, M., Jentsch, M., Holzhammer, U., 2011. Energiewirtschaftliche und ökologische Bewertung eines Windgas-Angebotes. Gutachten für Greenpeace Energy. Fraunhofer IWES, Kassel.
United Nations, 2015. Paris Agreement [WWW Document]. http://unfccc.int/files/essential_background/convention/application/pdf/english_paris_agreement.pdf. (Accessed 29 December 2017).
Williamson, I., 2010. H2-Mobility—Towards Commercialisation of Fuel Cell Vehicles. Presentation to US HTAC. https://www.hydrogen.energy.gov/pdfs/4_williamson_0610.pdf.
World Energy Council, 2013. World Energy Scenarios. Composing Energy Futures to 2050. PSI, United Kingdom.

FURTHER READING

Dagdougui, H., 2011a. Decision Support Systems for Sustainable Renewable Energy Systems and Hydrogen Logistics: Modelling, Control and Risk Analysis. UNIGE; CRC, MINES ParisTech, Italy; Paris.

Chapter 2

Hydrogen Supply Chain Design: Key Technological Components and Sustainable Assessment

Jesus Ochoa Robles, Sofía De-León Almaraz and Catherine Azzaro-Pantel
Laboratoire de Génie Chimique, Université de Toulouse, CNRS, Toulouse, France

ABBREVIATIONS

CCS	carbon capture and storage
FCEV	fuel cell electric vehicle
GHG	greenhouse gas
GWP	global warming potential
HAZID	hazard identification
HRS	hydrogen retail station
HSC	hydrogen supply chain
LCA	life cycle assessment
NGO	nongovernmental organization
OECD	Organisation for Economic Cooperation and Development
PEM	proton exchange membrane
PSA	pressure swing adsorption
R&D	research and development
RES	renewable energy sources
SC	supply chain
SMR	steam methane reforming
SOEC	solid oxide electrolyzer cell
SOFC	solid oxide fuel cell
WtW	well-to-wheel

Hydrogen Supply Chain. https://doi.org/10.1016/B978-0-12-811197-0.00002-6

2.1 INTRODUCTION

The plethora of sources for hydrogen production, along with the variety of methods for its extraction, distribution, and storage, makes hydrogen a very promising fuel in both industrialized and developing countries. This wide variety of options also makes the development of its associated supply chain a challenging task. Supply chain management (SCM) generally uses a set of approaches to efficiently integrate energy source supply, hydrogen production, conditioning, storage, and distribution, so that hydrogen is produced and distributed at the right quantities, to the right locations, and at the right time, in order to minimize system global cost while satisfying service level requirements (Papageorgiou, 2009) and taking into account other constraints. As previously explained in Chapter 1, the path toward a hydrogen economy, and more particularly here, toward the use of hydrogen as an energy vector, must address the economic, social, and environmental pillars of sustainability. This explains why SCM models can be considered as useful to design improved business pathways, which could result in reduced environmental impacts that satisfy local regulations and international treaties for greenhouse gas emissions, while being also economically achievable (De León Almaraz, 2014). A supply chain (SC) can be defined as a set of three or more entities (organizations or individuals) directly involved in the upstream and downstream flows of products, services, finances, and/or information from a source to a customer (Mentzer et al., 2001).

A hydrogen supply chain involves:

- *Multiple echelons* from the choice of energy source, production, storage, transportation, and distribution that interact along the chain.
- *Multiple stakeholders* covering industry, academia, government, energy sector, etc. The transition to a hydrogen economy is a global issue and requires extensive consultation with industry (covering the various echelons of the HSC), energy sector, national and local Government, academia, NGOs, and the finance sector. The importance of public-private partnerships in the achievement of a hydrogen economy is emphasized in several roadmaps (Tomei, 2009). The involvement and support of local communities is also of crucial importance. The current energy system requires collaboration between countries. A review of hydrogen roadmaps also emphasizes the importance of consumer acceptance and awareness of hydrogen technologies in driving the markets for hydrogen (Tomei, 2009).
- *Multiple periods* that reflect the dynamic process from deployment to maturity and the gradual approach to infrastructure buildup, that may be initially concentrated around existing hydrogen clusters, or user centers.
- *Multiple objectives* that reflect the economic, environmental, and social pillars of sustainable development.
- *Multiple users*, including mobility, heat, and industrial applications; this aspect has been tackled in the previous chapter.

The flexibility of hydrogen systems creates many opportunities for system design. A combination of equipment can enable interactions with different sectors. The purpose of the following analysis is to establish which configurations should be considered and explored according to the desired goal. A big challenge is then to assess if hydrogen produced from renewable energy sources can be competitive with current fuels and to deploy an infrastructure of hydrogen SC (HSC) for new applications.

The remainder of this chapter is organized as follows; in Section 2.2, the HSC for industrial and mobility use is introduced. The system is represented by nodes and connections, and a generic HSC framework is displayed. The different echelons (source, production, storage, transportation, etc.) involved in an HSC are then presented. Hydrogen can be produced from both fossil and renewable resources (Section 2.3). Section 2.4 reviews the different technologies related to the various sources. A highlight is that hydrogen can provide storage options for intermittent renewable technologies, such as solar and wind, and might thus facilitate their large-scale introduction into the energy mix. The degree of centralization, determined by the plant capacity, is introduced, and processes, such as steam methane reforming, gasification, and electrolysis, are presented. Finally, issues related to CO_2 capture and storage are mentioned. Hydrogen conditioning and storage are the core of Section 2.5. Section 2.6 is devoted to transportation, with different modes, such as pipelines, tube trailers, and tanker trucks, to supply H_2 fuel to the refueling stations, which are discussed in Section 2.7. These key technological components have to be integrated in an HSC, and criteria of sustainable assessment are presented in Section 2.8.

2.2 HYDROGEN SUPPLY CHAINS

Because the design of the supply chain may vary according to the desired goal, there is no unique hydrogen supply chain, and thus it is difficult to embed in one generic formulation all the various possibilities that may be encountered. Several options that can be captured to build a hydrogen supply chain network framework are collected in Fig. 2.1.

Two cases that may be considered are covered in the following sections.

2.2.1 H_2 Supply Chain as a Feedstock for Industrial Uses

The chemical industry is the largest producer and consumer of hydrogen (see Tables 2.1 and 2.2) as one of the key starting materials used (Ausfelder and Bazzanella, 2016). It is a fundamental building block for the manufacture of ammonia, and hence fertilizers, of methanol, used in the manufacture of many polymers, and also of the refining industry (for hydrotreatments by hydrogenation of unsaturated hydrocarbons and hydrosulfuration).

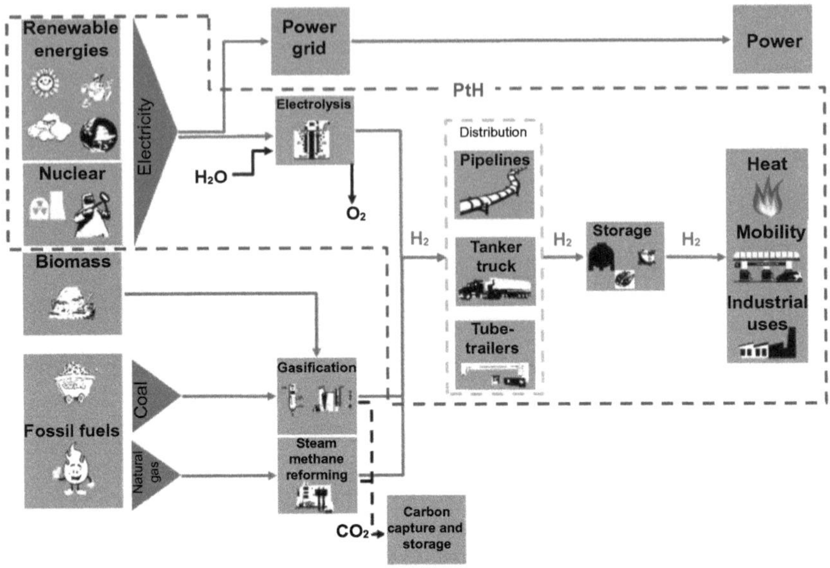

FIG. 2.1 Various pathways involved in hydrogen supply chains.

TABLE 2.1 Worldwide Hydrogen Consumption in Industry (Le Duigou and Miguet, 2010)

Use	Consumption (Mt)	%
Refinery	26.4	44
Ammonia production	22.8	38
Other chemical products	4.8	8
Others	6	10

TABLE 2.2 Hydrogen Consumption in Industry in France (Le Duigou and Miguet, 2010)

Use	Consumption (kt)	%
Refinery	544	59
Ammonia and fertilizers	240	26
Chemical industry (methanol)	92	10
Metal industry	9.2	1
Others	36.8	4

In the chemical industry, hydrogen is generally produced via syngas processes from fossil feedstocks, including natural gas, oil-, and coal-based sources. The syngas process variant applied depends strongly on the available feedstock and the downstream processes. Steam reforming of natural gas is by far the dominant process. Electrolytic production of hydrogen plays a minor role; it is not competitive with steam reforming and is only applied when relatively small amounts of pure hydrogen are required. Hydrogen is also produced as a significant side stream of other reactions, such as dehydrogenation processes and chlorine production.

In current schemes of industrial distribution, hydrogen is produced in centralized plants or produced onsite for captive uses. H_2 is obtained mainly from steam methane reforming (SMR) and coal gasification, and also obtained as a byproduct from chloralkali electrolysis plants. It is used onsite or transported over short distances by pipelines. The option for hydrogen to be liquefied or compressed and then transported via tube trailers or tanker trucks also exists but is less frequent. Ammonia production plants, oil processing units, and methanol and metal industries are customers with high demand volume.

As highlighted in Ausfelder and Bazzanella (2016) there are a lot of opportunities and challenges for the chemical industry in the face of large-scale production of renewable hydrogen.

2.2.2 H_2 Supply Chain as a Fuel

The hydrogen supply chain for H_2 as fuel is defined as a system of activities from suppliers to customers, including energy source, production, storage, transportation, and dispensation of hydrogen to refueling stations. Unlike most other fuel infrastructure, hydrogen can be produced either centrally (similar to existing gasoline supply chains) or in a distributed mode (using small-scale units that can produce H_2 close to the point of use in small quantities) at forecourt refueling stations, so that the distribution cost can be strongly reduced.

In this context, several questions arise:

- What is the environmental impact of the energy source used to produce hydrogen?
- Which production option is more cost effective?
- Is centralized production or decentralized production (small-scale production at local fueling stations) more cost-effective?
- What are the most cost-effective transportation modes and pathways to connect hydrogen demand with its supply?
- Which quantitative criteria can be used to evaluate sustainability of the hydrogen supply chain?

2.3 MULTIPLE SOURCES TO HYDROGEN

Hydrogen can be produced by two pathways, either power-to-hydrogen (from nuclear or from renewable sources, such as hydro, wind, solar, and geothermal) or non-power-to-hydrogen options (biomass or fossil fuels associated with carbon capture and storage, for instance).

The local market conditions and availability of regional primary energy feedstock have a large impact on the selection of supply chain pathways. As previously explained, hydrogen has the benefit of improving the security of fuel supply because it can be produced from diverse primary energy sources, such as fossil fuels (natural gas, naphtha, heavy oil, and coal) and renewable energy sources (RES), such as wind, biomass, water, and solar energy. Currently, almost 95% of hydrogen is produced from fossil fuels via steam reforming, gasification, and partial oxidation processes (Heracleous, 2011). According to Murthy Konda et al. (2011), feedstock remains the biggest contributor to the cost of H_2 fuel with around 40% share.

RES are playing an ever increasing role in worldwide electricity generation (EIA, 2017); their contribution has increased from 21.5% in 1973 to 23.6% in 2015. Hydropower plays by far the most important role in electricity generation, producing 16% of the energy from RES in 2016. Nevertheless, the importance of RES other than hydro has grown considerably, generating 7.1% of the total energy in 2015. Renewables are the fastest growing source of energy for electricity generation, with average increases of 2.8% per year expected from 2015 to 2040. Renewable resources other than hydropower are the fastest-growing energy sources for new generation capacity, and this phenomenon is valid for both the OECD and non-OECD regions. Renewables other than hydropower accounted for 7% of total world generation in 2015, with an expected share in 2040 of 15% (EIA, 2017), with more than half of the growth coming from wind power. After renewable energy sources, natural gas and nuclear power are the next fastest growing sources of energy used to generate electricity (Fig. 2.2).

Several scenarios can also be found in the literature. Some of them are more optimistic in the rate projected for RES use, such as WWF (2011) and EREC (2010), that consider a 100% production rate for RES by 2050. For instance, some examples of different energy scenarios developed in French reports include the RES roadmap for 2020 and 2030 (presented by the Renewable

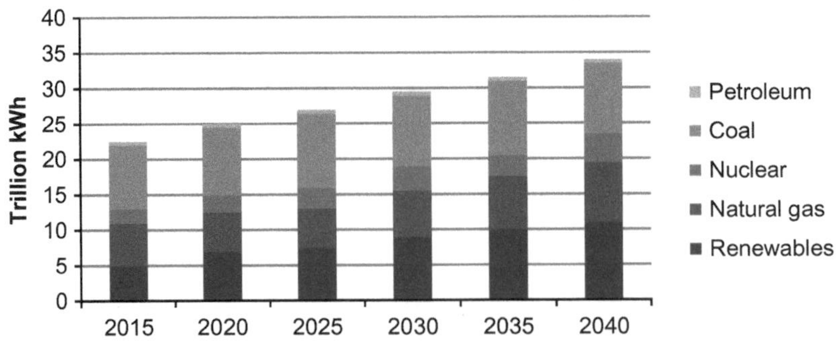

FIG. 2.2 World net electricity generation by energy source (EIA, 2017).

Energy Association) (Syndicat des énergies renouvelables, 2012) and the French energy report 2050 (Percebois and Mandil, 2012), in which some scenarios (e.g., CEA, AREVA, Global chance, RTE) regarding nuclear and RES pathways are thoroughly analyzed.

2.3.1 Coal

Coal is a sedimentary rock consisting of both organic and inorganic material. It consists of carbon, hydrogen, oxygen, and, in lower amounts, sulfur and nitrogen. Coal is the most abundant fossil fuel in the world (Miller, 2004). Coal is the world's most abundant and widely distributed fossil fuel with reserves for all types of coal estimated to be about 990 billion metric tons, enough for 150 years at current consumption (BGR, 2009). Coal makes up 42% of electricity generation and is likely to remain a key component of the fuel mix for power generation to meet electricity demand, especially the growing demand in developing countries. As pointed out by the European Directive of 2009, a key factor in reducing GHG emissions in the short term is to improve energy efficiency in end uses in different sectors: construction, transport, and industry. In this regard, using coal as an energy source is not without problems because coal combustion releases carbon dioxide and other pollutants. Coal gasification and carbon sequestration (see also Section 2.4.6) may be able to minimize pollutants and greenhouse gases while changes in mining practices can reduce the environmental concerns.

2.3.2 Natural Gas

Natural gas is a mixture of several hydrocarbon gases, including methane (between 70% and 90%), ethane, propane, butane, and pentane, as well as carbon dioxide, nitrogen, and hydrogen sulfide. The composition of natural gas can vary widely, depending on the gas field. Natural gas is produced through the decay of organic materials. Most natural gas is obtained from wells, although it can be produced from organic waste through the use of digesters (Bartels et al., 2010). A third way to obtain natural gas is by abiogenic processes. In general, extremely deep under the Earth's crust, hydrogen-rich gases and carbon molecules exist, interacting with minerals in the absence of oxygen that combined with the high pressure, form methane deposits (Bahadori, 2014).

The remaining resources of natural gas are abundant and can satisfy the projections of global demand growth included in all three of the IEA *World Energy Outlook* (IEA, 2011) scenarios to 2040 and well beyond. Proven reserves have been estimated at 216 trillion cubic meters at the end of 2014, equal to more than 60 years of production at current rates, yet proven reserves are only a fraction of the total remaining technically recoverable resources.

2.3.3 Biomass

Biomass, one of the most abundant renewable resources, is formed by fixing carbon dioxide in the atmosphere during the process of plant photosynthesis. It is therefore carbon neutral in its lifecycle. Biomass comprises any organic matter, either plant or animal in origin. Biomass energy refers to the stored energy (solar, carbon and hydrogen) that is available within organic matter and can take a variety of forms, with woody biomass being the most used (Macqueen and Korhaliller, 2011).

Biomass resources can be divided into four categories (Ni et al., 2006):

- Energy crops: herbaceous, woody industrial, agricultural, and aquatic.
- Agricultural residues and waste: crop and animal waste.
- Forestry waste and residues: mill wood, logging, trees, and shrub residues.
- Industrial and municipal wastes: municipal solid and industry waste, sewage sludge.

From the point of view of quality, biomass can be categorized according to its content of humidity, lignocellulose, sugar, starch, and oil (Orecchini and Bocci, 2007).

Much research has focused on sustainable and environmentally friendly energy from biomass to replace conventional fossil fuels. Thermochemical and biological processes are the two main pathways to obtain energy from biomass. To produce hydrogen, the most used processes are fast pyrolysis and gasification (thermochemical).

One issue with the increasing use of biomass for energy purposes (Bartels et al., 2010) is related to its impact on land use and its implications for a whole range of sustainable development issues, including social development and environmental impacts associated with land use change.

Another limitation is that the processes of hydrogen production from biomass are still in the development stage and require a strong effort in terms of R&D and demonstration activities (Balat and Balat, 2009). The characteristics of biomass are very important because they can vary greatly from location to location, and also seasonally and yearly, so that hydrogen production via the biomass route may not be competitive with hydrogen production with fossil fuels (Dagdougui, 2011a). However, the waste-to-energy application has received much attention due to its potential to become a major hydrogen source.

Bioenergy (Sawin et al., 2017) is the largest contributor to global renewable energy supply. Total primary energy supplied from biomass in 2016 was approximately 62.5 exajoules (EJ) (2.5% increase per year since 2010). The bioenergy share in total global primary energy consumption has remained relatively steady since 2005, at around 10.5%, despite a 21% increase in overall global energy demand over the last 10 years. The contribution of bioenergy to final energy demand for heat in buildings and industry far outweighs its use for electricity and transport combined. Europe is the largest consumer of bioheat following the mandatory national targets under the Renewable Energy Directive (Sawin et al., 2017).

2.3.4 Solar Energy

The ubiquitous solar energy is a clean, renewable source with average around 120,000 TW irradiation at the Earth's surface (Hosseini and Wahid, 2016). The estimated potential of the direct capture of solar energy is enormous. When solar energy strikes the Earth's atmosphere, approximately 30% is reflected. After reflection by the atmosphere, Earth's surface receives about $3 \times 9\ 10^{24}$ MJ incident solar energy per year, which is almost 10,000 times more than current global energy consumption. Thus, the harvesting of <1% of photonic energy would serve all human energy needs. Although solar energy is the largest energy source of the planet, it produces only 1% of all electricity used globally. However, the global installed capacity for solar-powered electricity has seen exponential growth, reaching about 227 GWe at the end of 2015 (World Energy Council, 2016).

There are two ways solar energy is used to generate electricity; photovoltaic cells directly convert sunlight to electricity, while solar thermal power plants or concentrating solar power systems focus sunlight with mirrors, heating water and producing high-pressure steam that drives turbines.

Photovoltaic cells only absorb a portion of the solar spectrum, but they can generate electricity from both direct and diffuse sunlight. Solar thermal power plants can use more wavelengths of the solar spectrum, but they can only operate in direct sunlight, limiting them to sun-rich areas. Moreover, the highest conversion efficiencies reported so far for solar thermal power plants are significantly less than those for photovoltaic cells.

2.3.5 Wind

Wind mills and horizontal- and vertical-axis turbines are used to convert the kinetic energy of the wind into electricity. Wind energy is one of the more cost-effective forms of renewable energy today. Wind turbines are typically constructed in large groups of individual wind turbines in order to form a large wind farm (Bartels et al., 2010). Onshore wind has now reached a certain maturity, even if the intermittent nature of the production makes it difficult to integrate into power grid systems, and the technical improvements are mostly incremental. However, offshore wind power has emerged recently (the first field of 500 MW in 2003 in Denmark), but faces some technical barriers, such as keeping equipment in harsh environments (saline, weather) and connection to the power grid (Direction générale de la compétitivité, de l'industrie et des services, 2011).

Wind-to-hydrogen may allow wind energy to be harnessed in areas without electrical transmission capacity, or it could provide an energy storage medium for the intermittent wind resource in order to provide a more constant renewable electricity supply (Bartels et al., 2010).

World wind power generation capacity has reached 435 GW at the end of 2015, around 7% of total global power generation capacity (World Energy Council, 2016).

2.3.6 Hydropower

Hydropower is currently the world's largest renewable power source for electricity generation, supplying 71% of all renewable electricity. Reaching 1064 GW of installed capacity in 2016, it generated 16.4% of the world's electricity from all sources (World Energy Council, 2016). Large-scale hydropower plants store water in a reservoir behind a dam, and then regulate the flow according to electricity demand. Hydropower can provide a relatively reliable source of power on demand. The Ecofys scenario reflects a relatively small increase in hydropower. In this scenario, hydropower would provide 12% of the electricity in 2050 compared with 15% today (WWF, 2011). Nowadays, highly efficient turbines that generate electricity by spinning water are installed. Small hydropower, defined by installed capacity of up to 10 MW, is the backbone of electricity production in many countries in the European Union (EREC, 2010). Small hydropower takes advantage of the kinetic energy and pressure from falling water or rivers for instance.

Regarding the type of operation, hydropower can be classified as *run-of-river* or *pumped hydro*. Run-of-river stations have no reservoir capacity and provide a continuous supply of electricity (base load) with some flexibility of operation for daily fluctuations in demand through water flow that is regulated by the facility. Pumped hydro produces electricity to supply high peak demands by moving water between reservoirs at different elevations.

2.3.7 Geothermal

Geothermal energy for hydrogen production is considered as a sustainable option for those countries with abundant geothermal energy resources. The technologies of hydrogen production and use can be easily integrated with geothermal sources and stand-alone energy systems (Balta et al., 2010). Although many studies (e.g., Bockris and Veziroglu, 2007; Costogue and Yasui, 1977; Eisenstadt and Cox, 1975; Maack and Skulason, 2006; Yilanci et al., 2009) have been carried out on hydrogen production from solar and wind in the dedicated literature, hydrogen production from geothermal resources has received far less attention. These works highlight some demonstration projects and their technical details, feasibility studies, and planning.

Geothermal energy produces less than 1% of the world's electricity generation. There were 315 MW of new geothermal power capacity installed in 2015, raising the total capacity to 13.2 GW (World Energy Council, 2016).

2.3.8 Uranium and Nuclear

Nuclear power could produce hydrogen by either electrolysis of water, or by thermal decomposition of water using heat from high-temperature reactors (thermochemical cycles must be implemented). No greenhouse gas emissions are created directly during the generation of nuclear power. The assessments of global uranium resources show that total identified resources have grown by about 70% over the last 10 years. The total identified resources of uranium are considered sufficient for over 100 years of supply based on current requirements in 2015 (World Energy Council, 2016).

Global nuclear power capacity reached 390 GWe at the end of 2015, generating about 11% of global electricity. In 2015, 65 reactors were under construction (6 more than in July 2012) with a total generating capacity of 64 GW. The increasing need to moderate the local pollution effects of fossil fuel use means that nuclear is increasingly seen as an option in order to add large-scale baseload power generation while limiting the amount of GHG emissions.

2.4 MULTIPLE HYDROGEN PRODUCTION MODES

Three main categories of hydrogen production technologies are considered here: (1) gasification and pyrolysis from coal or biomass; (2) reforming, from natural gas, ethanol, biomass, or heavy fuel oil; and (3) electrolysis with alkaline and proton exchange membrane (PEM) and solide oxide (SOE) electrolyzers.

In this section, the centralization/decentralization level of hydrogen production is discussed and the production technologies are briefly analyzed.

The hydrogen supply chain may be classified as either centralized or decentralized (on site), depending on the degree of centralization.

2.4.1 Centralized Versus Distributed Hydrogen Production

Hydrogen can be produced: (1) at or near the site of use in distributed production, or (2) at large facilities and then delivered to the point of use in central production (Office of Energy Efficiency and Renewable Energy, n.d.).

The centralized production option (750,000 kg/day of hydrogen) would be analogous to current gasoline supply chains, for which the economies of scale are capitalized upon within an industrial context and large quantities are produced at a central site and then distributed (Hugo et al., 2005; Murthy Konda et al., 2011). Centralized plants promise higher hydrogen production efficiency but require more capital investment and a substantial hydrogen transport and delivery infrastructure.

Decentralized production consists of small regional plants, or even local filling stations, that generate hydrogen. While the hydrogen generation efficiency for decentralized production is lower than that for centralized plants, losses in hydrogen transport can make such a scheme more efficient (Kim et al., 2008; Haeseldonckx and D'haeseleer, 2011). There is a tendency in the literature to argue that decentralized production plants could overcome many of the infrastructural barriers facing a transition to hydrogen (Ball and Wietschel, 2008). Most studies consider the decentralized route as the key to bypassing the infrastructural problem (Haeseldonckx and D'haeseleer, 2011). A decentralized approach often results in higher costs as efficiencies are generally lower and because on-site production facilities are often dimensioned to cover peak demand (especially when no storage is foreseen or possible). However, a further increase in demand will require larger pipelines, and consequently additional investment costs (Haeseldonckx and D'haeseleer, 2011). Two distributed hydrogen production technologies have been identified: (1) reforming natural gas or liquid fuels, including renewable liquids, such as ethanol and bio-oil, and (2) small-scale water electrolysis.

Large hydrogen production facilities currently exist in or near petroleum refineries because hydrogen is used in petroleum processing. A small fraction of this hydrogen may be transported to nearby refueling stations during the transition phase of hydrogen deployment (Office of Energy Efficiency and Renewable Energy, n.d.).

2.4.2 Steam Reforming of Natural Gas (SMR)

Most hydrogen (97%) is produced by steam reforming of natural gas, also known as SMR (Koroneos et al., 2004). SMR is used in the chemical and petrochemical industries; it is currently the cheapest production method and has the lowest CO_2 emissions of all fossil production routes (Ball and Wietschel, 2008). This process is the most efficient in large-scale production. Fig. 2.3 presents the typical features of the SMR process.

The main steps in the production of hydrogen from natural gas are (Hajjaji, 2011): (1) production of the synthesis gas, (2) conversion of carbon monoxide to hydrogen (water shift gas); and (3) purification (Scipioni et al., 2017).

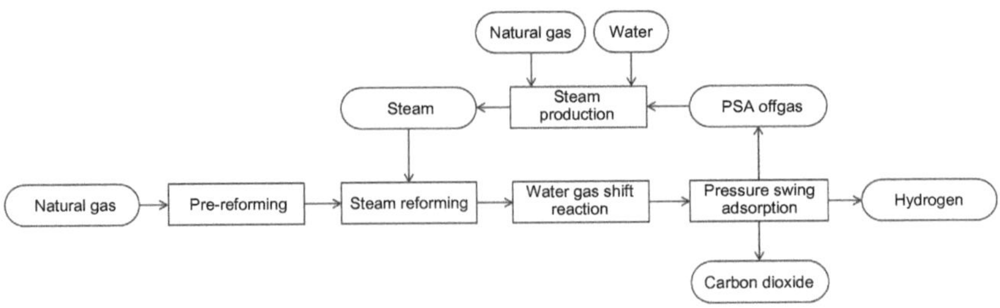

FIG. 2.3 SMR process. *(From Wulf, C., Kaltschmitt, M., 2012. Life cycle assessment of hydrogen supply chain with special attention on hydrogen refuelling stations. Int. J. Hydrogen Energy, Advances in Hydrogen Production (Selected papers from ICH2P-2011) 37, 16711–16721. https://doi.org/10.1016/j.ijhydene.2012.03.028.)*

$$CH_4 + H_2O \xrightarrow[\;]{\substack{800^\circ C \\ 15-20\,\text{bar}}} CO + 3H_2 \quad \Delta H = \pm 206\,\text{kJ} \tag{2.1}$$

$$CO + H_2O \xrightarrow[\;]{\substack{400^\circ C \\ 15-18\,\text{bar}}} CO_2 + 3H_2 \quad \Delta H = -41.2\,\text{kJ} \tag{2.2}$$

Governing equation

$$CH_4 + H_2O \rightarrow 4H_2 \quad \Delta H = 164.8\,\text{k} \tag{2.3}$$

The process starts with desulfurized natural gas below the ppm-level (the catalyst of the process is very sensitive to sulfur). Then, natural gas is reformed with the addition of water to a hydrogen- and carbon monoxide-rich gas at 800°C and 3 MPa. The gas is transformed into hydrogen by the water-gas-shift reaction, in which carbon monoxide reacts with a catalyst to produce carbon dioxide and more hydrogen. The hydrogen so produced must be cleaned by removing CO_2 by pressure swing absorption, leaving essentially pure hydrogen (Wulf and Kaltschmitt, 2012).

During hydrogen production, some heat is produced that can be recovered in an industrial furnace, making the process most efficient.

2.4.3 Electrolysis

Electrolysis is a process based on the generation of hydrogen and oxygen by applying a direct electric current to water to dissociate it. Hydrogen obtained with this technology has a high purity that can reach 99.999 vol% once hydrogen has been dried up and oxygen impurities have been removed. Electrolyzers consist of an anode and a cathode separated by an electrolyte (see Fig. 2.4).

Electrolysis needs electrical energy to convert water into oxygen and hydrogen. After hydrogen is obtained, it must be dried and purified by a deoxo dryer. Oxygen is released to the air. The governing reaction is shown in Eq. 2.4 (Scipioni et al., 2017).

$$2H_2O \rightarrow 2H_2 + O_2 \tag{2.4}$$

The different electrolyzers, operated in slightly different ways according to the different type of electrolyte material involved, are described in the following paragraphs.

2.4.3.1 Polymer Electrolyte Membrane Electrolyzer

In a polymer electrolyte membrane (PEM) electrolyzer, the electrolyte is a solid specialty plastic material. The water reacts at the anode to form oxygen and positively charged hydrogen ions (protons). Then, the electrons flow through an external circuit and the hydrogen ions selectively move across the PEM to

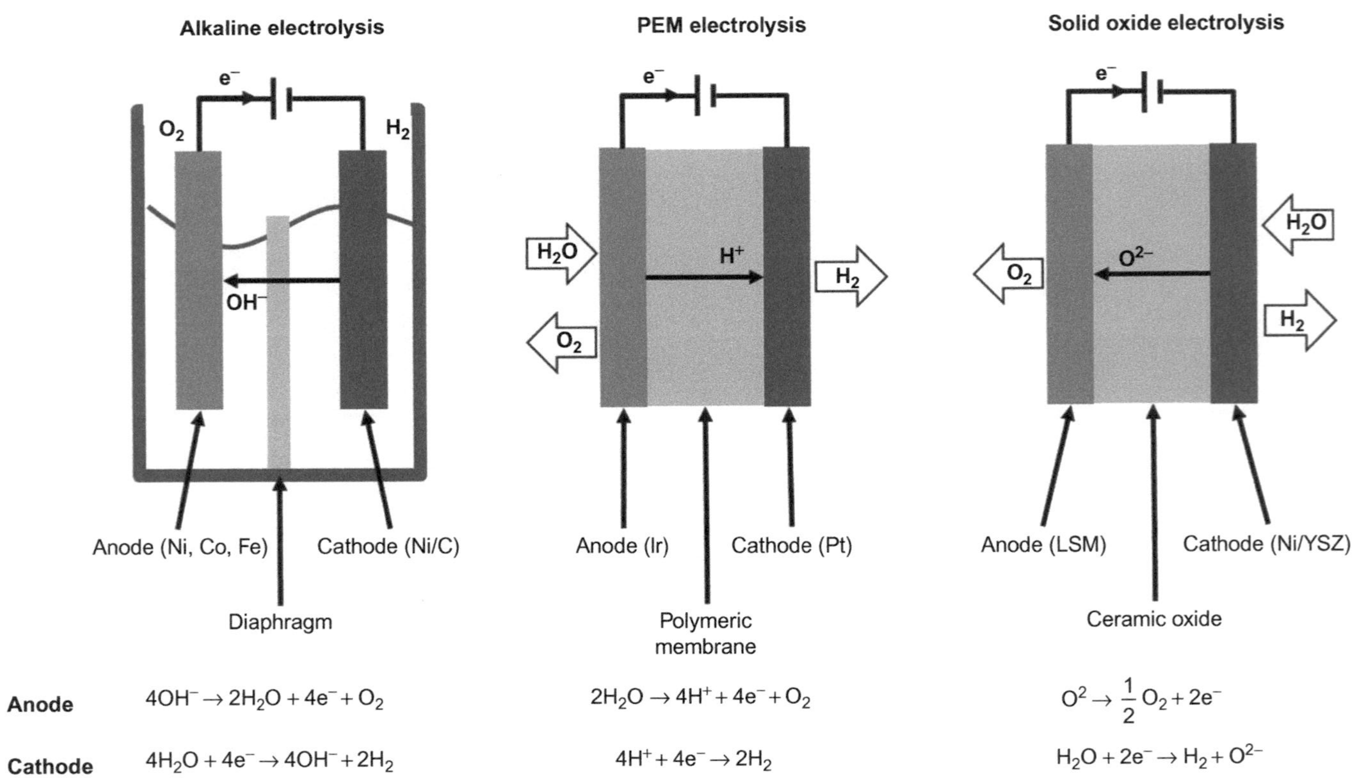

FIG. 2.4 Operation principles of an alkaline, PEM, and solid oxide electrolyzer. *(From Sapountzi, F.M., Gracia, J., Weststrate, C., Fredriksson, H., Niemantsverdriet, H., 2017. Electrocatalysts for the generation of hydrogen, oxygen and synthesis gas. https://doi.org/10.1016/j.pecs.2016.09.001.)*

the cathode. At the cathode, hydrogen ions combine with electrons from the external circuit to form hydrogen gas, with the following reactions:

$$\text{Anode reaction}: 2H_2O \rightarrow O_2 + 4H^+ + 4e^- \tag{2.5}$$

$$\text{Cathode reaction}: 4H^+ + 4e^- \rightarrow 2H_2 \tag{2.6}$$

2.4.3.2 Alkaline Electrolyzers

Alkaline electrolyzers operate via transport of hydroxide ions (OH^-) through the electrolyte from the cathode to the anode with hydrogen being generated on the cathode side. Electrolyzers using a liquid alkaline solution of sodium or potassium hydroxide as the electrolyte have been commercially available for many years. Newer approaches using solid alkaline exchange membranes as the electrolyte are showing promise at the lab scale.

2.4.3.3 Solid Oxide Electrolyzers

Solid oxide electrolyzers, which use a solid ceramic material (zirconium dioxide doped with yttrium oxide) as the electrolyte that selectively conducts negatively charged oxygen ions (O^{2-}) at elevated temperatures, generate hydrogen in a slightly different way. Water at the cathode combines with electrons from the external circuit to form hydrogen gas and negatively charged oxygen ions. The oxygen ions pass through the solid ceramic membrane and react at the anode to form oxygen gas and generate electrons for the external circuit.

Solid oxide electrolyzers must operate at temperatures high enough for the solid oxide membranes to operate properly (about 700–800°C, compared to PEM electrolyzers, which operate at 70–90°C, and commercial alkaline electrolyzers, which operate at 100–150°C). The solid oxide electrolyzers can effectively use heat available at these elevated temperatures (from various sources, including nuclear energy) to decrease the amount of electrical energy needed to produce hydrogen from water. This system has the advantage of being reversible and able to work either as electrolyzer or as solid oxide fuel cell (SOFC). It is thus possible to operate in electrolyzer mode when the price of electricity is low and in fuel cell mode producing electricity during electrical demand peaks. This process is still in the research phase (Cueugniet et al., 2015).

An electrolysis flow diagram is shown in Fig. 2.5, although some differences may occur according to the process. For example, PEM units will not require a KOH mixing tank, as no electrolytic solution is needed for these electrolyzers. Water purification equipment may also vary, because water quality requirements differ across electrolyzers.

All systems have a hydrogen generation unit (electrolysis stack, gas purification, and heat removal). Hydrogen and oxygen are produced in the generation

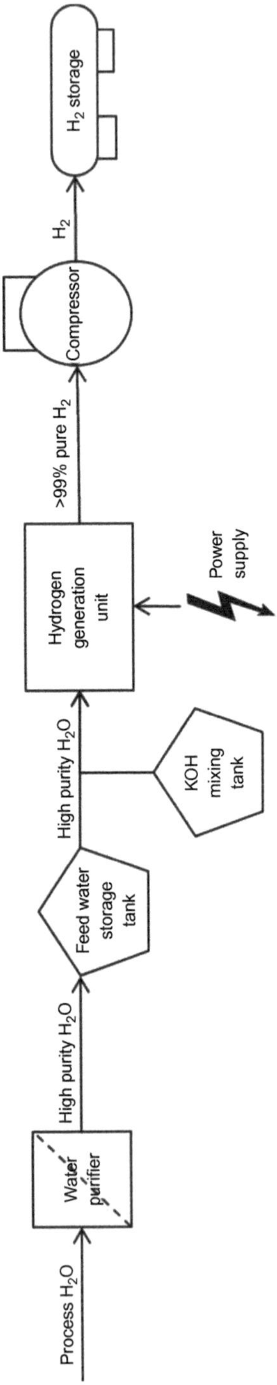

FIG. 2.5 Electrolysis flow diagram. (*Adapted from Ivy, J., 2004. Summary of electrolytic hydrogen production: milestone completion report (No. NREL/MP-560-36734). National Renewable Energy Lab., Golden, CO (US).*)

unit. At the end of the system, a compressor and hydrogen storage can be added (Ivy, 2004).

While PEM electrolysis is less proven and costlier in terms of capital expenditure (CAPEX) than alkaline electrolysis, it is more compact and suitable for dynamic load balancing of electricity grids needed with the use of intermittent renewable energy.

In a central electrolyzer, production can reach a capacity of 50,000 kg/day. The plant design capacity is 51,020 kg/day with a 98% operating efficiency (Table 2.3).

From a technological point of view, the development of alkaline electrolyzers is currently sufficiently advanced to start the production of renewable hydrogen at significant rates. However, it is generally agreed that the massive hydrogen production required by the hydrogen economy will need electrolysis units with production capacities much higher than current ones. Polymer electrolyte membrane (PEM) electrolyzers are commercially available, but their cost is high, and the technology seems more suitable for small-scale

TABLE 2.3 Specifications of the Different Types of Electrolyzers Summarized From Dincer and Acar (2015)

Specification	Alkaline	PEM	SOE
Technology maturity	State of the art	Demonstration	R&D
Cell temperature (°C)	60–80	50–80	900–1000
Cell pressure (bar)	<30	<30	<30
Current density (A/cm^2)	0.2–0.4	0.6–2.0	0.3–1.0
Cell voltage (V)	1.8–2.4	1.8–2.2	0.95–1.3
Power density (W/cm^2)	Up to 1.0	Up to 4.4	–
Specific system energy consumption (kWh/Nm$_3$)	4.5–7.0	4.5–7.5	2.5–3.5
Hydrogen production (Nm^3/h)	<760	<30	–
Stack lifetime (h)	<90,000	<20,000	<40,000
System lifetime (y)	20–30	10–20	–
Hydrogen purity (%)	>99.8	99.999	–
Cold start up time (min)	15	<15	>60

Data from Bhandari, R., Trudewind, C.A., Zapp, P., 2014. Life cycle assessment of hydrogen production via electrolysis – a review. J. Clean. Prod., Special Volume: Making Progress Towards More Sustainable Societies through Lean and Green Initiatives 85, 151–163. DOI: 10.1016/j.jclepro.2013.07.048.

applications. Solid oxide electrolyzers (SOEs) for hydrogen production from steam are at the R&D stage. They show enormous potential because the use of high-temperature heat reduces their electricity usage.

2.4.4 Coal Gasification

Coal gasification is economically attractive, but has significant CO_2 emissions, even if CO_2 can be captured and stored. Nowadays, hydrogen production from coal gasification is not much used, except in those places where the other sources are very expensive (Dagdougui, 2012). Fig. 2.6 shows the standard process of producing hydrogen with the coal gasification method.

The governing reaction of gasification is presented in Eq. 2.7 (Scipioni et al., 2017).

$$C + 2H_2O \rightarrow CO_2 + 2H_2 \tag{2.7}$$

In this process, hydrogen is produced by the added water, and the coal serves merely as a reducing agent and an energy source. The first step is to pyrolyze the coal to crack carbon and obtain minuscule molecules while water is added as the synthesis gas production starts. A water-gas-shift reaction follows gasification. Depending on the ingredients of the hard coal, a pressure swing adsorption may not be enough to clean hydrogen. Hydrogen sulfide (H_2S) produced from the sulfur content of the coal must also be removed (Wulf and Kaltschmitt, 2012; Xiang et al., 2010).

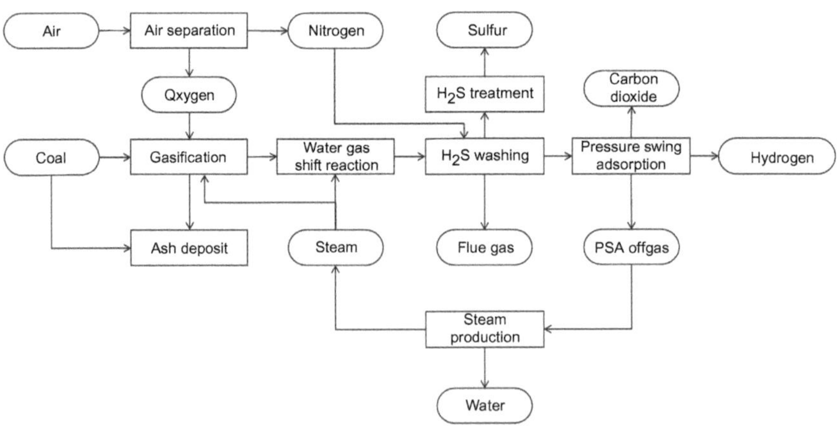

FIG. 2.6 Coal gasification to produce hydrogen. *(From Wulf, C., Kaltschmitt, M., 2012. Life cycle assessment of hydrogen supply chain with special attention on hydrogen refuelling stations. Int. J. Hydrogen Energy, Advances in Hydrogen Production (Selected papers from ICH2P-2011) 37, 16711–16721. https://doi.org/10.1016/j.ijhydene.2012.03.028.)*

2.4.5 Biomass

Biomass gasification is still in the development stage and competes with other uses of biomass. There are two main ways to produce hydrogen from biomass (Orecchini and Bocci, 2007), either biochemical or thermochemical. Fig. 2.7 shows a general view of the conversion from biomass to hydrogen.

Biochemical methods convert biomass into energy through the use of enzymes, mushrooms, and microorganisms formed from the biomass. Four processes can be found, namely, aerobic digestion, anaerobic digestion, oil extraction, and alcoholic fermentation (production of ethanol).

Aerobic digestion uses aerobic microorganisms, decomposing the organic matter and producing heat, CO_2, and water. This process is used to depurate sewage and is not suitable for hydrogen production.

Anaerobic digestion produces biogas with the help of bacteria in an anaerobic environment, producing a gas mixture of 65%–70% CH_4 and 30%–35% CO_2. The produced biogas can then be used to obtain hydrogen by the SMR method.

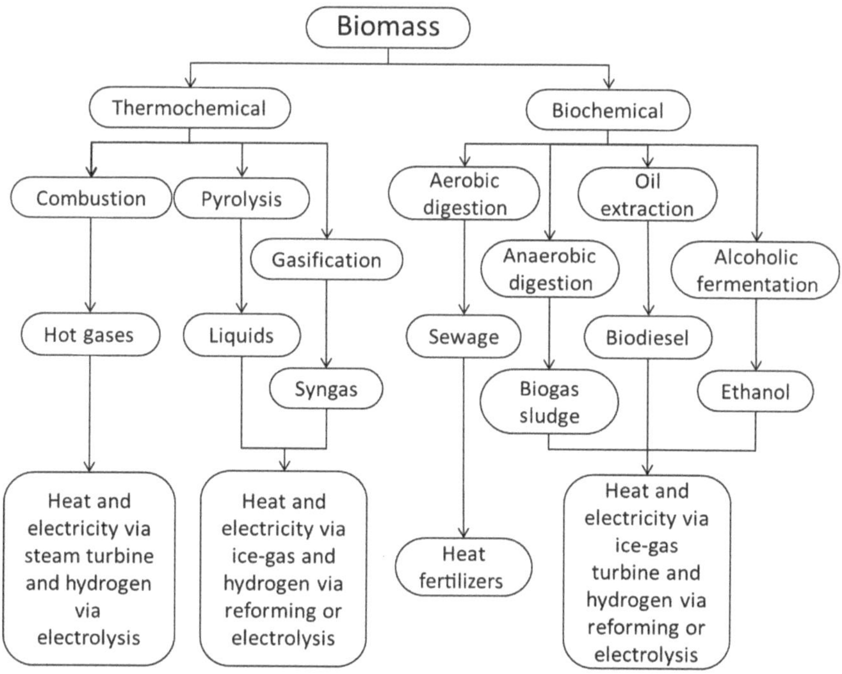

FIG. 2.7 Biomass to hydrogen processes. *(From Orecchini, F., Bocci, E., 2007. Biomass to hydrogen for the realization of closed cycles of energy resources. Energy, Third Dubrovnik Conference on Sustainable Development of Energy, Water and Environment Systems 32, 1006–1011. DOI: 10.1016/j.energy.2006.10.021.)*

In the oil extraction process, oil can be extracted from biomass via pressure. Then it can be processed with alcohol through esterification, in order to obtain biodiesel and hydrogen.

Alcoholic fermentation in controlled air conditions can directly produce hydrogen or ethyl alcohol, from which hydrogen can be obtained by steam reforming, but the energetic and economic reliability of the process has not been demonstrated.

Oil extraction is a mechanical-chemical conversion process that can be used to produce oil from seeds, and that can be followed by esterification to obtain biodiesel and hydrogen. The disadvantage of this pathway is its low energy ratio (between 0.3 and 3).

On the thermochemical side, three processes can be highlighted: combustion, pyrolysis, and gasification.

Biomass combustion produces hot gases (800–1000°C), which is generally possible with a moisture level of <50%. Biomass pyrolysis generates liquid oils, solid charcoal, and gaseous compounds by heating the biomass at a temperature of 650–800 K at 0.1–0.5 MPa. It occurs in the total absence of oxygen, except in cases in which partial combustion is allowed to provide the thermal energy needed for the process (Orecchini and Bocci, 2007).

The interested reader can refer to state-of-the-art reviews of hydrogen production technologies (Dincer and Acar, 2015; Nikolaidis and Poullikkas, 2017).

2.4.6 Carbon Capture and Storage (CCS)

Hydrogen supply chains offer the possibility of capturing most of the CO_2 emissions (~90%) and sequestering them (Murthy Konda et al., 2011) because they originate mostly from point sources (production facilities using nonrenewable sources). CCS is an energy intensive and costly process involving several steps: CO_2 capture, pressurization, transportation, and final disposal into geological formations or in aquifers (Direction générale de la compétitivité, de l'industrie et des services, 2011; Haeseldonckx and D'haeseleer, 2011). There are three main technology options for CO_2 capture and storage: precombustion capture, postcombustion capture, and oxy fuel combustion.

2.4.6.1 Precombustion Capture

Precombustion capture processes can be used in coal or natural gas plants. The fuel is reacted first with oxygen and/or steam and then further processed in a shift reactor to produce a mixture of H_2 and CO_2. CO_2 is captured from a high-pressure gas mixture that contains between 15% and 40% CO_2 (Haeseldonckx and D'haeseleer, 2011; Direction générale de la compétitivité, de l'industrie et des services, 2011).

2.4.6.2 Postcombustion Capture

The postcombustion process is commercially applied to produce high-purity CO_2 from the exhaust of coal and gas fired boilers, furnaces, and turbines. CO_2 is captured typically through the use of solvents and subsequent solvent regeneration, sometimes in combination with membrane separation (Direction générale de la compétitivité, de l'industrie et des services, 2011; Haeseldonckx and D'haeseleer, 2011). The basic technology, using mono ethanolamine, has been used on an industrial scale for decades, but the challenge is the massive scaleup for power plants and to recover CO_2 with a minimum energy penalty and at acceptable cost. At present, the largest operating unit has a capacity of 800t CO_2/day. To put this into perspective, large coal-fired units produce up to roughly 10,000t CO_2/day (Haeseldonckx and D'haeseleer, 2011).

2.4.6.3 Oxygen Combustion

In this technology, the fuel is combusted using (nearly) pure oxygen, which is produced by a cryogenic ASU, although new technologies, such as ion transport membranes, are investigated. The flue gas, containing primarily CO_2, is partially recycled to the boiler to control the combustion temperature. The main advantage of oxy fuel combustion is that it enables nearly 100% CO_2 capture (Haeseldonckx and D'haeseleer, 2011).

After CO_2 is captured via physical absorption, it is compressed to supercritical pressure at 15MPa, which permits efficient pipeline transmission of CO_2. Energy use and CO_2 emissions from CO_2 sequestration are assumed to be predominately associated with compression (Johnson et al., 2008).

2.4.7 Other Hydrogen Production Methods

Other methods exist, such as water splitting by high-temperature heat, photoelectrolysis, and biological processes. These methods have been extensively presented in some review papers (Acar and Dincer, 2014; Dincer, 2012; Hallenbeck and Benemann, 2002), and include plasma arc decomposition, water thermolysis, biological processes, thermochemical water splitting, biophotolysis, and photofermentation among others (Dincer, 2012).

Table 2.4 shows the advantages and disadvantages of the hydrogen production methods described in this chapter.

2.4.8 Key Parameters of Some Hydrogen Production Technologies

Selected hydrogen production technologies are listed in Table 2.5 with their typical features.

TABLE 2.4 Key Benefits and Critical Challenges of the Hydrogen Production Methods

	Fossil Fuel Reforming	Coal and Biomass Gasification	Water Electrolysis
Critical challenges	High capital costs	High reactor costs	Low system efficiency
	Design	System efficiency	High capital costs
	High operation and maintenance costs	Feedstock impurities	System integration
		Carbon capture and storage	Design issues
Major R&D needs	Efficiency and cost	Low-cost and efficient purification	Durable and cheap materials
	Low-cost and efficient purification	Cofed gasifiers	Corrosive-resistant membranes
	Feedstock pretreatment	Carbon capture and storage	Durable, active, and cheap catalysts
	Optimization	Hydrogen quality	Large-scale applications
	Automated process control	Cost and feedstock preparation	Storage and production rate
	Reliability	Tolerance of impurities	Reliability
Key benefits	Most viable approach	Low-cost syngas production	No pollution with renewable energy sources
	Lowest current cost	Abundant and cheap feedstock	Existing infrastructure
	Existing infrastructure		Integration with fuel cells

Adapted from Dincer, I., Acar, C., 2015. Review and evaluation of hydrogen production methods for better sustainability. Int. J. Hydrogen Energy 40, 11094–11111. DOI: 10.1016/j.ijhydene.2014.12.035.

TABLE 2.5 Key Parameters for Hydrogen Production in the HSC

				kg CO_2eq/kg (H_2)[a]	Capital Cost	Operational Cost ($/kg)	Capacity	Primary Energy[a]
Centralized production	LH_2 production	Electrolysis[a]	Wind	1.03	4420–6421.05 $/(kg/day)[a]	4.69–4.59[a]	300–150,000 kg/day[a]	52.49 kWh/kg (H_2)
			Solar	6.21				
			Hydro	2.07				
			Nuclear	3.10				
		SMR		10.1	903 M$[b]	1.12[b]	10,000–960,000 kg/day[b]	4.02–3.16 kg (NG)/kg(H_2)
					201.16–431.85 €/kW[c]		1530–33 MW[c]	
					29–903 M€[a]	3.36–1.43[a]	300–960,000 kg/day[a]	
		Coal gasification			1611 M$[b]	1.43 $/kg[b]	10,000–960,000 kg/day[b]	
					462.46–573.37 €/kW[c]		1667–434 MW[c]	
		Biomass gasification		3.10	1836 M$[b]	2.38[b]	10,000–960,000 kg/day[b]	
					2637.55–4101.10 €/kW[c]		33–0.7 MW[c]	
					575–1836 M€[a]	3.52–2.15[a]	10,000–960,000 kg/day[a]	
Distributed production	LH_2 production	Electrolysis[a]	Wind	1.03	4.03–20.20 M€[a]	6.24–4.94[a]	50–2500 kg/day[a]	52.49 kWh/kg(H_2)
			Solar	6.21				
			Hydro	2.07				
			Nuclear	3.10				

[a] *De-León Almaraz et al. (2014).*
[b] *Almansoori and Betancourt-Torcat (2016).*
[c] *Sgobbi et al. (2016).*

2.5 HYDROGEN CONDITIONING AND STORAGE

Hydrogen is a very low density gas (0.08988 kg/Nm^3 at 273 K) (Patay, 2008) and can be stored in three ways (Florida Solar Energy Center, n.d.):

- as a compressed gas in high-pressure tanks.
- as a liquid in dewars or tanks (requires cryogenic temperatures because the boiling point of hydrogen at one atmosphere pressure is −252.8°C).
- as a solid by either absorbing or reacting with metals or chemical compounds.

The physical limits for the storage density of compressed and liquid hydrogen have more or less been reached, while there is still potential in the development of solid materials for hydrogen storage, such as systems involving metal hydrides (Ball and Wietschel, 2008). Designing tanks that are compact, lightweight, safe, and cheap is crucial because this would open the possibility of making hydrogen storage particularly attractive compared to electricity (CEA, 2013).

The hydrogen storage pathways from production to onboard storage in vehicles is presented in Fig. 2.8.

2.5.1 Gaseous Hydrogen (GH_2)

The most commonly used method for hydrogen storage is compressed hydrogen tanks at high pressure (>200 bar). Compressed hydrogen in hydrogen tanks at 350 bar and 700 bar is used in fuel cell hydrogen vehicles (Krishna et al., 2012). This is currently one of the simplest, most common, and efficient storage

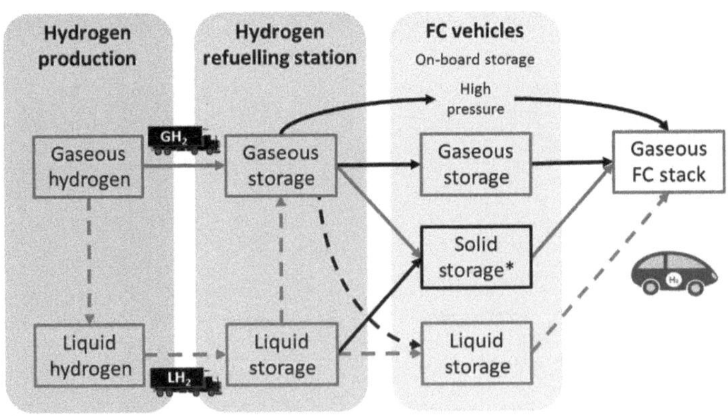

FIG. 2.8 Hydrogen storage pathways from production to onboard storage in vehicles. *(Adapted from Mori, D., Hirose, K., 2009. Recent challenges of hydrogen storage technologies for fuel cell vehicles. Int. J. Hydrogen Energy, 2nd World Hydrogen Technologies Convention 34, 4569–4574. https://doi.org/10.1016/j.ijhydene.2008.07.115.)*

technologies in use. High-pressure hydrogen is stored in thick-walled tanks (mainly of cylindrical or quasiconformable shape) made of high-strength materials to ensure durability (Krishna et al., 2012; Tzimas et al., 2003).

According to James (2008), for stationary hydrogen storage, GH_2 also offers the advantages of simplicity and stable storage (no boil-off losses) but at a considerably greater volume than LH_2. Even accounting for compression costs, high-pressure gaseous hydrogen is cheaper than LH_2. However, except for pipeline transmission, GH_2 lacks the bulk transportability of LH_2. Consequently, GH_2 will mostly be used for storage of limited hydrogen quantities, for long term storage, or when the cost of liquefaction is prohibitive. Some issues for GH_2 are its safety perception and the high cost of the pressure vessels and hydrogen compressors.

2.5.2 Liquid Hydrogen (LH_2)

Hydrogen in liquid form has a considerably higher energy density than in gaseous form, making it an attractive storage medium (Tzimas et al., 2003). For example, the theoretical volumetric capacity of hydrogen increases from 24 or 40 g/L (for compressed H_2 at 350 or 700 bar at 300 K) to 70 g/L (for liquid H_2 at 1 atm and 20 K). When hydrogen is stored as liquid at 1 atm, it must be maintained below its boiling point (−253°C or 20 K) (Krishna et al., 2012). Effective thermal insulation is essential to maximize the efficiency of the liquid hydrogen (LH_2) tank. Typical LH_2 tanks consist of metallic double-walled containers, in which the inner and outer walls are separated by vacuum for thermal insulation purposes (Züttel, 2003).

The energy requirements of liquefaction are high, typically 30% of the hydrogen's heating value, leading to relatively high hydrogen cost as compared to gaseous hydrogen. The loss of hydrogen by evaporation effects during storage periods are further disadvantages of liquid hydrogen (LH_2) storage systems (Hake et al., 2006).

In the initial phases of hydrogen infrastructure development, the transportation of cryogenic liquid hydrogen via trucking or rail could play a significant role (James, 2008). Hydrogen is typically liquefied at the production site in large quantities (10–30 tons/day) and then trucked cross country in LH_2 tankers with no boil-off losses.

Another option is based on cryo-compressed storage; studies of physical hydrogen storage have currently shifted to cryo-compressed H_2, which combines compression and cryogenic storage (Hwang and Varma, 2014).

2.5.3 Solid Hydrogen

The storage of hydrogen in solid materials has the potential to become a safe and efficient way to store energy, both for stationary and mobile applications.

Solid hydrogen may be stored in two ways, either by metal hydrides or by carbon adsorption.

Metal hydrides store hydrogen by chemically bonding the hydrogen to metal or metalloid elements and alloys. Some hydrides can adsorb hydrogen at or below atmospheric pressure, then release the hydrogen at significantly higher pressure when heated (Tzimas et al., 2003). For example, $LaNi_5H_6$ can release hydrogen under PEM fuel cell operating conditions (1–10 atm and 25–1008°C), but its gravimetric capacity is too low (1.4 wt%) and its cost too high for vehicle applications (Hwang and Varma, 2014).

Each alloy has different performance characteristics (cycle life, heat of reaction). Metal hydrides offer the advantages of lower pressure storage, conformable shapes, and reasonable volumetric storage efficiency. However, they suffer from weight penalties and thermal management issues. Improving resistance to gaseous contaminants and increasing system cycle life are still obstacles to overcome.

Carbon-based materials/porous structures (single-walled nanotubes and graphite nanofibers) can also be used to store significant amounts of hydrogen at room temperature due to their high surface area and abundant pore volume. There are still challenges to overcome, such as the understanding of the adsorption/desorption mechanism and the volumetric capacity of porous structures (Tzimas et al., 2003).

2.5.4 Key Factors

All the aforementioned hydrogen storage options (compressed, liquid, metal hydrides, and porous structures) have their advantages and drawbacks with respect to weight, volume, energy efficiency, refueling time, cost, and safety.

Some key parameters of the liquid and gaseous options are presented in Table 2.6.

2.6 HYDROGEN TRANSPORTATION

Conceptually, transportation is divided into two parts: transmission and distribution. Transmission refers to H_2 transportation from a plant to other regions without plant units and distribution refers to H_2 transportation to the refueling stations from a plant or regional conditioning center in any region (Murthy Konda et al., 2011).

There are various methods for transporting hydrogen, but the choice of the best transportation mode and natural state (compressed gas trucks, cryogenic liquid trucks, or gas pipelines) depends mostly on the geographic and market characteristics of the zone in which the transport occurs (market penetration, population density, demand, state of the roads, and infrastructure available) (Yang and Ogden, 2007).

TABLE 2.6 Key Parameters for Hydrogen Storage in the HSC

	kg CO_2eq/kg (H_2)[a]	Capital Cost	Operational Cost	Capacity (kg)	Efficiency[a] kWh/kg (H_2)
Liquid H_2	0.704	106 M€[b]	0.0043 €/(kg/day)[b]	540,000[b]	8.42
		0.802–122 M€[a]	0.064–0.005 $/(kg/day)[a]	50–540,000[a]	
		122 M$[c]	0.005 $/kg[c]	10,000–540,000[c]	
Gaseous H_2	0.349	1645 M€[b]	0.066 €/(kg/day)[b]	540,000[b]	4.17
		1894 M$[c]	0.076 $/kg[c]	10,000–540,000[c]	

[a]*De-León Almaraz et al. (2014).*
[b]*Sabio et al. (2010).*
[c]*Almansoori and Betancourt-Torcat (2016).*

FIG. 2.9 Hydrogen pipelines and hydrogen tube trailers.

Hydrogen may be transported as either a gas or a liquid. Gaseous hydrogen can be transported by high-pressure pipelines or by tube trailers (see Fig. 2.9). Liquefied hydrogen can be transported in tankers (Dagdougui, 2012).

2.6.1 Hydrogen Pipelines

Pipelines are highly recommended for long distances and large quantities. They have been used for several years to deliver hydrogen to refineries and chemical plants (Ball and Weeda, 2015).

From the economic point of view, pipelines have very low operating costs, but very high capital costs due to the high-quality material that must be used in their construction that assures a lack of risk. Of course, their cost can be reduced if the natural gas infrastructure could be adapted to hydrogen (Ball and Wietschel, 2008).

Pipelines for medium and large fueling stations may become relevant once a significant market penetration of hydrogen vehicles has been achieved, but these are mostly used for local distribution in highly populated areas and for large-scale interregional energy transport according to the HyWays roadmap (European Commission, 2008). In Ball and Wietschel (2008), a pipeline network could be constructed after 2030 when the potential demand can justify the high investment.

2.6.2 Hydrogen Tube Trailers

Compressed gaseous hydrogen trailers (at 200 bar) can be a very suitable option when hydrogen must be transported over short distances, mostly up to 200 km. Recently, the pressure level for these trailers has been increased from 200 bar to 500 bar, increasing the payload from 400 kg to more than 1000 kg for distances up to 500 km (Ball and Weeda, 2015). Liquid hydrogen is recommended for smaller volumes and longer distances (Ball and Wietschel, 2009).

2.6.3 Tanker Trucks

From the liquefaction unit, LH_2 is transported by tanker trucks (cryogenic liquid hydrogen tankers) (Patay, 2008). This transportation mode is the most economic pathway for medium market penetration (Dagdougui, 2011b). Tanker trucks could transport relatively large amounts of hydrogen and reach markets located throughout large geographic areas. Forty ton trucks can carry 3500 kg of LH_2. Thus, the transport of liquid hydrogen is limited by volume, not by weight (Bossel, 2006).

2.6.4 Key Parameters of Some Hydrogen Transportation Modes

Some key parameters of various hydrogen transportation modes are summarized in Table 2.7.

Hydrogen storage, along with distribution, is a key technology associated with the wide utilization of hydrogen. Both of these technologies represent limiting factors, currently and in the near future, to the deployment of hydrogen in the energy system

2.7 HYDROGEN REFUELING STATIONS

Hydrogen refueling stations (HRS) are not only associated with the market but also to the expectations and needs of FCEV owners. Major progress has been made in hydrogen refueling by developing equipment standards and refueling protocols for high pressure (700 bar) and fast dispensing ($<$3 min) in cars (Ball and Weeda, 2015).

Patay (2008) identified four types of refueling stations according to their main mode of supply, allowing the filling station vehicles equipped with tanks for hydrogen gas at high pressure to about 700 bar[1] (a priori these tanks will be dominant in the market):

- Type 1: Stations receiving liquid hydrogen require evaporation and compression facilities.
- Type 2: Stations receiving compressed hydrogen (200 bar) from tube trailers require a high-compression unit (400–700 bar).
- Type 3: Stations with a piped gaseous H_2 supply require a high-compression unit (from 100 to 400–700 bar).
- Type 4: Stations with onsite H_2 production by small electrolysis units require compression (from 15 to 400–700 bar).

1. In cases where the FCEV has 350 bar onboard storage tanks, H_2 has to be compressed up to 400 bar before being distributed (Murthy Konda et al., 2011).

TABLE 2.7 Key Parameters for Hydrogen Transportation

		g CO_2eq/ metric ton/ km	Capital Cost (€)[a]	Operational Cost[a]	Capacity[a] (kg/ day)	Avg Speed[a] (km/h)	Un/Loading Time[a] (h)	Efficiency (Fuel)[a] (km/L)
LH_2	Road	62[b]	434,236	40.56 €/h	4082–960,000	55	2	3.58
	Rail		434,236	23.62 €/h	9072–960,000	45	12	10.13
	Ship			0.00115 €/h/ kg	4082–960,000	16	48	18.75
GH_2	Road		217,118	40.56 €/h	181–960,000	55	2	3.58
	Rail		260,541	23.62 €/h	454–960,000	45	12	10.13
	Pipeline		708,673 €/km	0.05767 €/kg/ day	10–960,000 kg/ day			

[a]*Sabio et al. (2010).*
[b]*De-León Almaraz et al. (2014).*

2.7.1 Key Parameters of Hydrogen Refueling Stations

The key parameters of the refueling stations are summarized in Table 2.8.

The choice between central and onsite production and storage of hydrogen, and of the transport mode between sites, depends on the demand of either gaseous or liquid hydrogen. The HSC design offers multiple possibilities to explore.

2.8 MULTIPLE OBJECTIVES IN HSC SUSTAINABLE ASSESSMENT

Three categories of objectives related to economic, environmental, and social issues are of major interest for HSC design. Sustainable development usually refers to the simultaneous achievement of economic prosperity, environmental cleanness, and social responsibility, namely the so-called "triple bottom line" (Jiménez-González and Woodley, 2010; Othman et al., 2010). Table 2.9 presents the set of criteria that have been studied in the dedicated HSC literature.

2.8.1 Economic Assessment

The principal and the most-treated objective is the economic one (Agnolucci et al., 2013; Almansoori and Betancourt-Torcat, 2016; Almansoori and Shah, 2006, 2009, 2012; Gondal and Sahir, 2013; Kamarudin et al., 2009; Kim et al., 2008; Woo et al., 2016), mainly based on the total network cost.

These criteria have been generally embedded in optimization strategies for hydrogen supply chains.

Almansoori (2015) presents a MILP model for the optimal design and operation of an HSC under CO_2 emission constraints. This model includes carbon capture and storage (CCS) methods and a carbon tax as CO_2 mitigation strategies. The optimization objective consists of the minimization of the total network cost, both in terms of capital and operating expenditures, under techno-economic and environmental constraints, as in Almansoori and Betancourt-Torcat (2016). The model determines the most suitable delivered product form (gaseous or liquid) into the market.

Two scenarios were studied to determine the optimal HSC configuration of the country (with and without the application of a carbon tax). A case study of a future HSC in the United Arab Emirates (UAE) was analyzed. The production technology selected was based on steam methane reforming, as similarly in (Almansoori and Betancourt-Torcat, 2016; Agnolucci et al., 2013; (Moreno-Benito et al., 2016).

The total annualized cost to plan the HSC under emission constraints has also been studied in (Almansoori and Betancourt-Torcat, 2016). A case study for Germany in 2030 was treated considering CCS decisions.

Other works present techno-economic inputs and technological background information for hydrogen delivery pathways (Bolat and Thiel, 2014a, b).

TABLE 2.8 Key Parameters for Hydrogen Refueling Stations in the HSC

	Tank Pressure[a] (bar)	Storage Pressure (MP)[a]	Capacity[a]	Efficiency[a] kWh/kg(H_2)	Refueling Time[b] (min)	Reliability[a]	Life Duration[a] (Years)
High pressure	700	85	750 kg (H_2)/day	3.0	<3	<95%	50 (building)
Low pressure	350	40	280 hg (H_2) Storage				10 (comp.)

[a]Wulf and Kaltschmitt (2012).
[b]Ball and Weeda (2015).

TABLE 2.9 Sustainable Assessment Criteria for HSC

Multiobjective Assessment		
Economic	*Environmental*	*Social*
Total network cost (Agnolucci et al., 2013; Almansoori and Betancourt-Torcat, 2016; Almansoori and Shah, 2006, 2009, 2012; Kim et al., 2008)	Global warming potential (De-León Almaraz et al., 2013)	Hazard identification (HAZID) (Oyama et al., 2016)
Total investment cost (Woo et al., 2016)	Life cycle assessment (LCA): Eco-Indicator99 (EI-99) (Guillén-Gosálbez et al., 2010)	Total relative risk (Han et al., 2013; Kim and Moon, 2008)
	Cumulative WtW (well-to-wheel) life cycle GHG emissions (Hugo et al., 2005; Li et al., 2008)	Social effects (Cantuarias-Villessuzanne et al., 2016; Creti et al., 2015)

2.8.2 Environmental Assessment

Life cycle assessment (LCA) has been extensively used for energy supply chain environmental assessment (Curran, 2000). LCA is an established and internationally accepted method that is defined in the ISO standards ISO 14040 and ISO 14044. Even if hydrogen is considered to be a clean fuel in its use phase, its production still has impacts on the environment. Examining resource consumption, energy requirements, and emissions from a life cycle point of view (production to distribution) gives a complete picture of the environmental burdens associated with the hydrogen supply chain. Some LCA-based approaches for hydrogen production processes are listed in Table 2.10.

To limit the global mean temperature increase to 2 °C by the middle of this century, roughly 80% of the global greenhouse gas (GHG) emissions must be reduced by 2050 using the 1990 baseline. This explains why the global warming potential (GWP) is perhaps the most used indicator for energy supply chains, and for hydrogen ones in particular.

De-León Almaraz et al. (2013) uses the GWP as an indicator of the overall effect of the process related to the heat radiation absorption of the atmosphere due to emissions of greenhouse gases (CO_2-equiv) of the network. The total GWP is the cumulative value of the GWP associated with the different echelons of the supply chain, namely, production, storage, and distribution.

TABLE 2.10 Reviews of the LCA of Hydrogen Production

Authors	Title
Boyano et al. (2011)	Exergo-environmental analysis of a steam methane reforming process for hydrogen production
Cetinkaya et al. (2012)	Life cycle assessment of various hydrogen production methods
Dufour et al. (2012)	Life cycle assessment of alternatives for hydrogen production from renewable and fossil sources
Hacatoglu et al. (2012)	Comparative life cycle assessment of hydrogen and other selected fuels
Koroneos et al. (2008)	Hydrogen production via biomass gasification. A life cycle assessment approach
Lee et al. (2010)	Life cycle environmental and economic analyses of a hydrogen station with wind energy
Ozbilen et al. (2011)	A comparative life cycle analysis of hydrogen production via thermochemical water splitting
Spath and Mann (2004)	Life cycle assessment of renewable hydrogen production via wind/electrolysis
Utgikar and Thiesen (2006)	Life cycle assessment of high-temperature electrolysis for hydrogen production via nuclear energy
Wulf and Kaltschmitt (2012)	Life cycle assessment of hydrogen supply chain with special attention on hydrogen refueling stations

Based on Bhandari, R., Trudewind, C.A., Zapp, P., 2012. Life cycle assessment of hydrogen production methods—a review, STE research report. Forschungszentrum Jülich, Institute of Energy and Climate Research.

Guillén-Gosálbez et al. (2010) uses the Eco-indicator 99 framework, focusing their attention on only one of its impact categories, that is, damage to human health caused by climate change. The units of this metric are disability adjusted life years (DALYs).

Hugo et al. (2005) and Li et al. (2008) propose the analysis and reduction of the environmental performance of the competing hydrogen infrastructures using the cumulative WtW (well-to-wheel) life cycle GHG emissions that result from delivering hydrogen to the FCEV consumer.

2.8.3 Social Assessment

Social aspects are perhaps the most difficult to formalize. The social aspects involved in the development of the hydrogen economy have been addressed from different perspectives and at different levels. Firstly, several studies have been conducted to evaluate hydrogen sustainability (Afgan et al., 2007; Chang et al., 2011; Hsu, 2013; Markert et al., 2016; Ren et al., 2013a, 2013b). Secondly, other works have focused on the analysis of strategies to narrow the gap toward a hydrogen economy (Keles et al., 2008; Moliner et al., 2016; Qadrdan et al., 2008). Lastly, several assessments have been performed to evaluate the potential of a hydrogen economy compared with other sustainable alternatives, and its acceptance in the future (Ball and Weeda, 2015; Ricci et al., 2008; Sgobbi et al., 2016).

One approach from a social point of view is the real case presented by Oyama et al. (2016) and developed by Kawasaki Heavy Industries, Ltd., focused on inexpensive Australian brown coal, which boasts tremendous reserves. Oyama et al. (2016) have developed the concept of the "CO_2-Free Hydrogen Chain" in which a large amount of hydrogen is imported into Japan by liquefied hydrogen carriers after gasifying and refining the brown coal to produce hydrogen. This model takes safety into account using Hazard Identification (HAZID) for safety design.

Han et al. (2013) have proposed an HSC model optimizing a risk safety objective considering the cumulative risk of production sites, storage sites, and of transportation.

In addition to economic objectives, Kim and Moon (2008) have also considered a risk index. The index is based on three subrisk indexes according to the types of hydrogen activities: hydrogen production, storage, and transportation. In addition, the index takes into account the characteristics of the regions (e.g., the population density) in which hydrogen activities are performed.

More recently, other studies (Cantuarias-Villessuzanne et al., 2016; Creti et al., 2015) have also investigated the social effects of developing a hydrogen economy, for electromobility applications. Cantuarias-Villessuzanne et al. (2016) have quantified the societal benefits for the reduction of greenhouse gas emissions and social costs for the increase of platinum consumption in the manufacture of fuel cells. According to their study, the social balance is positive, including also platinum depletion, generating net savings for Europe.

2.9 CONCLUSIONS

In this chapter, the key components of an HSC have been presented. An HSC involves a variety of energy sources, production, storage, transportation options, and supply of hydrogen to the refueling stations. These individual components constitute the building blocks that need to be integrated to form the global supply chain. The resulting network will heavily depend on country- and region-specific conditions.

HSC deployment and design are clearly a problem with multiple objectives, in a context of energy transition with multiple stakeholders and multiple end users.

Even though most of the aspects of the HSC infrastructure have been covered in the literature, an open question that remains is how the hydrogen infrastructure will back the transition toward long-term sustainable hydrogen economies while guaranteeing short- and mid-term system economic acceptability, starting from the current carbon-based economy.

This issue has been tackled in several roadmaps, but the formulation for optimizing simultaneously the range of production technologies, scales, transportation modes, and CCS elements across time and space for solving the transition toward a low-carbon energy system, and in particular toward a hydrogen economy remains. Such an approach will be required to manage a gradual penetration of hydrogen demand.

All these activities need to be integrated in a systematic modeling framework that allows a more precise approach than roadmaps and planning scenarios (as previously explained in Chapter 1). This analysis emphasizes that achieving the potential benefits of a hydrogen system requires careful integration of production, storage, and end-use components with minimized cost and maximized efficiency, and a strong understanding of the environmental impact, reliability, and opportunities, positioning this book in a system modeling approach based on a multiobjective optimization framework.

REFERENCES

Acar, C., Dincer, I., 2014. Comparative assessment of hydrogen production methods from renewable and non-renewable sources. Int. J. Hydrogen Energy 39, 1–12.

Afgan, N.H., Veziroglu, A., Carvalho, M.G., 2007. Multi-criteria evaluation of hydrogen system options. Int. J. Hydrogen Energy 32, 3183–3193.

Agnolucci, P., Akgul, O., McDowall, W., Papageorgiou, L.G., 2013. The importance of economies of scale, transport costs and demand patterns in optimising hydrogen fuelling infrastructure: an exploration with SHIPMod (spatial hydrogen infrastructure planning model). Int. J. Hydrogen Energy 38, 11189–11201. https://doi.org/10.1016/j.ijhydene.2013.06.071.

Almansoori, A., 2015. Design and operation of a hydrogen supply chain considering CO2 mitigation strategies—a case study of the United Arab Emirates.2015 IEEE International Conference on Smart Energy Grid Engineering (SEGE). Presented at the 2015 IEEE International Conference on Smart Energy Grid Engineering (SEGE)pp. 1–5. https://doi.org/10.1109/SEGE.2015.7324585.

Almansoori, A., Betancourt-Torcat, A., 2016. Design of optimization model for a hydrogen supply chain under emission constraints—a case study of Germany. Energy 111, 414–429. https://doi.org/10.1016/j.energy.2016.05.123.

Almansoori, A., Shah, N., 2006. Design and operation of a future hydrogen supply chain: snapshot model. Chem. Eng. Res. Des. 84, 423–438. https://doi.org/10.1205/cherd.05193.

Almansoori, A., Shah, N., 2009. Design and operation of a future hydrogen supply chain: multi-period model. Int. J. Hydrogen Energy 34, 7883–7897. https://doi.org/10.1016/j.ijhydene.2009.07.109.

Almansoori, A., Shah, N., 2012. Design and operation of a stochastic hydrogen supply chain network under demand uncertainty. Int. J. Hydrogen Energy 37, 3965–3977. Portable Fuel Cells – Fundamental and Applications (ISPFC2010). https://doi.org/10.1016/j.ijhydene.2011.11.091.

Ausfelder, F., Bazzanella, A., 2016. Hydrogen in the chemical industry. In: Stolten, D., Emonts, B. (Eds.), Hydrogen Science and Engineering: Materials, Processes, Systems and Technology. Wiley-VCH Verlag GmbH & Co. KGaA, Weinheim, Germany, pp. 19–40. https://doi.org/10.1002/9783527674268.ch02.

Bahadori, A., 2014. Natural Gas Processing: Technology and Engineering Design. Gulf Professional Publishing, Oxford.

Balat, M., Balat, M., 2009. Political, economic and environmental impacts of biomass-based hydrogen. Int. J. Hydrogen Energy 34, 3589–3603. https://doi.org/10.1016/j.ijhydene.2009.02.067.

Ball, M., Weeda, M., 2015. The hydrogen economy – vision or reality?1. Int. J. Hydrogen Energy 40, 7903–7919. https://doi.org/10.1016/j.ijhydene.2015.04.032.

Ball, M., Wietschel, M., 2008. The future of hydrogen – opportunities and challenges. Int. J. Hydrogen Energy 34, 615–627. https://doi.org/10.1016/j.ijhydene.2008.11.014.

Ball, M., Wietschel, M., 2009. The future of hydrogen – opportunities and challenges. Int. J. Hydrogen Energy 34, 615–627. https://doi.org/10.1016/j.ijhydene.2008.11.014.

Balta, M.T., Dincer, I., Hepbasli, A., 2010. Potential methods for geothermal-based hydrogen production. Int. J. Hydrogen Energy 35, 4949–4961. Novel Hydrogen Production Technologies and Applications. https://doi.org/10.1016/j.ijhydene.2009.09.040.

Bartels, J.R., Pate, M.B., Olson, N.K., 2010. An economic survey of hydrogen production from conventional and alternative energy sources. Int. J. Hydrogen Energy 35, 8371–8384. https://doi.org/10.1016/j.ijhydene.2010.04.035.

BGR, 2009. Reserves, Resources and Availability of Energy Resources – Annual Report 2009. Bundesanstalt für Geowissenschaften und Rohstoffe – Federal Institute for Geosciences and Natural Resources, Hannover, Germany.

Bockris, J.O., Veziroglu, T.N., 2007. Estimates of the price of hydrogen as a medium for wind and solar sources. Int. J. Hydrogen Energy 32, 1605–1610. https://doi.org/10.1016/j.ijhydene.2007.04.037.

Bolat, P., Thiel, C., 2014a. Hydrogen supply chain architecture for bottom-up energy systems models. Part 1: Developing pathways. Int. J. Hydrogen Energy 39, 8881–8897. https://doi.org/10.1016/j.ijhydene.2014.03.176.

Bolat, P., Thiel, C., 2014b. Hydrogen supply chain architecture for bottom-up energy systems models. Part 2: Techno-economic inputs for hydrogen production pathways. Int. J. Hydrogen Energy 39, 8898–8925. https://doi.org/10.1016/j.ijhydene.2014.03.170.

Bossel, U., 2006. Why a hydrogen economy doesn't make sense [WWW Document]. http://phys.org/news85074285.html (Accessed 1.6.14).

Boyano, A., Blanco-Marigorta, A.M., Morosuk, T., Tsatsaronis, G., 2011. Exergoenvironmental analysis of a steam methane reforming process for hydrogen production. Energy 36, 2202–2214. 5th Dubrovnik Conference on Sustainable Development of Energy, Water & Environment Systems. https://doi.org/10.1016/j.energy.2010.05.020.

Cantuarias-Villessuzanne, C., Weinberger, B., Roses, L., Vignes, A., Brignon, J.-M., 2016. Social cost-benefit analysis of hydrogen mobility in Europe. Int. J. Hydrogen Energy 41, 19304–19311.

CEA, 2013. CEA – Jeunes – Thèmes – Les énergies renouvelables – L'hydrogène – Caractéristiques de l'hydrogène [WWW Document] http://www.cea.fr/jeunes/themes/les-energies-renouvelables/l-hydrogene/caracteristiques-de-l-hydrogene?_tmc=evrEoS_mLKqhnmJQmo92PCxU0EkakD5e21f5S2yiQKo (Accessed 7.10.13).

Cetinkaya, E., Dincer, I., Naterer, G.F., 2012. Life cycle assessment of various hydrogen production methods. Int. J. Hydrogen Energy 37, 2071–2080. 2010 AIChE Annual Meeting Topical Conference on Hydrogen Production and Storage Special Issue. https://doi.org/10.1016/j.ijhydene.2011.10.064.

Chang, P.-L., Hsu, C.-W., Chang, P.-C., 2011. Fuzzy Delphi method for evaluating hydrogen production technologies. Int. J. Hydrogen Energy 36, 14172–141792010 Asian/APEC BioH2.

Costogue, E.N., Yasui, R.K., 1977. Performance data for a terrestrial solar photovoltaic water electrolysis experiment. Sol. Energy 19, 205–210. https://doi.org/10.1016/0038-092X(77)90060-3.

Creti, A., Kotelnikova, A., Meunier, G., Ponssard, J.-P., 2015. A cost benefit analysis of fuel cell electric vehicles, Research Report, Report 1:1–41.

Cueugniet, J., Legait, B., Durville, J.-L., Gazeau, J.-C., Nataf, J.-M., 2015. Filière hydrogène-énergie [WWW Document]. Portail Ministères Économiques Financ. https://www.economie.gouv.fr/cge/filiere-hydrogene-energie (Accessed 12.31.17).

Curran, M.A., 2000. Life cycle assessment. In: Kirk-Othmer Encyclopedia of Chemical Technology. John Wiley & Sons, Inc., New York. https://doi.org/10.1002/0471238961.lifeguin.a01.pub2.

Dagdougui, H., 2011a. Models, methods and approaches for the planning and design of the future hydrogen supply chain. Int. J. Hydrogen Energy 37, 5318–5327. https://doi.org/10.1016/j.ijhydene.2011.08.041.

Dagdougui, H., 2011b. Decision Support Systems for Sustainable Renewable Energy Systems and Hydrogen Logistics: Modelling, Control and Risk Analysis. UNIGE, Italy; CRC, MINES ParisTech, France.

Dagdougui, H., 2012. Models, methods and approaches for the planning and design of the future hydrogen supply chain. Int. J. Hydrogen Energy 37, 5318–5327. Optimization Approaches to Hydrogen Logistics. https://doi.org/10.1016/j.ijhydene.2011.08.041.

De León Almaraz, S., 2014. Multi-objective optimisation of a hydrogen supply chain [WWW Document]. http://ethesis.inp-toulouse.fr/archive/00002723/ (Accessed 1.13.15).

De-León Almaraz, S., Azzaro-Pantel, C., Montastruc, L., Pibouleau, L., Senties, O.B., 2013. Assessment of mono and multi-objective optimization to design a hydrogen supply chain. Int. J. Hydrogen Energy 38, 14121–14145. https://doi.org/10.1016/j.ijhydene.2013.07.059.

De-León Almaraz, S., Azzaro-Pantel, C., Montastruc, L., Domenech, S., 2014. Hydrogen supply chain optimization for deployment scenarios in the Midi-Pyrénées region, France. Int. J. Hydrogen Energy 39, 11831–11845. https://doi.org/10.1016/j.ijhydene.2014.05.165.

Dincer, I., 2012. Green methods for hydrogen production. Int. J. Hydrogen Energy 37, 1954–1971. 10th International Conference on Clean Energy 2010. https://doi.org/10.1016/j.ijhydene.2011.03.173.

Dincer, I., Acar, C., 2015. Review and evaluation of hydrogen production methods for better sustainability. Int. J. Hydrogen Energy 40, 11094–11111. https://doi.org/10.1016/j.ijhydene.2014.12.035.

Direction générale de la compétitivité, de l'industrie et des services, 2011. Technologies clés 2015 [WWW Document]. http://www.dgcis.gouv.fr/politique-et-enjeux/technologies-cles-2015-mars-2011 (Accessed 9.9.13).

Dufour, J., Serrano, D.P., Gálvez, J.L., González, A., Soria, E., Fierro, J.L.G., 2012. Life cycle assessment of alternatives for hydrogen production from renewable and fossil sources. Int. J. Hydrogen Energy 37, 1173–1183. 10th International Conference on Clean Energy 2010. https://doi.org/10.1016/j.ijhydene.2011.09.135.

EIA, 2017. International Energy Outlook 2017 (no. DOE/EIA-0484(2017)). U.S. Energy Information Administration.

Eisenstadt, M.M., Cox, K.E., 1975. Hydrogen production from solar energy. Sol. Energy 17, 59–65. https://doi.org/10.1016/0038-092X(75)90017-1.

EREC, 2010. Rethinking 2050: a 100% renewable energy vision for the European Union [WWW Document]. http://setis.ec.europa.eu/newsroom-items-folder/news-container/2018rethinking-2050-a-100-renewable-energy-vision-for-the-european-union2019-conv (Accessed 4.8.13).

European Commission, 2008. HyWays the European Hydrogen Roadmap (No. EUR 23123).

Florida Solar Energy Center, n.d. Hydrogen basics – storage [WWW Document]. http://www.fsec.ucf.edu/en/consumer/hydrogen/basics/storage.htm (Accessed 10.3.17).

Gondal, I.A., Sahir, M.H., 2013. Model for biomass-based renewable hydrogen supply chain. Int. J. Energy Res. 37, 1151–1159. https://doi.org/10.1002/er.2970.

Guillén-Gosálbez, G., Mele, F.D., Grossmann, I.E., 2010. A bi-criterion optimization approach for the design and planning of hydrogen supply chains for vehicle use. AICHE J. 56, 650–667. https://doi.org/10.1002/aic.12024.

Hacatoglu, K., Rosen, M.A., Dincer, I., 2012. Comparative life cycle assessment of hydrogen and other selected fuels. Int. J. Hydrog. Energy 37, 9933–9940. Proceedings of the Symposium on Hydrogen Production and Applications at the 240th American Chemical Society National Meeting, August 22–26, 2010, Boston, Massachusetts, USA. https://doi.org/10.1016/j.ijhydene.2012.04.020.

Haeseldonckx, D., D'haeseleer, W., 2011. Concrete transition issues towards a fully-fledged use of hydrogen as an energy carrier: methodology and modelling. Int. J. Hydrogen Energy 36, 4636–4652. https://doi.org/10.1016/j.ijhydene.2011.01.113.

Hajjaji, N., 2011. Analyse de cycle de vie exergétique de systèmes de production d'hydrogène (Doctorat). INPL.

Hake, J.-F., Linssen, J., Walbeck, M., 2006. Prospects for hydrogen in the German energy system. Energy Policy 34, 1271–1283. Hydrogen. https://doi.org/10.1016/j.enpol.2005.12.015.

Hallenbeck, P.C., Benemann, J.R., 2002. Biological hydrogen production: fundamentals and limiting processes. Int. J. Hydrogen Energy 27, 1185–1193.

Han, J.-H., Ryu, J.-H., Lee, I.-B., 2013. Multi-objective optimization design of hydrogen infrastructures simultaneously considering economic cost, safety and CO_2 emission. Chem. Eng. Res. Des. 91, 1427–1439. Special Issue: Computer Aided Process Engineering (CAPE) Tools for a Sustainable World. https://doi.org/10.1016/j.cherd.2013.04.026.

Heracleous, E., 2011. Well-to-Wheels analysis of hydrogen production from bio-oil reforming for use in internal combustion engines. Int. J. Hydrogen Energy. In Press, Corrected Proof. https://doi.org/16/j.ijhydene.2011.06.052.

Hosseini, S.E., Wahid, M.A., 2016. Hydrogen production from renewable and sustainable energy resources: promising green energy carrier for clean development. Renew. Sust. Energy Rev. 57, 850–866. https://doi.org/10.1016/j.rser.2015.12.112.

Hsu, C.-W., 2013. Constructing an evaluation model for hydrogen application pathways. Int. J. Hydrogen Energy 38, 15836–15842.

Syndicat des énergies renouvelables, 2012. Le Livre Blanc des énergies renouvelables [WWW Document]. http://www.enr.fr/. Accessed 4 August 2013.

Hugo, A., Rutter, P., Pistikopoulos, S., Amorelli, A., Zoia, G., 2005. Hydrogen infrastructure strategic planning using multi-objective optimization. Int. J. Hydrogen Energy 30, 1523–1534. https://doi.org/10.1016/j.ijhydene.2005.04.017.

Hwang, H.T., Varma, A., 2014. Hydrogen storage for fuel cell vehicles. Curr. Opin. Chem. Eng. 5, 42–48. Energy and environmental engineering/Reaction engineering. https://doi.org/10.1016/j.coche.2014.04.004.

IEA, 2011. World energy outlook. Resume 2011.

Ivy, J., 2004. Summary of electrolytic hydrogen production: milestone completion report (no. NREL/MP-560-36734). Golden, CO, National Renewable Energy Lab.

James, B.D., 2008. Overview of hydrogen storage technologies. Prepr. Symp., Div. Fuel Chem., Am. Chem. Soc.568–569.

Jiménez-González, C., Woodley, J.M., 2010. Bioprocesses: modeling needs for process evaluation and sustainability assessment. Comput. Chem. Eng. 34, 1009–1017. Process Modeling and

Control in Drug Development and Manufacturing. https://doi.org/10.1016/j.compchemeng.2010.03.010.

Johnson, N., Yang, C., Ogden, J., 2008. A GIS-based assessment of coal-based hydrogen infrastructure deployment in the state of Ohio. Int. J. Hydrogen Energy 33, 5287–5303. https://doi.org/10.1016/j.ijhydene.2008.06.069.

Kamarudin, S.K., Daud, W.R.W., Yaakub, Z., Misron, Z., Anuar, W., Yusuf, N.N.A.N., 2009. Synthesis and optimization of future hydrogen energy infrastructure planning in Peninsular Malaysia. Int. J. Hydrogen Energy 34, 2077–2088. https://doi.org/10.1016/j.ijhydene.2008.12.086.

Keles, D., Wietschel, M., Most, D., Rentz, O., 2008. Market penetration of fuel cell vehicles? Analysis based on agent behaviour. Int. J. Hydrogen Energy 33, 4444–4455.

Kim, J., Moon, I., 2008. Strategic design of hydrogen infrastructure considering cost and safety using multiobjective optimization. Int. J. Hydrogen Energy 33, 5887–5896. https://doi.org/10.1016/j.ijhydene.2008.07.028.

Kim, J., Lee, Y., Moon, I., 2008. Optimization of a hydrogen supply chain under demand uncertainty. Int. J. Hydrogen Energy 33, 4715–4729. https://doi.org/10.1016/j.ijhydene.2008.06.007.

Koroneos, C., Dompros, A., Roumbas, G., Moussiopoulos, N., 2004. Life cycle assessment of hydrogen fuel production processes. Int. J. Hydrogen Energy 29, 1443–1450. https://doi.org/16/j.ijhydene.2004.01.016.

Koroneos, C., Dompros, A., Roumbas, G., 2008. Hydrogen production via biomass gasification—a life cycle assessment approach. Chem. Eng. Process. Process Intensif. 47, 1261–1268. https://doi.org/10.1016/j.cep.2007.04.003.

Krishna, R., Titus, E., Salimian, M., Okhay, O., Rajendran, S., Rajkumar, A., Sousa, J.M.G., Ferreira, A.L.C., Gil, J.C., Gracio, J., 2012. Hydrogen storage for energy application, Intech Open, Europe. https://doi.org/10.5772/51238

Le Duigou, A., Miguet, M., 2010. Les marchés de l'hydrogène industriel français. Actual. Chim. 347, 52–57.

Lee, J.-Y., An, S., Cha, K., Hur, T., 2010. Life cycle environmental and economic analyses of a hydrogen station with wind energy. Int. J. Hydrogen Energy 35, 2213–2225. https://doi.org/10.1016/j.ijhydene.2009.12.082.

Li, Z., Gao, D., Chang, L., Liu, P., Pistikopoulos, E.N., 2008. Hydrogen infrastructure design and optimization: a case study of China. Int. J. Hydrogen Energy 33, 5275–5286. https://doi.org/10.1016/j.ijhydene.2008.06.076.

Maack, M.H., Skulason, J.B., 2006. Implementing the hydrogen economy. J. Clean. Prod. 14, 52–64. https://doi.org/10.1016/j.jclepro.2005.05.027.

Macqueen, D., Korhaliller, S., 2011. Bundles of Energy: The Case for Renewable Biomass Energy. IIED, London.

Markert, F., Marangon, A., Carcassi, M., Duijm, N.J., 2016. Risk and sustainability analysis of complex hydrogen infrastructures. Int. J. Hydrogen Energy 42 (11), 7698–7706.

Mentzer, J.T., DeWitt, W., Keebler, J.S., Min, S., Nix, N.W., Smith, C.D., Zacharia, Z.G., 2001. Defining supply chain management. J. Bus. Logist. 22, 1–25.

Miller, B.G., 2004. Coal Energy Systems. Elsevier, Amsterdam.

Moliner, R., Lázaro, M.J., Suelves, I., 2016. Analysis of the strategies for bridging the gap towards the hydrogen economy. Int. J. Hydrogen Energy 41, 19500–19508.

Moreno-Benito, M., Agnolucci, P., Papageorgiou, L.G., 2016. Towards a sustainable hydrogen economy: optimisation-based framework for hydrogen infrastructure development. Comput. Chem. Eng. https://doi.org/10.1016/j.compchemeng.2016.08.005.

Murthy Konda, N.V.S.N., Shah, N., Brandon, N.P., 2011. Optimal transition towards a large-scale hydrogen infrastructure for the transport sector: the case for the Netherlands. Int. J. Hydrogen Energy 36, 4619–4635. https://doi.org/10.1016/j.ijhydene.2011.01.104.

Ni, M., Leung, D.Y.C., Leung, M.K.H., Sumathy, K., 2006. An overview of hydrogen production from biomass. Fuel Process. Technol. 87, 461–472. https://doi.org/10.1016/j.fuproc.2005.11.003.

Nikolaidis, P., Poullikkas, A., 2017. A comparative overview of hydrogen production processes. Renew. Sust. Energy Rev. 67, 597–611. https://doi.org/10.1016/j.rser.2016.09.044.

Office of Energy Efficiency & Renewable Energy, n.d. Central Versus Distributed Hydrogen Production | Department of Energy [WWW Document] https://energy.gov/eere/fuelcells/central-versus-distributed-hydrogen-production (Accessed 1.28.18).

Orecchini, F., Bocci, E., 2007. Biomass to hydrogen for the realization of closed cycles of energy resources. Energy 32, 1006–1011. Third Dubrovnik Conference on Sustainable Development of Energy, Water and Environment Systems. https://doi.org/10.1016/j.energy.2006.10.021.

Othman, M.R., Repke, J.-U., Wozny, G., Huang, Y., 2010. A modular approach to sustainability assessment and decision support in chemical process design. Ind. Eng. Chem. Res. 49, 7870–7881. https://doi.org/10.1021/ie901943d.

Oyama, S., Satoh, S., Sakanaka, S., 2016. HAZID for CO_2-free hydrogen supply chain FEED (front end engineering design). Int. J. Hydrog. Energy. https://doi.org/10.1016/j.ijhydene.2016.07.023.

Ozbilen, A., Dincer, I., Rosen, M.A., 2011. A comparative life cycle analysis of hydrogen production via thermochemical water splitting using a Cu–Cl cycle. Int. J. Hydrogen Energy 36, 11321–11327. International Conference on Hydrogen Production (ICH2P)-2010. https://doi.org/10.1016/j.ijhydene.2010.12.035.

Papageorgiou, L.G., 2009. Supply chain optimisation for the process industries: advances and opportunities. Comput. Chem. Eng. 33, 1931–1938. https://doi.org/10.1016/j.compchemeng.2009.06.014.

Patay, E., 2008. Déploiement de la chaîne logistique de l'hydrogène pour le marché des carburants en 2050: Conception et développement d'un outil d'aide à la décision pour l'analyse de scénarios. Ecole Centrale Paris-Laboratoire Génie Industriel, Paris.

Percebois, J., Mandil, C., 2012. Rapport – Énergies 2050. Centre d'analyse stratégique.

Qadrdan, M., Saboohi, Y., Shayegan, J., 2008. A model for investigation of optimal hydrogen pathway, and evaluation of environmental impacts of hydrogen supply system. Int. J. Hydrogen Energy 33, 7314–7325.

Ren, J., Manzardo, A., Toniolo, S., Scipioni, A., 2013a. Sustainability of hydrogen supply chain. Part I: Identification of critical criteria and cause–effect analysis for enhancing the sustainability using DEMATEL. Int. J. Hydrogen Energy 38, 14159–14171.

Ren, J., Manzardo, A., Toniolo, S., Scipioni, A., 2013b. Sustainability of hydrogen supply chain. Part II: Prioritizing and classifying the sustainability of hydrogen supply chains based on the combination of extension theory and AHP. Int. J. Hydrogen Energy 38, 13845–13855.

Ricci, M., Bellaby, P., Flynn, R., 2008. What do we know about public perceptions and acceptance of hydrogen? A critical review and new case study evidence. Int. J. Hydrogen Energy 33, 5868–5880. https://doi.org/10.1016/j.ijhydene.2008.07.106.

Sabio, N., Gadalla, M., Guillén-Gosálbez, G., Jiménez, L., 2010. Strategic planning with risk control of hydrogen supply chains for vehicle use under uncertainty in operating costs: a case study of Spain. Int. J. Hydrogen Energy 35, 6836–6852. ISMF-09 International Symposium on Multiphase Flow, Heat Mass Transfer and Energy Conversion. https://doi.org/10.1016/j.ijhydene.2010.04.010.

Sawin, J.L., Sverrisson, F., Seyboth, K., Adib, R., Murdock, H.E., Lins, C., Edwards, I., Hullin, M., Nguyen, L.H., Prillianto, S.S., Satzinger, K., Appavou, F., Brown, A., Chernyakhovskiy, I., Logan, J., Milligan, M., Zinaman, O., Epp, B., Huber, L., Lyons, L., Nowak, T., Otte, P., Skeen, J., Sovacool, B., Witkamp, B., Musolino, E., Brown, A., Williamson, L.E., Ashworth, L., Mastny, L., 2017. Renewables 2017 Global Status Report—REN21. 48(27).

Scipioni, A., Manzardo, A., Ren, J., 2017. Hydrogen Economy: Supply Chain, Life Cycle Analysis and Energy Transition for Sustainability. Academic Press, London.

Sgobbi, A., Nijs, W., De Miglio, R., Chiodi, A., Gargiulo, M., Thiel, C., 2016. How far away is hydrogen? Its role in the medium and long-term decarbonisation of the European energy system. Int. J. Hydrogen Energy 41, 19–35. https://doi.org/10.1016/j.ijhydene.2015.09.004.

Spath, P.L., Mann, M.K., 2004. Life Cycle Assessment of Renewable Hydrogen Production Via Wind/Electrolysis: Milestone Completion Report. National Renewable Energy Laboratory, Golden, CO.

Tomei, J., 2009. Planning for a transition to a hydrogen economy: a review of roadmaps. UKSHEC Working Paper.

Tzimas, E., Filiou, C., Peteves, S.D., Veyret, J.-B., 2003. Hydrogen Storage: State of the Art and Future Perspective, EU Commission, JRC Petten, EUR 20995EN.

Utgikar, V., Thiesen, T., 2006. Life cycle assessment of high temperature electrolysis for hydrogen production via nuclear energy. Int. J. Hydrogen Energy 31, 939–944. https://doi.org/10.1016/j.ijhydene.2005.07.001.

Woo, Y., Cho, S., Kim, J., Kim, B.S., 2016. Optimization-based approach for strategic design and operation of a biomass-to-hydrogen supply chain. Int. J. Hydrogen Energy 41, 5405–5418. https://doi.org/10.1016/j.ijhydene.2016.01.153.

World Energy Council, 2016. World Energy Resources | 2016. World Energy Council, United Kingdom.

Wulf, C., Kaltschmitt, M., 2012. Life cycle assessment of hydrogen supply chain with special attention on hydrogen refuelling stations. Int. J. Hydrogen Energy 37, 16711–16721. Advances in Hydrogen Production (Selected papers from ICH2P-2011). https://doi.org/10.1016/j.ijhydene.2012.03.028.

WWF, 2011. The Energy Report – 100% renewable energy by 2050 [WWW Document]. http://www.wwf.org.uk/wwf_articles.cfm?unewsid=4565. Accessed 4 August 2013.

Xiang, W., Chen, S., Xue, Z., Sun, X., 2010. Investigation of coal gasification hydrogen and electricity co-production plant with three-reactors chemical looping process. Int. J. Hydrogen Energy 35, 8580–8591. https://doi.org/10.1016/j.ijhydene.2010.04.167.

Yang, C., Ogden, J., 2007. Determining the lowest-cost hydrogen delivery mode. Int. J. Hydrogen Energy 32, 268–286.

Yilanci, A., Dincer, I., Ozturk, H.K., 2009. A review on solar-hydrogen/fuel cell hybrid energy systems for stationary applications. Prog. Energy Combust. Sci. 35, 231–244. https://doi.org/10.1016/j.pecs.2008.07.004.

Züttel, A., 2003. Materials for hydrogen storage. Mater. Today 6, 24–33. https://doi.org/10.1016/S1369-7021(03)00922-2.

FURTHER READING

Bhandari, R., Trudewind, C.A., Zapp, P., 2012. Life cycle assessment of hydrogen production methods – a review, STE research report. Forschungszentrum Jülich, Institute of Energy and Climate Research.

Bhandari, R., Trudewind, C.A., Zapp, P., 2014. Life cycle assessment of hydrogen production via electrolysis – a review. J. Clean. Prod. 85, 151–163. Special Volume: Making Progress Towards More Sustainable Societies through Lean and Green Initiatives. https://doi.org/10.1016/j.jclepro.2013.07.048.

Mori, D., Hirose, K., 2009. Recent challenges of hydrogen storage technologies for fuel cell vehicles. Int. J. Hydrogen Energy 34, 4569–4574. 2nd World Hydrogen Technologies Convention. https://doi.org/10.1016/j.ijhydene.2008.07.115.

Sapountzi, F.M., Gracia, J., Weststrate, C., Fredriksson, H., Niemantsverdriet, H., Electrocatalysts for the generation of hydrogen, oxygen and synthesis gas, *Prog. Energy Combust. Sci.* **58**, 2017, 1–35. https://doi.org/10.1016/j.pecs.2016.09.001

Chapter 3

Assessment of Selected Hydrogen Supply Chains—Factors Determining the Overall GHG Emissions

Anne Rödl*, Christina Wulf[†] and Martin Kaltschmitt*

*Hamburg University of Technology (TUHH), Institute of Environmental Technology and Energy Economics (IUE), Hamburg, Germany, [†]Institute of Energy and Climate Research—Systems Analysis and Technology Evaluation (IEK-STE), Forschungszentrum Julich, Julich, Germany

3.1 INTRODUCTION AND SCOPE

Passenger transport volumes in cars, buses, and powered two-wheelers, as well as volumes of inland freight transports, have increased in Europe since the late 1990s (EEA, 2016c,d). Today the mobility sector is the largest emitter of greenhouse gas emissions (GHG) within the European Union (EU), even without taking shipping and aviation into account (EEA, 2016b). As shown in Fig. 3.1 road transport has the highest share of GHG emissions in the transport sector (EPSC, 2016). The mobility sector is also a major contributor to overall NO_x emissions and contributes at a lower share to total CO, particle, and NMVOC emissions within the EU (EEA, 2016a).

In 2016 the European Commission published a strategy for low-emission mobility (European Commission, 2016) that, among other things, focusses on the transition toward zero-emission vehicles, such as hydrogen-powered fuel cell vehicles. By the middle of the 21st century 60% of the 1990 GHG emissions from the transport sector should be reduced (European Commission, 2011). Also other air pollution emitted from the transport sector should be reduced significantly (European Commission, 2016).

Thus, growing mobility demands on the one hand and increasingly stringent targets for GHG emissions and other airborne pollutants on the other hand require a shift toward low-carbon and low-emission mobility. Existing mobility options have to be improved considerably and new concepts for road transport systems have to be developed and deployed.

Hydrogen Supply Chain. https://doi.org/10.1016/B978-0-12-811197-0.00003-8

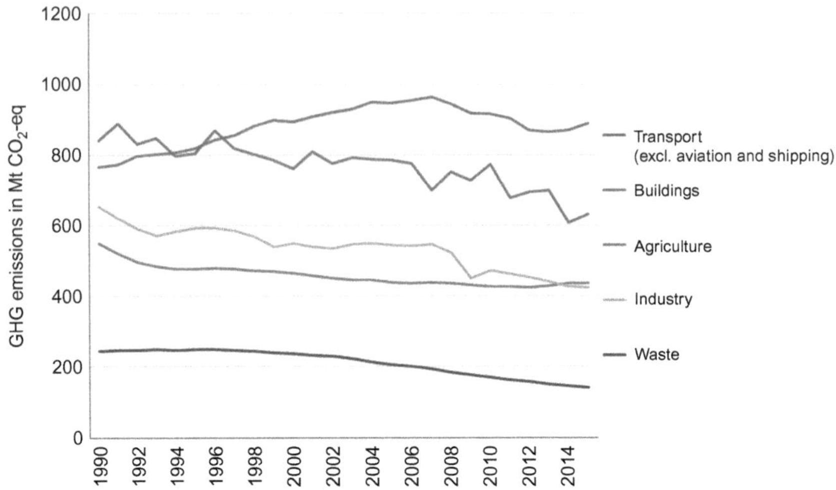

FIG. 3.1 Historic trends of GHG emissions between 1990 and 2015. *(Data from EEA, 2016b. GHG Emission Trends and Projections. Data Visualization. European Environment Agency. Available at: https://www.eea.europa.eu/data-and-maps/daviz/ghg-emission-trends-and-projections#tab-chart_3 (Accessed 12 May 2017).)*

Fuel cell vehicles powered by hydrogen from renewable sources could be such an innovative solution (IEA, 2015). This approach might help to solve problems with GHG emissions and other locally effective emissions, such as NOx, CO, and particles. The latter group is mostly responsible for toxic effects on humans and the natural environment. By introducing fuel cell vehicles, particle emissions are shifted from the vehicle operation stage to the fuel production stage, or will be completely avoided if renewable energy sources are used (Ahmadi and Kjeang, 2015).

At present hydrogen is mainly produced from fossil energy sources (da Silva Veras et al., 2017). However, due to the legal requirements outlined above, alternatives for the provision of "green" hydrogen from renewable energy sources are being explored. There is a variety of possible renewable supply chains that could be used to implement hydrogen as an alternative transport fuel in the market. These options need to be assessed from an environmental point of view to allow the identification of the most promising supply chains with minimal environmental impacts.

Against this background the overall goal of this chapter is to analyze the environmental impacts of different hydrogen production options and their respective transport needs. An assessment of the most important elements of the hydrogen provision chains should be considered when designing a future hydrogen infrastructure for sustainable supply of the transport sector. To this end, this chapter provides information on non-renewable energy consumption and GHG emissions of different hydrogen provision chains and their various

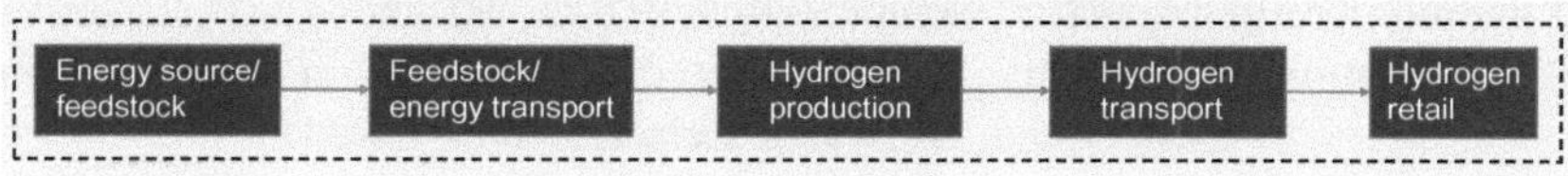

FIG. 3.2 Overview of the analyzed hydrogen supply chains.

system elements. These elements, such as feedstock provision, hydrogen generation, and transportation, as illustrated exemplarily in Fig. 3.2, are analyzed in relation to their contribution to overall non-renewable energy consumption and GHG emissions.

The hydrogen supply chains are investigated based on a well-to-tank approach, including energy for feedstock provision, hydrogen production technologies, transport, and the sale of the final product. The influence of hydrogen production infrastructure design on the overall emissions and energy consumption of hydrogen provision is assessed in detail. The greenhouse gas emission saving potential of the different renewable pathways of hydrogen provision is compared to that of hydrogen provision based on fossil fuels.

3.2 BACKGROUND

In the following the status and future plans for hydrogen infrastructure development in Europe and Germany are discussed in detail.

Although fuel cell vehicles were developed in the 1960s, most of the vehicles operated today are fueled by fossil fuels, such as gasoline and diesel. Nevertheless, hydrogen is produced in considerable quantities and used mainly within the chemical and the crude oil refinery industry (IEA, 2015).

In Europe only 4% of the overall new vehicle registrations have been cars with engines based on alternative fuels (Eurostat, 2015) in 2013. The majority of these vehicles are hybrid electric vehicles; they sum up to 2.1% of the overall newly registered cars (ACEA, 2017). However, there are no official statistics on the number of fuel cell vehicles in use; statistics available in Europe (Eurostat, 2015; ACEA, 2017; KBA, 2017b) just distinguish between conventional fossil and alternative fuel engines. The latter can be fueled by electricity, LPG (liquefied petroleum gas), CNG (compressed natural gas), alcohols (B85, B100), hydrogen, or biodiesel (B100).

Globally, roughly 550 fuel cell electric vehicles (including cars and buses) were in use in 2014 (IEA, 2015). Thus, it is likely that in Europe only a few hundred of such vehicles are in operation. This is an extremely small number compared to the total number of 1.2 billion vehicles in use worldwide and roughly 250 million passenger cars operated within the EU-28 in 2014 (ACEA, 2017; Eurostat, 2015).

So far, Europe, the United States, and Japan have been the most active in promoting hydrogen use within the transportation sector and in developing the necessary refueling infrastructure (Hydrogen Council, 2017). In Europe the hydrogen refueling infrastructure is already quite advanced with 22 active refueling stations in Germany, 15 in the United Kingdom, 11 in Denmark, 9 in France, and 6 in Switzerland. In Scandinavia (Finland, Sweden, Norway) 6 hydrogen refueling stations can be found, 4 each in Austria and Spain, 3 in Belgium, 3 in the Netherlands, 2 in Italy, 1 in the Czech Republic, and also 1 in Slovenia (status early 2017). In the United States, hydrogen refueling stations have been mainly installed on the East Coast, the Great Lakes Area, and in California (TÜV SÜD, 2017).

This infrastructure development goes hand in hand with the planned expansion of the fuel cell vehicles market. It is planned to have 520,000 fuel cell vehicles on the streets globally in 2020 (IEA, 2015). Depending on the development scenario, the share of fuel cell vehicle sales will range between 4% and 11% in 2050 (Pasaoglu et al., 2016). This requires an efficient hydrogen refueling infrastructure.

In Germany, about 215 fuel cell vehicles have been registered at the beginning of 2016. Most of them are used within demonstration projects initiated and sponsored by the National Innovation Program (NIP) for Hydrogen and Fuel Cell Technology (BMVI, 2016). However, due to the high cost of these vehicles and the still only scarcely developed fuel provision infrastructure, market development is only very limited so far. Only 49 new fuel cell passenger cars have been registered throughout the year 2016 (Gerster and Wimmelbücker, 2017).

Therefore, it is planned to significantly expand the network of refueling stations in the coming years to achieve much better coverage, and thus clearly improved fuel availability. Based on the 22 hydrogen refueling stations currently in use in Germany, it is planned to have around 400 active refueling stations in 2025 (H2 Mobility, 2017). Until 2020 the installation of 100 refueling stations will be supported by governmental institutions regardless of the number of registered vehicles (BMVI, 2016).

Investigations have shown that hydrogen could provide 40% of the German end energy demand within the transportation sector (H2 Mobility, 2017; Robinius, 2015). For 2050 an final energy demand of the overall German mobility sector of 2412 PJ/a has been projected (Blanck et al., 2013). About 965 PJ/a (roughly 40%) could be provided by hydrogen (Joest et al., 2009).

Assuming such a prosperous development and an unchanged number (about 14,500) of refueling stations (Statista, 2017), each station would have to provide >2 t/d of hydrogen. Currently, the largest hydrogen refueling station in operation can deliver 1 t/d of hydrogen. To reach sufficient coverage, at least 10 hydrogen refueling stations within each major urban center and in a 90 km distance along major motorways between the urban centers should be erected (Altmann et al., 2014). Most likely, the capacity of these refueling stations will be adapted according to the expected demand; at the earliest in the 2020-years it is likely that there will be a mixture of small refueling stations (212 kg/d), medium-sized stations (420 kg/d), and large ones (1000 kg/d). An overall hydrogen demand of about 45,300 t/a in 2023 can be expected if the market develops as discussed (calculated on the basis of: Altmann et al., 2014).

3.3 HYDROGEN PRODUCTION AND TRANSPORTATION

There are a variety of possible strategies for hydrogen production, distribution, storage, and refueling. From these various options only the environmental performance of proven hydrogen production technologies will be analyzed here.

Established technologies for hydrogen production are steam reforming of natural gas, coal gasification, and water electrolysis (Sgobbi et al., 2016). In the future, further opportunities might become available through biomass-based H_2 production technologies, solar steam reforming of methane by CSP (concentrated solar power) in countries with high solar radiation, and also by coal gasification with a subsequent water-gas-shift-reaction with carbon capture and storage (Sgobbi et al., 2016). Further options based on bacterial hydrogen production, such as dark fermentation and photofermentation, as well as thermochemical cyclic processes, are still in an early demonstration stage or even just available on a lab scale (Nikolaidis and Poullikkas, 2017).

Based on these conversion technologies, hydrogen can be produced in small-scale facilities onsite close to the refueling station or offsite in large-scale, optimized facilities with huge hydrogen production capacities. If hydrogen is produced offsite in large-scale plants it has to be delivered by truck or pipeline to the refueling station (Altmann et al., 2014). This can be realized in a gaseous form at high pressure or in a liquid state at very low temperatures. At the refueling station, hydrogen can be stored also pressurized or in a liquid form. Consequently, different hydrogen provision concepts require different means of transportation.

Initially the feedstock (e.g., biomass, coal) and the energy (e.g., electricity) have to be transported to the small-scale or large-scale production plant. Feedstock is transported by truck or pipeline and electricity is conducted via electrical lines. Finally, the hydrogen needs to be transported to the refueling station via pipeline or truck.

Table 3.1 gives an overview of the various hydrogen production technologies that are assessed in this chapter. In the following these technologies will be explained in detail.

3.3.1 Steam Methane Reforming (SMR)

Currently around 95% of worldwide commercial hydrogen production is based on fossil fuels. The largest part of it is produced via SMR from natural gas (Hydrogen Council, 2017; Santos, 2015; Abe, 2008). Essentially all hydrogen-rich gases, biomethane for example, can be used for hydrogen production via such a process. However, natural gas is the currently the method of choice because of its easy availability and low costs, mainly in large-scale production facilities (HPTT, 2009). Thus, at present SMR is, besides coal gasification, the cheapest hydrogen production option and has the lowest CO_2 emissions compared with other fossil fuel-based hydrogen production options

TABLE 3.1 Conversion Efficiencies of Different Hydrogen Production Technologies

Hydrogen Production Technologies	Plant Capacity (MW)	Conversion Efficiency		References
		Min (%)	Max (%)	
Large-scale steam methane reforming (SMR)	150–300	70	85	3; 4; 2; 5
Small-scale steam methane reforming (SMR)	0.15–15	51	80	3; 2
Biogas reforming	~100	22	32	6
Alkaline electrolysis (large scale)	Up to 150	65	82	4; 2
PEM electrolysis	Up to 1	65	78	3; 2; 5
Biomass gasification	4–160	35	53	1; 2; 5, 6; 7

1, Hosseini and Wahid, 2016; 2, IEA, 2015; 3, Nilsen et al., 2007; 4, Altmann et al., 2014; 5, US Department of Energy, 2015b; 6, Gellert, 2013; 7, Vakkilainen et al., 2013.

(Sharma and Ghoshal, 2015). Additionally, some small-scale reforming units exist on a demonstration level (IEA 2015). Such small-scale reformers would allow a distributed (decentralized) natural gas reforming directly at fueling stations and is therefore considered as a near-term option for relatively cheap hydrogen provision (HPTT, 2009; Schjolberg et al., 2015).

Natural gas reforming consists of two coupled process steps. First the hydrocarbon is converted together with steam to a synthesis gas containing hydrogen (H_2) and carbon monoxide (CO). In the respective exothermic chemical reaction methane is partly oxidized to CO by reducing water to H_2. The product gas is then treated within a second process step in which additional water is reduced to hydrogen within a catalytically controlled conversion process by oxidizing the carbon monoxide to carbon dioxide (Sharma and Ghoshal, 2015). Afterwards the produced gas is purified in a final cleaning step by pressure swing adsorption to provide hydrogen with a purity of nearly 100%. On average SMR plants show conversion efficiencies between 74% and 85% (Nikolaidis and Poullikkas, 2017).

3.3.2 Electrolysis

Another common way of producing hydrogen is electrolysis. This process splits water into hydrogen and oxygen by applying electricity; electrical energy is the driving force of the chemical reaction and provides the energy to split water into its components. Thus, it is an endothermic reaction characterized by a high energy (here: electricity) demand and the provided products (hydrogen and oxygen) are more energy rich than the reactant (water). The energy input-output ratio amounts to approximately 1.5 (Nikolaidis and Poullikkas, 2017). Only 4% of the world's commercial hydrogen production is based on electrolysis (IEA, 2015). However, a great future is predicted for this option because electricity from volatile renewable sources of energy (e.g., wind, solar) can be converted into a storable gas to be used as an energy carrier or as a raw material for the chemical industry.

Alkaline water electrolyzers using alkaline electrolytes, such as potassium hydroxide, are currently the most mature technology used for hydrogen provision via this technical approach (Ursua et al., 2012; IEA, 2015). Commercial-scale alkaline electrolyzers operate with an overall efficiency of 65%–80% (see Table 3.1). Electrolysis via PEM (Polymer Electrolyte Membrane) is a variation where the electrolyte membrane is a plastic material (US Department of Energy, 2017). Large-scale water electrolysis plants can produce 50,000 to 750,000 kg of hydrogen per day. In comparison, distributed small-scale systems located, e.g., at refueling stations, have much lower capacities and can provide up to 1500 kg of hydrogen per day.

In most cases electricity from the public grid is used for hydrogen production within an electrolyzer operated at commercial scale. Water electrolysis using only electricity provided by wind energy (from onshore, and especially offshore wind parks) or solar radiation (e.g., large-scale photovoltaic power plants) is seen as an important future technology that might reach full market maturity within the next 5–10 years. This time schedule is realistic because PEM electrolysis demonstration systems already exist at selected locations (Miller, 2011).

Hydrogen production from electrolysis is considered as a possibility to store surplus electricity from wind or solar in order to facilitate the integration of their fluctuating supply characteristics into the energy system and to provide a "clean" and flexible fuel for several applications. Thus, hydrogen is often regarded as the most important solution to realize an energy system based fully on renewable energies with fluctuating characteristics; in a figurative sense this is also true for the mobility sector.

3.3.3 Solid Biomass Gasification

Gasification is defined as the thermochemical conversion of solid biomass within different mediums, such as air, oxygen, or steam, into a synthesis gas (Nikolaidis and Poullikkas, 2017; Kaltschmitt et al., 2016). Within such a gasifier the solid organic matter (e.g., wood chips) reacts chemically at high temperatures (700–1400 °C), among others, with an oxidation agent to hydrogen. If water steam is used as an oxidation agent, the relative amount of hydrogen per unit biomass is higher compared with the use of oxygen, because with water additional hydrogen is introduced into the system. Such a steam gasification can reach an efficiency of 44%, on average. It requires a primary energy input of 2.3 MJ per MJ of hydrogen. Typically, such systems are realized within an intermediate size. Economies of scale prevent very small systems from being profitable because the costs of producing one unit is higher the smaller facility. The size limit is defined by the biomass logistic costs, which are increasing with the transportation distance.

3.3.4 Hydrogen Transportation

3.3.4.1 Pipelines

Hydrogen pipelines are constructed in a similar manner to pipelines for natural gas. There are already some examples of such hydrogen networks in Germany, the United States, France, and Belgium (Krieg, 2012). They are made of steel or tough cast iron and do have significant requirements on the weld seams.

3.3.4.2 Trucks

Hydrogen can be transported by truck in compressed or liquid form. Hereafter the advantages and disadvantages are highlighted briefly.

Compressed transportation

Compressed hydrogen can be transported by trucks in gas cylinders or gas tubes with pressures between 200 and 500 bar. Usually several cylinders or tubes are bundled to modules in a 20′ or 40′ container that is mounted on a trailer (tube trailer).

The transport capacities and tank weights are important variables in assessing the GHG emissions and non-renewable energy consumption in hydrogen transportation via truck. Typically, the high weight of the cylinders or tubes limit the maximum hydrogen load that can be transported. A tube trailer with steel cylinders can store up to 25,000 liters of hydrogen compressed to 200 bar (Wystrach GmbH, 2017a), which amounts to around 420 kg of hydrogen.

Currently lighter tank materials (composite materials for gas cylinders or gas tubes) that can be operated at higher pressure are under development in order to increase the hydrogen transport quantities per trailer. For example, superlight cylinder materials consisting of carbon fiber over high-density polyethylene liners (Wystrach et al., 2012) have been investigated. Trailers with such composite cylinders can carry up to 39,600 liters of hydrogen with a pressure level of 200 bar equivalent to about 666 kg of hydrogen (Wystrach GmbH, 2017b).

Recently a jumbo trailer was released that can carry 13,000 m^3 of hydrogen compressed with 500 bar (Linde Group, 2013), which amounts to a transported hydrogen weight of about 1100 kg.

Liquid transportation

If hydrogen is transported in liquid form, it has to be cooled down to a temperature of at least −253 °C using a very energy consuming process. Typically,

10 kWh of electricity are needed per kg of liquid hydrogen; this represents around 30% of the energy content of the hydrogen (Krieg, 2012). The latest research has shown that energy consumption for liquefaction can be reduced to 6.7–7.5 kWh/kg H_2 (Cardella et al., 2017; Stolzenburg et al., 2013). Because of the higher energy density resulting from liquefaction, the transport capacities for liquid hydrogen per supply task are higher than for compressed hydrogen. Furthermore, about 1.65% of the hydrogen is lost during the liquefaction process and around 0.3% of the liquefied hydrogen is "boiled-off" per day during transportation and storage (Stolzenburg and Mubbala, 2013).

3.4 ASSESSMENT OF HYDROGEN SUPPLY CHAINS

3.4.1 Method

In this part of the chapter life cycle assessment (LCA) results on different hydrogen supply chains based on a literature review are presented. Further, literature data is used for calculating GHG emissions and energy consumption of hydrogen transport processes.

In general, a LCA looks at potential environmental impacts during the lifetime of a product (DIN EN ISO, 2006). In this approach, all inputs and outputs of manufacturing, and the use and disposal of products and services, including all related upstream processes, are summarized. Thus, all data assessed below use the LCA method to determine the environmental impacts of the different hydrogen provision pathways.

The functional unit chosen here is 1 MJ contained in the final product hydrogen at an ambient temperature of 15 °C, 1.5 MPa, and a purity of 99.999%, required to operate fuel cell vehicles successfully. It is assumed that 1 kg of hydrogen contains 120.1 MJ. The assessment concentrates on the situation in Central Europe. Assumptions for the parameters that have been taken into account for own calculations are presented in the subsequent sections.

3.4.2 Supply Chains

The following hydrogen supply chains have been assessed here:

(1) *SMR from natural gas*

- Natural gas (German mix available from the gas grid) → large-scale SMR unit → pipeline or truck transportation (liquid or compressed) → compression retail
- Natural gas (German mix available from the gas grid) → small-scale[1] SMR unit → compression retail

1. Regional distribution at the refueling station.

(2) *SMR from biogas*

- Biogas from maize (energy crops) and from animal manure ➔ large-scale SMR unit ➔ transportation via pipeline or truck (compressed or liquid) ➔ compression retail

(3) *Electrolysis based on electrical energy from wind power*

- Offshore wind park ➔ electricity distribution ➔ large-scale electrolysis unit ➔ pipeline or truck transportation (liquid or compressed) ➔ compression retail
- Onshore wind park ➔ electricity distribution ➔ large-scale electrolysis unit ➔ pipeline or truck transportation (liquid or compressed) ➔ compression retail
- Locally distributed wind power generation ➔ small-scale electrolysis unit[1]➔ compression retail

(4) *Electrolysis based on electricity from solar radiation (PV)*

- Solar park ➔ electricity distribution ➔ large-scale electrolysis unit ➔ pipeline or truck transportation (liquid or compressed) ➔ compression retail

(5) *Electrolysis based on grid power*

- Electricity provision and distribution ➔ large-scale electrolysis unit ➔ pipeline or truck transportation (liquid or compressed) ➔ compression retail
- Electricity provision and distribution ➔ small-scale electrolysis unit[1] ➔ compression retail

(6) *Biomass gasification*

- Farmed wood (short rotation plantation) ➔ large-scale gasification unit ➔ pipeline or truck transportation (compressed or liquid) ➔ compression retail

Fig. 3.3 gives an overview of the different hydrogen supply chains considered here. Depending on the assessed hydrogen production option, the analyzed chains start with feedstock provision or electricity generation.

3.4.3 Assumptions and Data

The assessment of the above defined hydrogen supply chains is based on various data and assumptions that will be explained in detail below.

3.4.3.1 Feedstock and Energy Resources

The most important data related to the implemented feedstock and the necessary energy resources are presented in the following.

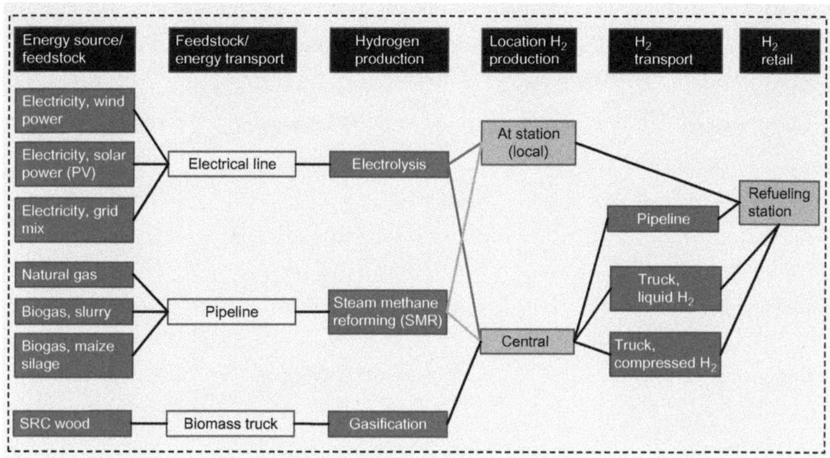

FIG. 3.3 Overview of the hydrogen supply chains considered.

Electricity. Three options for electricity provision are considered for electrolytic hydrogen production, namely, from wind and solar power, and from the German electricity mix as a reference. The LCA data for the German electricity mix have been taken from a commercially available database (Ecoinvent, 2016). The used dataset represents the consumption mix for the year 2015, and includes, among others, electricity inputs produced in Germany and from imports, the losses within the transmission network, as well as direct emissions to the ambient air. Table 3.2 contains the considered data sources for modelling electricity provision for the electrolytic hydrogen production.

Methane. Natural gas supply for the large-scale and small-scale steam methane reformers (SMR) is realized via the German natural gas pipeline network. According to Table 3.3, the average LCA data representing the mix of gas imports from domestic production in Germany is found in Ecoinvent (2016).

For the cases in which biomethane is used for SMR, biogas produced from two different types of feedstock have been taken into consideration:

- biomethane from cultivated biomass—maize, whole crops, grass.
- biomethane from residues and byproducts—agricultural and industrial organic waste streams.

The system boundaries for biomethane production from cultivated biomass include the cultivation, transportation, and storage of the energy crops, whereas the assessment of biomethane production from biowaste starts with the collection and transportation of the organic waste materials. For both chains, biomass fermentation, fermentation residue storage, and their transportation and application are considered, as well as the upgrading of the biogas to biomethane and its feed-in into the local gas network (Westerkamp et al., 2014).

TABLE 3.2 LCA Data Used for Modelling the Impacts of Energy Provision for the Respective Hydrogen Supply Chains

Energy Source	Dataset Name/Parameters	Reference
Electricity grid mix (large-scale application)	Market for electricity, high voltage—DE[a]	Ecoinvent (2016), v.3.3
Electricity grid mix (small-scale application)	Market for electricity, medium voltage—DE[a]	Ecoinvent (2016), v.3.3
Wind energy, onshore	Energy generation onshore (3,5 MW, 5,5 m/s)	Kaltschmitt et al. (2013b)
Wind energy, offshore	Energy generation offshore (6 MW, 9,5 m/s)	Kaltschmitt et al. (2013b)
Solar radiation, PV	Energy generation, polysilicon cells, solar park	Kaltschmitt et al. (2013a)

[a]*DE—data representing German Electricity Mix.*

TABLE 3.3 LCA Data Used for Modeling Methane Sources for Hydrogen Supply Chains From SRM (Steam Methane Reforming)

Feedstock	Dataset Name/Parameters	Reference
German natural gas mix	Market for natural gas, high pressure—DE[a]	Ecoinvent (2016), v.3.3
Natural gas from low-pressure network	Market for natural gas, low pressure—ROW[b]	Ecoinvent (2016), v.3.3
Biomethane from cultivated biomass	Cultivation maize (82%), whole crops (8%), grass (10%)—fermentation, covered—upgrading—feed-in	Westerkamp et al. (2014)
Biomethane from organic wastes	Collection of biogenic wastes from agricultural (17%) and industry (83%), fermentation, covered—upgrading—feed-in	Westerkamp et al. (2014)

[a]*DE—data representing German Electricity Mix.*
[b]*ROW—data representing a worldwide average.*

Solid biofuels. The biomass gasification process is fueled by poplar wood chips from a short rotation plantation in Germany (Table 3.4). The data considered include cultivation with fertilizer application and the harvesting of trees with special choppers that chip the trees and blow them into a tractor transport container. Transportation to the gasification plant is realized by truck (Roedl, 2010).

TABLE 3.4 LCA Data Used for Modeling of the Solid Biofuels

Feedstock	Dataset Name/Parameters	Reference
Wood chips from short rotation coppice (SRC)	SRC poplar, fertilized, chipped, transported	Roedl (2010)

3.4.3.2 Hydrogen Production

Three different options for hydrogen production are considered here. These are SMR, electrolysis, and biomass gasification. They mark the currently most likely technology for small-scale (locally distributed at a refueling station) and large-scale hydrogen production. The following framework has been defined.

SMR. Hydrogen production is assumed to be realized from natural gas via SMR in two different plant sizes: a large-scale plant and a small-scale system at a retail site. Biomethane reforming is also taken into consideration, in addition to natural gas. This can be realized from two different types of feedstock, cultivated biomass and biowaste. The following assumptions are made on the efficiency of the reformers, regardless of the origin of their feed-in gases:

- Small-scale steam reformer typically used at refueling station: efficiency 69% (Edwards et al., 2014).
- Large-scale reformer typically used at large-scale production sites: efficiency 76% (Edwards et al., 2014).

The natural gas input for the small-scale reformer is assumed to be 1.4 MJ/MJ hydrogen and 1.3 MJ/MJ for the large-scale reformer (Edwards et al., 2014). These gas demands are also considered if biomethane is used as a feedstock. The average heating value of natural gas is assumed to be 35.7 MJ/m^3 and 35.9 MJ/m^3 for biomethane upgraded from biomass. The small difference results from the purity of the gas, which is slightly higher for biomethane than for natural gas (which is a bit more polluted with nitrogen and carbon dioxide).

Electrolysis. For hydrogen production via electrolysis a PEM electrolyzer is considered. This technological solution is characterized by an efficiency of 65% (Edwards et al., 2014), i.e., an electricity input of around 1.5 MJ/MJ hydrogen is needed regardless of the origin of the electricity (Edwards et al., 2014).

Gasification. For the hydrogen provision route via gasification of woody biomass, a large-scale gasifier with an efficiency of 44% is considered (Edwards et al., 2014). The required wood input amounts to 2.3 MJ/MJ hydrogen.

3.4.3.3 Hydrogen Transportation

The options for hydrogen transportation assessed here are discussed and defined in detail below. If the hydrogen is not produced in a facility close to the point of retail a transport distance of 250 km is taken into consideration.

Pipeline. Data on the emissions and energy requirements for the construction of a hydrogen pipeline have been transferred from the construction of high-pressure natural gas pipelines (Ecoinvent, 2016). This is justified by the fact that both pipeline systems are quite similar. On its way through the pipeline the hydrogen gas loses pressure and has to be recompressed regularly (Faist-Emmenegger et al., 2007). It is assumed that a gas recompression is necessary every 100 km. Energy consumption for recompression and further assumptions concerning the pipeline properties are shown in Table 3.5. Recompression units and pipeline networks have to be maintained on a permanent basis. As an adjustment, 2% of the initial GHG emissions resulting from pipeline construction have been assumed for maintenance work per year.

Truck (compressed). The assumptions outlined in Table 3.6 related to tank types, volumes, and weights, as well as storage capacities have been made.

TABLE 3.5 Data for the Pipeline Considered

Service Life Time	40	a
Maintenance	2	%/a
Mass flow	35	t/d
Length	250	km
Recompression every 100 km	0.02	kWh/kg hydrogen

TABLE 3.6 Capacity and Weights of Selected Tank Types for Compressed Hydrogen Transportation Used in the Assessment

Tank Type	Volume (L)	Pressure (bar)	H_2-Capacity (kg)	Tank Tare Weight (kg)
Steel cylinder container (SC)[a]	23,800	200	400	26,298
Steel tubes (ST)[b]	19,292	200	324	27,254
Composite super light container (CC)[c]	45,500	250	957	18,854
Composite (TITAN V) trailer (CT)[d]	44,200	250	979	21,810

[a]*Wystrach GmbH (2017a).*
[b]*CMW (2017).*
[c]*Wystrach GmbH (2017b).*
[d]*Hexagon Composites (2012).*

Before a truck can be loaded the hydrogen has to be compressed. Therefore, it is assumed that the hydrogen is generated at a 20-bar level. Values on the necessary energy consumption for compression vary in literature. However, the energy demand for compression primarily depends on the pressure level of the hydrogen to be compressed. For example, an energy demand of 0.5 kWh/kg hydrogen has been presented for compression to 160 bar (Krieg, 2012). Here it is assumed that hydrogen is compressed from 20 to 200 bar or 250 bar. For 1 kg of hydrogen, it is assumed that 0.7 kWh electricity is needed for a compression to 200 bar and 0.9 kWh/kg for a compression to 250 bar. Conventional high-voltage power from the public grid is used for this purpose (Ecoinvent, 2016).

Throughout the overall production, distribution, and retail chain of compressed hydrogen about 4% losses occur (Bond et al., 2011). Additionally, during discharge of the compressed hydrogen gas from the cylinders, a small amount of hydrogen remains within the transportation unit and cannot be used. This reduces the net amount of the compressed hydrogen to be transported.

In summary, during each trip almost 980 kg of compressed hydrogen can be delivered to a refueling station. To ensure a continuous supply at least two deliveries per day are needed if an average hydrogen demand of 1500 kg/d is assumed (Krieg, 2012).

Truck (liquefied). The assessment of the respective transport emissions and energy use is based on a transported amount of 4000 kg liquid hydrogen per supply task. This is a rather conservative assumption because the latest published results indicate a transport capacity of approximately 4500 kg of liquid hydrogen per truck (US Department of Energy, 2015a). The empty tank has a weight of 24,400 kg (Krieg, 2012). A standard refueling station with a demand of 1500 kg/d hydrogen would have to be supplied roughly every three days.

For the total liquefaction process a conservative assumption on electricity demand of 10 kWh/kg hydrogen is made. Modern high-end liquefaction facilities have been demonstrated to require less energy already. However, this value is chosen also to consider the energy demand for auxiliaries, flash gas management, losses, and boil-off during the liquefaction process (Stolzenburg and Mubbala, 2013).

Modern tanks for liquid hydrogen production reduce boil-off losses to a minimum (Linde AG, 2014). Nevertheless, it is assumed that roughly 1% of the liquefied hydrogen will be lost due to boil-off per day during hydrogen storage at the refueling station (Krieg, 2012). An average storage period of approximately three days is assumed.

For all truck transports the distance from the hydrogen production plant to the refueling station (250 km) and back to the production facility is considered. On the outward journey the truck is fully loaded, while it goes back with an empty tank. The vehicle considered is in each case a 40 t freight lorry fulfilling EURO 5 emission standards (KBA, 2017a).

Electricity transport. For transportation of electricity to the electrolysis unit, losses of 2% have been assumed (Edwards et al., 2014).

TABLE 3.7 Energy Demand for Compression or Vaporization at Refueling Station

Onsite hydrogen production (from 15 to 880bar)	3.1 kWh/kg hydrogen
Pipeline delivery (from 20 to 880bar)	3.1 kWh/kg hydrogen
Compressed truck delivery (200–880bar)	2.4 kWh/kg hydrogen
Compressed truck delivery (250–880bar)	2.3 kWh/kg hydrogen
Liquid truck delivery (compression and dispensing)	1.7 kWh/kg hydrogen

3.4.3.4 Conditioning at the Point of Retail

Depending on the mode of transport, hydrogen has to be compressed or vaporized at the refueling station to be ready for sale. The worldwide automotive industry, as well as gas and plant suppliers, have agreed on a standard for hydrogen fueling of light duty vehicles (SAE, 2016). It regulates the fuel delivery temperature, maximum fuel flow rate, and refueling pressure. Thus, hydrogen refueling stations for passenger cars operate with 700 bar compressed hydrogen, buses are refueled with 350 bar compressed hydrogen (BMVI, 2016). To fill vehicles in practice actually requires a higher pressure of 880 bar (Gardiner and Satyapal, 2009). The refueling of passenger cars takes 3 min with hydrogen cooled to −33 to −40 °C (H2ME, 2016). Energy requirements for compression of the delivered hydrogen at the refueling station are calculated according to the assumptions displayed in Table 3.7 (Edwards et al., 2014; Gardiner and Satyapal, 2009; Krieg, 2012).

3.5 RESULTS

The results presented below are based on the assumptions discussed above.

3.5.1 Energy Demand

The analysis of the cumulative energy input of non-renewable energy (fossil biogenic resources, such as crude oil, natural gas, and coal, as well as fossil mineral resources, such as uranium) per energy unit hydrogen is shown in Fig. 3.4. The presented results can be summarized as follows.

- The lowest non-renewable energy input is required for hydrogen provision chains based on wind energy followed by solar energy-based chains.
- The highest non-renewable energy input is required when hydrogen is produced from grid electricity.
- The lowest overall energy inputs can be realized if the produced hydrogen is transported via pipeline.

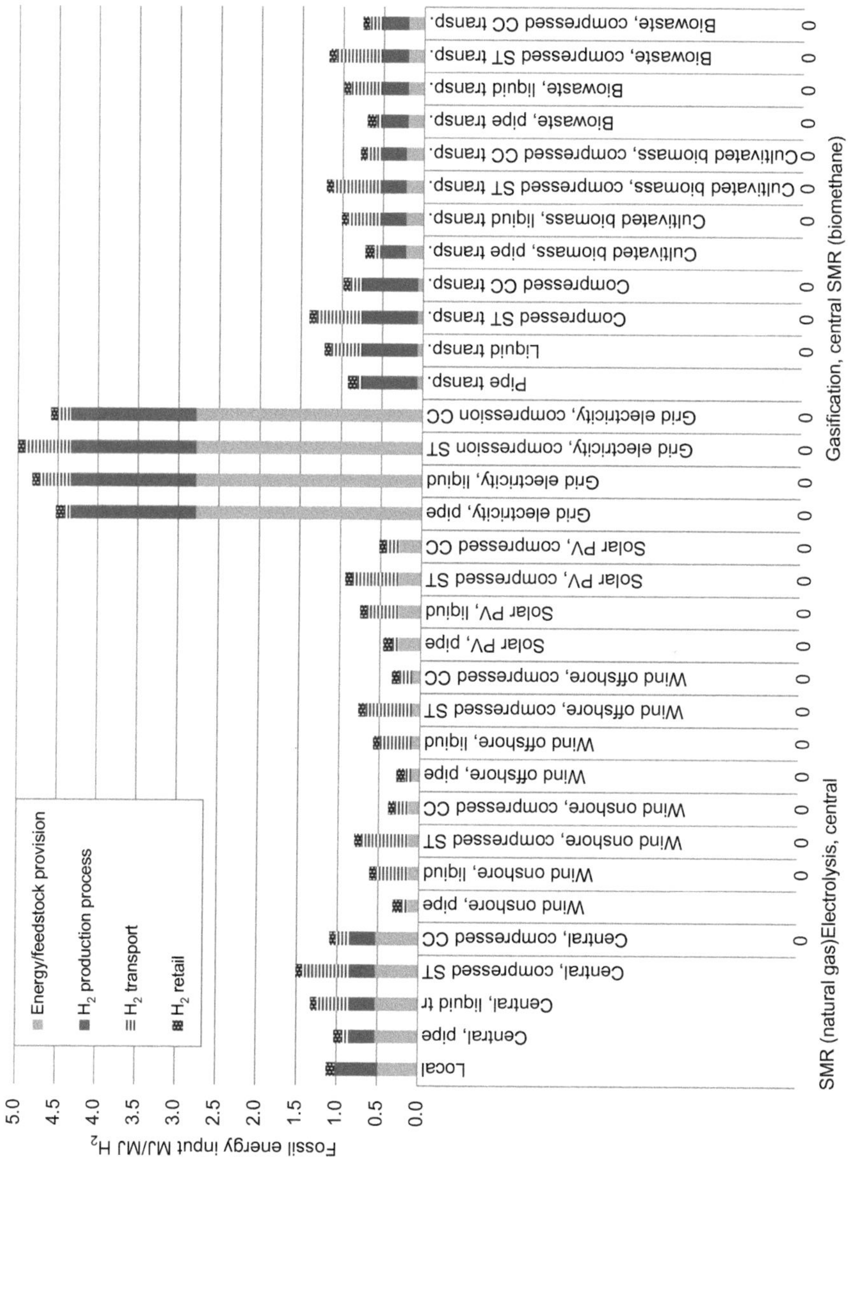

FIG. 3.4 Cumulative energy consumption of the assessed hydrogen supply pathways.

- Non-renewable energy demand for transportation has a high share on the overall energy demand of the whole chain if liquid hydrogen is transported or heavy steel tubes are used for compressed hydrogen transportation.
- The overall energy demand of hydrogen provision is rather low if biomethane from biowaste is used via SMR or farmed wood (SRC) is converted into hydrogen via gasification.

3.5.2 Greenhouse Gas Emissions

Below the greenhouse gas (GHG) emissions, expressed in CO_2-equivalents (CO_2e) of the overall hydrogen supply chain and its respective production steps, are discussed related to 1 MJ hydrogen. Fig. 3.5 shows a close relationship to the results presented in Fig. 3.4 for the demand for non-renewable energy. This is caused by the fact that the use of non-renewable energy, such as oil, natural gas, and coal, leads directly to GHG emissions. Typically, the use of natural gas shows relatively low GHG emissions, the use of crude oil intermediate GHG emissions, and the use of hard coal relatively high GHG emissions related to one unit of energy. Thus, the energy balances of the various hydrogen provision routes show that a substantial part of the non-renewable energy is needed to provide hydrogen at the refueling station. Only a small share of the overall GHG emissions occur, for example, from fermentation processes or from gas releases from soil or pipeline leaks.

Looking at the complete hydrogen supply chain, the lowest GHG emissions per MJ of hydrogen are produced if wind energy or solar radiation, that is, renewable sources of energy, are converted to hydrogen via electrolysis (25–89 g CO_2-eq/MJ). There are only emissions resulting from the upstream processes of the electricity provision due to the construction of the wind mills or the PV systems, as well as from the electricity distribution system. Due to the large-scale electrolysis assumed here, additional emissions are caused by hydrogen transportation. There are no emissions from the hydrogen production process itself due to the use of renewable energy as outlined above.

The situation is completely different if electricity provision from conventional sources, as in the current German electricity mix, is assumed as a feedstock for hydrogen production. According to Fig. 3.5, the hydrogen supply chains based on electrolysis from grid electricity (German electricity mix with roughly 40% of electricity coming from hard coal and lignite) have the highest overall GHG emissions (317–370 g CO_2-eq/MJ).

The calculated GHG emissions from gasification and subsequent hydrogen production from woody biomass are also rather low (27–80 g CO_2-eq/MJ). Like the production chains based on biomethane (0.5 g CO_2-eq/MJ), the hydrogen production process based on thermochemical gasification causes only low emissions of GHG (0.4 g CO_2-eq/MJ) and is almost not visible in Fig. 3.5. Biomethane provision causes higher GHG emissions than supply with woody biomass, especially if biomethane is produced from cultivated biomass.

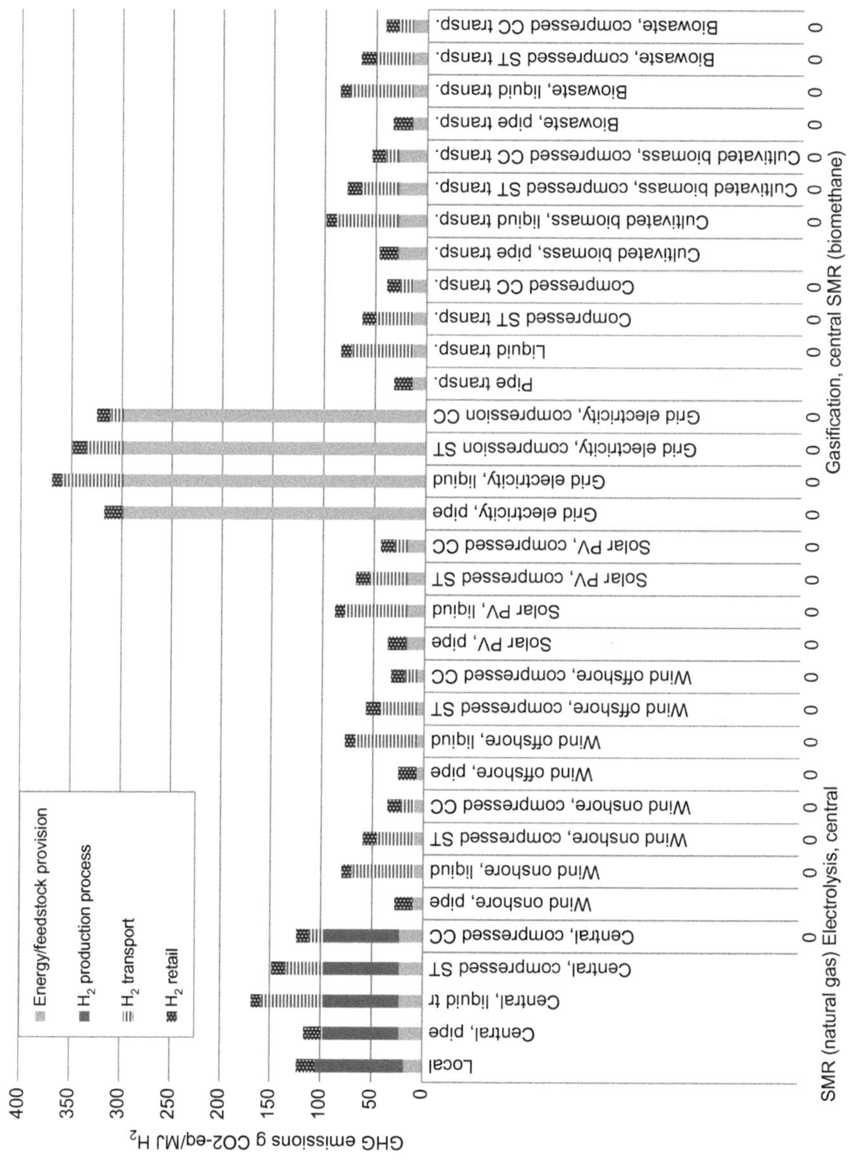

FIG. 3.5 Greenhouse gas (GHG) emissions of the assessed hydrogen supply pathways.

This is due to higher fertilizer inputs for agricultural crops compared with tree plantations and GHG emissions released during fermentation of the organic matter. The use of biowaste instead of cultivated biomass for biomethane production reduces GHG emissions of the overall hydrogen supply chains.

In comparison, hydrogen production based on natural gas releases significant amounts of GHG emissions during the feedstock provision phase as well as the hydrogen production phase. Therefore, the overall hydrogen provision chains from natural gas have the second highest GHG emissions per MJ of hydrogen (116–169 g CO_2-eq/MJ). Locally realized small-scale SMR causes slightly higher emissions (87 g CO_2-eq/MJ) compared to large-scale reforming (74 g CO_2-eq/MJ); this is mainly due to the lower conversion efficiencies of small-scale systems compared to large-scale units. However, if the hydrogen is produced at the place of retail, GHG emissions from hydrogen transportation do not occur so that the overall GHG emissions of the complete supply chain are lower for the distributed hydrogen production chains. Only if hydrogen from a large-scale SMR plant is transported by pipeline the total GHG emissions of the supply chain are lower than the overall emissions of locally realized hydrogen production. Higher production-related GHG emissions are not compensated by pipeline transport emissions (1.2 g CO_2-eq/MJ).

The contribution of the different production steps to the overall GHG emissions throughout the total hydrogen supply chains are displayed in Fig. 3.6. Thus, overall GHG emissions of hydrogen production chains from large-scale and small-scale SMR are dominated by GHG emissions from the hydrogen production process itself. GHG emissions of an electrolysis-based hydrogen production can mainly be traced back to energy requirements for compression at the retail site. As explained above, overall emissions are low because the production process itself causes almost no additional emissions and it is assumed that only electricity from renewable sources of energy, characterized by very low GHG emissions, is used. Therefore, the shares of conditioning and transportation are higher. Biomass-based hydrogen production pathways are mainly determined by the feedstock provision and the conditioning of the gas at the retail site. Thus, transport is a critical issue in the overall emissions balance.

The transport emissions in the studied supply chains (Fig. 3.5) are based on assumptions of a fixed transport distance. However, the results of the overall chains differ with the transported amounts of hydrogen and the transport distances. According to Fig. 3.7, GHG emissions due to truck transport of the hydrogen consist of two components. GHG emissions represented by the intercept with the y-axis are caused by the compression or liquefaction. With increasing distance of the transported hydrogen, the total GHG emissions rise because the truck releases GHG emissions from the combustion of diesel fuel during operation; this is especially true for transport modes with heavy hydrogen tanks. As apparent from Fig. 3.7 transport of liquefied hydrogen causes the highest GHG emissions per transported MJ of hydrogen.

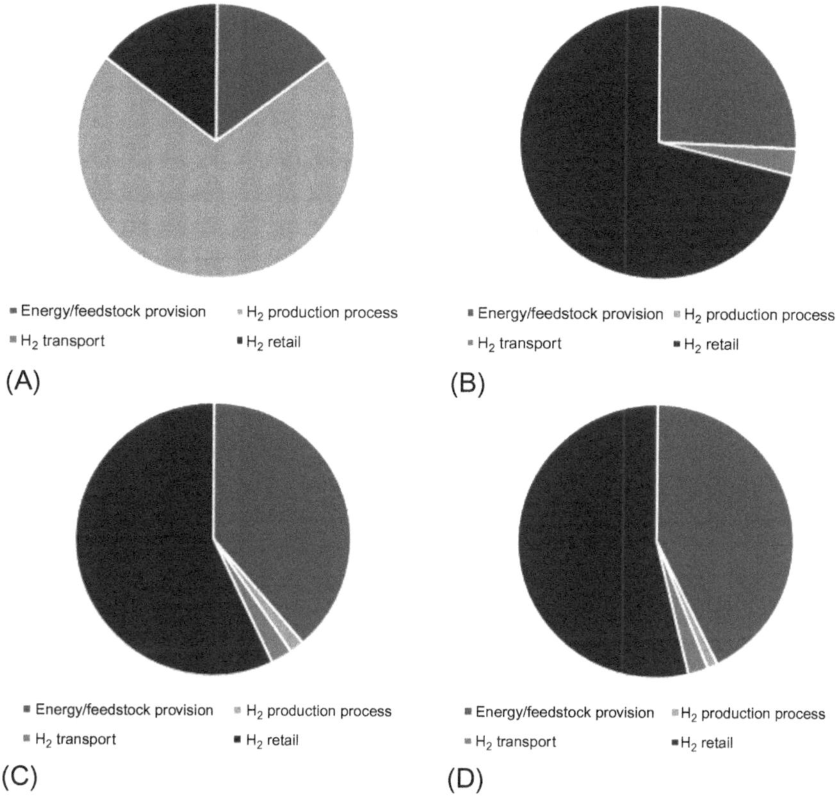

FIG. 3.6 (A–D) Contribution of the individual process steps to the overall GHG emissions of the different hydrogen production chains (exemplary for each hydrogen production option). (A) Steam methane reforming, natural gas, small scale. (B) Electrolysis wind offshore, pipeline transport. (C) Gasification, short rotation coppice wood, pipeline transport. (D) Steam methane reforming, biomethane from biowaste, pipeline transport.

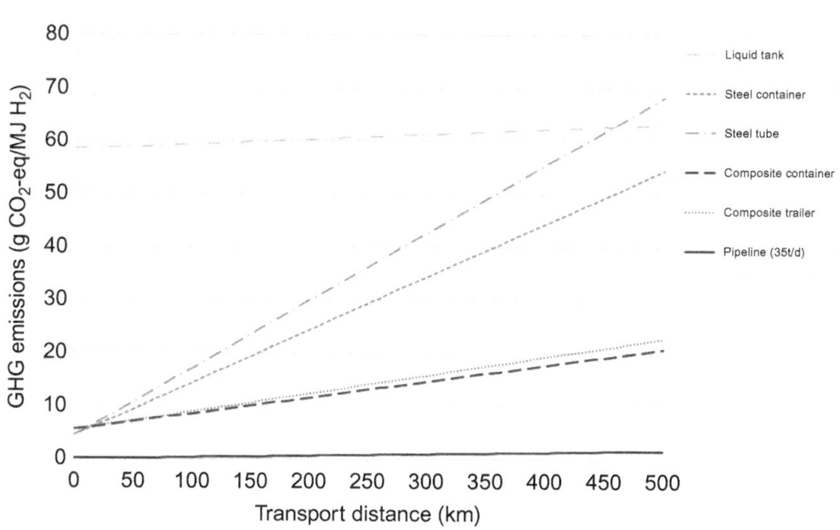

FIG. 3.7 Comparison of GHG emissions from the different hydrogen transportation options.

However, for large transport distances above 450km, the calculation results show that liquid hydrogen transportation can be more advantageous than compressed transportation in steel tubes. Although the liquid transportation causes higher initial GHG emissions due to the higher energy demand for liquefaction, GHG emissions do not increase as fast per transported kilometer because of the higher energy content of the transported liquid hydrogen. Therefore, the GHG emissions of compressed hydrogen transportation with heavy steel tubes exceed those of liquid transportation on long distances because of the higher fuel demand of the truck.

If hydrogen is transported as a compressed gas, the use of composite tank materials is advisable to reduce GHG emissions over long distances. The lighter the transport vessels, the lower the GHG emissions per transported energy unit. Especially over long distances, this effect increases. On short distances up to 20km, initial emissions from the compression outweigh the total GHG emissions per transported MJ of hydrogen. This is because larger amounts of hydrogen can be transported in composite vessels than in steel vessels. Initially the compression of these larger amounts requires more energy but with every traveled kilometer they are offset by lower fuel requirements because of the lighter specific weight of the composite vessels. Therefore, GHG emissions per inherent MJ are lowest if hydrogen is transported in composite vessels over distances above 20km.

Compared to truck transport, the specific GHG emissions of hydrogen transportation via pipeline are rather low. They increase only slightly with an increasing transport distance. If the GHG emissions of hydrogen transportation are displayed per t km (Fig. 3.8), the GHG emissions from truck transportation decrease strongly with every driven kilometer; again, the reason is that the GHG emissions of compression or liquefaction are allocated to a longer transport distance. In comparison, GHG emissions from pipeline transport increase slightly per t km because of the need for recompression every 100km.

GHG emissions from pipeline transport are not only dependent on the transportation distance but also on the total pipeline capacity because the overall GHG emissions of the pipeline construction have to be allocated to the overall amount of hydrogen passing through the pipeline during its entire service lifetime. The more hydrogen is transported within an existing pipeline, the lower the GHG emissions per unit of hydrogen. Fig. 3.9 shows that if a long-distance pipeline is only used for the transportation of a low amount of hydrogen, GHG emissions increase sharply if the pipeline diameter is assumed to be constant (the pressure level is increased). If the throughput of the pipeline has to be increased significantly, the pipeline itself needs to be enlarged with the result that the GHG emissions resulting from the pipeline construction also rise.

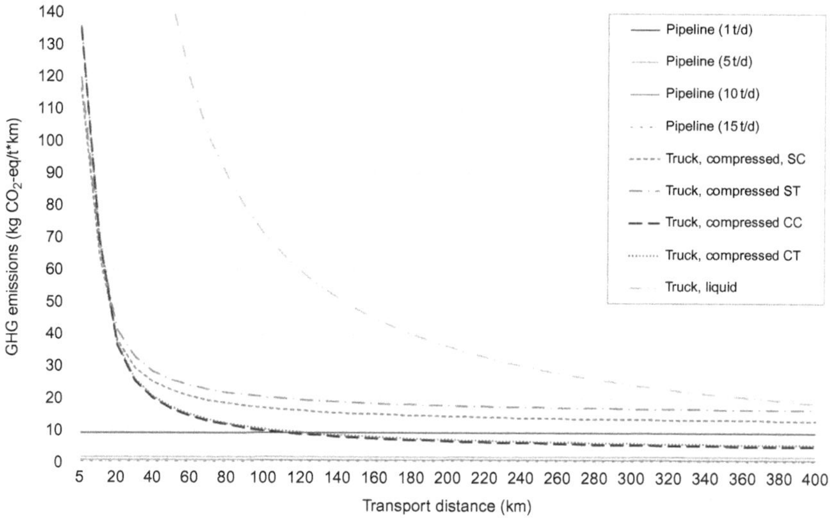

FIG. 3.8 Comparison of GHG emissions from different hydrogen transportation options via truck (SC, steel container; ST, steel tube; CC, composite container; CT, composite trailer) and pipeline.

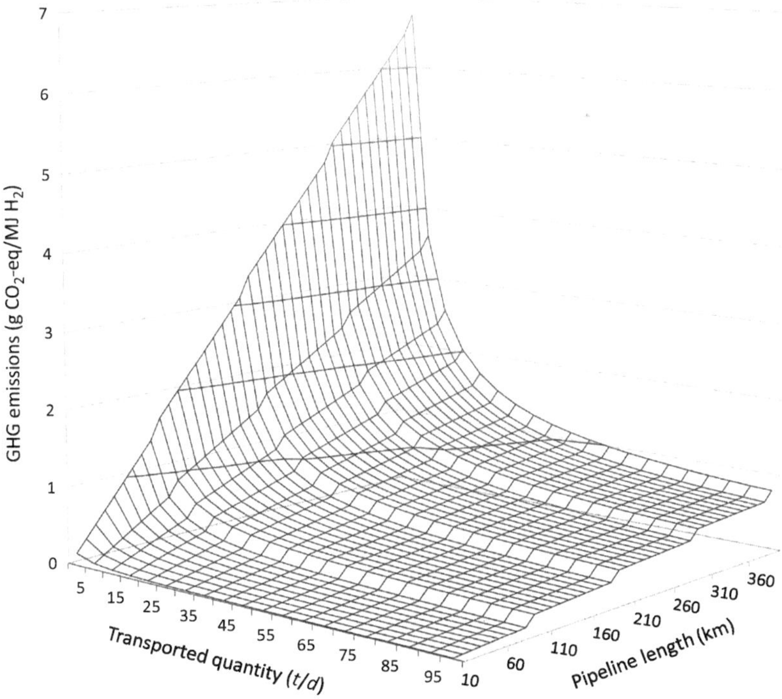

FIG. 3.9 GHG emissions due to pipeline transport in relation from pipeline length and total capacity.

3.6 FINAL CONSIDERATIONS

Hydrogen as a transportation fuel is seen as the great hope for climactically sound ground transportation. Within this context, the overall goal of this chapter has been to assess different possible hydrogen supply chains from a life cycle perspective related to the overall GHG emissions. Additionally, the sensitivity of these environmental effects to the design of the provision chain (e.g., transportation distances, location of the production facility) has been assessed. Based on these variations, promising hydrogen provision chains with minimized environmental impact have been identified for the mobility sector.

The most important results can be summarized as follows:

- Hydrogen provision chains based on wind or solar energy produce the least GHG emissions of all investigated pathways.
- Hydrogen production from natural gas via SMR is in each case more energy consuming and causes more GHG emissions than all other provision chains based on renewable feedstocks.
- The worst option of all investigated cases, in terms of non-renewable energy consumption and GHG emissions, is the use of electricity from the current grid, which is mostly based on non-renewable resources, such as coal and natural gas.
- With an increasing proportion of renewable electricity from the grid, it will become a more climate friendly feedstock for hydrogen production.
- If hydrogen is transported via pipelines from the production facility to the retail station, GHG emissions of the overall provision chain can be minimized.
- If pipeline transportation is not possible, hydrogen should be transported in vessels made of composite materials.
- For very long distances ($> 450\,\text{km}$), hydrogen transport in liquid form can be advantageous instead of in trucks with steel vessels.

The results summarized above can be compared to literature results. Fig. 3.10 displays the average GHG emissions from different studies published in the past. These results show great variations because of differences in the assumed system boundaries, the assumptions related to the hydrogen provision chain, as well as the database used. In general, it has also been found in the literature that emissions of hydrogen production from electrolysis with renewable electricity are the lowest, followed by gasification, and SMR from biomethane. The highest GHG emissions occur from SMR of natural gas. These findings support the results presented here.

In conclusion, it can be stated that large-scale hydrogen production combined with pipeline transport can contribute to reduced GHG emissions and non-renewable energy use of hydrogen provision chains. GHG emissions can be minimized if electricity from renewable sources is used for electrolytic hydrogen production. However, if pipelines are used, they should be utilized up to the installed transportation capacity. In this respect, it has to be kept in

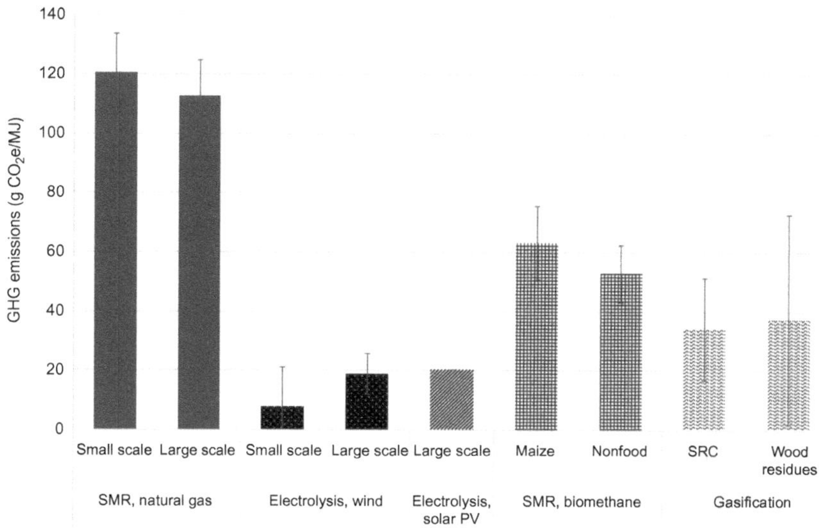

FIG. 3.10 Average GHG emissions of different hydrogen supply chains compiled from literature (Ruth et al., 2009; Altmann et al., 2004; Albrecht et al., 2015; Edwards et al., 2014; Wulf and Kaltschmitt, 2012, 2013; Suleman et al., 2016; Hajjaji et al., 2016; Cetinkaya et al. 2012).

mind that pipelines would mean erecting a new infrastructure beside the existing natural gas grid. For shorter transport distances, compressed delivery is suitable, especially if modern hydrogen transport tanks from composite materials are used. With longer distances, the transport of liquid hydrogen becomes increasingly more feasible in terms of GHG emissions and energy consumption.

REFERENCES

Abe, I., 2008. Statistics on hydrogen production and consumption. In: Ohta, T., Veziroglu, T.N. (Eds.), Energy Carriers and Conversion Systems With Emphasis on Hydrogen. In: Encyclopedia of Life Support Systems; Vol. 1. UNESCO-EOLSS, Paris, pp. 55–57.

ACEA, 2017. Trends in Fuel Type of New Cars Between 2015 and 2016, by Country. European Automobile Manufacturers Association, Brussels. Available at: http://www.acea.be/statistics/article/trends-in-fuel-type-of-new-cars-between-2015-and-2016-by-country (Accessed 31 May 2017).

Ahmadi, P., Kjeang, E., 2015. Comparative life cycle assessment of hydrogen fuel cell passenger vehicles in different Canadian provinces. Int. J. Hydrog. Energy 40 (38), 12905–12917.

Albrecht, U., Altmann, M., Barth, F., Bünger, U., Fraile, D., Lanoix, J.-C., Pschorr-Schoberer, E., Vanhoudt, W., Weindorf, W., Zerta, M. and Zittel, W., 2015. Study on Hydrogen From Renewable Resources in the EU-Final Report. Ludwig-Bölkow-Systemtechnik GmbH (LBST), Ottobrunn and Hinicio S.A., Brussels.

Altmann, M., Schmidt, P., Wurster, R., Zittel, W., 2004. Potential for Hydrogen as a Fuel for Transport in the Long Term (20202 – 2030)—Full Background Report. European Commission Joint Research Centre; Institute for Prospective Technological Studies, Seville, Spain.

Altmann, M., Bünger, U., Landinger, H., Pschorr-Schoberer, E., Raksha, T., Wurster, R., Zerta, M., 2014. e-mobil BW GmbH, Ludwig-Bölkow-Systemtechnik GmbH (LBST), (Eds.), Die Rolle von Wasserstoff in der Energiewende. Entwicklungsstand und Perspektiven. e-mobil BW GmbH, Stuttgart.

Blanck, R., Kasten, P., Hacker, F., Mottschall, M., 2013. Treibhausgasneutraler Verkehr 2050: Ein Szenario zur zunehmenden Elektrifizierung und dem Einsatz stromerzeugter Kraftstoffe im Verkehr. Öko-Institut e.V. -Institut für angewandte Ökologie, Freiburg.

BMVI, 2016. Nationaler Strategierahmen über den Aufbau der Infrastruktur für alternative Kraftstoffe. als Teil der Umsetzung der Richtlinie 2014/94/EU. Bundesministerium für Verkehr und digitale Infrastruktur, Berlin.

Bond, S., Alvarez, R., Reimann, S., Weilenmann, M., Buchmann, B., 2011. Hydrogen emissions to the atmosphere from industry and transportation. In: Wokaun, A., Wilhelm, E. (Eds.), Transition to Hydrogen: Pathways Toward Clean Transportation. Cambridge University Press, pp. 96–127.

Cardella, U., Decker, L., Sundberg, J., Klein, H., 2017. Process optimization for large-scale hydrogen liquefaction. Int. J. Hydrog. Energy 42 (17), 12339–12354.

Cetinkaya, E., Dincer, I., Naterer, G.F., 2012. Life cycle assessment of various hydrogen production methods. Int. J. Hydrog. Energy 37 (3), 2071–2080.

CMW, 2017. Multiple Element Gas Container (MEGC) Datasheet. City Mashine & Welding, Inc (CMW), Amarillo, TX. Available at: http://www.cmwelding.com/technical/megc-datasheet (Accessed 31 May 2017).

DIN EN ISO, 2006. Environmental Management—Life Cycle Assessment—Principles and Framework. ISO 14040, Beuth Verlag, Berlin.

Ecoinvent, 2016. Ecoinvent Database Version 3.3 [Software]. Ecoinvent, Zürich.

Edwards, R., Larive, J.-F., Rickeard, D., Weindorf, W., (KEINE ANGABE), 2014. Godwin, S. et al., (Ed.), Well-to-Tank Report Version 4.a. JEC Well-To-Wheels Analysis. Well-to-Wheels Analysis of Future Automotive Fuels and Powertrains in the European Context. European Commission, Joint Research Center (JRC), Luxembourg.

EEA, 2016a. Contribution of the Transport Sector to Total Emissions of the Main Air Pollutants. Data Visualization. European Environment Agency. Available at: https://tinyurl.com/lwbrjkt (Accessed 12 May 2017).

EEA, 2016b. GHG Emission Trends and Projections. Data Visualization. European Environment Agency. Available at: https://www.eea.europa.eu/data-and-maps/daviz/ghg-emission-trends-and-projections#tab-chart_3 (Accessed 12 May 2017).

EEA, 2016c. Inland Freight Transport Volumes and GDP. Data Visualization. European Environment Agency. Available at: https://www.eea.europa.eu/data-and-maps/daviz/freight-transport-volumes-and-gdp-4#tab-chart_4 (Accessed 12 May 2017).

EEA, 2016d. Passenger Transport Volume and Modal Split. Data Visualization. European Environment Agency. Available at: https://tinyurl.com/l6nb4m8 (Accessed 12 May 2017).

EPSC, 2016. Towards Low-Emission Mobility. Driving the Modernisation of the EU Economy. European Political Strategy Center, Brussels (EPSC Strategic Notes; Issue 18).

European Commission, 2011. White Paper. Roadmap to a Single European Transport Area—Towards a Competittive and Resource Efficient Transport System. COM (2011) 144 Final. Available at: http://eur-lex.europa.eu/legal-content/EN/ALL/?uri=CELEX:52011DC0144%3e.

European Commission, 2016. A European Strategy for Low-Emission Mobility. COM (2016) 501 Final. 2016.

Eurostat, 2015. Passenger Cars in the EU. Statistical Unit of the European Commission, Luxembourg. Available at: http://ec.europa.eu/eurostat/statistics-explained/index.php/Passenger_cars_in_the_EU#Buyers_prefer_new_petrol_cars_rather_than_diesel_cars_in_most_Member_States.2C_while_alternative_fuels_continue_to_play_a_minor_role (Accessed 31 May 2017).

Faist-Emmenegger, M., Heck, T., Jungbluth, N., 2007. Erdgas. Teil V, Data v2.0 (2007). In: Dones, R., Kägi, T., Spielmann, M. (Eds.), Sachbilanzen von Energiesystemen: Grundlagen für den ökologischen Vergleich von Eenrgiesystemen und den Einbezug von Energiesytemen in Ökobilanzen für die Schweiz. Swiss Centre for Life Cycle Inventories, Dübendorf (Final report ecoinvent; No. 6-V).

Gardiner, M., Satyapal, S., 2009. Energy Requirements for Hydrogen Gas Compression and Liquefation as Related to Vehicle Storage Needs. DOE Hydrogen and Fuel Cells Program Record. U.S. Department of Energy, Washington, DC.

Gellert, S., 2013. Thermochemische Herstellung von Wasserstoff aus Biomasse unter besonderer Berücksichtigung der Rohgasreformierung. DissertationTechnische Universität Hamburg-Harburg.

Gerster, M., Wimmelbücker, S., 2017. Wasserstofffahrzeuge in Deutschland: Warum bisher erst 314 Brennstoffzellenautos zugelassen wurden. *Automobilwoche*. 18 January 2017. Available at: http://www.automobilwoche.de/article/20170118/NACHRICHTEN/170119880/wasserstofffahrzeuge-in-deutschland-warum-bisher-erst–brennstoffzellenautos-zugelassen-wurden (Accessed 15 May 2017).

H2 Mobility, 2017. H2-Stationen, Roadmap der H2 Mobility, Berlin. Available at: http://h2-mobility.de/h2-stationen/.

H2ME, 2016. Hydrogen Mobility Europe. EU Funded Project for Hydrogen Mobility and Fuel Cell Electric Vehicles Deployment, London. Available at: http://h2me.eu/.

Hajjaji, N., Martinez, S., Trably, E., Steyer, J.-P., Helias, A., 2016. Life cycle assessment of hydrogen production from biogas reforming. Int. J. Hydrog. Energy 41 (14), 6064–6075.

Hexagon Composites, 2012. Move Gas- Not Steel. Leading Solutions for Bulk Gas Transportation. Hexagon Composites Group; Lincoln Composites, Alesund, Norway; Lincoln, NB. Available at: https://www.slideshare.net/vasisht123/bulk-trabnsportsept2012 (Accessed 31 May 2017).

Hosseini, S.E., Wahid, M.A., 2016. Hydrogen production from renewable and sustainable energy resources: promising green energy carrier for clean development. Renew. Sust. Energ. Rev. 57, 850–866.

HPTT, 2009. Hydrogen Production Roadmap. Technology Pathways to the Future. Hydrogen Production Technical Team (HPTT) of the FreedomCAR and Fuel Partnership, Washington, DC.

Hydrogen Council, 2017. How Hydrogen Empowers the Energy Transition. Hydrogen Council, Brussels.

IEA, 2015. Technology Roadmap. Hydrogen and Fuel Cells. International Energy Agency (IEA), Paris.

Joest, S., Fichtner, M., Wietschel, M., Bünger, U., Stiller, C., Schmidt, P., Merten, F., Bonhoff, K., 2009. Bundesminiserium für Verkehr, Bau und Stadentwicklung (BMVBS), (Ed.), Woher kommt der Wasserstoff in Deutschland bis 2050?. Deutsche Energieagentur (dena); Forschungszentrum Karlsruhe; Fraunhofer ISI; Ludwig-Bölkow-Systemtechnik GmbH (LBST); Wuppertaler Institut für Klima, Umwelt, Energie GmbH, Berlin.

Kaltschmitt, M., Lippitsch, K., Müller, J., Reichert, S., Schulz, D., Schwunk, S., 2013a. Photovoltaische Stromerzeugung. In: Kaltschmitt, M., Streicher, W., Wiese, A. (Eds.), Erneuerbare Energien. Systemtechnik, Wirtschaftlichkeit, Umweltaspekte. 5. Auflage Springer Vieweg, Berlin, Heidelberg, pp. 353–452.

Kaltschmitt, M., Schlüter, M., Schulz, D., Skiba, M., Özdirik, B., 2013b. Stromerzeugung aus Windenergie. In: Kaltschmitt, M., Streicher, W., Wiese, A. (Eds.), Erneuerbare Energien. Systemtechnik, Wirtschaftlichkeit, Umweltaspekte. 5. Auflage Springer Vieweg, Berlin, Heidelberg, pp. 453–553.

Kaltschmitt, M., Hartmann, H., Hofbauer, H. (Eds.), 2016. Energie aus Biomasse. Grundlagen, Techniken und Verfahren. Springer Vieweg, Berlin, Heidelberg.

KBA, 2017a. Fahrzeugzulassungen (FZ) Bestand an Kraftfahrzeugen nach Umwelt-Merkmalen. 1. Januar 2017. Kraftfahrtbundesamt, Flensburg (FZ 13) (Accessed 31 May 2017).

KBA, 2017b. Neuzulassungen von Pkw im Jahr 2016 nach ausgewählten Kraftstoffarten. Kraftfahrtbundesamt, Flensburg. Available at: http://www.kba.de/DE/Statistik/Fahrzeuge/Neuzulassungen/Umwelt/2016_n_umwelt_dusl.html;jsessionid=AC6CD6CA2AA4CB9F71A059A6A359B820.live21302?nn=652326 (Accessed 31 May 2017).

Krieg, D., 2012. Konzept und Kosten eines Pipelinesystems zur Versorgung des deutschen Straßenverkehrs mit Wasserstoff. Dissertation an der RWTH Aachen, Jülich. (Schriften des Forschungszentrum Jülich; Band/Volume 144).

Linde AG, 2014. The Cleanest Energy Carrier Ever. Hydrogen Solutions From Linde Gas. Linde AG, Munich (Accessed 31 May 2017).

Linde Group, 2013. Linde setzt neue Maßstäbe beim Wasserstofftransport, München. Available at: http://www.the-linde-group.com/de/news_and_media/press_releases/news_20130925.html. (Accessed 31 May 2017).

Miller, E., 2011. Hydrogen Production by Polymer Electrolyte Membrane (PEM) Electrolysis—Spotlight on Giner and Proton. Water Electrolysis Hydrogen Production Overview. Office of Energy Efficiency & Renewable Energy, Washington, DC.

Nikolaidis, P., Poullikkas, A., 2017. A comparative overview of hydrogen production processes. Renew. Sust. Energ. Rev. 67, 597–611.

Nilsen, S., Haugom, G.P., Markert, F., Verfondern, K., Schitter, C., Barthelemy, H., Edimbourg, M., Barth, F., Adams, P., Azkarate, I., Ezponda, E., Crutzen, H., 2007. Hydrogen Technologies, 1.2th ed. HySafe Safety of Hydrogen as an Energy Carrier, Eggenstein-Leopoldshafen. (Biennal Report on hydrogen safety). Available at: http://www.hysafe.org/brhs%3e.

Pasaoglu, G., Harrison, G., Jones, L., Hill, A., Beaudet, A., Thiel, C., 2016. A system dynamics based market agent model simulating future powertrain technology transition: scenarios in the EU light duty vehicle road transport sector. Technol. Forecast. Soc. Chang. 104, 133–146.

Robinius, 2015. Strom- und Gasmarktdesign zur Versorgung des deutschen Straßenverkehrs mit Wasserstoff. DissertationFakultät für Maschinenwesen. Rheinisch-Westfälische Technische Hochschule Aachen.

Roedl, A., 2010. Production and energetic utilization of wood from short rotation coppice—a life cycle assessment. Int. J. Life Cycle Assess. 15 (6), 567–578.

Ruth, M., Laffen, M., Timbario, T.A., 2009. Hydrogen pathways: cost, Well-to-Wheels energy use, and emissions for the current technology status of seven hydrogen production, delivery, and distribution scenarios: National Renewable Energy Laboratory. (Technical Report; NREL/TP-6A1-46612).

SAE, 2016. Fueling Protocols for Light Duty Gaseous Hydrogen Surface Vehicles. SAE J2601. SAE International, Warrendale, PA. Available at: http://standards.sae.org/j2601_201612/%3e.

Santos, S., 2015. Understanding the Potential of CCS in Hydrogen Production. Review of Current State-of-the-Art. Process Industry CCS Workshop, Joint IEAGHG & IETS Meeting, Lisbon.

Schjolberg, I., Hulteberg, C., Lieftink, D., 2015. Small-scale reforming for in-site hydrogen supply. In: Stolten, D., Emonts, B. (Eds.), Hydrogen Science and Engineering: Materials, Processes, Systems and Technology. In: Vols. 1 and 2. Wiley-VHC, Weinheim, pp. 217–236.

Sgobbi, A., Nijs, W., de Miglio, R., Chiodi, A., Gargiulo, M., Thiel, C., 2016. How far away is hydrogen? Its role in the medium and long-term decarbonisation of the European energy system. Int. J. Hydrog. Energy 41 (1), 19–35.

Sharma, S., Ghoshal, S.K., 2015. Hydrogen the future transportation fuel: from production to applications. Renew. Sust. Energ. Rev. 43, 1151–1158.

da Silva Veras, T., Mozer, T.S., da Costa Rubim Messeder dos Santos, D., da Silva César, A., 2017. Hydrogen: trends, production and characterization of the main process worldwide. Int. J. Hydrog. Energy 42 (4), 2018–2033.

Statista, 2017. Anzahl der Tankstellen in Deutschland 1950 bis 2016. Statista GmbH, Hamburg. Available at: https://de.statista.com/statistik/daten/studie/2621/umfrage/anzahl-der-tankstellen-in-deutschland-zeitreihe/ (Accessed 13 June 2017).

Stolzenburg, K., Mubbala, R., 2013. Hydrogen Liquefaction Report. Integrated Design for Demonstration of Efficient Liquefation of Hydrogen (IDEALHY). Fuel Cells and Hydrogen Joint Undertaking (FCH JU). PLANET Energie udn Technik GbR, Oldenburg (WP 3.16, Whole Chain Assessment).

Stolzenburg, K., Berstad, D., Decker, L., Elliott, A., Haberstroh, C., Hatto, C., Klaus, M., Mortimer, N.D., Mubbala, R., Mwabonje, O., Neksa, P., Quack, H., Rix, J.H.R., Seemann, I., Walnum, H.T., 2013. Efficient liquifaction of hydrogen: results of the IDEALHY Project. XXth energie—Symposium, Stralsund, 2013.

Suleman, F., Dincer, I., Agelin-Chaab, M., 2016. Comparative impact assessment study of various hydrogen production methods in terms of emissions. Int. J. Hydrog. Energy 41 (19), 8364–8375.

TÜV SÜD, 2017. Hydrogen filling stations worldwide. Ludwig-Bölkow-Systemtechnik GmbH (LBST); TÜV SÜD, Munich. Available at: https://www.netinform.net/H2/H2Stations/Default.aspx (Accessed 15 May 2017).

Ursua, A., Gandia, L.M., Sanchis, P., 2012. Hydrogen production from water electrolysis: current status and future trends. Proc. IEEE 100 (2), 410–426.

US Department of Energy, 2015a. Fuel cell technologies office multi-year research, development, and demonstration plan. 3.2 Hydrogen delivery. US Department of Energy Office of Energy Efficiency & Renewable Energy, Washington, DC. Available at: https://energy.gov/eere/fuelcells/downloads/fuel-cell-technologies-office-multi-year-research-development-and-22 (Accessed 14 June 2017).

US Department of Energy, 2015b. Hydrogen production energy conversion efficiencies. US Department of Energy, Fuel Cell Technologies Office, Washington, DC. Available at: http://hydrogen.pnl.gov/hydrogen-data/hydrogen-production-energy-conversion-efficiencies%3e.

US Department of Energy, 2017. Hydrogen Production: Electrolysis. US Department of Energy Office of Energy Efficiency & Renewable Energy, Washington, DC. Available at: https://energy.gov/eere/fuelcells/hydrogen-production-electrolysis (Accessed 18 May 2017).

Vakkilainen, E., Kuparinen, K., Heinimö, J., 2013. Large Industrial Users of Energy Biomass. IEA Bioenergy Task 40: Sustainable International Bioenergy Trade. Lappeenranta University of Technology, Lappeenranta.

Westerkamp, T., Reinelt, T., Oehmichen, K., Ponitka, J., Nauman, K., 2014. KlimaCH4. klimaeffekte von Biomethan. Deutsches Biomasseforschungszentrum (DBFZ), Leipzig (DBFZ Report; Nr. 20).

Wulf, C., Kaltschmitt, M., 2012. Life cycle assessment of hydrogen supply chain with special attention on hydrogen refuelling stations. Int. J. Hydrog. Energy 37 (21), 16711–16721.

Wulf, C., Kaltschmitt, M., 2013. Life cycle assessment of biohydrogen production as a transportation fuel in Germany. Bioresour. Technol. 150, 466–475.

Wystrach GmbH, 2017a. Gastransport-Fahrzeuge, Weeze. Available at: http://www.wystrach-gmbh.de/produkt-wystrach-gastransportfahrzeuge.html (Accessed 31 May 2017).

Wystrach GmbH, 2017b. Transportmodule für Gase—CNG/H2, Weeze. Available at: http://www.wystrach-gmbh.de/produkt-wystrach-gascontainer-cng.html (Accessed 31 May 2017).

Wystrach, J., Wolter, W., Valks, B., 2012. Wystrach News. Wystrach GmbH, Weeze (Ausgabe Dezember 2012).

Chapter 4

Hydrogen Production From Biogas Reforming: An Overview of Steam Reforming, Dry Reforming, Dual Reforming, and Tri-Reforming of Methane

Doan Pham Minh*, Tan Ji Siang†, Dai-Viet N. Vo†, Thanh Son Phan*, Cyrille Ridart‡, Ange Nzihou* and Didier Grouset*
**Université de Toulouse, IMT Mines Albi, UMR CNRS 5302, Centre RAPSODEE, Campus Jarlard, Albi, France, †Faculty of Chemical and Natural Resources Engineering, University Malaysia Pahang, Lebuhraya Tun Razak, Gambang, Kuantan, Malaysia, ‡HERA—ALBHYON, Technopole Innoprod, rue Pierre Gilles de Gennes, Albi, France*

4.1 INTRODUCTION

The present context of climate change, fossil resource scarcity, and environmental issues imposes the development of renewable energy, such as solar, wind, geothermal, and biomass. In France, 24.5 million metric tons of oil equivalent of renewable energy were produced in 2013, which corresponds to 17.6% of the total energy amount produced this year (ADEME, 2014a). According to the last report of ADEME (agence de l'environnement et de la maîtrise de l'énergie, France) on the "*lightened environmental footprint*" scenario, the objective is to reduce by 17% the carbon footprint in France by 2030 (ADEME, 2014b). This energy transition dynamic is also being conducted in many other countries, such as Germany, the United Kingdom, the United States, Sweden, Norway, Switzerland, and China (IEA, 2017).

In the transport domain, much effort has been devoted to the development of electric vehicles. In parallel, the development of hydrogen vehicles is also very attractive. Up-to-date, hydrogen vehicles present some major advantages compared to electric vehicles, namely, a shorter charging time and a higher autonomy. The use of hydrogen as a clean fuel starts now, and hydrogen can become a strategic energy vector in the near future.

Hydrogen Supply Chain. https://doi.org/10.1016/B978-0-12-811197-0.00004-X

Nowadays, around 96% of hydrogen is produced from fossil resources, and particularly from natural gas, oil and derivatives, and coal. Chemicals synthesized from these fossil resources are usually obtained via synthetic gas, which is produced by steam reforming of natural gas. Details about steam reforming will be discussed later. Scheme 4.1 below illustrates the most common products that can be obtained from syngas, including hydrogen, methanol, liquid fuels, synthetic natural gas (SNG), ammonia, and heat and power.

Hydrogen from water electrolysis accounts only for about 4% of total hydrogen production. Naturally, water electrolysis using electricity from renewable resources, such as solar, wind, or hydro is a promising way for green hydrogen production. However, the process still needs to be improved in order to reach the competitive cost required by the hydrogen market.

Another way to produce green hydrogen is the valorization of bioresources, such as biogas. This last one is obtained from biomass, residues, or wastes by anaerobic digestion. In general, two types of biogas are distinguished: (1) digested gas from anaerobic digester, which is commonly called biogas, and (2) landfill gas from landfill sites (IEA, 2009).

The production of hydrogen from biogas and landfill gas can be achieved via a syngas route as shown in Scheme 4.2 below. Syngas is obtained from biogas or landfill gas by different reforming processes, which will be detailed later. Then a mixture of H_2 and CO_2 can be obtained from syngas by the water-gas-shift (WGS) reaction. Finally, hydrogen can be separated from the mixture with CO_2 by using a pressure swing adsorption (PSA) or an equivalent process.

Both biogas/landfill gas reforming and WGS are catalytic processes. The reforming step can be achieved with catalytic steam reforming process, which is already commercialized for natural gas. However, the energy balance of the steam reforming is not optimized because of a large excess of water vapor, which is required with current catalysts. Because the composition of

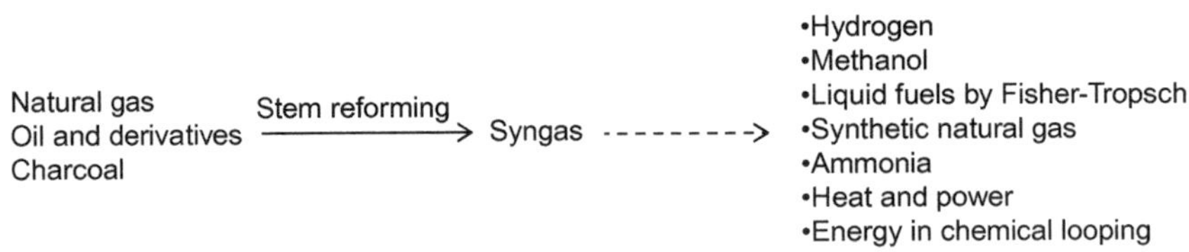

SCHEME 4.1 Chemicals and energy from fossil resources via steam reforming.

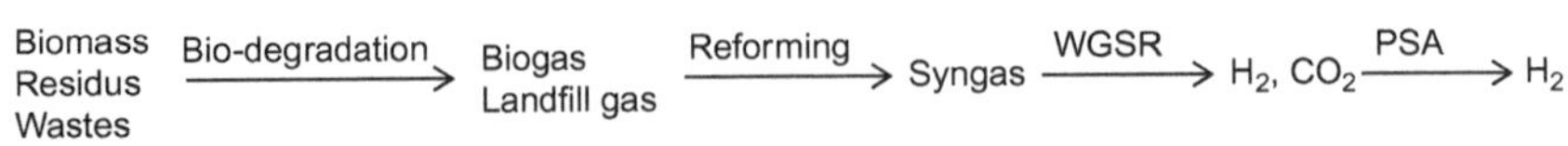

SCHEME 4.2 Main steps for the production of hydrogen from biogas or landfill gas via syngas route.

biogas/landfill gas varies as shown in Table 4.1, other reforming processes can be considered, as listed below:

- Dry reforming of methane (DRM) with CO_2.
- Dual reforming of methane with a mixture of CO_2 and H_2O.
- Tri-reforming of methane (TRM) with a mixture of CO_2, H_2O and O_2.

For a given mixture of CH_4 with at least one of the three oxidants (CO_2, H_2O, and O_2), different chemical equilibriums can take place. Table 4.2 summarizes the most important chemical reactions usually found in methane reforming processes. The dependence of the Gibbs free energy change per mole of reaction ($\Delta_r G$) on the reaction temperature has been previously reported in the literature, and some of them have been calculated from the handbook of chemistry data. FactSage software (Factsage, n.d.) (6.3.1 version) is used for the thermodynamic equilibrium calculation of a given mixture at well-defined conditions of temperature and pressure. The calculation is based on the principle of the Gibbs free energy minimization. Standard enthalpies of reaction ($\Delta_r H^{\circ}_{298}$) are calculated from the handbook of chemistry (Lide, 2003–2004).

TABLE 4.1 Main Characteristics of Digested Gas and Landfill Gas in Comparison With Some Natural Gases

Compounds	Biogas	Landfill gas	Natural Gas (Danish)	Natural Gas (Dutch)
Methane (vol%)	60–70	35–65	89	81
Other hydrocarbons (vol%)	0	0	9.4	3.5
Hydrogen (vol%)	0	0–3	0	–
Carbon dioxide (vol %)	30–40	15–50	0.67	1
Nitrogen (vol%)	≈0.2	5–40	0.28	14
Oxygen (vol%)	0	0–5	0	0
Hydrogen sulfide (ppm)	0–4000	0–100	2.9	–
Ammonia (ppm)	≈100	≈5	0	–
Lower heating value (kWh/Nm^3)	6.5	4.4	11.0	8.8

Adapted from IEA, 2009. IEA bioenergy, Task 37—Energy from biogas and landfill gas.

TABLE 4.2 The Main Chemical Reactions That Occur During the Reforming of Methane With At Least One of the Following Oxidants: CO_2, H_2O, and O_2

Reaction	$\Delta_r G$ function (P = 1 bar) (kJ/mol)	$\Delta_r H^\circ_{298}$ (kJ/mol)	Equations
$CH_4+CO_2 \rightarrow 2CO+2H_2$	$\Delta_r G = 61{,}770 - 67.3 \times T$	+247	(4.1)
$2CO \rightarrow C+CO_2$	$\Delta_r G = -39{,}810 + 40.9 \times T$	−172	(4.2)
$CO+H_2O \rightarrow CO_2+H_2$	$\Delta_r G = -39{,}802 + 37.673 \times T$	−41	(4.3)
$CH_4 \rightarrow C+2H_2$	$\Delta_r G = 21{,}960 - 26.5 \times T$	+75	(4.4)
$CH_4+H_2O \rightarrow CO+3H_2$	$\Delta_r G = 210{,}359 - 233.9 \times T$	+206	(4.5)
$C_s+H_2O \rightarrow CO+H_2$	$\Delta_r G = 132{,}184 - 138.8 \times T$	+131	(4.6)
$2CH_4+O_2 \rightarrow 2CO+4H_2$	$\Delta_r G = -653.9 - 369 \times T$	−71	(4.7)
$CH_4+2O_2 \rightarrow CO_2+2H_2O$	$\Delta_r G = -803{,}508 + 13 \times T - 0.018 \times T^2 + 8 \times 10^{-6} \times T^3$	−802.5	(4.8)
$C+O_2 \rightarrow CO$	$\Delta_r G = -110{,}872 - 89.4 \times T$	−110.5	(4.9)
$C+O_2 \rightarrow CO_2$	$\Delta_r G = -393{,}647 - 2.5 \times T$	−393.5	(4.10)
$CH_4+2H_2O \rightarrow CO_2+4H_2$	$\Delta_r G = 170{,}557 - 196.29 \times T$	+165	(4.11)

(4.1) Dry reforming of methane (DRM); (4.2) Boudouard reaction; (4.3) water-gas-shift (WGS) reaction; (4.4) methane cracking; (4.5) steam reforming of methane; (4.6) steam reforming of carbon; (4.7) partial oxidation of methane; (4.8) methane combustion; (4.9) partial oxidation of carbon; (4.10) carbon combustion; and (4.11) Methane reforming with large excess of steam. All molecules are in the gas state except carbon (C), which is in the solid state.

4.2 METHANE REFORMING WITH STEAM (SMR) AND WITH A MIXTURE OF STEAM/CARBON DIOXIDE (DUAL-MR)

4.2.1 Thermodynamic Equilibrium of Steam Methane Reforming

The ideal reaction of steam methane reforming (SMR) to syngas is given in Eq. (4.5) (Table 4.2). One mole of H_2O is needed to transform 1 mol of CH_4. In reality, SMR is usually carried out with a large excess of steam (molar steam/carbon ratio or S/C close to 3). Fig. 4.1 shows the thermodynamic equilibrium of a CH_4-H_2O mixture with two different molar ratios of H_2O/CH_4. Only products present at significant amounts are presented. They are all in

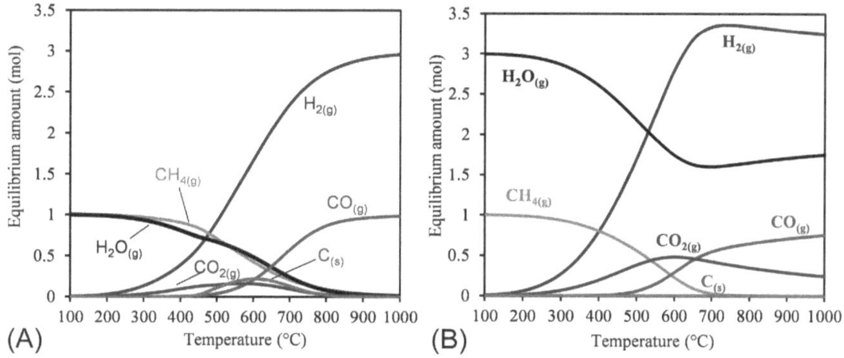

FIG. 4.1 Thermodynamic equilibrium as a function of the temperature at atmospheric pressure of the mixture containing (A) 1 mol CH_4 and 1 mol H_2O and (B) 1 mol CH_4 and 3 mol H_2O.

the gas state except for solid carbon ($C_{(s)}$). Other hydrocarbons, such as ethylene, ethane, methanol, and acetone, can be formed in very small quantities (approximately 10^{-6} to 10^{-14} mol) and are not presented.

With the equimolar ratio of H_2O/CH_4 (Fig. 4.1A), hydrogen is formed above 180°C and its content increases with the temperature. Carbon dioxide and solid carbon are formed in the temperature ranges of around 250–800°C and 450–800°C, respectively. This makes the process difficult because solid carbon formation causes catalyst deactivation, and carbon dioxide is an undesirable byproduct. Under these conditions, syngas formation is really favorable above 800°C from the thermodynamic point of view.

For the molar ratio of H_2O/CH_4 equal to 3/1 (Fig. 4.1B), solid carbon is potentially eliminated and methane can be completely consumed above around 720°C. This is the main advantage of using a large molar ratio of H_2O/CH_4. However, a large excess of water (>50%) strongly burdens the energy balance of the process. Also, carbon dioxide is omnipresent above 200°C because of the WGS reaction (Eq. 4.3 in Table 4.2).

Fig. 4.2 shows CH_4 conversion and coke selectivity as a function of the molar ratio of H_2O/CH_4 (or S/C) and reaction temperature. Increasing the molar ratio of H_2O/CH_4 leads to the increase of the CH_4 conversion at a given reaction temperature, and to the decrease of the reaction temperature at a given CH_4 conversion (Fig. 4.2A). Solid carbon can be formed at a low molar ratio of H_2O/CH_4 of 1/1 and 1.2/1 and can be avoided above 1.5/1 (Fig. 4.2B).

In Fig. 4.3A, as previously observed for CH_4 conversion, at a given reaction temperature, the amount of hydrogen increases with the molar ratio of H_2O/CH_4. At a high molar ratio of H_2O/CH_4 (2/1 to 3/1), the amount of hydrogen reaches a maximum due to the presence of WGS. On Fig. 4.3B, CO can be obtained above 400°C, and its amount increases with the reaction temperature. The molar ratio of H_2O/CH_4 slightly influences the CO amount up to around 600°C. Then, above this temperature, WGS is favored by increasing the molar ratio of H_2O/CH_4, and thus the amount of CO decreases. WGS explains also the

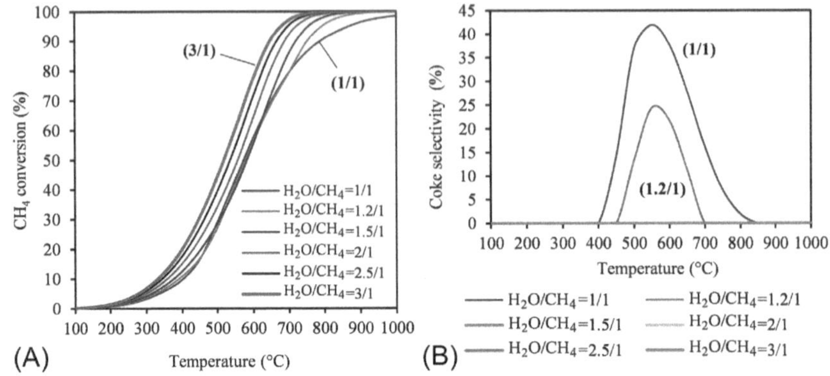

FIG. 4.2 Influence of the molar ratio of H_2O/CH_4 and the temperature on (A) CH_4 conversion and (B) coke selectivity at the thermodynamic equilibrium; Initial mixture: 1 mol of CH_4 and 1 to 3 mol of H_2O.

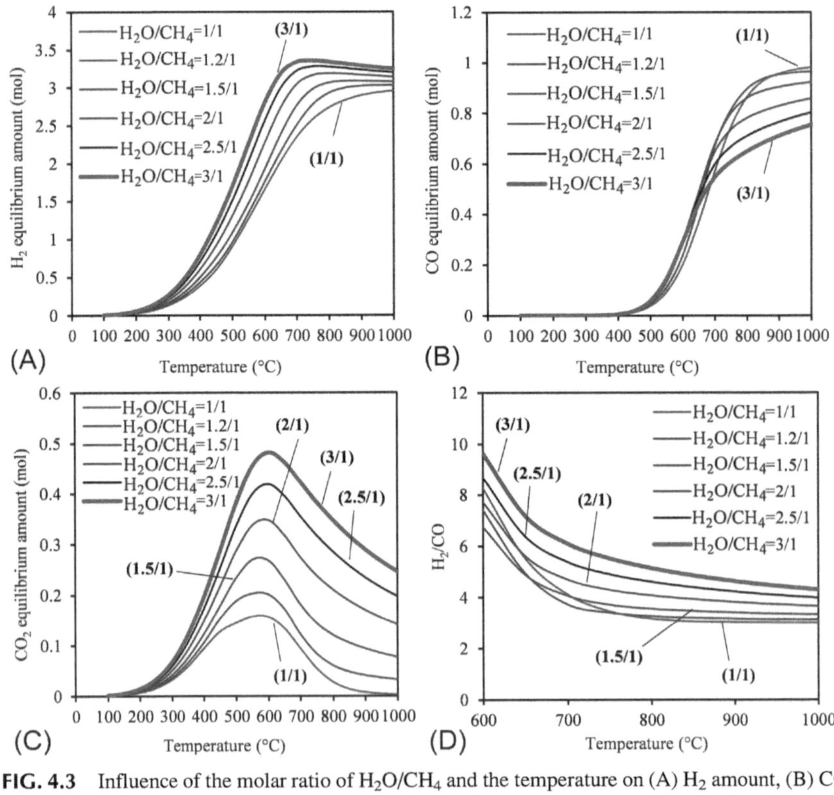

FIG. 4.3 Influence of the molar ratio of H_2O/CH_4 and the temperature on (A) H_2 amount, (B) CO amount, (C) CO_2 amount, and (D) molar ratio of H_2/CO at the thermodynamic equilibrium; Initial mixture: 1 mol of CH_4 and 1 to 3 mol of H_2O.

evolution of the amount of CO_2 in Fig. 4.3C. At a given reaction temperature, the amount of CO_2 increases with an increase in the molar ratio of H_2O/CH_4. Also, for each molar ratio of H_2O/CH_4, the amount of CO_2 reaches a maximum around 600°C. The molar ratio of H_2/CO is always higher than 3 and is favorable for all downstream uses as listed in Scheme 4.1, with an eventual regulation of this ratio by WGS.

From these results, it appears that, at atmospheric pressure, SMR is highly favored above 700°C using a molar ratio of $H_2O/CH_4 \geq 1.5/1$.

SMR is not favored at high pressure. However, for different reasons, it can be carried out at 30 bar pressure or even higher in large units. Fig. 4.4 shows the influence of the pressure on the amounts of the main species in SMR at 700, 800, and 900°C using a mixture of 1 mol CH_4 and 3 mol H_2O. Under these conditions, the behavior of the system seems to be similar from each other. As expected, increasing the pressure of the reaction strongly burdens the transformation of methane. At each reaction temperature, the amount of CH_4 increases with an increase of the reaction pressure, which is evidenced in Fig. 4.4D. This leads to a strong decrease in the amount of hydrogen with an increase of the

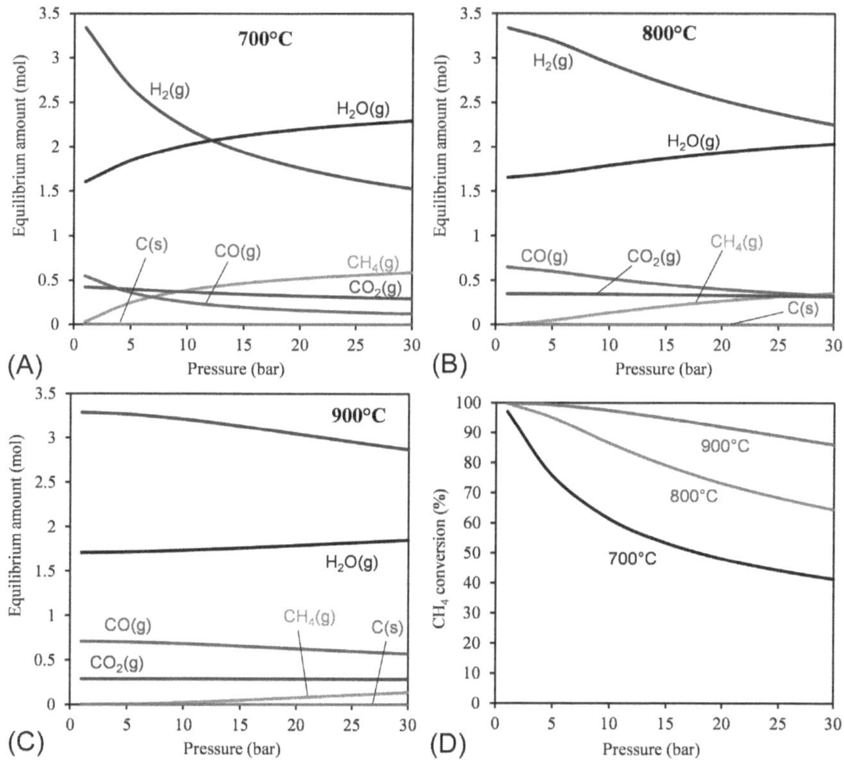

FIG. 4.4 Influence of the reaction pressure in steam methane reforming in the temperature range of 700–900°C for a mixture containing 1 mol of CH_4 and 3 mol of H_2O: (A) 700°C, (B) 800°C, (C) 900°C, and (D) CH_4 conversion at different temperatures.

reaction pressure. Increasing the reaction temperature to 900°C does not allow the complete transformation of CH_4 above 5 bar (Fig. 4.4D. No solid carbon is observed under these reaction conditions, as explained previously by a large excess of steam. Because of the WGS, CO_2 is omnipresent in large quantities in the temperature range of 700–900°C.

4.2.2 Thermodynamic Equilibrium of Dual Methane Reforming (Dual-MR)

Dual methane reforming (DMR) refers to the transformation of a CH_4-CO_2-H_2O mixture into syngas. This system is particularly interesting for different configurations, such as steam reforming of biogas and also carbon dioxide valorization by its incorporation into the reforming of methane or natural gas. Thus, both CO_2 and H_2O play the role of oxidants. The simultaneous presence of CO_2 and H_2O influences also the WGS equilibrium. Hereafter, we consider two cases: (1) an initial mixture of CH_4 and CO_2 is kept constant, then H_2O is added in various amounts; and (2) an initial mixture of CH_4 and H_2O is kept constant, then CO_2 is added in various amounts.

Fig. 4.5 shows the thermodynamic equilibrium of mixtures containing 1 mol of CH_4, 1 mol of CO_2 and 0.5 to 2.0 mol of H_2O at 1 bar. The equimolar mixture of CH_4 and CO_2 can represent the composition of a purified landfill gas. Adding water to this mixture is not beneficial for methane conversion under 700°C. In fact, as shown in Fig. 4.5A, the addition of water decreases the methane conversion below 600°C because of the formation of water from the CO_2 and CH_4 mixture by the WGS reaction. Then, above 700°C, the increase in the amount of water leads to a slight increase in CH_4 conversion. Thus, it is possible to reach 100% CH_4 conversion. Also, CO_2 conversion decreases with an increase in the amount of water added to the system. This may be explained by the WGS. On the other hand, the amount of water added to the system strongly influences the amounts of solid carbon, hydrogen, and CO. In Fig. 4.5C, the coke selectivity curve is displaced on the left-hand side, which signifies that coke can be efficiently limited by the addition of water. For hydrogen formation, the addition of water is favorable because of the gasification of the coke. The CO amounts are different from each other above 700°C, as explained by the equilibrium of the WGS. Finally, the molar ratio of H_2/CO is only slightly affected by the amount of water above 700°C, as highlighted by the insert in Fig. 4.5F.

Fig. 4.6 shows the thermodynamic equilibrium of mixtures containing 2 mol of CH_4, 1 mol of H_2O, and 0.5 to 2.0 mol of CO_2 at 1 bar. In this case, the initial amounts of CH_4 and H_2O are fixed while the amount of CO_2 varies. In Fig. 4.6A, increasing CO_2 helps to notably increase the CH_4 conversion below about 700°C, but this is much less important above this temperature. In Fig. 4.6B and D, the initial amount of CO_2 slightly influences the CO_2 conversion and the amount of H_2. However, the initial amount of CO_2 has a strong impact on the formation of coke (Fig. 4.6C) and CO (Fig. 4.6E) and the molar ratio of

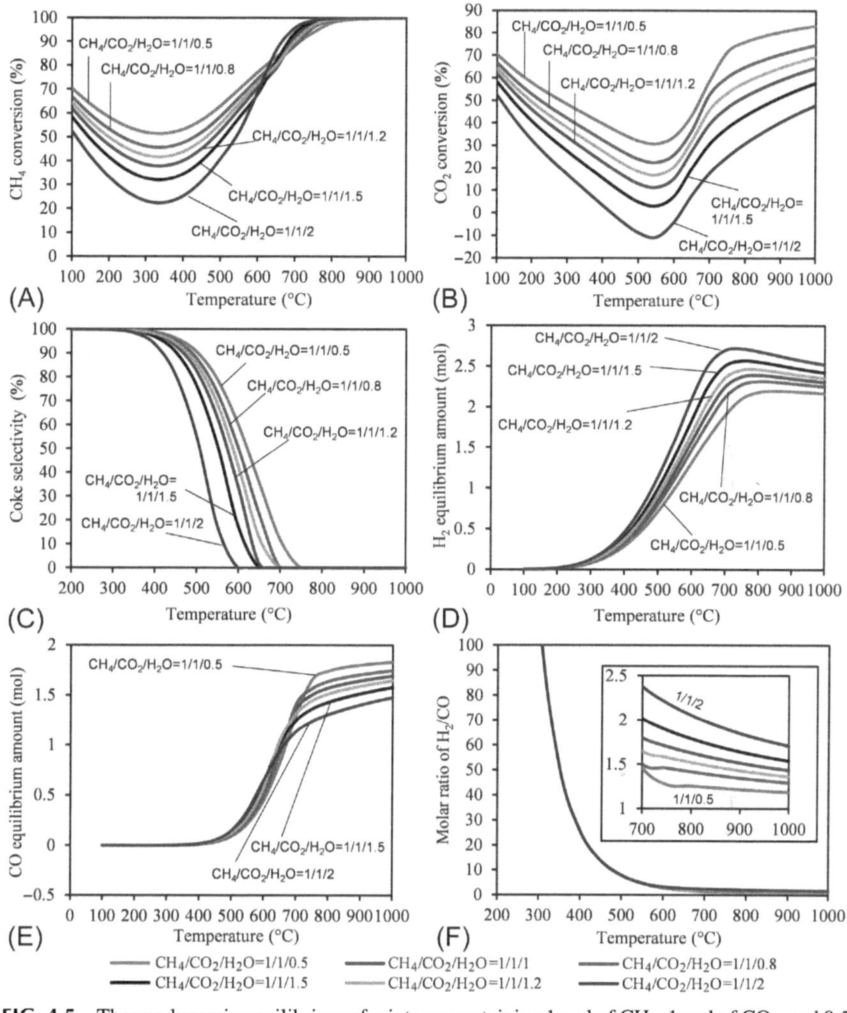

FIG. 4.5 Thermodynamic equilibrium of mixtures containing 1 mol of CH_4, 1 mol of CO_2, and 0.5 to 2.0 mol of H_2O at 1 bar: (A) CH_4 conversion, (B) CO_2 conversion, (C) coke selectivity, (D) H_2 amount, (E) CO amount, (F) molar ratio of H_2/CO.

H_2/CO (Fig. 4.6F). Above 500°C, coke selectivity is highly limited by an increasing amount of CO_2, and this is explained by the gasification of coke by CO_2 to form CO. This helps to explain also the positive effect of the initial amount of CO_2 on CO formation (Fig. 4.6E). Finally, the H_2/CO molar ratio decreases with the increase of the initial amount of CO_2 (Fig. 4.6F), explained by the fact that H_2 formation is less influenced, while CO formation is strongly affected, by the initial amount of CO_2.

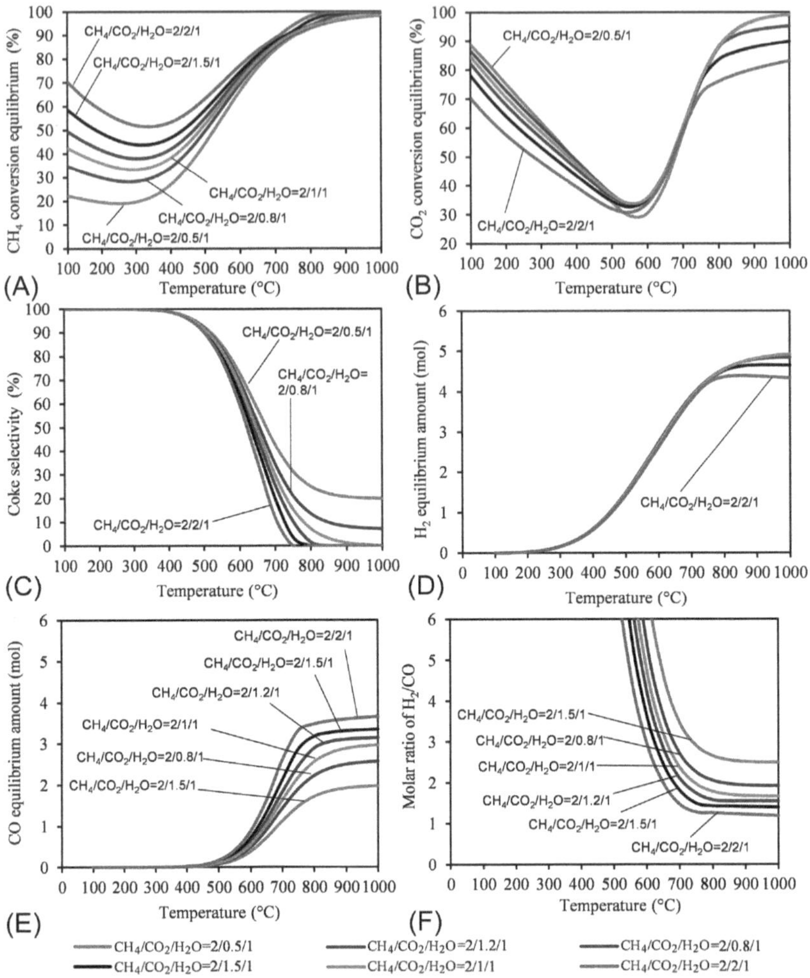

FIG. 4.6 Thermodynamic equilibrium of mixtures containing 2 mol of CH_4, 1 mol of H_2O, and 0.5 to 2.0 mol of CO_2 at 1 bar: (A) CH_4 conversion, (B) CO_2 conversion, (C) coke selectivity, (D) H_2 amount, (E) CO amount, (F) molar ratio of H_2/CO.

4.2.3 Steam Methane Reforming: The SMR Process

As previously mentioned, hydrogen is mostly produced from fossil resources, including natural gas. Using natural gas as feedstock, the hydrogen production process is composed of the following steps:

- Natural gas pretreatment: removal of H_2S and sulfur-containing compounds using a zinc oxide bed, which has high affinity for these pollutants.
- Generation of steam: production of steam from demineralization water. Natural gas can be used as fuel, together with waste heat recovered from the reforming reactor and waste combustible gases from H_2 separation.

- Steam reforming (Eq. 4.5 in Table 4.2): reaction of CH_4 with H_2O to give principally CO and H_2. Other byproducts are also formed, such as CO_2 and solid carbon. This is a catalytic process that generally needs a heterogeneous catalyst. As shown in Table 4.2, this reaction is highly endothermal ($\Delta_r H^\circ_{298} = +206\,kJ/mol$). The thermodynamic study in Section 4.2.1 shows that the reaction is favorable at high temperature (typically 500–900°C) and high steam-to-carbon (S/C) ratio (typically 2.5–3). According to Fig. 4.2A, a high temperature is needed to get high CH_4 conversion and high chemical kinetic, while a high S/C ratio allows limiting the catalyst deactivation by coke deposition. The reactor for the SMR process is composed of several vertical tubes filled with solid catalysts. These tubes are placed inside a furnace that is heated (using a gas burner for example) to the desired temperature for the SMR reaction (Linde, n.d.).
- WGS: Syngas from the SMR reactor is cooled down to around 300–500°C and then fed into the WGS reactors. Waste heat from this step is recovered and used for steam production. A WGS reactor is classified as either a high-temperature shift reactor (HTSR) or a low-temperature shift reactor (LTSR). Because of the exothermicity of the WGS (Eq. 4.3 in Table 4.2), a LTSR allows a high CO conversion to be obtained. However, its kinetic is slow. On the other hand, a HTSR allows reaching high kinetic, but low CO conversion. In general, the combination of both a HTSR and a LTSR leads to higher performance than using only one shift reactor (Chen et al., 2008).
- H_2 separation: Hydrogen of high purity can be separated from a H_2/CO_2 mixture left from a WGS reactor by using the PSA technique (Pressure Swing Adsorption). This technique involves the difference in adsorption capacity of H_2 compared to other molecules, as well as the fact that adsorption and desorption can be controlled by the increase or decrease of the process pressure.

Concerning the energy balance of hydrogen production from natural gas at an industrial scale, Table 4.3 shows the typical performance figures for a steam reforming–based hydrogen plant of 50,000 Nm^3/h, according to Linde—a world leader in the hydrogen production field (Linde, n.d.). A fuel consumption of 744.4 GJ/h for a production 50 kNm^3/h represents a specific consumption of 14.88 MJ/Nm^3 H_2 and as 1 Nm^3 H_2 contains 10.82 MJ/Nm^3, the hydrogen production efficiency is 72.7%; but 31 t/h extra steam are exported, so it is a specific steam production of 0.62 kg/Nm^3 H_2 with a heat content of 3.2 MJ/kg at 390°C and an extra 2.03 MJ/Nm^3 H_2, leading finally to an overall efficiency of 86.2%., including the steam credit. For comparison, hydrogen plants by Air Liquide Engineering & Construction have an energy consumption of 12.3 to 13.2 MJ/Nm^3 H_2, which is comparable to Linde plants (Air Liquide, n.d.).

The utilization of a reforming catalyst is mandatory for the transformation of methane and oxidants to syngas. Nickel-based catalysts appear as the best choice for both SMR and Dual-MR processes, taking into account their catalytic

TABLE 4.3 Typical Performance Figures of a Steam Reforming-Based Hydrogen Plant Using Natural Gas as Feedstock

	Flow Rate	**Nm^3/h**	**50,000**
	Pressure	***bar***	***25***
Product hydrogen	*Purity*	*mol.%*	*99.9*
Export steam	Flow rate	T/h	31
	Steam temperature	°C	390
	Steam pressure	bar	40
Feed and fuel consumption		GJ/h	744.4
Energy consumption (including steam credit)		GJ/ 1000 Nm^3 H_2	12.853
Utilities	Demineralized water	T/h	55.6
	Cooling water	T/h	160
	Electrical energy	kW	850
Design flexibility	Export steam production	T/ 1000 Nm^3 H_2	0.5–1.2
	Fuel consumption	GJ/ 1000 Nm^3 H_2	0.9–3.5

Adapted from Linde, n.d. http://www.linde-engineering.in/en/process_plants/hydrogen_and_synthesis_gas_plants/gas_products/hydrogen/index.html (Accessed on December 13, 2017).

performance and their low cost compared to noble metals, such as Ru, Pt, and Rh (LeValley et al., 2014; Baysal et al., 2017; Ferreira-Aparicio and Benito, 2005). Alumina is also the most appropriate support for these processes (Baysal et al., 2017; Ferreira-Aparicio and Benito, 2005). Alumina-supported nickel catalysts are generally well shaped for use under industrial conditions. The recent review on catalysts for the SMR process by Angeli et al., 2014 confirmed this observation again. The lifetime of a given alumina-supported nickel catalyst is required to be about 5 years under continuous operation (Liu et al., 2010). SMR catalysts can be supplied by different companies, such as Johnson Matthey, BASF, Haldo Topsoe, Thermo Fisher GMBH, and Süd-Chemie (Ferreira-Aparicio and Benito, 2005; Liu et al., 2010).

There are several producers of hydrogen via the SMR process at a large industrial scale: Linde, Haldo Topsoe, Howe-Baker, Foster Wheeler, Sanyo Electric, Osaka Gas Co, Air Liquide, Thyssenkrupp, and others (Linde, n.d.; Ferreira-Aparicio and Benito, 2005; Liu et al., 2010; Shah, 2017). As an example, Linde has built more than 200 plants for hydrogen production with capacities from 1000 to over 100,000 Nm^3/h (Linde, n.d.). Air Liquide Engineering & Construction provides solutions for hydrogen plants with capacities in the range of 10,000–200,000 Nm^3/h of hydrogen (Air Liquide, n.d.).

4.2.4 Steam Biogas Reforming: The SBR Process

Most of the features of the SMR process presented in the previous section are still valid for the steam biogas reforming (SBR) process. One of the most important differences is the capacity of the units: biogas production plants have a capacity ranging from 100 Nm^3/h for small agricultural or agro-food waste digesters to a few 1000 Nm^3/h for large municipal waste landfills; moreover, in some cases all of the biogas is not converted to hydrogen and other biogas valorization can coexist on the site. Thus, biogas steam reforming processes have to be designed for capacities ranging from 50 to 1000 Nm^3 H_2/h. The example presented in this section is relative to a 50 Nm^3 H_2/h production unit to be realized in 2018 by ALBHYON HERA's office under the VABHYOGAZ project on the site of TRIFYL's bioreactor, which produces about 1500 Nm^3/h of biogas, 40 km from the city of Albi (Occitanie region, France). This unit has a capacity about 1000 times smaller than the industrial SMR process presented in the previous section. Yet it produces more than 100 kg H_2/day, enough to travel 10,000 km/day on hydrogen. It is adapted to supply several tens of cars per day, refueling each with a few kg H_2, and this is the size of fuel cell car fleets for the next few years. As biogas sources are located in many places all over the territory, this underlines an important feature of biogas steam reforming: it concerns distributed hydrogen production.

A second difference with respect to large SMR processes is the integration of the SBR process on the site, which is radically different in that there is no need for steam on biogas production sites: in most cases, SBR has to be a standalone process and so efficiency optimization is more difficult. When another biogas valorization process coexists on the biogas production site, some interaction between the two processes can be implemented in order to improve efficiency or simplify the process and lower the investment cost; an example is shown here.

Then, of course, an important difference comes from the nature of the feedstock in that biogas contains more CO_2 than natural gas. The typical composition of the biogas feeding the SBR process at the TRIFYL bioreactor is given in Table 4.4.

Until now, the choice of designers for biogas reforming has been mainly to add a preliminary step in the process in order to extract the CO_2 from the

TABLE 4.4 Composition of the Biogas Feeding the Steam Biogas Reforming Process

Biogas Composition	Molar Fraction	Mass Fraction
CH_4	59.75%	35.05%
CO_2	40.0%	64.60%
N_2	0.20%	0.20%
O_2	0.04%	0.05%
H_2S	725 ppmv	900 ppm

mixture and then to reform the methane in a classical SMR. This preliminary step for CO_2 separation, also called enrichment of the biogas, can be done through PSA, or amines absorption, or membrane permeation. This case in which SBR is Enrichment + SMR has the disadvantage of increasing the complexity and the investment cost of the overall process. "Direct SBR" is considered here, without preliminary enrichment and with steam injected into the reforming reactor. Thus, it is a form of DMR, and is, in fact, not new as a commercial process for DMR, in that SPARG was developed 30 years ago by Haldo-Topsoe (Rostrup-Nielsen, 2002). Here the feeding ratio CH_4/CO_2 is 1/0.67, in the range explored in Section 4.2.2. Yet, for a given CH_4 flow rate feeding the process, the total volume flow rate is increased by a factor of 1.67 and the total mass flow rate is increased by a factor of 2.85 due to the presence of CO_2. Thus, the aerodynamic, thermal, and chemical behaviors of the mixture are considerably modified in the equipment used for the SBR process as compared to the SMR process. The process developed under the VABHYOGAZ project has been designed so that, through its control system, it can deal with these different behaviors and manage mixtures with CH_4/CO_2 ratios ranging from 1/0 to 1/1. The nature of the CO_2 must be emphasized: SMR, when applied to natural gas, produces large amounts of CO_2 with a fossil origin. On the other hand, SBR also produces CO_2, but this CO_2 comes from the biodegradation of green waste or of agro-food waste; it has no fossil origin as it was previously captured by plants from the atmosphere. Thus, the SBR process does not add any fossil CO_2 or greenhouse gas to the atmosphere and produces renewable hydrogen.

The successive steps of the SBR process are similar to those of the SMR process, with the necessary adaptations to accommodate the different features, as explained above. These steps are sketched in Fig. 4.7 and presented here, together the main operating parameters:

- Biogas cleaning: H_2S and other sulfur compounds are removed by adsorption on an active carbon-fixed bed.
- Compression of the biogas: the PSA requires pressure to be efficient and it costs less energy to compress the biogas and liquid water before reforming

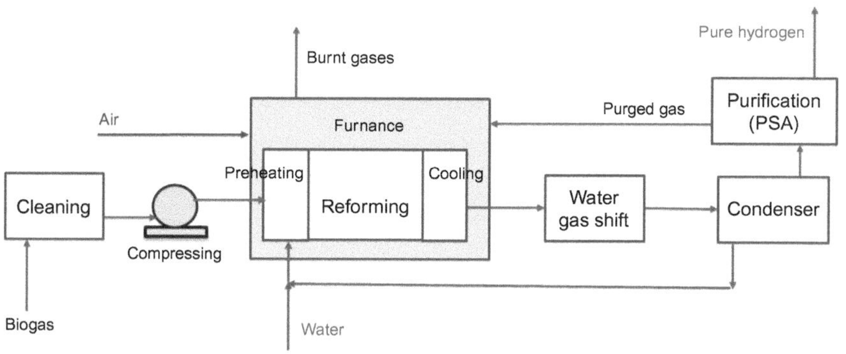

FIG. 4.7 Block diagram of the SBR process in the VABHYOGAZ project.

than compressing the gas mixture after reforming. The pressure level here is 15 bar. As seen in the previous section, increasing the pressure has a negative effect on CH_4 conversion, but at this level, the residual CH_4 is low.

- Heating of biogas, vaporization and overheating of water: this is mainly realized in a specific zone inside the reforming tubes through heat recovery from the reformate gas before it leaves the tubes and through exchange with the combustion gases.
- SBR (Eq. 4.5) on a catalytic fixed bed inside the reforming tubes. The heat for this highly endothermal reaction ($\Delta_r H^\circ_{298} = +206$ kJ/mol) is provided through exchange with the combustion gases inside the furnace where the tubes are placed. The reaction temperature is 860°C and the S/C ratio is 3; for a production of 50 Nm^3 H_2/h, the reformed biogas flow rate is 31.7 Nm^3/h and the water flow rate is 45.6 kg/h. With these operating parameters, a chemical equilibrium calculation shows no carbon formation and a 100% CH_4 conversion. Nevertheless, as indicated in Table 4.5, some residual CH_4 is assumed, corresponding to an imperfect catalytic effect.
- Cooling of the reformate gas mixture down to 300°C before leaving the tubes.
- WGS catalytic reactor to convert CO and water into H_2 and CO_2 (Eq. 4.3). The heat from this exothermal reaction ($\Delta_r H^\circ_{298} = -41$ kJ/mol) is valorized for preheating and partial vaporization of water. WGS is conducted in 2 steps: first, HTSR around 300°C, and then LTSR around 200°C. The CO level is decreased to 1.5%.
- Cooling the reformate mixture down to 10°C through exchange with cold water. Most of the water contained in the reformate mixture condenses and can be recycled as reforming water.
- H_2 separation using a PSA. The purity achieved for hydrogen is 99.995%. At 15 bar, the PSA efficiency is 79%, which means that 21% of the produced hydrogen leaves the PSA with the purged gases, which also contain CO_2 and residual CO and CH_4.
- The pure hydrogen can then be compressed and stored at high pressure.

TABLE 4.5 Gas Composition in the 50 Nm^3 H_2/h SBR Process

Volume Fractions	Reformed Biogas	Reforming Water	Reformer Output	WGS Output	Condenser Output	PSA Output	PSA Purged Gas	Burnt Gases
H_2	0		40.2%	52.3%	66.6%	>99.995%	29.5%	
H_2O	0	100%	33.6%	21.5%	0.1%		0.2%	16%
N_2	0,2%		0.1%	0.1%	0.1%		0.1%	56.5%
CO	0		13.6%	1.5%	1.9%	<1 ppm	4.0%	
CO_2	40%		10.3%	22.5%	28.6%		60.3%	24.5%
CH_4	59.7%		2.2%	2.2%	2.8%		5.9%	
Other	O_2, H_2S							O_2 3%
Flow rate (kg/h)	38.5	45.6	84.1	84.1	63.3	4.47	58.3	195.8

- Heat production in the furnace surrounding the reforming tubes comes from the combustion of the purged gases. The energy needed is 72.3 kW to balance the SMR endothermicity and to complete the vaporizing and overheating of water and biogas. The energy available in the purged gases is 74 kW and taking into account the energy losses in the exhaust gases, a complement must be provided by some biogas injected into the furnace; 3.7 Nm^3/h of biogas are added to supply 22 kW. It can be noticed here that as long as the energy contained in the purged gas is lower than the energy to be burnt in the furnace, there is no need to improve the PSA efficiency, nor the CH_4 conversion in the reforming tube, nor the CO conversion in the WGS reactor as any decrease in the H_2, CH_4 or CO content of the purged gas should be balanced by an extra injection of biogas in the furnace.

A better way to valorize the purged gases may be considered when another valorization process coexists on site (electricity production in gas engines for example), especially when the flow of reformed biogas is small with respect to the total available biogas flow, as in the present case. Then, the purged gases can be returned and mixed with the flow of biogas feeding the other process for electricity production. Heat production in the furnace is then realized with biogas only: 14 Nm^3/h of biogas supply 84 kW to the furnace. With this integration, thanks to the interaction between the two processes, the global efficiency, including the purged gas credit, is increased from 71% to 82%, as presented in Table 4.6.

It must be pointed out that the SBR process, similar to SMR process, is an energy-intensive process, and high-temperature thermal integration is the key to reach good efficiency. For a 150 kW hydrogen output, the reforming reaction heat (206 kJ/mol CH_4) is 48.3 kW (32.2% of the hydrogen energy content) and this heat

TABLE 4.6 Energy Efficiency of a 50 Nm^3H_2/h SBR Process With Two Options, a Standalone Process or an Integrated Process With Purged Gas Credit

Parameter	Stand-Alone Process	Integrated Process
Reformed biogas flow rate (Nm^3/h)	31.7	31.7
Combustion biogas flow rate (Nm^3/h)	3.7	14
Purged gas energy returned to the main biogas flow (kW)	0	74
Energy in the produced hydrogen (kW)	150	150
Hydrogen energy efficiency (% LHV)	71.3	55.7
Energy efficiency with purged gas credit (% LHV)	71.3	82.4

TABLE 4.7 Energy Repartition of the SBR Process With 150 kW Hydrogen Output

Energy needs	Energy sources
Reaction heat: 48.3 kW (32.2%)	Combustion of purges gas and extra biogas 72.3 kW
Preheating of the reactants (water and biogas): 71.3 kW (47.5%)	Cooling of reformate mixture 33.5 kW
	Cooling WGS reactor: 13.8 kW
Total : 119.6 kW (79.7% of hydrogen energy content)	

has to be brought through an external exchange at very high temperature to reach the 860°C reacting temperature. Together with preheating, vaporizing, and overheating of water and biogas, it is 119.6 kW (79.7% of the hydrogen energy content) to be brought to the reactants, among which 13.8 kW come from the WGS reactors (11.6%), 33.5 kW come from the cooling of the reformate mixture (28%), and the remaining 72.3 kW come from the furnace (60.4%) (Table 4.7).

For this small-capacity unit, the electric consumption of the auxiliaries is estimated to be 12.6 kW and the main parts are for biogas compression (5.7 kW) and for refrigeration (3.6 kW). This can be compared to the 750 kW for the 1000 times larger SMR process, in which natural gas compression is not included, as natural gas is delivered at a sufficient pressure from the pipe. It appears the small-scale effect is sensible: excluding compression, auxiliary consumption is divided by 100 (from 750 to 7.5 kW) when capacity is divided by 1000 (from 50,000 to 50 Nm^3 H_2/h).

4.2.5 Kinetic of Steam Methane Reforming

Kinetic data on the SMR process are rich after several decades of research and development. Because the process is complex, with numerous intermediate steps, and because of the nature of kinetic parameters, which strongly depend on experimental conditions, it is difficult to achieve a general agreement on kinetic models. Discrepancies, even contradictions, exist as reported in the literature (Liu et al., 2010; Nguyen et al., 2014). The most common kinetic laws considered for the SMR process are: Langmuir-Hinshelwood, Temkin, Eley-Rideal, power law, and microkinetic analysis (Soria et al., 2012; Temkin, 1979). Table 4.8 summarizes some selected models and results reported in the literature.

TABLE 4.8 Selected Kinetic Models for SMR

Catalyst	Condition	Kinetic Model and Parameters	Equations	Reference
$Ni/MgAl_2O_4$	773–848 K 3–15 bar	$r_1 = \frac{\frac{k_1}{p_{H_2}^{2.5}}\left(p_{CH_4}p_{H_2O} - \frac{p_{H_2}^3 p_{CO}}{K_1}\right)}{(DEN)^2}$ $DEN = \frac{1 + K_{CO}p_{CO} + K_{H_2}p_{H_2} + K_{CH_4}p_{CH_4} + K_{H_2O}p_{H_2O}}{p_{H_2}}$ $k_1 = 0.2088$, 0.5254 and 2.069 kmol $bar^{0.5}\ kg_{cat}^{-1}\ h^{-1}$ at 773, 798 and 823 K, respectively $A\ (k_1) = 4.225E+15$ $E_1 = 240.1$ kJ/mol	(4.12)	Xu and Froment (1989)
NiO/Al_2O_3	573–973 K 1 bar	$R_1 = \frac{k_1}{p_{H_2}^{2.5}}\left(p_{CH_4}p_{H_2O} - \frac{p_{H_2}^3 p_{CO}}{K_I}\right)\left(\frac{1}{\Omega^2}\right)$ $\Omega = 1 + K_{CO}p_{CO} + K_{H_2}p_{H_2} + K_{CH_4}p_{CH_4} + K_{H_2O}\frac{p_{H_2O}}{p_{H_2}}$ $k_{o,1} = 5.19E+9$ mol $bar^{0.5}\ g^{-1}\ s^{-1}$ $E_1 = 257.01$ kJ/mol	(4.12)	Abbas et al. (2017)
Ni/YSZ	773–1013 K 1 bar	$r_{r,\ Arr} = k \cdot p_{CH_4}^{\alpha} \cdot p_{H_2O}^{\beta}$ $k = 40 \pm 1\ s^{-1}$ at 773 K and $393 \pm 48\ s^{-1}$ at 1013 K when S/C = 2 $E_a = 62 \pm 1$ kJ/mol	(4.13)	Nguyen et al. (2016)
Ru/SiO_2	723–823 K 1 bar	$r_1 = \frac{k_1}{p^{2.5}}\theta^2 p_{CH_4}p_{H_2O}(1-\beta_1)$ $k_1 = 4.6$ mol $kg_{cat}^{-1} 1 s^{-1} Pa^{0.5}$ at 773 K; $E_a = 140$ kJ/mol	(4.14)	Soria et al. (2012)

Continued

TABLE 4.8 Selected Kinetic Models for SMR—cont'd

Catalyst	Condition	Kinetic Model and Parameters	Equations	Reference
Ru/Al_2O_3	823–1023 K 1–5 bar	$r_f = k \times P_{CH_4}$ $E_a = 91$ kJ/mol	(4.15)	Wei and Iglesia (2004a, b)
Rhodium- and Nickel-based catalysts	723–1023 K 1 bar	$r_f = k \times P_{CH_4}$ $E_a = 69.1$ kJ/mol and $k_{0a} = 2.88E+6\ s^{-1}$ for Rh catalyst $E_a = 96.1$ kJ/mol and $k_{0a} = 7.26E+7\ s^{-1}$ for Ni catalyst	(4.15)	Zeppieri et al. (2010)

The model proposed by Xu and Froment (1989) is based on Langmuir-Hinshelwood methodology and is usually considered as the most widely used model for the SMR process. In this model, various chemical reactions were considered in order to explain the formation of different products, including H_2, CO, CO_2, H_2O, and solid carbon (C). The following assumptions were made to build reaction schemes of the SMR process (Xu and Froment, 1989):

- Water reacts with surface nickel atoms, yielding adsorbed oxygen and gaseous hydrogen.
- Methane is adsorbed on surface nickel atoms. The adsorbed methane either reacts with the adsorbed oxygen or is dissociated to form chemisorbed radicals.
- The concentrations of the carbon-containing radicals are much lower than the total concentration of the active sites.
- The adsorbed oxygen and the carbon-containing radicals react to form oxygen-containing radicals.
- The hydrogen formed is directly released into the gas phase and/or the gaseous hydrogen is in equilibrium with adsorbed atomic or molecular hydrogen.
- All reaction schemes have a rate-determining step among the following reactions: SMR, WGS, and the direct formation of CO_2 and H_2 from CH_4 and H_2O (reactions 4.3, 4.5, 4.11 in Table 4.2).

After the work reported by Xu and Froment (1989), other research confirmed the validity of this model for the SMR process, such as the recent work of Abbas et al. (2017).

Nguyen et al. (2016) studied the kinetic of the SMR process by expressing the reaction rate as a function of the exponential of methane and steam pressure (power law). They highlighted also that the reaction order (for methane and steam) varies significantly, as reported by several studies. The value of α (reaction order with respect to partial pressure of methane) was typically found in the range of 0.85–1.4, while the value of β (reaction order with respect to partial pressure of steam) can be negative, zero, or positive depending strongly on the ratio of steam/carbon (S/C). Small S/C ratios yield positive β, S/C ratios on the order of 2 yield β close to zero, and high S/C ratios yield negative β. In their study, Nguyen et al. (2016) found the value of α and β to be equal to 2 and 1, respectively, for S/C ratios of 2–3, and an activation energy, E_a, equal to 62 ± 1 kJ/mol.

As previously assumed by Xu and Froment (1989) and other researchers and based on a thermodynamic analysis of the SMR process under the studied conditions, three reactions build the kinetic mechanism of the process, which are the reactions (4.3), (4.5), (4.11) in Table 4.2. Thus, methane is dissociated on nickel sites. In the study of Soria et al. (2012), the SMR process was carried out at 723–823 K and 1 bar over a Ru/SiO_2 catalyst. Their kinetic models were built by supposing that both methane and steam react as adsorbed species. Thus, steam can be adsorbed on the catalyst surface with or without dissociation, while methane presents more possibilities for dissociation, forming C or the intermediate specie CH_2. The combination of these possibilities yields a total of six kinetic mechanisms. Among these, the best kinetic model was found by fitting with the experimental data, and the reaction rate equation is shown in Table 4.8 (Eq. 4.14). For this kinetic model, the following assumptions were made:

- Steam is dissociated into gaseous H_2 and adsorbed O when it is adsorbed on the catalyst surface.
- Methane is adsorbed to form the intermediate $CH_{2(ads)}$, which later reacts with oxygen to form $CHO_{(ads)}$.
- $CHO_{(ads)}$ species react in parallel to form CO and CO_2. These surface reactions that produce CO and CO_2 are considered to be the rate controlling steps.
- Also, WGS takes place.
- Desorption of adsorbed CO, CO_2 species, and a combination of adsorbed H species to form gaseous CO, CO_2 and H_2.

The activation energy was found to be 140kJ/mol for the SMR reaction over a Ru/SiO_2 catalyst (Soria et al., 2012).

In their earlier study, Wei and Iglesia (2004a, b) experimentally studied the SMR process using a Ru/Al_2O_3 catalyst at 823–1023K and 1–5bar. They found that the reaction rate depends only on the partial pressure of methane and not on the partial pressure of steam. Thus, the global kinetic equation is simple, as shown in Eq. (4.15) in Table 4.8. They also obtained a similar result for the CO_2 reforming of methane in this study. This kinetic mechanism was later confirmed by Zeppieri et al. (2010), using rhodium- and nickel-based catalysts. Note that in these two papers, both research groups used a very small quantity of catalyst. Wei and Iglesia (2004a, b) used 5mg of catalyst (250–425μm) diluted in 500mg of acid-washed quartz powder (250–425μm) to fill a quartz or steel tube (8mm inner diameter). Zeppieri et al. (2010) used 5–30mg of catalyst samples diluted with 500–1500mg of acid-washed silicon dioxide to fill a quartz tube reactor of 8mm inner diameter. The absence of preferential passage of reactants through the catalytic bed should be confirmed, considering the small quantities of catalysts versus the reactor inner diameter of 8mm.

As a partial conclusion, there have been several kinetic studies of the SMR process. Because the process is complex, with the implication of different chemical equilibriums, different kinetic models have been built and discussed. The validation of a given model is generally done by fitting with the experimental data. Thus, it is only meaningful to retain a kinetic mechanism taking into account the experimental conditions. More effort is still needed to determine the intrinsic kinetic of the SMR process.

4.3 DRY REFORMING OF METHANE

4.3.1 Thermodynamic Equilibrium Aspect

DRM relates to the reaction between CH_4 and CO_2 to form CO and H_2 (Eq. 4.1 in Table 4.2). According to Eq. (4.1), 1 mol of CO_2 is needed to reform 1 mol of CH_4. Fig. 4.8 shows the thermodynamic equilibrium of an equimolar mixture of CH_4 and CO_2 at atmospheric pressure. Only the main species are presented. Methane conversion increases above 350°C and reaches more than 98% at 900°C. Solid carbon and water are highly selected below 600°C. Above this temperature, CO and H_2 become the main products. Taking into account the conversion of CH_4 and CO_2, and the selectivity of the reaction, this process should be carried out at 850°C or above.

Fig. 4.9 shows the influence of the molar ratio of CO_2/CH_4 on the thermodynamic equilibrium of a CO_2 and CH_4 mixture. Increasing this ratio increases

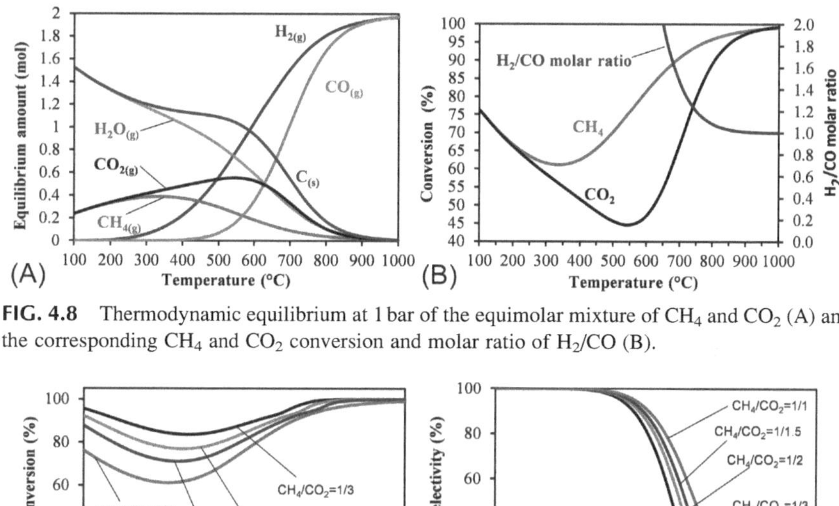

FIG. 4.8 Thermodynamic equilibrium at 1 bar of the equimolar mixture of CH_4 and CO_2 (A) and the corresponding CH_4 and CO_2 conversion and molar ratio of H_2/CO (B).

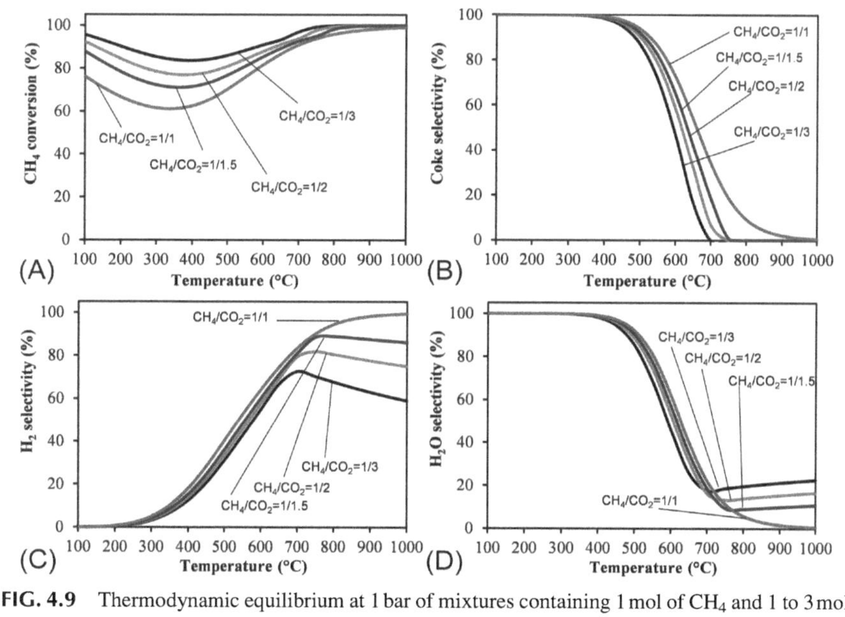

FIG. 4.9 Thermodynamic equilibrium at 1 bar of mixtures containing 1 mol of CH_4 and 1 to 3 mol of CO_2: (A) CH_4 conversion, (B) coke selectivity, (C) H_2 selectivity, and (D) H_2O selectivity.

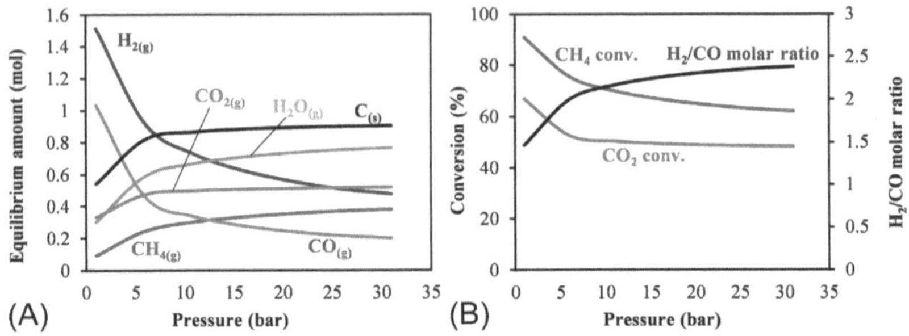

FIG. 4.10 Influence of the reaction pressure on the equilibrium of an equimolar mixture of CH_4 and CO_2 at 700°C.

methane conversion at a given temperature and decreases coke selectivity. At a ratio of CO_2/CH_4 equal to 3, coke can be avoided above 700°C. However, adding more CO_2 influences also WGS (Eq. 4.3 in Table 4.2). Consequently, the amount of H_2 decreases and the amount of H_2O increases. Thus, the choice of CO_2/CH_4 molar ratio must integrate all the factors, i.e., reaction conversion, reaction selectivity, and energy consumption, in order to optimize the DRM process, in particular for H_2 production.

High pressure is generally not favorable for the DRM process. Fig. 4.10 presents the influence of the reaction pressure on the thermodynamic equilibrium of an equimolar mixture of CH_4 and CO_2 at 700°C. Methane conversion decreases quickly from 91 to 62% when the pressure increases from 1 to 30 bar. In parallel, the amounts of CO and H_2 drastically decrease while the amounts of H_2O and solid carbon strongly increase with the increase of the reaction pressure. Thus, DRM at high pressure is not recommended.

4.3.2 Catalysts for Methane Dry Reforming

Because DRM is an extremely endothermic reaction, one of the main challenges in the heterogeneous catalytic DRM process is the considerable loss of catalytic activity with time-on-stream (TOS) due to the formation of carbonaceous species on the catalyst surface from CH_4 cracking and metal sintering (Argyle and Bartholomew, 2015; Usman et al., 2015). Thus, numerous studies have been conducted to seek for ideal and cost-effective solutions to overcome the limitations in catalytic stability during DRM. In general, the employment of noble metal catalysts (namely, Pt, Ru, and Rh) could solve the above-mentioned problems with high catalytic activity and coke resistance (Jones et al., 2008; Pakhare and Spivey, 2014). However, the high cost and limited availability of noble metals have limited their use in industrial applications. For these reasons, catalyst development has tended to shift toward the design of metal catalysts other than noble metals, such as Ni- and Co-based catalysts, which are globally available and more economical. In order to achieve catalyst activity and stability comparable to those of precious metals, tuning catalytic properties by modification of supports and addition of promoters has become a main focus area in many studies (Wang et al., 2017; Qian et al., 2014; Zarei et al., 2016; Djinović and Pintar, 2017). The recent efforts in designing promoted and supported nonnoble metal catalysts for DRM are comprehensively discussed in this section.

4.3.2.1 Catalyst Supports

The active metals (including noble metals, Ni, and Co) for DRM are normally dispersed on various single-metal oxides possessing basic (CeO_2 and La_2O_3 according to Barroso-Quiroga and Castro-Luna, 2010), acidic (SiO_2, amphoteric γ-Al_2O_3 (Barroso-Quiroga and Castro-Luna, 2010; Selvarajah et al., 2016), and ZrO_2 (Rong-jun et al., 2015; Lou et al., 2017)) attributes. Metal oxide supports play an important role in catalytic DRM performance. Apart from providing mechanical strength, they increase specific surface area of the catalyst and act as a host for facilitating metal dispersion, thus increasing active metal dispersion (Wang et al., 2018; Abdullah et al., 2017).

Barroso-Quiroga and Castro-Luna (2010) investigated the effect of different ceramic oxide supports, such as Al_2O_3, CeO_2, La_2O_3, and ZrO_2, on the catalytic activity of Ni metal using the incipient wetness impregnation technique for catalytic synthesis. Among supported Ni catalysts, 10% Ni/CeO_2 exhibited the highest CH_4 conversion followed by Ni/ZrO_2 catalyst. However, Ni/ZrO_2 catalyst was the best catalyst in terms of catalytic stability, with negligible deactivation observed. Lou et al. (2017) also designed a stable ZrO_2-supported Ni nanocatalyst system with metal loading of 1–5 wt% using air-stable Ni colloids. Ni/ZrO_2 catalyst having a 1.1 nm particle dimension reportedly exhibited excellent catalytic stability with TOS and maintained up to 90% initial activity beyond 60 h on-stream. The location of nearly all of the Ni atoms at the interface and perimeter of the ZrO_2 support could result in easier access to the oxygen generated from activated CO_2 at the Ni-ZrO_2 interface. Thus, the labile oxygen could facilitate the gasification of surface carbon to CO and prevent Ni/ZrO_2 catalyst deactivation. The interaction between NiO and various semiconductor metal oxide supports, including SiO_2, Al_2O_3, MgO, TiO_2, and ZrO_2, was the focus in the DRM study by Rong-jun et al. (2015). SiO_2, TiO_2, and ZrO_2 weakly interact with NiO and hence ease NiO reduction. However, the weak metal support interaction induced quick catalytic deactivation owing to metal agglomeration. TiO_2-supported Ni catalyst had the lowest reactant conversions of $<5\%$ during 20 h on-stream because of low surface area and weak interaction with NiO, whereas the basic nature of the MgO support attracted acidic CO_2 adsorption, which in turn gasified surface carbon. Thus, a gradual decrease in the activity of the Ni/MgO catalyst was observed within 50 h and then the catalyst appeared to be stable for 100 h on-stream (Rong-jun et al., 2015).

Ordered mesoporous silica (such as SBA-16 (Zhang et al., 2013), MCM-41 (Xie et al., 2015), and SBA-15 (Omoregbe et al., 2017)) has attracted significant attention from academics and appears to be a promising support because its large specific surface area, thick framework walls, high thermal stability, and well-ordered hexagonal structure with a narrow pore size distribution yields a supported catalyst with a uniform particle size and high metal dispersion (Zhang et al., 2013; Xie et al., 2015; Omoregbe et al., 2017). Wang et al. (2013) compared different types of silica-based supports on the DRM performance of $LaNiO_3$ perovskite catalysts, including SiO_2, mesoporous SBA-15, and MCM-41 materials. Supported catalysts showed higher catalytic activity and stability than those of bare $LaNiO_3$ catalyst due to a strong metal-support interaction and greater metal dispersion. The initial CH_4 conversion decreased in the following order: $LaNiO_3$/MCM-41 > $LaNiO_3$/SBA-15 > $LaNiO_3/SiO_2$ > $LaNiO_3$ catalysts. However, the longevity test revealed that the SBA-15 supported $LaNiO_3$ catalyst was free from graphitic carbon and possessed the highest catalytic stability within 60h at 973K owing to the greater anchoring effect of the SBA-15 mesostructure, resulting in well-dispersed active Ni species. Additionally, the mesoporous structure SBA-15 support was largely undamaged after 60h on-stream, while structural collapse was observed for the MCM-41 support. Cai et al. (2014) also reported that the anchoring effect of mesoporous silica not only contributed to great Ni particle dispersion, inducing remarkable catalytic activity, but also significantly stabilized metallic Ni nanoparticles inside mesoporous channels and prevented them from agglomerating.

Owing to the outstanding oxygen storage capacities and distinctive redox properties of rare earth metal oxides, namely, CeO_2 and La_2O_3, these metal oxides have been utilized as supports for reducing carbon deposition (Barroso-Quiroga and Castro-Luna, 2010; Ayodele et al., 2016). Ay and Üner (2015) synthesized Ce-supported Ni, monometallic Co, and bimetallic Ni-Co catalysts using the incipient wetness impregnation approach for DRM evaluation. Both Ni/CeO_2 and $Ni\text{-}Co/CeO_2$ catalysts had excellent interaction between the Ni metal and the CeO_2 support as evidenced in HRTEM analysis, and hence they had higher catalytic activities compared to the Co/CeO_2 catalyst. In the assessment of DRM over Ni/La_2O_3 catalyst, Li et al. (2017) found that Ni/La_2O_3 catalyst exhibited outstanding catalytic stability with comparable CH_4 and CO_2 conversions of about 75% at 973K and atmospheric pressure for 50h on-stream. This noteworthy observation was attributed to improved Ni dispersion due to strong metal-support interaction, which thus inhibited Ni sintering and suppressed carbon deposition on the catalyst surface. Additionally, they reported that a complete transformation cycle between the La_2O_3 and $La_2O_2CO_3$ phases that occurred during DRM was responsible for deposited carbon removal. In particular, CO_2 could adsorb on La_2O_3 form to generate the intermediate $La_2O_2CO_3$ phase, which in turn reacted with neighboring active surface carbon species to produce CO product and La_2O_3 form.

Besides single metal oxide supports, mixed metal oxides, such as CeO_2-Al_2O_3, La_2O_3-Al_2O_3 (Charisiou et al., 2016), and MgO-Al_2O_3 (Zhang et al., 2016), have been recently utilized as a support in DRM catalysts in order to receive synergistic benefits from these compounds. In a study of the DRM reaction over Ni supported on mixed metal oxides between Al_2O_3 and CeO_2 or La_2O_3, Charisiou et al. (2016) noticed that Ni/CeAl, Ni/LaAl, and Ni/CeLaAl catalysts appreciably improved reactant conversions compared with an unmodified Ni/Al catalyst owing to the high oxygen storage capacity of CeO_2 and La_2O_3 alleviating surface carbon oxidation. Additionally, a longevity test revealed that modified alumina catalysts could constantly retain their catalytic activity over 20 h on-stream while the unmodified Ni/Al catalyst experienced catalytic deterioration within 5 h.

Zhang et al. (2016) examined a series of Ni/MgO-Al_2O_3 catalysts (with varying Al_2O_3 content from 0 to 45 wt%) synthesized by a two-step hydrothermal approach. NiO reportedly dissolved into the MgO support to form a solid solution, which was difficult to reduce to the active Ni° metallic phase. They found that the addition of Al_2O_3 to the MgO support could weaken the formation of Mg-Ni-O solid solution and hence alleviate the H_2 reaction. Nevertheless, the generation of a more spinel-like $NiAl_2O_4$ form at high Al_2O_3 loading beyond 15 wt% required a greater reduction temperature than for Ni/MgO catalyst. Additionally, basic site concentration improved with increasing Al_2O_3 content in the support mixture and attained an optimum at 15 wt%, which was also the best Al_2O_3 loading for the DRM reaction in terms of CH_4 conversion (52%) and stability due to the formation of a stable and basic $MgAl_2O_4$ phase instead of the undesirable $MgNiO_2$ and $NiAl_2O_4$ forms. In the study of Ni/$MgAl_2O_4$ spinel catalysts with various Ni contents from 2.5 to 10% for DRM, Habibi et al. (2016) also found that 5% Ni/$MgAl_2O_4$ exhibited the highest catalytic activity because of high Ni dispersion on the mesoporous $MgAl_2O_4$ support, large BET surface area, low-temperature reducibility, and the strong interaction between the Ni and $MgAl_2O_4$ spinel support.

4.3.2.2 Promoters

Apart from adjusting operational conditions or modifying the structure or composition of catalyst supports, adding small amounts of promoters is an effective way to suppress catalyst deactivation from carbon deposition and metal sintering. Although precious metals exhibit excellent catalytic performance and stability, these metals are preferably employed as promoters due to their high cost and low availability (Pakhare and Spivey, 2014; Abdullah et al., 2017). In a comparative study of noble metal effect on catalytic activity and carbon deposition for DRM, Arandiyan et al. (2014) prepared various precious metal–doped LaAlNi catalysts (including Pt, Pd, Ru, Rh, and Ir) via the sol-gel method. The addition of noble metal promoters eased the H_2 reduction of cationic nickel owing to the H_2-spillover effect in agreement with the findings from Jabbour et al. (2014). Arandiyan et al. (2014) reported the order of CH_4 conversion as: $La_{0.4}Rh_{0.6}Al_{0.2}Ni_{0.8}O_3 > La_{0.4}Ru_{0.6}Al_{0.2}Ni_{0.8}O_3 > LaAl_{0.2}Ni_{0.8}O_3 > La_{0.4}Ir_{0.6}Al_{0.2}Ni_{0.8}O_3 \geq La_{0.4}Pt_{0.6}Al_{0.2}Ni_{0.8}O_3 > La_{0.4}Pd_{0.6}Al_{0.2}Ni_{0.8}O_3$. The enhancement of the catalytic performance and stability was ascribed to the low reduction temperature and the high content of oxygen adspecies on noble metal promoters accelerating surface carbon gasification.

Mahoney et al. (2014) examined the promotional effect of Pt (0.2 wt%) on 15% Ni/Al_2O_3 and 15% $Ni/CeZrO_2$ catalysts in a fixed-bed reactor at a stoichiometric CH_4:CO_2 ratio = 1:1 and 1073 K. The addition of Pt did not significantly affect CH_4 and CO_2 conversions of 15% Ni/Al_2O_3 catalyst. However, the superior catalytic activity and stability of 0.2% Pt–15% $Ni/CeZrO_2$ catalyst were attributed to the facilitated oxygen spillover from Pt atoms to the oxygen vacancies in the ceria-zirconia support. Based on density functional theory (DFT) calculations, Gould et al. (2015) found that Pt-promoted Ni catalyst possessed a great resistance to carbon because the presence of Pt atoms weakened the carbon adsorption energy at step edges and increased the carbon diffusion barrier on terraces.

Besides noble metals, rare earth metal oxides (such as CeO_2 and La_2O_3) have been widely employed as promoters for Ni-based catalysts due to their basic property and high oxygen storage capacity, which hinders carbonaceous deposition on the catalyst surface and enhances catalytic activity (Usman et al., 2015; Omoregbe et al., 2016). Shamskar et al. (2017) scrutinized the influence of CeO_2 and La_2O_3 dopants on 25% Ni/Al_2O_3 catalyst prepared by the ultrasonic-assisted coprecipitation approach. Temperature-programmed oxidation analysis shows that the addition of promoter reduced the amount of accumulated carbon and the 5% CeO_2–25% Ni/Al_2O_3 catalyst was stable during 15 h on-stream. The promotional effect of a series of rare-earth elements, namely, Sc, Y, Ce, and Pr on NiMgAl catalysts produced from layered double hydroxides for the DRM reaction was also investigated by Cao et al. (2016). The addition of promoter increased the CH_4 conversion in the order: Ce-promoted > Pr-promoted > Y-promoted > Sc-promoted > unpromoted catalysts. The amount of carbon deposition decreased from 47.22% (unpromoted catalyst) to 33.76% (Ce-doped catalyst). The increased reactant conversion and coke resistance with the addition of a rare-earth promoter were attributed to enhanced surface basicity, the great redox properties, and well-dispersed Ni particles on the catalyst surface. Notably, Ce- and Pr-modified catalysts had higher catalytic activity and carbon resilience compared with other promoted and unpromoted catalysts owing to their fast redox cycling that involved the reversible formation of Ce^{3+}/Ce^{4+} and Pr^{3+}/Pr^{4+} redox pairs.

Sarkar et al. (2016) investigated the addition of 0.2 wt% Gd dopant to 5% Ni/ZSM-5 catalyst and found that Gd promotion improved the oxygen atom supply to the adjacent Ni sites and hence oxidized accumulated coke on the catalyst surface. As a result, Gd-doped Ni/ZSM-5 catalyst exhibited a high CH_4 conversion of 83% without noticeable carbon deposition and sintering at 1023 K after 100 h on-stream. Al-Fatesh (2017) also examined the influence of Gd loading from 1 to 5 wt% Gd on the catalytic performance of 10% Ni/Y_2O_3 catalyst for the DRM reaction. Both CH_4 and CO_2 conversions reportedly increased with increasing Gd loading from 1 to 3 wt% owing to the synergetic effect between

Gd and Ni metals. Metal dispersion was also improved with Gd addition, and the optimal value was obtained at 3 wt% Gd because Gd metal incorporated itself into the pores of the catalyst and inhibited Ni agglomeration. Thus, 3% Gd–10% Ni/Y_2O_3 catalyst appeared to be the best catalyst in terms of reactant conversions, H_2 yield, and carbon resistance. Boron promoter has recently gained considerable attention for coke resistance in the DRM reaction because, according to DFT calculations from Xu and Saeys (2007), boron atoms effectively suppress surface carbon diffusion into the Ni lattice by preoccupying the octahedral sites of the initial Ni (111) subsurface layer. In the study of boron-promoted Ni/Al_2O_3 catalyst, Fouskas et al. (2014) found that low boron content (0.6–5.6 wt%) improved Ni dispersion in an Al_2O_3 support without any changes in textural attributes. Additionally, the total amount of carbonaceous deposits on a B-promoted catalyst was about 74%–86% lower than on unpromoted catalyst.

Table 4.9 summarizes the catalytic performance of various nonnoble metal-based catalysts employing various support and promoter types. In order to facilitate the comparison among the catalysts used, the experimental conditions, including feed composition, reaction temperature, and gas hourly space velocity (GHSV), are also listed. In addition, the initial and final reactant conversions are given to justify catalytic stability. In fact, the estimated degree of catalyst deactivation based on CH_4 conversion is also summarized, together with the H_2/CO ratio.

4.3.2.3 DRM Kinetic Models

Fundamentally understanding the intrinsic mechanistic DRM pathways and determining the kinetic parameters based on the best-fit kinetic model are essential for optimizing catalyst synthesis and subsequent reactor design for commercialized applications in industry. The typical kinetic models commonly implemented in the DRM reaction are the power law, Eley Rideal (ER), and Langmuir Hinshelwood-Hougen Watson (LHHW) models. Among these main kinetic models, the empirical power law model is regarded as the simplest one capable of generating approximately estimated values for associated reaction parameters. The empirical power law model for the DRM reaction is given in Eq. (4.16).

$$-r_{CH_4} = kP_{CH_4}^{\alpha} P_{CO_2}^{\beta} \tag{4.16}$$

where P_i is the partial pressure of component i (i: CH_4 or CO_2) and k is the apparent rate constant. Additionally, α and β are reaction orders varying with the type of catalyst while $-r_{CH_4}$ is the reaction rate of CH_4.

Although the power law model has been widely employed in many studies (Pakhare and Spivey, 2014; Kathiraser et al., 2015) for roughly computing the kinetic parameters because of its simplicity and because there is no requirement to understand the mechanistic reaction pathways (see summary in Table 4.10),

TABLE 4.9 Bibliographic Listing of Catalytic Performance for Recently Employed Catalysts in DRM Reaction

				Initial Performance				Final Performance				
Catalysts	CH_4/CO_2 Ratio	*T* (K)	GHSV (L g_{cat}^{-1} h^{-1})	CH_4 Conversion (%)	CO_2 Conversion (%)	H_2/CO Ratio	TOS[a] (h)	CH_4 Conversion (%)	CO_2 Conversion (%)	H_2/CO Ratio	D[b] (%)	References
Supported catalysts												
NiO/SiO_2	1.0/1.0	1023	–	86.9	89.8	–	100	74.7	80.4	–	7.5	Zhang et al. (2015)
NiO/TiO_2	1.0/1.0	1023	–	4.1	3.2	–	20	1.4	4.3	–	65.9	Zhang et al. (2015)
NiO/Al_2O_3	1.0/1.0	1023	–	77.5	57.8	–	30	76.8	53.3	–	0.1	Zhang et al. (2015)
NiO/ZrO_2	1.0/1.0	1023	–	88.5	88.8	–	30	16.7	21.8	–	81.1	Zhang et al. (2015)
NiO/MgO	1.0/1.0	1023	–	88.1	90.1	–	100	74.3	76.5	–	15.7	Zhang et al. (2015)
Ni/CZ/SBA-15	1.0/1.0	873	20.0	49.6	66.4	0.76	28	41.9	58.0	0.24	15.5	Albarazi et al. (2013)
Ni/CeO_2-Al_2O_3	–	973	–	79.4	–	0.93	80	76.9	–	0.91	3.1	Wang et al. (2014)
Ni/$MgAl_2O_4$	1.0/1.0	973	18.0	74.4	83.3	–	15	70.6	81.2	–	5.1	Habibi et al. (2016)

$Ni/MgAl_2O_4$	1.0/1.0	973	12.0	57.7	–	–	5	56.3	–	–	2.4	Zarei et al. (2016)
$Ni\text{-}MgO\text{-}Al_2O_3$	1.0/1.0	973	180.0	76.1	82.3	0.96	12	71.4	79.8	0.92	6.2	Akbari et al. (2017)
$Ni/MgO\text{-}Al_2O_3$	1.0/1.0	923	60.0	46.5	57.1	0.64	10	40.5	52.0	0.63	12.9	Zhang et al. (2016)
$Mg_{0.8}Ni_{0.2}Al_2O_4$	1.0/1.0	973	18.0	74.0	82.5	–		67.2	78.6	–	9.2	Habibi et al. (2017)
Promoted catalysts												
$Pt\text{-}Ni/CeZrO_2$	0.5/0.5	1073	120.0	94.2	96.2	0.85	15	93.5	95.8	0.91	0.7	Mahoney et al. (2014)
Gd-promoted Ni/ZSM-5	1.0/1.0	1023	50.0	85.0	–	0.98	1	–	–	–	–	Sarkar et al. (2016)
Sm-promoted Ni/ZSM-5	1.0/1.0	1023	50.0	84.0	–	0.92	1	–	–	–	–	Sarkar et al. (2016)
Gd-promoted Ni/Y_2O_3	1.0/1.0	973	8.0	85.7	83.6	–	6	83.4	81.1	–	2.7	Al-Fatesh (2017)
Ce-promoted Ni/Al_2O_3	1.0/1.0	973	12.0	59.9	78.2	0.78	5	58.9	78.3	0.77	1.7	Shamskar et al. (2017)
La-promoted Ni/Al_2O_3	1.0/1.0	973	12.0	56.4	76.7	0.76	5	54.6	76.6	0.75	3.2	Shamskar et al. (2017)
Zr-promoted Ni/Al_2O_3	1.0/1.0	973	12.0	53.4	74.0	0.76	5	51.8	74.2	0.74	3.0	Shamskar et al. (2017)

Continued

TABLE 4.9 Bibliographic Listing of Catalytic Performance for Recently Employed Catalysts in DRM Reaction—cont'd

Catalysts	CH_4/CO_2 Ratio	T (K)	GHSV (L g_{cat}^{-1} h^{-1})	Initial Performance			TOS (h)	Final Performance			D (%)	References
				CH_4 Conversion (%)	CO_2 Conversion (%)	H_2/CO Ratio		CH_4 Conversion (%)	CO_2 Conversion (%)	H_2/CO Ratio		
Mo-promoted Ni/Al_2O_3	1.0/1.0	873	200.0	25.6	33.8	0.64	7	22.1	30.2	0.60	13.7	Yao et al. (2017)
B-promoted Ni/Al_2O_3	1.0/1.0	973	30.0	–	–	–	24	51.4	61.9	–	–	Fouskas et al. (2014)
Bimetallic catalysts												
Co-Zr/AC	1.0/1.0	873	7.2	91.7	91.2	–	45	88.9	89.1	–	3.1	Zhang et al. (2015)
Co-Ni/Al_2O_3-MgO	1.0/1.0	1123	24.0	97.3	99.1	0.97	20	92.3	95.4	0.97	5.2	Abdollahifar et al. (2016)
Pd-Ni/Al_2O_3	1.0/1.0	1023	8.0	87.0	94.2	1.06	100	84.7	83.3	1.03	2.6	Ma et al. (2016)
Co_3Mo_3N	1.0/1.0	1073	6.0	96.9	98.9	0.93	50	86.5	98.4	0.83	10.7	Fu et al. (2017)
Co-Ni/Al_2O_3	1.0/1.0	873	–	23.4	30.9	0.58	180	15.7	17.7	0.46	32.9	Ray et al. (2017)

[a] *Time-on-stream (TOS) is defined as the period of stability measurements at a fixed reaction temperature.*
[b] *Degree of catalyst deactivation, D (%) = [1 − (Final CH_4 conversion/Initial CH_4 conversion)] × 100%.*

TABLE 4.10 Summary of Kinetic Parameters Derived From Power Law Models of DRM Over Different Catalysts

Catalysts	Temperature (K)	Activation Energy, E_a (kJ/mol)	Reaction Order		References
			α	β	
		CH_4	CH_4	CO_2	
0.3%Pt–10%Ni/Al_2O_3	853–893	112.5	1.0	0.9	Özkara-Aydınoğlu and Aksoylu (2013)
0.2%Pt–15%Ni/Al_2O_3	853–893	111.3	1.1	1.4	Wang et al. (2014)
Co/La_2O_3	923–1023	96.4	3.7	0.4	Ayodele et al. (2016)
$SmCoO_3$	973–1073	41.0	1.2	0.3	Osazuwa et al. (2017)
Co/Sm_2O_3	923–1023	88.6	3.1	0.2	Ayodele et al. (2017)
$La_{2-x}Sr_xNiO_4$	633–713	41.8	0.9–0.4	0.2–0.1	Pichas et al. (2010)
13.5%Ni–2%K/CeO_2-Al_2O_3	823–1073	113.8	0.9–1.0	0.3–0.5	Pechimuthu et al. (2006)

the power law model may not be applicable and accurate for a wide range of reactant partial pressure (Pakhare and Spivey, 2014; Kathiraser et al., 2015). Nevertheless, this empirical model can be effectively used for initially guessing the overall activation energy and power constants of each reactant for comparison among catalysts used in the literature.

Table 4.10 summarizes the estimated CH_4 activation energy and reaction orders obtained from power law models recently reported in the literature for several catalysts. Özkara-Aydınoğlu and Aksoylu (2013) fitted the experimental data for DRM on Pt-Ni/Al_2O_3 catalysts to a power law model and found that the reaction orders of CH_4 and CO_2 varied from 1.0 to 1.1 and 0.9 to 1.4 in this order depending on the Pt and Ni loading. For 0.2% Pt–15% Ni/Al_2O_3 catalyst, the higher reaction order for CO_2 than for CH_4 suggests a greater reliance on the CO_2 partial pressure for surface carbon gasification. However, at higher Pt content (0.3% Pt–10% Ni/Al_2O_3), the CO_2 dependency was reduced and CH_4 appeared to be the dominant component affecting the reaction rate, because higher Pt content enhanced the oxygen utilization for removing deposited carbon. As seen in Table 4.10, the CH_4 reaction order was superior to that of CO_2 for most catalysts, suggesting that CH_4 adsorption on the catalyst surface dominated the DRM reaction.

As discussed above, the power law models do not reflect the intrinsic DRM reaction steps occurring on the catalyst surface. Thus, in order to thoroughly understand various complex mechanistic DRM schemes, ER and LHHW models are normally employed, together with statistical criteria and thermodynamic consistency (Abdullah et al., 2017), for assessing the adequacy of the fitting of experimental data to the mechanistic-based kinetic models. Özkara-Aydınoğlu and Aksoylu (2013) investigated Pt-Ni/Al_2O_3 catalyst for the DRM reaction and fitted their experimental data on the CH_4 consumption rate into eight ER and LHHW kinetic models derived from various mechanistic pathways with different rate-determining steps (RDS). Based on the squared error from nonlinear regression, they deduced that the ER model (see Eq. 4.17 in Table 4.11) in which the reaction between molecularly adsorbed CO_2 and CH_4 in the gas phase is the RDS, was the model that provided the best fit.

In kinetic studies of Pt- and Ru-doped TiO_2 catalysts, Singh and Madras (2016) performed an in situ FTIR analysis during the DRM reaction and found that CH_4 adsorbed on the catalyst surface while the absence of adsorbed CO_2 species on the metal or support surfaces was evident. Hence, an ER model (see Eq. 4.18 in Table 4.11) in which CH_4 adsorbs molecularly, followed by CH_4 dissociative adsorption as RDS (rate-determining step) while CO_2 exists in the gas phase, was proposed to forecast the experimental reaction rate.

Ayodele et al. (2017) examined the CH_4 reaction rate over Co/La_2O_3 catalyst in six different LHHW models derived from both single-site (adsorption of CH_4 and CO_2 reactants on the same active site) and dual-site (CH_4 and CO_2 favorably adsorbed on two nonidentical sites on active metal and support) mechanisms. They discovered that the experimental data on the CH_4

TABLE 4.11 Representative ER and LHHW Kinetic Rate Models for DRM Over Different Catalysts

Catalysts	Assumption	Mechanism	Kinetic Model	Equations	References
Pt-Ni/ Al_2O_3	Reaction of molecularly adsorbed CO_2 with CH_4 in gas phase as RDS[a]	$CO_2 + X \overset{K_1}{\leftrightarrow} CO_2 - X$ $CO_2 - X + CH_4 \overset{k_2}{\rightarrow} 2CO + 2H_2 + X$ (RDS*)	$-r_{CH_4} = \frac{k_2 K_1 P_{CH_4} P_{CO}}{(1 + K_1 P_{CO_2})}$	(4.17)	Özkara-Aydınoğlu and Aksoylu (2013)
Pt/TiO_2	Single-site CH_4 molecular adsorption with CH_4 dissociation as RDS	$CH_4 + X \overset{K_1}{\leftrightarrow} CH_4 - X$ $CH_4 - X \overset{k_2}{\rightarrow} C - X + 2H_2$ (RDS) $X - C + CO_2 + X \overset{K_3}{\leftrightarrow} 2X - CO$ $2X - CO \overset{K_4}{\leftrightarrow} 2X + 2CO$	$-r_{CH_4} = \frac{k_2 K_1 K_3 K_4 P_{CH_4} P_{CO_2}}{(1 + K_1 P_{CH_4}) K_3 K_4 P_{CO_2} + K_3 \sqrt{K_4} P_{CO} + P_{CO}^2}$	(4.18)	Singh and Madras (2016)
Co/La_2O_3	Dual-site CH_4 and CO_2 associative adsorption with bimolecular surface reaction as RDS	$CH_4 + X_1 \overset{K_1}{\leftrightarrow} CH_4 - X_1$ $CO_2 + X_2 \overset{K_2}{\leftrightarrow} CO_2 - X_2$ $CH_4 - X_1 + CO_2 - X_2 \overset{k_3}{\rightarrow} 2H_2 + 2CO + X_1 + X_2$ (RDS)	$-r_{CH_4} = \frac{k_3 P_{CH_4} P_{CO_2}}{(1 + K_1 P_{CH_4})(1 + K_2 P_{CO_2})}$	(4.19)	Ayodele et al. (2016)

Continued

TABLE 4.11 Representative ER and LHHW Kinetic Rate Models for DRM Over Different Catalysts—cont'd

Catalysts	Assumption	Mechanism	Kinetic Model	Equations	References
Ni-Rh/ Al_2O_3	Dual-site mechanism and surface reaction between adsorbed CH_3 and CO_2 species as RDS	$CH_4 + X_1 \overset{K_1}{\leftrightarrow} X_1 - CH_3 + \frac{1}{2}H_2$ $CO_2 + X_2 \overset{K_2}{\leftrightarrow} CO_2 - X_2$ $X_1 - CH_3 + CO_2 - X_2 \overset{k_3}{\rightarrow} 2CO + \frac{3}{2}H_2 + X_1 + X_2$ (RDS)	$-r_{CH_4} = \frac{k_3 K_1 K_2 \left(\frac{P_{CH_4} P_{CO_2}}{P_{H_2}^{0.5}} - \frac{P_{H_2}^{1.5} P_{CO_2}^2}{K_{ref}} \right)}{\left(1 + \frac{P_{CH_4}}{P_{H_2}^{0.5} K_1} + P_{CO_2} K_2 \right)^2}$	(4.20)	Barroso-Quiroga and Castro-Luna (2007)
Co-Ni/ Al_2O_3	Dual-site LHHW mechanism in which surface reaction between adsorbed carbon and oxygen species is RDS	$CH_4 + 2X_1 \leftrightarrow CH_3\text{-}X_1 + H\text{-}X_1$ $CH_3\text{-}X_1 + X_1 \leftrightarrow CH_y\text{-}X_1 + H\text{-}X_1;\quad 1 \le y \le 3$ $CH\text{-}X_1 + X_1 \leftrightarrow C\text{-}X_1 + H\text{-}X_1$ $CO_2 + 2X_2 \leftrightarrow CO\text{-}X_2 + O\text{-}X_2$ $C - X_1 + O - X_2 \overset{k_{rxn}}{\rightarrow} CO - X_2 + X_1$ (RDS) $2H\text{-}X_1 \leftrightarrow H_2 + 2X_1$ $CO\text{-}X_2 \leftrightarrow CO + X_2$ $H\text{-}X_1 + O\text{-}X_2 \leftrightarrow OH\text{-}X_2 + X_1$ $H\text{-}X_1 + OH\text{-}X_2 \leftrightarrow H_2O + X_1 + X_2$	$-r_{CH_4} = \frac{k_{rxn} \sqrt{P_{CH_4}} \sqrt{P_{CO_2}}}{\left(1 + \sqrt{K_{CH_4} P_{CH_4}}\right)\left(1 + \sqrt{K_{CO_2} P_{CO_2}}\right)}$	(4.21)	Foo et al. (2011)

[a] *RDS is a rate-determining step with X being available active site.*

consumption rate at different reactant partial pressures from 5 to 50 kPa, and different reaction temperatures from 923 to 1023 K, were sufficiently captured by a dual-site LHHW model (see Eq. 4.19 in Table 4.11) with surface reaction between molecularly adsorbed CH_4 and CO_2 species as RDS. This model satisfied both statistical and thermodynamic criteria with a high correlation coefficient (>0.95). Other dual-site LHHW models with detailed mechanistic steps are also summarized in Table 4.11.

4.3.2.4 Conclusions and Outlook

DRM is a promising green technology to produce a syngas mixture appropriate for downstream production of synthetic long-chained hydrocarbons via Fischer-Tropsch synthesis. However, the DRM technology is still impractical for implementation at an industrial scale because the highly endothermic nature of DRM requires excessive energy consumption, leading to high operating and investment costs. In addition, the inevitably severe catalyst deactivation induced by carbon deposition and metal sintering at high reaction temperature is other limitation of DRM. The synergistic effect of promoters and supports on catalytic attributes, including metal dispersion, extent of metal-support interaction, and crystallite size, has been extensively studied to improve catalytic performance and stability. The use of appropriate supports or promoters possessing a mesoporous structure and basic or redox properties help prevent the catalyst from deteriorating by carbonaceous deposition and sintering.

Power law, ER, and LHHW models are the three standard DRM kinetic models. The empirical power law models do not explain the complex and multistep pathways of the DRM reaction, but the reaction orders could indicate the dominance or important role of reactant partial pressure on overall catalytic activity. A large number of mechanistic-based ER and LHHW models derived from single-site or dual-site mechanisms with various RDS have been proposed for capturing the CH_4 consumption rate.

Apart from utilizing suitable supports and promoters to formulate carbon-resistant catalysts, carbonaceous deposition may be suppressed by cofeeding a small amount of oxidizing reactants, namely, H_2O and O_2, to the CH_4 and CO_2 reactant mixture. In fact, bi-reforming (or combined steam and DRM) and oxidative DRM (referring to DRM with the corresponding addition of H_2O and O_2 reactants) have recently been considered as alternative approaches to inhibit coke deposition. In the presence of these oxidizing agents, surface carbon can be easily gasified to gaseous products, thus keeping the catalyst surface free from carbonaceous species. Additionally, these reforming processes are capable of generating syngas with desirable and flexible H_2/CO ratios for downstream processing via manipulation of reactant composition. Future investigations into bi-reforming and oxidative DRM in terms of catalyst recipe and kinetic modeling could be considered to optimize the efficiency of reforming processes for large-scale production.

4.4 TRI-REFORMING OF METHANE

4.4.1 Thermodynamic Equilibrium Aspect

In a given landfill gas, some quantity of oxygen is usually present, possibly up to 5 vol%. It is meaningful to be able to reform a purified landfill containing again oxygen. Fig. 4.11 shows the thermodynamic equilibrium of a mixture containing 51.3% CH_4, 25.6% CO_2, 20.5% H_2O, and 2.6% O_2 (molar percentage, or 2 mol of CH_4, 1 mol of H_2O, 1 mol of CO_2, and 0.1 mol of O_2) at atmospheric pressure. This mixture corresponds to a stoichiometric reaction of CH_4 with the three oxidants, according to Eqs. (4.1), (4.5), (4.7) in Table 4.2. Oxygen is completely consumed in the temperature range studied. Methane is gradually consumed above 350°C and it practically disappears above 850°C (>98%). Solid carbon can be nearly eliminated above 800°C and it is very important for avoiding the catalyst deactivation by coke deposition. Water is permanently present as a byproduct due to the WGS. Finally, H_2 and CO are formed as the main products, with high selectivity above 800°C, and the molar ratio of H_2/CO is close to 1.65 within 800–1000°C, which is interesting for hydrocarbon synthesis via the Fisher-Tropsch process. Thus, it is recommended to perform tri-reforming above 800°C.

Fig. 4.12 shows the thermodynamic equilibrium of a mixture containing 43.5% CH_4, 21.7% CO_2, 32.6% H_2O, and 2.2% O_2 (molar percentage, or

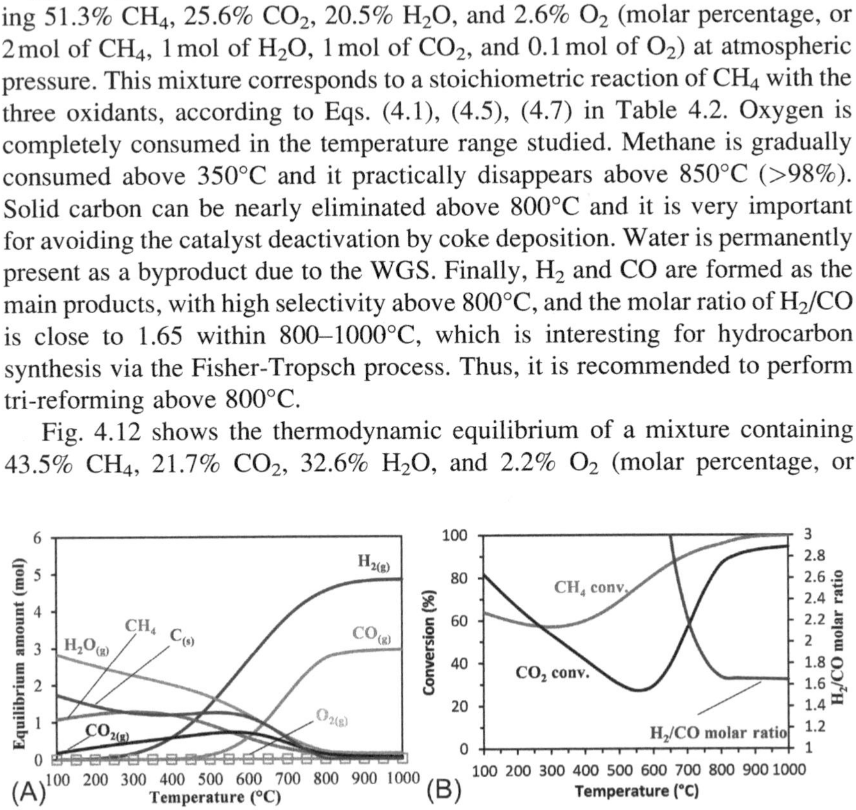

FIG. 4.11 Thermodynamic Equilibrium of the Mixture Containing 2 mol of CH_4, 1 mol of H_2O, 1 mol of CO_2, and 0.1 mol of O_2.

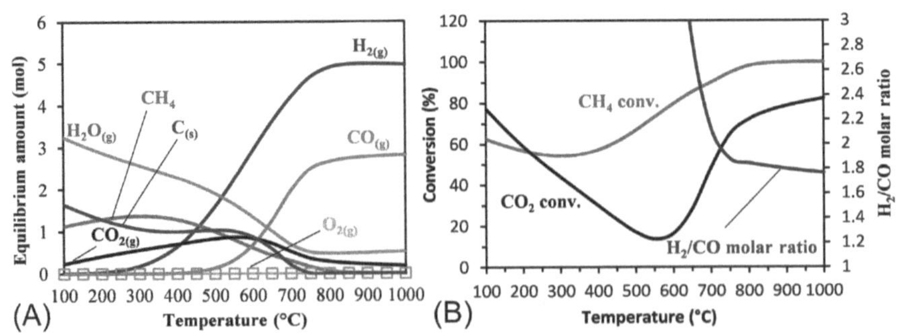

FIG. 4.12 Thermodynamic equilibrium of the mixture containing 2 mol of CH_4, 1.5 mol of H_2O, 1 mol of CO_2 and 0.1 mol of O_2.

2 mol of CH_4, 1.5 mol of H_2O, 1 mol of CO_2, and 0.1 mol of O_2) at atmospheric pressure. This mixture contains a slight excess of oxidants to methane, or the molar ratio of $(H_2O+CO_2+O_2)/CH_4$ of this mixture is equal to 2.1/2. The behavior of this system is similar to the system presented in Fig. 4.11. It is worth noting that adding more water avoids early solid carbon formation as of 750°C. At 800°C, the methane conversion can reach 98.1% with high selectivity in H_2 and CO, and the molar ratio of H_2/CO can reach 1.87. However, both H_2O and CO_2 are present in large excess and can negatively impact the final energy balance of the reforming process.

4.4.2 Catalysts for Methane Tri-Reforming

The TRM process has been regarded as a synergetic combination technique consisting of endothermic steam reforming of methane, DRM, and exothermic partial oxidation of methane. As a heterogeneous catalytic reforming process, Ni-based catalysts are widely used in TRM due to their low cost, high availability, and capability of C-H bond cleavage. Support and promoter are the main components of industrial Ni-based catalysts and substantially influence the catalytic performance in terms of reactant conversions, selectivity, and stability in harsh TRM reaction conditions. The effect of various supports and promoters on TRM catalysts will be discussed thoroughly in the following sections.

4.4.2.1 Catalyst Supports

As in steam reforming and dry reforming processes, the utilization of supports for dispersing active metals and providing mechanical strength for catalysts also plays a crucial role in TRM in terms of catalytic activity, product selectivity, and stability. A wide range of supports has been examined for TRM based on several main selection standards and their intrinsic nature (Song and Pan, 2004; García-Vargas et al., 2014). For instance, because CO_2 is one of the reactants in TRM, supports having a basic character or high oxygen storage capacity could enhance CO_2 adsorption and hence increase CO_2 conversion (Song and Pan, 2004).

Song and Pan (2004) investigated the effect of both single-metal oxides, namely, CeO_2, ZrO_2, MgO, and Al_2O_3, and mixed-metal oxides, including CeO_2-ZrO_2 with a Ce:Zr atomic ratio = 3, on the TRM catalytic performance of Ni catalysts prepared via the wet impregnation method. They observed that high CH_4 ($\geq$97%) and CO_2 (about 80%) conversions were attained on supported Ni catalysts with desirable H_2/CO ratios ranging from 1.5 to 2.0 at a reaction temperature of 1073–1123 K and atmospheric pressure. In addition, unlike DRM reaction, TPO measurements for the deposited carbon content on the catalysts used indicate that, except for the Ni/ZrO_2 catalyst with a small quantity of carbonaceous carbon (1.34 wt%), carbon deposition was not present on other Ni-based catalysts. They also found that the capability of CO_2 consumption over different supported catalysts was strongly reliant on the extent of interaction between metal and support. The order of CO_2 conversion in TRM was reported as: Ni/MgO > Ni/MgO/CeZrO > Ni/CeO_2 ≈ Ni/ZrO_2 ≈ Ni/Al_2O_3 > Ni/CeZrO. Among the supported Ni catalysts, the greater CO_2 conversion on the Ni/MgO and Ni/MgO/CeZrO catalysts was attributed to the stronger interaction of MgO and CO_2, as well as more Ni and MgO interface arising from the formation of NiO/MgO solid solution.

In the study of support influence on Ni catalysts for TRM, García-Vargas et al. (2014) synthesized five different Ni catalysts dispersed on various supports, including γ-Al_2O_3, yttria-stabilized zirconia (YSZ), yttria-stabilized zirconia calcined in an oxygen-poor environment (YSZ-O_2), CeO_2, and silicon carbide (β-*SiC*) (García-Vargas et al., 2014). Gamma-alumina was selected as a support due to its mechanical strength and thermal stability under the high reaction temperature required for TRM, while ZrO_2-based supports reportedly possessed high thermal resistance and ionic conductivity for reforming processes because of their crystal surface defects in which oxygen reactant could easily be activated, and hence enhance reactivity (Bellido and Assaf, 2009). Apart from high thermal conductivity and mechanical strength, *SiC* has chemical inertness and low specific weight, which are appropriate properties for being supports of the endothermic TRM reaction (García-Vargas et al., 2014). García-Vargas et al. (2014) reported that CeO_2 and β-*SiC* materials appeared to be the best catalytic supports for the TRM process. In fact, CeO_2- and β-*SiC*-supported Ni catalysts exhibited the highest CH_4 reaction rate without considerable deactivation. Ni/CeO_2 catalyst yielded the lowest H_2/CO ratio and this observation was attributed to the high basicity of the CeO_2 support, which enhanced CO_2 adsorption to initiate more of the methane dry reforming reaction. Among supported catalysts, Ni/Al_2O_3 catalyst showed the lowest CH_4 and CO_2 conversions owing to the formation of the $NiAl_2O_4$ phase. In addition, the catalytic performance of Ni/YSZ-O_2 catalyst was superior to that of Ni/YSZ catalyst because the calcination conditions of Ni/YSZ-O_2 catalyst resulted in a better NiO reducibility and greater number of oxygen vacancies on the support surface.

Based on the aforementioned important characteristics of supports for TRM catalyst design, composite oxides, such as MgO-TiO_2 (Jiang et al., 2007) and MgO- ZrO_2 (Walker et al., 2012), have been developed as promising supports as these materials also have high surface oxygen vacancies and great basic site concentration, which could contribute to increase carbonaceous gasification. Jiang et al. (2007) investigated $Mg_xTi_{1-x}O$ composite oxides (with $0 \leq x \leq 1$) as supports for impregnation of nickel nitrate. TRM longevity runs were conducted in a continuous-flow fixed-bed reactor at 1123 K and 1 MPa for 50 h. The activity of Ni/MgO, Ni/$Mg_{0.25}Ti_{0.75}O$ and Ni/TiO_2 catalysts declined within 6 h on-stream while high catalytic stability was observed for the Ni/$Mg_{0.75}Ti_{0.25}O$ and Ni/$Mg_{0.5}Ti_{0.5}O$ catalysts. Catalyst deactivation was reportedly induced by the re-oxidation of the Ni metallic phase to the inactive NiO phase or support phase transformation at high temperature. There was no graphitic carbon or carbon fiber detected in XRD and SEM measurements for $Ni/Mg_{0.75}Ti_{0.25}O$ and $Ni/Mg_{0.5}Ti_{0.5}O$ catalysts. The authors deduced that the moderate metal-support interaction and the right ability to be reduced contributed to the high stability of $Ni/Mg_{0.75}Ti_{0.25}O$ and $Ni/Mg_{0.5}Ti_{0.5}O$ catalysts.

Walker et al. (2012) studied the effect of support composition, metal loading, and preparation method on the catalytic performance of Ni-MgO-(Ce,Zr)O_2 composite catalysts for TRM. They discovered that implementing even more Ce:Zr ratio could yield an increase in oxygen mobility and redox attributes, which in turn allow the movement of oxygen atoms to suitable sites for removing surface carbonaceous species. In addition, a greater H_2 yield and production rate were observed during TRM over catalyst with a Ni:Mg ratio of 1:1. As the Ni:Mg metal ratio was about unity, a greater number of interfaces between the Ni and Mg phases were formed. Because MgO has a basic nature, preferably attracting CO_2 adsorption, more Ni and Mg interfaces could facilitate CO_2 dissociative adsorption, and consequently, the oxygen mobility could oxidize the adsorbed carbon species originating from CH_4 decomposition on the Ni surface.

Singha et al. (2016) modified CeO_2-ZrO_2 nanoporous composite with the addition of MgO for synthesizing a Ni nanocluster catalyst for TRM. MgO addition was reportedly coke-resistant and sintering-resistant in TRM reaction conditions. The 5% Ni-MgCeZr composite catalyst showed a highly stable catalytic activity, with high CH_4 conversion (95%) and an ideal H_2/CO ratio of about 2 for TRM, without any deactivation over 100h on-stream. They attributed this outstanding catalytic performance to the synergetic effects of the various metal oxides employed. In fact, ZrO_2 enhanced the thermal stability of the catalyst while the interaction between MgO and $Ce_xZr_{1-x}O_2$ metal oxides improved the dispersion of Ni particles. In addition, a CeO_2-ZrO_2 solid solution could increase the oxygen storage capacity, facilitating the carbon removal process due to high surface oxygen vacancies and great oxygen mobility in the lattice. Pino et al. (2014) also reported similar findings for TRM over Ni-based catalysts supported on La-Ce-O mixed oxides. In this study, a series of La-Ce-O mixed oxide—supported Ni catalysts with various Ni loadings was prepared using the combustion technique. The highest catalytic activity was evident for $Ce_{0.7}La_{0.2}Ni_{0.1}O_{2-\delta}$ catalyst, with no detectable carbon deposition observed in a SEM image after 150h on-stream. After the prereduction step, the formation of LaO_x species increased the d-electron density of Ni atoms, which in turn hindered carbonaceous deposition on the catalyst surface. The authors concluded that the high catalytic activity was associated with the interaction between nickel-lanthana and surface oxygen vacancies of CeO_2, which resulted in rising metal dispersion. However, at high active metal and La_2O_3 contents, the observed catalytic deterioration with TOS was attributed to re-oxidation of the Ni^0 metallic phase, metal sintering, and surface coverage by excessive lanthana species.

4.4.2.2 Promoters

Because the TRM process is a complex heterogeneous catalytic reaction involving the concomitant presence of three main catalytic reactions (including methane steam reforming, methane dry reforming, and methane partial oxidation), apart from metal oxide supports, the role of promoter addition to a TRM catalyst is essential for maximizing catalytic performance, product yield, and catalytic stability with TOS in drastic operation conditions. As a typical reforming process, rare-earth metal oxides are also employed as promoters for Ni-based catalysts in the TRM process in order to gain the benefits of their basic and redox attributes. In an investigation of La-doping on Ni-CeO_2 catalysts, Pino et al. (2011) reported that CH_4 and CO_2 conversions considerably improved with the addition of 10 wt% La, from 93% to 96% and 83% to 86.5%, respectively, with the absence of carbonaceous deposition. Nevertheless, a significant decline in both CO_2 and CH_4 conversions was observed at excessive La loading beyond 10 wt% La while the H_2/CO ratio appeared to be unaffected at about 1.62–1.65. The strong interaction within the Ni-La_2O_3-surface oxygen vacancies of CeO_2 was considered as the driving force for enhancing catalytic activity by improving the dispersion of Ni particles and inducing the formation of Ce^{3+} sites. Additionally, La_2O_3 promotion was able to create a large number of intermediate and strong basic sites, enabling a stronger binding interaction between the catalyst surface and the CO_2 reactant. As a result, the increasing CO_2 adsorption was able to enhance the reforming activity. The promotional effects of CeO_2 and La_2O_3 addition on the catalytic performance of Ni/Al_2O_3 catalyst for TRM were also examined by Solov'ev et al. (2012) at various feed compositions and with reaction temperatures of 814–1123 K. They reported that the addition of CeO_2 or La_2O_3 promoter provided additional pathways for CO_2 activation. Thus, the formation and decomposition of surface carbonates could yield active oxygen species reacting with surface carbon from CH_4 dissociation on Ni metallic phase.

García-Vargas et al. (2014) examined the influence of alkaline (including Na and K) and alkaline earth (namely, Mg and Ca) promoters on the performance of Ni/β-*SiC* catalyst. Although Na and K metal oxides were reported as appropriate promoters in other reforming processes, these metal oxides were not useful for TRM. This was assigned to their ability to enhance the oxidation rate of β-*SiC* during calcination to generate an α-cristobalite form. A similar behavior was also observed for Ca promoter with high loading. However, the promotion of Mg reduced the Ni metal particle size and increased the basicity of the catalyst resulting in increased activity and stability of the Mg-promoted catalyst. The authors also assessed the effect of Mg loading and found that an increase in the Mg/Ni ratio induced a decline in the crystallite size and in the reducibility of NiO particles, most likely due to NiO-MgO solid solution formation.

In order to facilitate the comparison between catalysts and easily identify the influence of support and promoter types on the TRM catalytic performance,

Table 4.12 summarizes the TRM reaction conditions and catalytic performance of various promoted and supported catalysts. The degree of catalyst deactivation is also estimated based on initial and final CH_4 conversions. Additionally, the H_2/CO ratio is given to justify the suitability of the produced syngas mixture for downstream processing.

4.4.3 Tri-reforming of Methane: Kinetic Model

Kinetic studies are generally conducted to determine the most appropriate reaction rate model that is fundamentally derived from mechanistic reaction pathways in order to capture the experimental reactant reaction rate and the product formation rate with the best fit. In comparison with other reforming processes (namely, methane dry reforming and methane steam reforming reactions), kinetic studies of TRM are relatively fewer and TRM kinetic investigations could become an essential focus in the near future. In kinetic studies of Ni catalysts supported on various metal oxides and mixed oxides, Song and Pan (2004) used a simplified power law model (Eq. (4.22)) to capture both the CH_4 and CO_2 reaction rates. In order to simplify the kinetic model for TRM, they kept the partial pressure of CH_4 and O_2 constant. Thus, the effect of the CH_4 and O_2 partial pressures could be incorporated into other associated constants.

$$r_i = A\exp\left(\frac{-E_{\text{app},i}}{RT}\right)(P_{CO_2})^{\alpha,i}(P_{H_2O})^{\beta,i} \tag{4.22}$$

where r_i is reaction rate of CH_4 or CO_2 while A and E_{app} are the preexponential factor and apparent activation energy, respectively. Reaction orders for CO_2 and H_2O are α and β in this order, whereas P_{CO_2} and P_{H_2O} are the corresponding partial pressures of CO_2 and H_2O. The activation energies and reaction orders for CH_4 and CO_2 reaction rates were estimated from Eq. (4.22) for various supported Ni catalysts and are summarized in Table 4.13.

Based on the estimated activation energies and reaction orders in Table 4.13, Song and Pan (2004) deduced that, because the reaction orders with respect to the H_2O partial pressure for the CO_2 consumption rate were negative for all catalysts, CO_2 and H_2O could compete with each other to convert CH_4 to CO and H_2. Thus, increasing P_{H_2O} may result in a reduction in the CO_2 reaction rate. Apart from the competition effect, the presence of the water gas shift reaction during TRM may partially contribute to the decline of the CO_2 reaction rate in a H_2O-rich reactant mixture.

As seen in Table 4.13, the TRM kinetic investigation shows that for both the CH_4 and CO_2 reaction rates, the reaction orders with respect to the CO_2 and H_2O partial pressures were observably different for various catalyst types. Additionally, power law models do not actually comply with intrinsically mechanistic reaction pathways, although they are widely employed to sufficiently capture reactant reaction rates in reforming processes to acquire apparent activation

TABLE 4.12 Bibliographic List for Catalytic Performance of Recently Employed Catalysts in TRM Reaction

Catalysts	$CH_4/CO_2/H_2O/O_2$ Ratio	*T* (K)	GHSV ($L\ g_{cat}^{-1}\ h^{-1}$)	Initial Performance			TOS[a] (h)	Final Performance			D^b (%)	References
				CH_4 Conversion (%)	CO_2 Conversion (%)	H_2/CO Ratio		CH_4 Conversion (%)	CO_2 Conversion (%)	H_2/CO Ratio		
Supported catalysts												
Ni/Al_2O_3	1.0/0.9/0.7/0.3	983	40	19.0	–	–	–	–	–	–	–	Solov'ev et al. (2012)
Ni/SiO_2	1.0/0.5/0.5/0.1	823	–	25.3	10.3	3.44	4	23.5	3.9	4.09	7.1	Majewski and Wood (2014)
Ni/β-*SiC*	1.0/0.5/0.5/0.1	1073	60	87.1	29.8	2.00	24	75.2	22.8	2.06	13.7	García-Vargas et al. (2014)
NiMoC	1.0/0.4/0.3/0.2	1073	46	97.0	99.9	–	–	–	–	–	–	Zou et al. (2016)
Ni/ZrO_2^{HTc}	5.0/1.0/2.1/1.0	1073	80	97.9	98.5	–	25	96.4	96.9	–	1.5	Singha et al. (2016)
Ni/ZrO_2^{Impd}	5.0/1.0/2.1/1.0	1073	80	92.7	94.2	–	25	14.5	17.7	–	84.4	Singha et al. (2016)
$Ni/ZrO_2^{ImpCome}$	5.0/1.0/2.1/1.0	1073	80	52.3	42.3	–	25	23.5	20.6	–	7.1	Singha et al. (2016)

Ni/ $Mg_{0.5}TiO_{0.5}O$	1.0/0.5/0.5/0.1	1123	–	98.0	82.3	–	50	93.6	79.4	–	4.5	Jiang et al. (2007)
Ni/ $Mg_{0.75}TiO_{0.25}O$	1.0/0.5/0.5/0.1	1123	–	86.7	65.2	–	50	84.1	62.3	–	3.0	Jiang et al. (2007)
Ni-Mg/ $Ce_{0.6}Zr_{0.4}O_2$	1.0/0.7/0.5/0.2	1073	61	99.6	65.6	2.20	4	–	–	–	–	Walker et al. (2012)
Promoted catalysts												
La-Ni/Al_2O_3	1.0/0.6/1.0/0.2	998	40	100.0	46.0	2.06	–	–	–	–	–	Solov'ev et al. (2012)
Ce-Ni/Al_2O_3	1.0/0.7/0.7/0.2	978	40	99.0	43.0	1.83	–	–	–	–	–	Solov'ev et al. (2012)
Ca-Ni/β-*SiC*	1.0/0.5/0.5/0.1	1073	60	81.1	12.0	2.56	24	72.8	7.0	2.75	11.0	García-Vargas et al. (2014)
Mg-Ni/β-*SiC*	1.0/0.5/0.5/0.1	1073	60	91.2	35.2	1.88	24	88.7	26.5	2.10	2.7	García-Vargas et al. (2014)
Mg-NiMoC	1.0/0.4/0.3/0.2	1073	46	89.9	84.2	–	–	–	–	–	–	Zou et al. (2016)
K-NiMoC	1.0/0.4/0.3/0.2	1073	46	18.0	0.5	–	–	–	–	–	–	Zou et al. (2016)
Co-NiMoC	1.0/0.4/0.3/0.2	1073	46	93.7	99.9	–	–	–	–	–	–	Zou et al. (2016)

Continued

TABLE 4.12 Bibliographic List for Catalytic Performance of Recently Employed Catalysts in TRM Reaction—cont'd

Catalysts	$CH_4/CO_2/H_2O/O_2$ Ratio	T (K)	GHSV (L g_{cat}^{-1} h^{-1})	Initial Performance			TOS (h)	Final Performance			D (%)	References
				CH_4 Conversion (%)	CO_2 Conversion (%)	H_2/CO Ratio		CH_4 Conversion (%)	CO_2 Conversion (%)	H_2/CO Ratio		
Ce-NiMoC	1.0/0.4/0.3/0.2	1073	46	92.9	99.9	–	–	–	–	–	–	Zou et al. (2016)
La-NiMoC	1.0/0.4/0.3/0.2	1073	46	93.8	88.0	–	–	–	–	–	–	Zou et al. (2016)

[a] *Time-on-stream (TOS) is defined as the period of stability measurements at a fixed reaction temperature.*
[b] *Degree of catalyst deactivation,* D *(%)* = *[1 − (Final CH_4 conversion/Initial CH_4 conversion)] × 100%.*
[c] *Catalyst was prepared via conventional impregnation method.*
[d] *Catalyst was prepared via hydrothermal synthesis.*
[e] *Catalyst was prepared via conventional impregnation method using a commercial ZrO_2 support.*

TABLE 4.13 Activation Energies and Reaction Orders for TRM on Different Supported Ni Catalysts (Song and Pan, 2004)

Catalysts	Temperature (K)	Apparent Activation Energy E_{app} (kJ/mol)		Reaction Order			
				CH_4		CO_2	
		CH_4	CO_2	α	β	α	β
Ni/Al_2O_3	1123	69.1	247.0	0.79	−0.06	1.90	−1.57
Ni/MgO	1123	219.6	160.1	−0.87	−0.64	0.53	−2.59
Ni/MgO/CeZrO	1123	67.4	165.7	0.00	0.03	0.98	−1.08

energy and reaction orders. Thus, in order to examine the interaction between CO_2 and the catalyst surface, Song and Pan (2004) used a simplified Langmuir-Hinshelwood model based on the DRM reaction to avoid the complexity of TRM. The mechanistic steps and rate of reaction for this simplified model are summarized in Table 4.14. Based on the derived TRM kinetic model, they reported that the CO_2 adsorption equilibrium constant, K_{CO_2}, depended strongly on the interaction of CO_2 with the catalyst surface. Because CO_2 interacted strongly with the catalyst surface, molecularly adsorbed CO_2 species became the most abundant reaction intermediates. Thus, most of active sites could be occupied by associatively adsorbed CO_2 species at increasing CO_2 partial pressure, hence leaving fewer available sites for CH_4 adsorption. Therefore, a decline in the CH_4 reaction rate with increased CO_2 partial pressure was observed and the reaction order with respect to CO_2 reactant became negative. However, in the case of weak CO_2 interaction with the catalyst, the opposite results were evident and the reaction order for CO_2 appeared to be positive.

Maciel et al. (2010) carried out a kinetic evaluation of Ni/γ-Al_2O_3 catalyst for the TRM reaction and proposed a four-step reaction mechanism as summarized in Table 4.15. Step I indicates the overall combined reactions, including

TABLE 4.14 Reaction Mechanism of TRM Suggested by Song and Pan (2004)

Mechanistic Steps	
Reactant adsorption	$CH_4 + * \overset{K_{CH_4}}{\leftrightarrow} CH_4^*$
	$CO_2 + * \overset{K_{CO_2}}{\leftrightarrow} CO_2^*$
Surface reaction	$CH_4^* + CO_2^* \rightarrow 2H_2 + 2CO + 2^*$
Rate of CH_4 reaction	$-r_{CH_4} = \frac{kK_{CH_4}K_{CO_2}P_{CH_4}P_{CO_2}}{(1 + K_{CH_4}P_{CH_4} + K_{CO_2}P_{CO_2})^2}$

Note: * *is an available active site on catalyst surface and* k *is the rate constant of rate-determining step.*

TABLE 4.15 The Four-Step Reaction Mechanism for the TRM Reaction Proposed by Maciel et al. (2010)

Reaction Steps	Rate Models
Step I: $CH_4 + \frac{5}{8}O_2 \leftrightarrow CO + \frac{7}{4}H_2 + \frac{1}{4}H_2O$	$r_I = \frac{k_1 K_{CH_4}P_{CH_4}K_{O_2}P_{O_2}}{(1 + K_{CH_4}P_{CH_4} + K_{O_2}P_{O_2})^2}$
Step II: CH_4 cracking $CH_4 \rightarrow C + 2H_2$	$r_{II} = \frac{k_2 K_{CH_4}P_{CH_4}}{1 + K_{CH_4}P_{CH_4}}$
Step III: Boudouard reaction $2CO \rightarrow C + CO_2$	$r_{III} = k_3 P_{CO}^2$
Step IV: Reverse water gas-shift reaction $CO_2 + H_2 \rightarrow CO + H_2O$	$r_{IV} = k_4\left(P_{H_2}P_{CO_2} - \frac{P_{CO}P_{H_2O}}{K_{eq}}\right)$

methane dry reforming, methane steam reforming, methane partial oxidation, and methane oxidation, whereas step II is the methane cracking reaction to form solid carbon and H_2. Besides these main reactions, the Boudouard reaction and the reverse water gas-shift reaction were also considered in the corresponding steps III and IV. Based on the rate laws of the four-step TRM reaction mechanism, the reactant consumption rate ($-r_{CH_4}$ and $-r_{CO_2}$) and the product formation rate (r_{H2} and r_{CO}) were expressed as follows: $-r_{CH4}=-r_I-r_{II}$; $-r_{CO2}=r_{III}-r_{IV}$; $r_{H2}=(7r_I/4)+2r_{II}-r_{IV}$; and $r_{CO}=r_I-2r_{III}+r_{IV}$. By evaluating the TRM kinetic model, the authors deduced that methane decomposition yielded carbon deposition and hydrogen during the TRM reaction while the high CO production rate was attributed to the concomitant presence of the reverse water gas-shift reaction and carbon gasification by carbon dioxide and oxygen reactants.

Garcia-Vargas et al. (2015) examined the effect of temperature and feed composition on the catalytic behavior of Ni-Mg/β-*SiC* catalyst for the TRM reaction. They observed that oxygen was not virtually detected in the effluent stream from the outlet of the reactor, indicating that methane partial oxidation rapidly achieved full conversion during the TRM process. Thus, experimental data were simply fitted to kinetic expressions for individual reactions, including methane steam reforming, methane dry reforming, and water gas-shift reaction. In fact, kinetic equations proposed by Wei and Iglesia (2004a, b) and De la Osa et al. (2011) for the corresponding reforming reactions (namely, steam and DRM) and water gas-shift reaction were employed for assessing the fitness of the experimental reaction rates of the TRM reaction.

Table 4.16 summarizes the reaction rates and associated kinetic parameters involved in the TRM reaction from the findings of Garcia-Vargas et al. (2015). In that study, the kinetic parameters were estimated using the Arrhenius' equation, $k_i=k_i^0e^{(-E_{ai}/RT)}$ with k_i^0, E_{a_i}, R, and T being the corresponding preexponential factor for reaction i, activation energy, the universal gas constant, and the reaction temperature. A good agreement was obtained between the modeled results and the experimental data, with average errors of 10.8% and 19.4% for CH_4 and CO_2 molar flows, respectively. Additionally, the estimated activation energies (see Table 4.16) for each main reaction during the TRM process were consistent with the range of values reported in the literature for steam reforming and DRM (Garcia-Vargas et al., 2015).

4.4.4 Conclusions and Outlook

The catalytic TRM process appears to be a promising substitution for other conventional reforming processes for syngas production at an industrial scale as it is a synergistic combination of methane steam reforming, methane dry reforming, and methane partial oxidation. Particularly, coupling methane dry reforming and methane steam reforming can provide the flexible tuning of H_2/CO ratios by adjusting the reactant feed composition, while the simultaneous presence of

TABLE 4.16 Kinetic Expressions and Associated Parameters for Each Reaction Involved in the TRM Process (Garcia-Vargas et al., 2015)

Rate Models	Associated Kinetic Parameters
For methane steam reforming: $r_{SR} = k_1 P_{CH_4}\left(1 - \frac{P_{CO}P_{H_2}^3}{P_{CH_4}P_{H_2O}K_{SR}}\right)$ $K_{SR} = 1.198 \times 10^{17} e^{\left(\frac{-26830}{T}\right)}$	$k_1^0 = 85.77\,\text{mol s}^{-1}\,\text{kPa}^{-1}$ $E_{a_1} = 74.72\,\text{kJ mol}^{-1}$
For methane dry reforming: $r_{DR} = k_2 P_{CH_4}\left(1 - \frac{P_{CO}^2 P_{H_2}^2}{P_{CH_4}P_{CO_2}K_{DR}}\right)$ $K_{DR} = 6.780 \times 10^{18} e^{\left(\frac{-31230}{T}\right)}$	$k_2^0 = 70.99\,\text{mol s}^{-1}\,\text{kPa}^{-1}$ $E_{a_2} = 77.82\,\text{kJ mol}^{-1}$
For water gas-shift reaction: $r_{WGS} = k_3\left(\frac{P_{CO}P_{H_2O}}{P_{H_2}} - \frac{P_{CO_2}}{K_{WGS}}\right)$ $K_{WGS} = 10^{\left(\frac{2078}{T} - 2.029\right)}$	$k_3^0 = 149.92\,\text{mol s}^{-1}\,\text{kPa}^{-1}$ $E_{a_3} = 54.26\,\text{kJ mol}^{-1}$

H_2O, CO_2, and O_2 oxidizing agents in the TRM reaction substantially reduce coke formation, and hence enhance the catalytic activity and prolong the catalytic lifespan.

However, bibliographic knowledge about the TRM reaction is still little-known in both academic and industrial realms due to the complexity of this reaction. Additionally, catalyst deactivation caused by carbonaceous deposition as well as thermal sintering is always an unavoidable problem of reforming processes. Hence, studies related to carbon-resistant catalyst development have been extensively conducted by employing different groups of metals and supports to examine their synergetic effect on catalytic performance and stability over the TRM reaction.

Additionally, the use of advanced catalyst preparation approaches in order to maximize the combined benefits of promoters and supports could substantially affect the physicochemical attributes of catalysts, namely, the extent of interaction between metal and support, as well as surface metal dispersion. In order to effectively design a functional catalyst and optimize reactor design, the best-fit kinetic modeling derived from the intrinsic TRM mechanistic steps is crucial. Hence, future research on adequate catalyst synthesis methods and TRM kinetics should be considered to improve catalytic performance and stability for the industrial application of the TRM process.

4.5 GENERAL CONCLUSIONS

At the present time, SMR is still the predominant process for hydrogen production from fossil resources. However, green hydrogen can be obtained from biogas as a renewable resource using not only SMR but also dry, dual, and tri-reforming processes. These alternative solutions have the advantages of employing carbon dioxide, a major component of biogas, as an oxidant in the reforming reaction, and also minimizing the steam to carbon (S/C) ratio compared to actual SMR processes. However, catalyst deactivation is still identified as the main drawback for the deployment of these reforming technologies. Among numerous studies on the cataysts for dry, dual, and tri-reforming of methane, some of them showed very promising catalytic performance. The thermodynamic equilibrium, reaction pathway and kinetic of these reforming processes are discussed in this chapter. Significant progress has been made recently but effort is still needed to determine most kinetic and mechanistic aspects of these processes.

ACKNOWLEDGMENTS

This work is a part of the *VABHYOGAZ3* project supported by the French "*Programme d'Investissements d'Avenir*" under the supervision of *ADEME*, the French Energy and Environment Agency. The authors are grateful to the ADEME for its support for this project. This work is also the result of the collaboration between "Université de Toulouse, IMT Mines Albi, UMR CNRS 5302, Centre RAPSODEE," "Faculty of Chemical and Natural Resources Engineering, University Malaysia Pahang," and HERA-ALBHYON enterprise.

REFERENCES

Abbas, S.Z., Dupont, V., Mahmud, T., 2017. Kinetics study and modelling of steam methane reforming process over a NiO/Al_2O_3 catalyst in an adiabatic packed bed reactor. Int. J. Hydrog. Energy 42, 2889–2903.

Abdollahifar, M., Haghighi, M., Babaluo, A.A., Talkhoncheh, S.K., 2016. Sono-synthesis and characterization of bimetallic Ni-Co/Al_2O_3-MgO nanocatalyst: effects of metal content on catalytic properties and activity for hydrogen production via CO_2 reforming of CH_4. Ultrason. Sonochem. 31, 173–183.

Abdullah, B., Ghani, N.A.A., Vo, D.-V.N., 2017. Recent advances in dry reforming of methane over Ni-based catalysts. J. Clean. Prod. 162, 170–185.

ADEME (2014a) Climat, Air, Energie.

ADEME (2014b) Alléger l'empreinte environnementale de la consommation des Français en 2030. Rapport de synthèse.

Air Liquide, n.d. https://www.engineering-airliquide.com/fr/reformage-du-methane-vapeur-production-dhydrogene (Accessed on May 16, 2018).

Akbari, E., Alavi, S.M., Rezaei, M., 2017. Synthesis gas production over highly active and stable nanostructured Ni-MgO-Al_2O_3 catalysts in dry reforming of methane: effects of Ni contents. Fuel 194, 171–179.

Albarazi, A., Beaunier, P., Da Costa, P., 2013. Hydrogen and syngas production by methane dry reforming on SBA-15 supported nickel catalysts: on the effect of promotion by $Ce_{0.75}Zr_{0.25}O_2$ mixed oxide. Int. J. Hydrog. Energy 38 (1), 127–139.

Al-Fatesh, A.S., 2017. Promotional effect of Gd over Ni/Y_2O_3 catalyst used in dry reforming of CH_4 for H_2 production. Int. J. Hydrog. Energy 42 (30), 18805–18816.

Angeli, S.D., Monteleone, G., Giaconia, A., Lemonidou, A.A., 2014. State-of-the-art catalysts for CH_4 steam reforming at low temperature. Int. J. Hydrog. Energy 39, 1979–1997.

Arandiyan, H., Peng, Y., Liu, C., Chang, H., Li, J., 2014. Effects of noble metals doped on mesoporous LaAlNi mixed oxide catalyst and identification of carbon deposit for reforming CH_4 with CO_2. J. Chem. Technol. Biotechnol. 89 (3), 372–381.

Argyle, M.D., Bartholomew, C.H., 2015. Heterogeneous catalyst deactivation and regeneration: a review. Catalysts 5 (1), 145–269.

Ay, H., Üner, D., 2015. Dry reforming of methane over CeO_2 supported Ni, Co and Ni-Co catalysts. Appl. Catal. B Environ. 179, 128–138.

Ayodele, B.V., Khan, M.R., Lam, S.S., Cheng, C.K., 2016. Production of CO-rich hydrogen from methane dry reforming over lanthania-supported cobalt catalyst: kinetic and mechanistic studies. Int. J. Hydrog. Energy 41 (8), 4603–4615.

Ayodele, B.V., Khan, M.R., Cheng, C.K., 2017. Greenhouse gases abatement by catalytic dry reforming of methane to syngas over samarium oxide-supported cobalt catalyst. Int. J. Environ. Sci. Technol. 14 (12), 2769–2782.

Barroso-Quiroga, M.M., Castro-Luna, A.E., 2007. Kinetic analysis of rate data for dry reforming of methane. Ind. Eng. Chem. Res. 46, 5265–5270.

Barroso-Quiroga, M.M., Castro-Luna, A.E., 2010. Catalytic activity and effect of modifiers on Ni-based catalysts for the dry reforming of methane. Int. J. Hydrog. Energy 35 (11), 6052–6056.

Baysal, M., Gunay, M.E., Yildirim, R., 2017. Decision tree analysis of past publications on catalytic steam reforming to develop heuristics for high performance: a statistical review. Int. J. Hydrog. Energy 42, 243–254.

Bellido, J.D.A., Assaf, E.M., 2009. Effect of the Y_2O_3–ZrO_2 support composition on nickel catalyst evaluated in dry reforming of methane. Appl. Catal. A Gen. 352, 179–187.

Cai, W., Ye, L., Zhang, L., Ren, Y., Yue, B., Chen, X., He, H., 2014. Highly dispersed nickel-containing mesoporous silica with superior stability in carbon dioxide reforming of methane: the effect of anchoring. Mater. 7 (3), 2340–2355.

Cao, Y., Li, H., Zhang, J., Shi, L., Zhang, D., 2016. Promotional effects of rare earth elements (Sc, Y, Ce, and Pr) on NiMgAl catalysts for dry reforming of methane. RSC Adv. 6 (113), 112215–112225.

Charisiou, N.D., Siakavelas, G., Papageridis, K.N., Baklavaridis, A., Tzounis, L., Avraam, D.G., Goula, M.A., 2016. Syngas production via the biogas dry reforming reaction over nickel supported on modified with CeO_2 and/or La_2O_3 alumina catalysts. J. Nat. Gas Sci. Eng. 31, 164–183.

Chen, W.H., Lin, M.R., Jiang, T.L., Chen, M.H., 2008. Modeling and simulation of hydrogen generation from high-temperature and low-temperature water gas shift reactions. Int. J. Hydrog. Energy 33, 6644–6656.

De la Osa, A.R., De Lucas, A., Romero, A., Valverde, J.L., Sánchez, P., 2011. Kinetic models discrimination for the high pressure WGS reaction over a commercial CoMo catalyst. Int. J. Hydrog. Energy 36 (16), 9673–9684.

Djinović, P., Pintar, A., 2017. Stable and selective syngas production from dry CH_4-CO_2 streams over supported bimetallic transition metal catalysts. Appl. Catal. B Environ. 206, 675–682.

Factsage, n.d., http://www.factsage.com/ (Accessed on May 16, 2018).

Ferreira-Aparicio, P., Benito, M.J., 2005. New trends in reforming technologies: from hydrogen industrial plants to multifuel microreformers. Catal. Rev. 47, 491–588.

Foo, S.Y., Cheng, C.K., Nguyen, T.H., Adesina, A.A., 2011. Kinetic study of methane CO_2 reforming on Co-Ni/Al_2O_3 and Ce-Co-Ni/Al_2O_3 catalysts. Catal. Today 164 (1), 221–226.

Fouskas, A., Kollia, M., Kambolis, A., Papadopoulou, C., Matralis, H., 2014. Boron-modified Ni/Al_2O_3 catalysts for reduced carbon deposition during dry reforming of methane. Appl. Catal. A Gen. 474, 125–134.

Fu, X., Su, H., Yin, W., Huang, Y., Gu, X., 2017. Bimetallic molybdenum nitride Co_3Mo_3N: a new promising catalyst for CO_2 reforming of methane. Cat. Sci. Technol. 7 (8), 1671–1678.

García-Vargas, J.M., Valverde, J.L., Dorado, F., Sánchez, P., 2014. Influence of the support on the catalytic behaviour of Ni catalysts for the dry reforming reaction and the tri-reforming process. J. Mol. Catal. A Chem. 395, 108–116.

Garcia-Vargas, J.M., Valverde, J.L., Diez, J., Dorado, F., Sanchez, P., 2015. Catalytic and kinetic analysis of the methane tri-reforming over a Ni-Mg/β-SiC catalyst. Int. J. Hydrog. Energy 40, 8677–8687.

Gould, T.D., Montemore, M.M., Lubers, A.M., Ellis, L.D., Weimer, A.W., Falconer, J.L., Medlin, J.W., 2015. Enhanced dry reforming of methane on Ni and Ni-Pt catalysts synthesized by atomic layer deposition. Appl. Catal. A Gen. 492, 107–116.

Habibi, N., Wang, Y., Arandiyan, H., Rezaei, M., 2016. Biogas reforming for hydrogen production: a new path to high-performance of Ni/$MgAl_2O_4$ spinel catalysts. ChemCatChem 8 (23), 3600–3610.

Habibi, N., Wang, Y., Arandiyan, H., Rezaei, M., 2017. Effect of substitution by Ni in $MgAl_2O_4$ spinel for biogas dry reforming. Int. J. Hydrog. Energy 42 (38), 24159–24168.

IEA, 2009. IEA bioenergy, task 37-energy from biogas and landfill gas. 2009.

IEA, 2017. IEA, renewables 2017, analysis and forecasts to 2022. 2017.

Jabbour, K., El Hassan, N., Casale, S., Estephane, J., El Zakhem, H., 2014. Promotional effect of Ru on the activity and stability of Co/SBA-15 catalysts in dry reforming of methane. Int. J. Hydrog. Energy 39 (15), 7780–7787.

Jiang, H., Li, H., Xu, H., Zhang, Y., 2007. Preparation of Ni/$Mg_xTi_{1-x}O$ catalysts and investigation on their stability in tri-reforming of methane. Fuel Process. Technol. 88 (10), 988–995.

Jones, G., Jakobsen, J.G., Shim, S.S., Kleis, J., Andersson, M.P., Rossmeisl, J., Abild-Pedersen, F., Bligaard, T., Helveg, S., Hinnemann, B., Rostrup-Nielsen, J.R., 2008. First principles calculations and experimental insight into methane steam reforming over transition metal catalysts. J. Catal. 259 (1), 147–160.

Kathiraser, Y., Oemar, U., Saw, E.T., Li, Z., Kawi, S., 2015. Kinetic and mechanistic aspects for CO_2 reforming of methane over Ni based catalysts. Chem. Eng. J. 278, 62–78.

LeValley, T.L., Richard, A.R., Fan, M., 2014. The progress in water gas shift and steam reforming hydrogen production technologies—a review. Int. J. Hydrog. Energy 39, 16983–17000.

Li, X., Li, D., Tian, H., Zeng, L., Zhao, Z.J., Gong, J., 2017. Dry reforming of methane over Ni/La_2O_3 nanorod catalysts with stabilized Ni nanoparticles. Appl. Catal. B Environ. 202, 683–694.

Lide, D.R., 2003–2004. Handbook of Chemistry and Physics, 84th ed. CRC Press, Boca Raton, FL.

Linde. n.d. http://www.lindeengineering.in/en/process_plants/hydrogen_and_synthesis_gas_plants/gas_products/hydrogen/index.html (Accessed on May 16, 2018).

Liu, K., Song, C., Subramani, V., 2010. Hydrogen and Syngas Production and Purification Technologies. John Wiley & Sons, Inc., Hoboken, New Jersey.

Lou, Y., Steib, M., Zhang, Q., Tiefenbacher, K., Horváth, A., Jentys, A., Liu, Y., Lercher, J.A., 2017. Design of stable Ni/ZrO_2 catalysts for dry reforming of methane. J. Catal. 356, 147–156.

Ma, Q., Sun, J., Gao, X., Zhang, J., Zhao, T., Yoneyama, Y., Tsubaki, N., 2016. Ordered mesoporous alumina-supported bimetallic Pd-Ni catalysts for methane dry reforming reaction. Cat. Sci. Technol. 6 (17), 6542–6550.

Maciel, L.J., de Souza, A.E., Cavalcanti-Filho, V.O., Knoechelmann, A., de Abreu, C.A., 2010. Kinetic evaluation of the tri-reforming process of methane for syngas production. React. Kinet. Mech. Catal. 101 (2), 407–416.

Mahoney, E.G., Pusel, J.M., Stagg-Williams, S.M., Faraji, S., 2014. The effects of Pt addition to supported Ni catalysts on dry (CO_2) reforming of methane to syngas. J. CO_2 Util. 6, 40–44.

Majewski, A.J., Wood, J., 2014. Tri-reforming of methane over Ni@SiO_2 catalyst. Int. J. Hydrog. Energy 39 (24), 12578–12585.

Nguyen, V.N., Blum, L., Peters, R., 2014. Operational behavior and reforming kinetics over Ni/YSZ of a planar type pre-reformer for SOFC systems. Int. J. Hydrog. Energy 39, 7131–7141.

Nguyen, V.N., Deja, R., Peters, R., Blum, L., 2016. Methane/steam global reforming kinetics over the Ni/YSZ of planar pre-reformers for SOFC systems. Chem. Eng. J. 292, 113–122.

Omoregbe, O., Danh, H.T., Abidin, S.Z., Setiabudi, H.D., Abdullah, B., Vu, K.B., Vo, D.-V.N., 2016. Influence of lanthanide promoters on Ni/SBA-15 catalysts for syngas production by methane dry reforming. Procedia Eng. 148, 1388–1395.

Omoregbe, O., Danh, H., Nguyen-Huy, C., Setiabudi, H., Abidin, S., Truong, Q., Vo, D.-V.N., 2017. Syngas production from methane dry reforming over Ni/SBA-15 catalyst: effect of operating parameters. Int. J. Hydrog. Energy 42 (16), 11283–11294.

Osazuwa, O.U., Setiabudi, H.D., Abdullah, S., Cheng, C.K., 2017. Syngas production from methane dry reforming over $SmCoO_3$ perovskite catalyst: kinetics and mechanistic studies. Int. J. Hydrog. Energy 42 (15), 9707–9721.

Özkara-Aydınoğlu, Ş., Aksoylu, A.E., 2013. A comparative study on the kinetics of carbon dioxide reforming of methane over Pt-Ni/Al_2O_3 catalyst: effect of Pt/Ni ratio. Chem. Eng. J. 215, 542–549.

Pakhare, D., Spivey, J., 2014. A review of dry (CO_2) reforming of methane over noble metal catalysts. Chem. Soc. Rev. 43 (22), 7813–7837.

Pechimuthu, N.A., Pant, K.K., Dhingra, S.C., Bhalla, R., 2006. Characterization and activity of K, CeO_2, and Mn promoted Ni/Al_2O_3 catalysts for carbon dioxide reforming of methane. Ind. Eng. Chem. Res. 45 (22), 7435–7443.

Pichas, C., Pomonis, P., Petrakis, D., Ladavos, A., 2010. Kinetic study of the catalytic dry reforming of CH_4 with CO_2 over $La_{2-x}Sr_xNiO_4$ perovskite-type oxides. Appl. Catal. A Gen. 386 (1), 116–123.

Pino, L., Vita, A., Cipitì, F., Laganà, M., Recupero, V., 2011. Hydrogen production by methane tri-reforming process over Ni-ceria catalysts: effect of La-doping. Appl. Catal. B Environ. 104, 64–73.

Pino, L., Vita, A., Laganà, M., Recupero, V., 2014. Hydrogen from biogas: catalytic tri-reforming process with Ni/LaCeO mixed oxides. Appl. Catal. B Environ. 148, 91–105.

Qian, L., Ma, Z., Ren, Y., Shi, H., Yue, B., Feng, S., Xie, S., 2014. Investigation of La promotion mechanism on Ni/SBA-15 catalysts in CH_4 reforming with CO_2. Fuel 122, 47–53.31.

Ray, K., Sengupta, S., Deo, G., 2017. Reforming and cracking of CH_4 over Al_2O_3 supported Ni, Ni-Fe and Ni-Co catalysts. Fuel Process. Technol. 156, 195–203.

Rong-jun, Z., Guo-fu, X., Ming-feng, L., Yu, W., Hong, N., Da-dong, L., 2015. Effect of support on the performance of Ni-based catalyst in methane dry reforming. J. Fuel Chem. Technol. 43 (11), 1359–1365.

Rostrup-Nielsen, J.R., 2002. Syngas in perspective. Catal. Today 71, 243–247.

Sarkar, B., Goyal, R., Pendem, C., Sasaki, T., Bal, R., 2016. Highly nanodispersed Gd-doped Ni/ZSM-5 catalyst for enhanced carbon-resistant dry reforming of methane. J. Mol. Catal. A Chem. 424, 17–26.

Selvarajah, K., Phuc, N.H.H., Abdullah, B., Alenazey, F., Vo, D.-V.N., 2016. Syngas production from methane dry reforming over Ni/Al_2O_3 catalyst. Res. Chem. Intermed. 42 (1), 269–288.

Shah, Y.T., 2017. Chemical Energy from Natural and Synthetic Gas. CRC Press, Boca Raton, FL.

Shamskar, F.R., Meshkani, F., Rezaei, M., 2017. Preparation and characterization of ultrasound-assisted co-precipitated nanocrystalline La-, Ce-, Zr-promoted Ni-Al_2O_3 catalysts for dry reforming reaction. J. CO_2 Util. 22, 124–134.

Singh, S.A., Madras, G., 2016. Sonochemical synthesis of Pt, Ru doped TiO_2 for methane reforming. Appl. Catal. A Gen. 518, 102–114.

Singha, R.K., Das, S., Pandey, M., Kumar, S., Bal, R., Bordoloi, A., 2016. Ni nanocluster on modified CeO_2-ZrO_2 nanoporous composite for tri-reforming of methane. Cat. Sci. Technol. 6 (19), 7122–7136.

Solov'ev, S.A., Gubareni, Y.V., Kurilets, Y.P., Orlik, S.N., 2012. Tri-reforming of methane on structured Ni-containing catalysts. Theor. Exp. Chem. 48 (3), 199–205.

Song, C., Pan, W., 2004. Tri-reforming of methane: a novel concept for catalytic production of industrially useful synthesis gas with desired H_2/CO ratios. Catal. Today 98 (4), 463–484.

Soria, M.A., Mateos-Pedrero, C., Marín, P., Ordonez, S., Guerrero-Ruiz, A., Rodríguez-Ramos, I., 2012. Kinetic analysis of the Ru/SiO_2-catalyzed low temperature methane steam reforming. Appl. Catal. A Gen. 413–414, 366–374.

Temkin, M.I., 1979. Eley, H.P.D.D., Paul, B.W. (Eds.), Advances in Catalysis. Academic Press, New York, pp. 173–291.

Usman, M., Daud, W.W., Abbas, H.F., 2015. Dry reforming of methane: influence of process parameters—a review. Renew. Sust. Energ. Rev. 45, 710–744.

Walker, D.M., Pettit, S.L., Wolan, J.T., Kuhn, J.N., 2012. Synthesis gas production to desired hydrogen to carbon monoxide ratios by tri-reforming of methane using Ni-MgO-(Ce, Zr)O_2 catalysts. Appl. Catal. A Gen. 445, 61–68.

Wang, C., Zhang, Y., Wang, Y., Zhao, Y., 2017. Comparative studies of non-noble metal modified mesoporous M-Ni-CaO-ZrO_2 (M = Fe, Co, Cu) catalysts for simulated biogas dry reforming. Chin. J. Chem. 35 (1), 113–120.

Wang, N., Yu, X., Wang, Y., Chu, W., Liu, M., 2013. A comparison study on methane dry reforming with carbon dioxide over $LaNiO_3$ perovskite catalysts supported on mesoporous SBA-15, MCM-41 and silica carrier. Catal. Today 212, 98–107.

Wang, N., Xu, Z., Deng, J., Shen, K., Yu, X., Qian, W., Wei, F., 2014. One-pot synthesis of ordered mesoporous NiCeAl oxide catalysts and a study of their performance in methane dry reforming. ChemCatChem 6 (5), 1470–1480.

Wang, Y., Yao, L., Wang, S., Mao, D., Hu, C., 2018. Low-temperature catalytic CO_2 dry reforming of methane on Ni-based catalysts: a review. Fuel Process. Technol. 169, 199–206.

Wei, J., Iglesia, E., 2004a. Reaction pathways and site requirements for the activation and chemical conversion of methane on Ru-based catalysts. J. Phys. Chem. B 108, 7253–7262.

Wei, J., Iglesia, E., 2004b. Isotopic and kinetic assessment of the mechanism of reactions of CH_4 with CO_2 or H_2O to form synthesis gas and carbon on nickel catalysts. J. Catal. 224 (2), 370–383.

Xie, T., Zhao, X., Zhang, J., Shi, L., Zhang, D., 2015. Ni nanoparticles immobilized Ce-modified mesoporous silica via a novel sublimation-deposition strategy for catalytic reforming of methane with carbon dioxide. Int. J. Hydrog. Energy 40 (31), 9685–9695.

Xu, J., Froment, G., 1989. Methane steam reforming, methanation and water-gas shift. 1. Intrinsic kinetics. AICHE J. 35, 88–96.

Xu, J., Saeys, M., 2007. First principles study of the coking resistance and the activity of a boron promoted Ni catalyst. Chem. Eng. Sci. 62 (18), 5039–5041.

Yao, L., Galvez, M.E., Hu, C., Da Costa, P., 2017. Mo-promoted Ni/Al_2O_3 catalyst for dry reforming of methane. Int. J. Hydrog. Energy 42 (37), 23500–23507.

Zarei, M., Meshkani, F., Rezaei, M., 2016. Preparation of mesoporous nanocrystalline Ni-$MgAl_2O_4$ catalysts by sol-gel combustion method and its applications in dry reforming reaction. Adv. Powder Technol. 27 (5), 1963–1970.

Zeppieri, M., Villa, P.L., Verdone, N., Scarsella, M., De Filippis, P., 2010. Kinetic of methane steam reforming reaction over nickel- and rhodium-based catalysts. Appl. Catal. A Gen. 387, 147–154.

Zhang, L., Zhang, Q., Liu, Y., Zhang, Y., 2016. Dry reforming of methane over Ni/MgO-Al_2O_3 catalysts prepared by two-step hydrothermal method. Appl. Surf. Sci. 389, 25–33.

Zhang, S., Muratsugu, S., Ishiguro, N., Tada, M., 2013. Ceria-doped Ni/SBA-16 catalysts for dry reforming of methane. ACS Catal. 3 (8), 1855–1864.

Zhang, S., Shi, C., Chen, B., Zhang, Y., Zhu, Y., Qiu, J., Au, C., 2015. Catalytic role of β-Mo_2C in DRM catalysts that contain Ni and Mo. Catal. Today 258, 676–683.

Zou, H., Chen, S., Huang, J., Zhao, Z., 2016. Effect of additives on the properties of nickel molybdenum carbides for the tri-reforming of methane. Int. J. Hydrog. Energy 41 (38), 16842–16850.

Chapter 5

Hydrogen Storage for Mobile Application: Technologies and Their Assessment

Lars Baetcke and Martin Kaltschmitt
Hamburg University of Technology (TUHH), Institute of Environmental Technology and Energy Economics (IUE), Hamburg, Germany

5.1 INTRODUCTION

For the development of modern transport systems fulfilling the demanding greenhouse gas (GHG) reduction goals stated in the Paris Agreement of late 2015, it is necessary to think about new technologies and innovative concepts to power vehicles. Besides the contractually agreed and accelerating necessity to contribute to a significant reduction of greenhouse gases, such a development would also help to substitute for finite fossil fuel energy, in which most of the countries around the globe are very much dependent on imports. Thus, fossil fuels cannot be a substantial part of a sustainable transportation system.

One promising technological concept that fulfills the demands outlined above is the use of hydrogen in combination with fuel cells and electric engines. Compared to battery electric vehicles, such an approach has the advantage that hydrogen cars could be operated for much longer distances without a refueling stop. But such a hydrogen-based mobility concept will only be successfully implemented within the existing mass market if hydrogen can be stored safely, quickly, and in a technically mature, economically efficient, and environmentally friendly way. For this reason, mobility concepts based on hydrogen depend directly on the availability of storage options fulfilling these requirements.

Thus, all frame conditions set by national governments, as well as the international community of states, point to a mobility system based on electric vehicles in the years to come. Therefore, the combustion engines used today within cars and trucks will most likely be phased out in the years to come due to this transformation within the mobility sector. For the resulting future mobility concepts characterized by much lower GHG emissions, the still open question is in which form the necessary energy will be transported. Promising approaches are

Hydrogen Supply Chain. https://doi.org/10.1016/B978-0-12-811197-0.00005-1

electrical energy stored in batteries or hydrogen as an energy carrier. The most important difference between these two technologies is the energy storage and, depending on the option chosen, the subsequent conversion into mechanical energy to realize motion.

In battery electric cars, the energy is stored in the form of electricity within an accumulator. Thus, the energy can be used directly by the electric power train of the car. In comparison, a fuel cell electric vehicle needs more energy conversion steps before the energy can be used for propulsion. First, hydrogen as a secondary energy carrier is produced within a stationary process from electrical energy. Hydrogen needs then to be stored, for example, at a hydrogen refueling station, until it is needed. From this stationary storage, hydrogen is transferred to storage located within the hydrogen vehicle. During vehicle motion, and thus depending on the given energy demand, this stored hydrogen is used within a fuel cell where it is again converted into electricity to provide the power demanded for the electric car.

Crucial for both options are the storage safety, storage density, storage capacity, storage losses, storage weight, storage costs, and other characteristic figures defining the storage of hydrogen or electricity. These aspects have to be solved in a competitive way on the market. In the longer term, systems are needed that are affordable by basically all customers and that can compete with the existing solutions based on fossil fuel energy. The system characterized by the lowest specific cost, the highest energy density, and other important technical, systemic, economic, environmental, and social points specific to the application, will be implemented within the mobility sector in the years to come to fulfill the GHG reduction goals. Thus, it may be possible that different solutions for different sub-segments could become market mature (e.g., battery electric vehicles for local personal mobility and fuel cell vehicles for supra-regional light duty traffic).

Against this background the overall goal of this chapter is it to present and to analyze the technical possibilities for hydrogen storage, especially for mobile applications (Fig. 5.1). Hydrogen storage is possible in pure form by applying physical effects (compression, liquefaction) or material-based effects (e.g., reversible chemical reaction with solid or liquid materials, physisorption with highly porous materials, for example, metal organic framework (MOF)). All the respective technologies are explained and categorized below. Finally, a short outlook on the possible use of each technology is given. By assessing and comparing the various solutions it will be seen that many of the described technologies are still within an early research and development (R&D) phase and, so far, they are not market ready. This fact makes a scientifically sound comparison very difficult.

5.2 HYDROGEN STORAGE IN PURE FORM

Storage of hydrogen in pure form includes all storage options in which hydrogen is stored directly without any interaction with other materials in any chemical

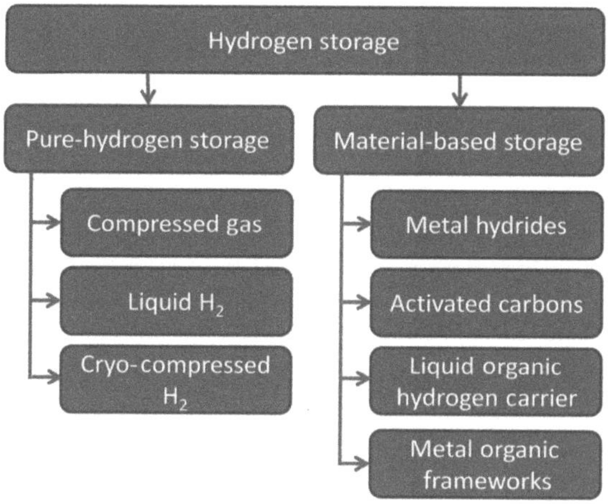

FIG. 5.1 Hydrogen storage options.

and/or physical way. Therefore, these options are typically relatively simple. The complexity, and thus the technical challenge, of these storage vessels lie in the tank design, and especially in the tank shell, as well as in the pretreatment of hydrogen. These kinds of storage options are mostly comparable with storage possibilities for many other liquids and/or gases.

So far three different hydrogen storage concepts exist for pure hydrogen: high-pressure storage, liquid storage, and cryo-compressed storage. Below these three options are explained in detail.

5.2.1 High-Pressure Storage

The storage of hydrogen in high-pressure vessels of varying size is currently the most widely used technological approach. In nearly all mobile applications that are commercially available, or that will be introduced into the market during the months and years to come, hydrogen is stored in high-pressure tank systems with a pressure level of up to 700 bar. Therefore, it is important to look more precisely at the design of such hydrogen high-pressure storage tanks.

Basically, high-pressure storage tanks for hydrogen are similar to other high-pressure gas storage tanks. Additionally, more or less all high-pressure hydrogen storage tanks available on the market or under development have basically the same main components. They have a valve, a tank shell made out of different materials to withstand the high pressure, and they have to be gas tight and have enough free space for containing hydrogen. High-pressure storage tanks are available in two different shapes (cylindrical and spherical); these shapes deal with the forces resulting from the high stresses inside the tank shell in a most efficient way.

One advantage of such high-pressure tanks is that no complex thermal management is necessary. Additionally, the weight of the storage tanks is not too high. Also, no other problems, such as chemical reactions inside the structure, typically occur. In addition, there are no losses during long parking times, because the storage tanks operate at ambient temperature (pressure does not change over time). The only significant disadvantage is the high operating pressure and the resulting demands on the strength of the tank materials. Nevertheless, experience shows the pressure is typically manageable from a technical point of view. However, such high pressure levels are also a severe safety problem. Therefore, a lot of effort is put into ways to protect the pressure vessel in case of an accident. All that effort has led to storage tanks in which an explosion is nearly impossible, even under very challenging conditions.

5.2.1.1 State of Technology

The state of the art for hydrogen storage tanks are CFRP-tanks (Carbon Fiber Reinforced Polymer) with a pressure level of 350 or 700 bar. The assembly of such a tank is explained in more detail below.

One very important design aspect for such high-pressure storage tanks is to ensure that hydrogen cannot diffuse out of the tank over time. For this reason, typically, a thin and absolutely gas-tight liner is placed inside of the CFRP shell. Thus, the CFRP shell only compensates the stresses resulting from the high pressure and does not guarantee a gas-tight system.

Such high-pressure tanks for hydrogen storage are typically divided into four different categories.

- The first category of tank (type I) is made out of steel only; that is, they can achieve a pressure level of around 200 to 300 bar.
- The second category of tank (type II) is made out of steel combined with carbon fiber reinforced plastic wrapped around the cylinder. This design approach makes the storage tank a bit lighter compared with type I. Additionally, the energy density, as well as the pressure of the storage tank, may be increased.
- The third category of tank (type III) is a cylinder made out of CFRP or GFRP (Glass Fiber Reinforced Polymer). This material is strong enough to withstand the stresses due to the high pressure. Inside the storage tank is a thin liner made from steel and/or aluminum to guarantee the gas-tightness. These storage tanks usually operate at a pressure of 350 bar.
- The fourth category of tank (type IV) is similar to the type III tanks. The difference is the liner inside the tank, which is made out of polymer, and the tank shell, which is always made out of CFRP. The consequence is that the storage tank is lighter again and can withstand a pressure of 700 bar. Thus, type IV tanks are usually used in modern vehicles (Mori and Hirose, 2009).

CFRP type IV high-pressure tanks are not a fully developed technology yet; a continuous improvement realized by different companies and research groups can be observed. Nevertheless, the solutions available already are quite advanced.

One example typical of several other approaches is presented in Fig. 5.2 (Nonobe, 2017), showing a pressure tank with a relatively high energy density for the overall tank system. The standard structure of such a CFRP type IV tank has three different layers. The inner layer is a plastic liner that seals the tank. After that the CFRP shell functions to withstand the stress that results from the pressure level inside the tank. This is also the layer in which weight reduction is realized by optimizing the structure of the single carbon fiber layer. The last layer of the storage tank consists of GFRP for safety reasons, as the function of this layer is to protect the storage tank from outside mechanical loads (e.g., events like a rock fall or other external impacts). Additional safety features for impacts or car accidents consist of two protectors that mainly protect the tank shell against dropping. In addition to the tank shell, there are also two aluminum fittings or bosses, which allow the implementation of valves required for the refilling and the safety blow-off.

There is also the possibility to implement sensors inside the tank to measure pressure and temperature (Nonobe, 2017). These sensors are necessary, because such high-pressure storage tanks show a clear temperature increase during loading with high pressure, especially at fast loading. According to international regulations and the material characteristics, the temperature limit for CFRP tanks is defined to be 85°C (Simonovski et al., 2015). To fulfill this requirement securely hydrogen is precooled at the fueling station to a temperature of −70°C. Under these conditions, the limits defined above can be met, thus allowing a comfortable refilling time.

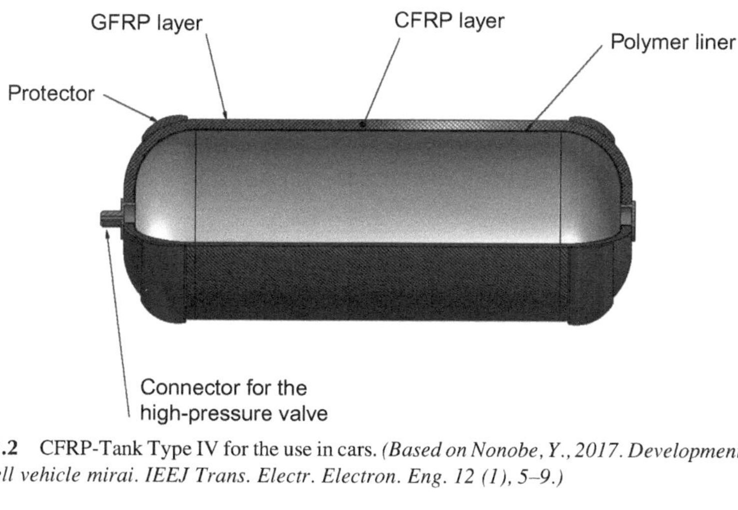

FIG. 5.2 CFRP-Tank Type IV for the use in cars. *(Based on Nonobe, Y., 2017. Development of the fuel cell vehicle mirai. IEEJ Trans. Electr. Electron. Eng. 12 (1), 5–9.)*

Besides lightweight design, the safety issues of hydrogen pressure storage tanks are an important issue. Therefore, different standards (e.g., ISO/TC 197) already exist that define and regulate selected parts of hydrogen supply, hydrogen storage, and other hydrogen-related components. All the tanks available so far do meet these challenging requirements.

If the pressure increases above the design pressure, safety valves (overpressure valves) react and blow off part of hydrogen in a controlled and predefined way. In comparison to "classical" fuel (e.g., gasoline) the fact that hydrogen is much lighter than air is a clear advantage. The consequence of this physical property is that hydrogen that gets outside the tank just disappears and does not lead to a dangerous atmosphere around and beneath the car. Also, in case of fire, hydrogen is much safer than liquid fossil fuels (e.g., gasoline, diesel fuel) because hydrogen released by the blow-off valve leads to one single flame of burning hydrogen. This is in contrast to "classical" liquid transportation fuels in which a big fire on the ground below the car occurs when gasoline or diesel tanks leak in an accident.

5.2.1.2 Characterization

The tank material used and the realized pressure level of the storage tank depend on the application it is designed for. For the mobile application assessed here, typically various types of compound materials (e.g., Glass Fiber Reinforced Polymer (GFRP) or Carbon Fiber Reinforced Polymer (CFRP)) are used for the tank shell. The two most common versions are CFRP storage tanks with 350 or 700 bar pressure, depending on the application.

The compression of hydrogen up to a pressure of 700 bar consumes around 18 MJ/kg_{H2}. This equals approximately 15% of the lower heating value of hydrogen. Thus, the cost of such high-pressure storage is typically relatively high, on account of losses as well as the compression. This results in an overall Well-to-Fuel Cell efficiency for the 700-bar tank of 54% and for the 350-bar tank of 56%. Therefore, this storage approach is already close to the ultimate DOE goal of 60%.

Storage costs also depend on the pressure level of the tank. On average a storage system with 700 bar costs roughly 20 US$/kWh. In comparison, a storage system with 350 bar is slightly cheaper, costing approximately 13.4 US$/kWh. The refueling rate sums up to 1.5–2 kg_{H2}/min for both pressure levels (Ahluwalia et al., 2010).

5.2.1.3 Markets and Perspectives

During implementation within the automotive market, a clear path for each of the two pressure levels outlined above can be seen. Currently cars are commonly equipped with storage tanks of approximately 700 bar because they show a higher energy density compared with storage tanks with a lower pressure. Such a high energy density is typically a clear advantage, especially in cars where space is very limited in most cases. In contrast, busses have mostly more available room. Therefore, it is easier to implement 350-bar storage tanks within such vehicles (e.g., on the roof). The resulting advantage is that such a "low-pressure" storage tank can be loaded more quickly and easily than the 700-bar tank. Additionally, a 350-bar storage tank shows overall better efficiency due to the lower pressure level. And, last but not least, the 350-bar storage tanks are typically slightly cheaper.

Thus, it is most likely that pressure tanks with the two pressure levels of 350 and 700 bar will remain on the market in the years to come. This is mainly due to the experience that has already been gained and the fact that the management of compressed gases is common within industry.

5.2.2 Liquid Hydrogen Storage

Another option to store hydrogen with high energy density is to store it in the liquid phase. For purely technical reasons there are a lot of benefits in dealing with liquids instead of compressed gases. However, the challenge with hydrogen is that it condenses at a very low temperature. Thus, such a storage concept can only be realized by guaranteeing sufficient cooling and re-gasification capacity at the entry into, and the exit from, the actual hydrogen storage tank. Even with these infrastructural challenges, some mobility concepts are based on liquid hydrogen. Therefore, this option is discussed in detail below.

5.2.2.1 State of Technology

Due to the physical characteristics outlined above, liquid hydrogen needs to be stored constantly below 20 to 30 K. Thus, one of the most important parts of such a liquid hydrogen tank is the insulation, which is needed due to the high temperature difference between inside and ambient conditions. Thus, for efficiency reasons, a tank design is chosen with a low surface-to-volume ratio because the thermodynamic losses depend directly on the surface area to the ambient environment. Therefore, nearly all liquid hydrogen storage tanks have a spherical shape. Additionally, some, especially smaller liquid storage tanks for mobile applications, have a shape that is a mix between a sphere and cylinder because this shape may be much better integrated within the design of a vehicle. Around the storage volume super insulation is assembled out of a combination of evacuated space and aluminum foil. The overall goal of such a design is to keep the convection losses, as well as the radiation losses, as low as possible. The possibilities are basically defined by the available technologies, as well as techno-economic constraints. Additionally, components, such as valves, heat exchangers, safety features, and measuring instruments, are needed. A system is also needed to take care of the boil-off losses. If such a tank is used within a vehicle that requires gaseous hydrogen as a fuel, there are some options to use the gas instead. But most times it is released to the atmosphere.

Thus, technological solutions are under development to make use of these boil-off losses. One option is to burn the gaseous hydrogen in a continuous process. Another possibility is to store this hydrogen gas within pressure storage that is able to deal with higher temperatures. There are also some concepts that use a part of the energy of the blow-off hydrogen to liquefy hydrogen again to minimize the losses (Schlapbach and Zuttel, 2001; Hirscher, 2010). So far, all these concepts are still very expensive and increase the overall cost of the storage.

Liquid storage is characterized by a considerably higher storage density than compressed hydrogen. Additionally, the volumetric storage density is in the same range, or even a higher range, compared to hydrogen storage in metal hydrides. This high energy density of liquid hydrogen is also the reason that it is still discussed as a storage option for some airplanes. But this clear advantage is realized at the expense of the need for around one-third of the stored hydrogen energy to be devoted to the operation of the cooling device to provide a liquid state. Another disadvantage is that, so far, there are no technically feasible possibilities to build a storage tank that can store the liquid hydrogen for more than a couple of days without considerable loss. This is due to the high temperature difference between the liquid hydrogen and the ambient temperature; this temperature difference might come to 275 K, or even more in the summer. Thus, some of hydrogen necessarily vaporizes, resulting in a rising pressure level within the tank. After reaching the maximum pressure, it is necessary to blow off the gaseous hydrogen to allow for ongoing safe storage. Currently this blow-off gas is released into the ambient atmosphere. Therefore, a vehicle equipped with such a liquid hydrogen storage tank is not allowed to be parked within a parking garage. The reason is that the blow-off gas can form an ignitable mixture with the surrounding air and there might be the danger of explosion. Another consequence of the permanent boil-off losses is that the storage tank will become empty after a time (depending on the storage concept and storage design). Thus, it is necessary to refill vehicles equipped with such a tank regularly, even if they are not being used. For this reason, this storage concept is most efficiently used for commercial applications with a constant demand and a clearly predictable use (e.g., commercially operated vehicles like planes, trains, ships, trucks) because these applications typically have a constant hydrogen need that can be planned ahead. Thus, this storage option is most suitable for applications characterized by a high energy demand and a need for high energy densities.

5.2.2.2 Characterization

The main consequence of the liquefaction of hydrogen is the significantly higher density of 70.8 kg/m^3 compared with hydrogen at standard conditions with 0.09 kg/m^3. Based on this significant difference, the resulting benefit for the required storage volume becomes obvious. But on the other hand, very challenging conditions are needed to provide hydrogen in a liquid phase. The temperature has to be between 20 and 30 K, because the triple point of hydrogen is 32 K (above this temperature hydrogen is always gaseous). The pressure typically realized for such a storage facility is in a range between ambient pressure and 6 bar (Schlapbach and Zuttel, 2001; Hirscher, 2010).

Liquefaction needs about 35% of the energy stored inside hydrogen; that is, the storage efficiency is around 22.3%. The system storage capacity has a gravimetric value of around 7 wt% and a volumetric value of 0.04 kg_{H2}/l. Liquid hydrogen can be transferred to a suitable tank with a speed of 1.5 to 2 kg_{H2}/min. But so far, no commercial application of this is available on the market. Thus, there are no reliable cost estimates, but it is expected that the cost for such storage (liquefaction, storage, re-gasification) exceeds the cost of a high-pressure storage, even at a pressure level of 700 bar and above (Hirscher, 2010; Paggiaro et al., 2010a; Emonts and Stolten, 2016a; Ahluwalia et al., 2010; Stolten et al., 2016).

5.2.2.3 Markets and Perspectives

The technology for the liquefaction of hydrogen has been proven in space applications (e.g., the Space Shuttle and Ariane rocket are fueled with liquid hydrogen). Therefore, it is possible to use this technology for some very specific applications (where costs do not play a major role). But for the time being, liquid hydrogen storage is not seen in the mass market for cars or other privately used technologies. Even if some automobile companies have developed such storage systems for passenger cars and performed many expensive tests, it is not expected that this storage option will be seen within the broader mobility market in the years to come. Additionally, the spread of this type of storage tank is also not supported by the fact that the refueling stations under operation and/or under construction are designed basically only for the provision of high-pressure hydrogen. So far, the most promising application for liquefied hydrogen is its use as a fuel for airplanes equipped with hydrogen turbines.

Overall there is limited scope for the storage of hydrogen in a liquid form. But when liquid hydrogen is used, most likely it will only be economically feasible for large amounts of hydrogen to minimize the boil-off losses (that is, to minimize the surface-volume ratio) (Klebanoff, 2013; Schlapbach and Zuttel, 2001).

5.2.3 Cryo-Compressed Hydrogen

Cryogenic hydrogen can also be stored under pressure; this option is called cryo-compressed storage. Besides the storage of liquid hydrogen, this is a storage option characterized by a relatively high volumetric energy density that does not change the chemical appearance of hydrogen.

5.2.3.1 State of Technology

In this storage concept, hydrogen is cooled down to the temperature of liquid hydrogen (20*K*) and additionally hydrogen is compressed to a pressure of up to 350 bar or even 700 bar. Under these challenging conditions, an energy density of maximum 7 wt% and maximum 0.07 kg_{H2}/l may be achieved. This is only possible at the expense of an enormous amount of energy and technological effort to cool and compress hydrogen. Additionally, the demands in the design of a storage vessel containing a fluid at such a low temperature level for a long time and at such a high pressure, are extremely high. To describe it in a simple way, usually a type IV pressure tank (Section 5.2.1) covered with the vacuum insulation typically used for liquefied hydrogen tanks (Section 5.2.2) is needed (Aceves et al., 2010; Brunner et al., 2016).

Due to these challenges, a technically less demanding compromise aims at realizing cyro-compressed hydrogen storage tanks characterized by around 4 bar and 50 K. This leads to a storage density of 5 wt% hydrogen inside the fully-loaded tank (Blagojević et al., 2012). Other concepts currently under development that are in various development stages use different storage parameters to try to optimize the amount of stored hydrogen in relation to the necessary technical effort.

5.2.3.2 Characterization

The cost of a cryo-compressed storage tank is in a range of 12–30 US$/kWh. Due to significant technical uncertainties, these values are more along the lines of the expected cost than the real cost for the few prototypes that exist. The storage efficiency is characterized by values somewhere in between the storage efficiency of liquid and compressed hydrogen; 41% is a realistic order of magnitude for the time being (Stolten et al., 2016; Ahluwalia et al., 2010).

5.2.3.3 Markets and Perspectives

Several attempts have been made to implement and test cryo-compressed storage for mobile applications. This is especially true for application in passenger cars because the high energy density is very attractive for this application. For example, a Toyota Prius has been equipped with a cryo-compressed storage tank and has achieved the longest driving distance so far (1050 km) with a single tank filling for a hydrogen-powered car (Aceves et al., 2010). This storage tank was a specially designed cyro-compressed storage tank filled either with liquid or compressed hydrogen; in the case it is filled with liquid or cryo-compressed hydrogen the storage tank contains 2–3 times more hydrogen than a storage tank of compressed hydrogen.

Cryo-compressed hydrogen storage has some advantages in comparison to other options (e.g., high storage capacity) and opens the technical possibility to fill the storage tank either with compressed, cryo-compressed, or liquid hydrogen. But so far cryo-compressed tanks are not available on the market and the longest possible time without losses if unused is seven days so far; compared to liquid hydrogen storage, this is around seven times longer (Brunner et al., 2016).

Because of these challenges it is unlikely that cryo-compressed storage will become a considerable part of the consumer market in the years to come. Additionally, all currently built refueling stations are designed for the distribution of compressed hydrogen. This might not be true for highly specialist fleets of company vehicles (e.g., trains, planes, ships), which might be refueled in only one place or a very limited number of places.

5.3 MATERIAL-BASED STORAGE

Besides the three options already discussed for the storage of pure hydrogen, there are also many other possibilities. These other options have one thing in common: without exception the hydrogen reacts with or accumulates on other materials based on reversible physical effects or chemical reactions. Such options are discussed below, limited to storage concepts that are the most promising based on the current state of knowledge. In interpreting the statements made below, it is important to keep in mind that all of the material-based hydrogen storage options are not market ready at this moment. So far, they are still in an advanced, or even an early, research and development (R&D) phase; some of them exist already in the demonstration phase, or even in small pilot applications, while others are still in at the early lab scale.

The border between the group of material-based storage options and the various options for storing hydrogen in a pure form is not always clearly defined, because there are numerous combinations of these different possibilities. Such combinations can be very promising because they might allow some system specific drawbacks of a single-storage option to be overcome. For example, in the most current applications of Metal Organic Frameworks (MOFs; Section 5.3.3) and in activated carbon (Section 5.3.4), hydrogen is stored also under cryogenic conditions.

The four storage options discussed below should not be considered a complete list of all approaches. There are many more options under investigation by public research organizations, such as universities and state-owned research organizations, as well as by small startup and large-scale companies globally; some examples are nonmetal hydrides, ammonia, and carbohydrates. Basically, however, the functionalities of these various possibilities are mostly very similar to the ones described below and the main difference is the material or material combination used. Nevertheless, research is ongoing in this area and it is expected that new ideas, other materials, innovative combinations, and alternative concepts, including the combination of different options, will come up.

5.3.1 Metal Hydride Storage

The storage of hydrogen in metal hydrides is a long-known technology. In the 1980s, there were some field tests with metal hydride hydrogen storage tanks for a car fleet. The storage tanks tested in those days had a weight of 140 kg and had a storage capacity of 1.4 kg of hydrogen (1 wt% of the overall tank mass). TiFe was used as an active material in these demonstration activities. This metal hydride allows a theoretical hydrogen load of 1.8 wt%. Since then a lot of progress has been made. Today a modern complex light-weight metal hydride (e.g., $LiBH_4$) is characterized by a theoretical hydrogen capacity of up to 18 wt %, but this storage capacity has not been verified experimentally so far (Wenger, 2009; Thiangviriya and Utke, 2016).

5.3.1.1 Principle

Metal hydride storage is based on chemical adsorption, that is, hydrogen atoms diffuse inside the metal lattice and react there with the metal to form metal hydride. This is the reason why metal hydride storage is often compared to a sponge. To understand the principle of metal hydride storage, it is essential to look into the reaction mechanism of the material. Due to the fact that the reaction takes place inside the metal hydride, it is very important to optimize the structure in a way that supports the diffusion of hydrogen inside the metal lattice.

The general reaction of metal hydrides is shown in Eq. (5.1).

$$H_2 + \text{Metal} \rightleftharpoons \text{Metal Hydride} + \Delta H_R;\quad \Delta H_R > 0 \qquad (5.1)$$

The chemical reaction of metal with hydrogen to metal hydride is exothermic. Thus, the backward reaction (the release of hydrogen from the metal) is endothermic. Therefore, in the storage of hydrogen in metal hydrides, there is always a heat flux that has to be taken into consideration during use.

This reaction, according to Eq. (5.1), proceeds in four steps (Fig. 5.3). At first the molecular hydrogen reaches the metal surface (Fig. 5.3A). There hydrogen dissociates into atomic hydrogen (Fig. 5.3B). Then the atomic hydrogen dissolves into the crystal lattice of the metal (Fig. 5.3C). Afterward, the atomic hydrogen reacts with the metal to form metal hydride with the release of heat, because the hydrogenation of metal is exothermic (Fig. 5.3D).

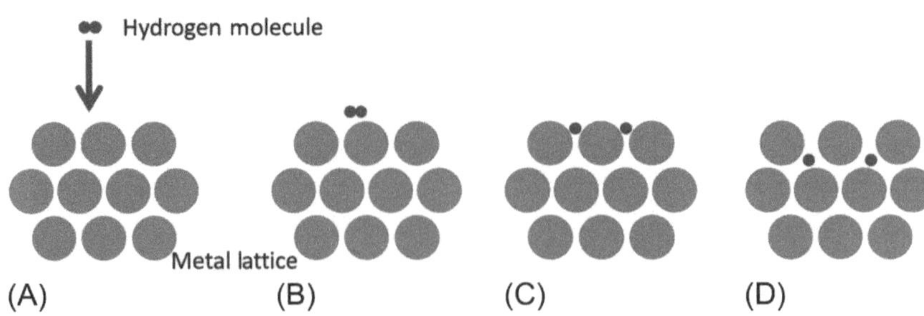

FIG. 5.3 Adsorption of hydrogen into a metal. (A) A hydrogen molecule reaches the metal surface; (B) hydrogen dissociates at the surface; (C) the atomic hydrogen dissolves into the metal lattice; (D) hydrogen reacts with the metal to metal hydride.

For dehydrogenation the same procedure takes place backward. But this time it is necessary to provide enough thermal energy to allow the reaction to take place, because dehydrogenation is an endothermic reaction.

While hydrogen reacts with the metal lattice, the lattice changes its volume. This means that it expands during absorption and contracts again during desorption. This results in a volume change of up to 30 to 40% that depends on the material and the amount of absorbed hydrogen (Klebanoff, 2013).

5.3.1.2 Characterization and Status

Nearly all elements can form hydrides. They are sorted into three different groups outlined below in order to give an overview of the differences between them. The main indicator of each group is the chemical bond of the element with hydrogen.

1. *Ionic or saline hydrides*. Here, all alkali metals and alkaline earth metals are included. These hydrides often have relatively similar physical properties to the original metal; that is, the crystal structures, the hardness, the brittleness, and the other properties do not change much due to the reaction with hydrogen. Hydrogen acts only as a negative loaded ion within these hydrides. Typical hydrides of this group are lithium hydride (LiH), sodium hydride (NaH), and calcium hydride (CaH_2) (Hirscher, 2010; Klebanoff, 2013).
2. *Covalent hydrides*. These hydrides are a combination of a nonmetal and hydrogen. Here the linking forces between the molecules are van der Waals forces (relatively weak forces). Therefore, most covalent hydrides are liquid or gaseous at room temperature. Some examples are water (H_2O), methane (CH_4), and aluminum borohydride ($Al(BH_4)_3$). This group does not form hydrides easily. Thus, most covalent hydrides cannot be built via a simple reaction of hydrogen and the element; a considerable effort is required for the synthesis of such complex chemical reactions, or as in the case of water, it is the reaction product of the combustion. This is the reason that there are no promising candidates for use in hydrogen storage or, in case of water, that it is thermodynamically impossible to store hydrogen in this form (Hirscher, 2010; Klebanoff, 2013).
3. *Metal hydrides*. In this kind of hydride, hydrogen acts as a metal so that there is a metal-hydrogen bond in the conduction band of the metal atom. These hydrides are formed by transition metals. In contrast to the ionic and saline hydrides, some physical properties of the metal hydride change significantly (e.g., they become brittle). But they still have a high thermal and electrical conductivity (Hirscher, 2010; Klebanoff, 2013). These metal hydrides can be sorted into two groups according to their stability. Group A contains metals that form easily stable hydrides (e.g., Li, Na, V). Group B metals, on the other hand, do not form stable hydrides. By combining these two different metal groups it is possible to create complex metal hydrides. Such a combination of both groups has the advantage that some of the material

properties can be modified to the range required by the application. Mostly the reaction temperature and pressure are adjusted by such modifications (Hirscher, 2010). Some of the currently most promising materials for this application are sodium alanate ($NaAlH_4$) and lithium boron hydride–magnesium hydride ($LiBH_4$-MgH_2). For example, $LiBH_4$ has a theoretical hydrogen capacity of 19wt%, which is the highest possible for metal hydride storage, but unfortunately the highest hydrogen loading realized in a lab has been 10wt%. Another challenge is that many metal hydrides only allow around 20cycles or less before the hydrogen concentration decreases rapidly.

The thermal management is very important in the technical realization of hydrogen storage based on metal hydride. The storage tank needs to be cooled during charging and heated up during discharging. Thus, it is helpful if the hydrogen consumer produces some waste heat that can be used for heating the storage tank. In this case, the temperature level is important because in most mobile applications a PEM-Fuel Cell is used. This fuel cell operates at temperatures around 100°C or less. Thus, in combination with a metal hydride storage tank, which needs much higher temperatures, an additional heating device is needed (e.g., part of the electricity produced by the fuel cell or a share of hydrogen is used to provide high temperature heat); this lowers the overall storage efficiency.

Metal hydride storage tanks have some advantages compared to other storage options. They are characterized by a relatively high volumetric energy density. Depending on the material used, such storage devices could operate at ambient pressure, and they are able to store hydrogen at ambient temperature when there is no hydrogen demand. Additionally, a very compact construction is possible. Also, hydrogen coming out of the metal hydride is ultrapure. But there are also some disadvantages. The gravimetric energy density is low, which leads to a very heavy storage tank. Some metal hydrides require very high temperatures for the reaction; the thermal management of metal hydride storage tanks is one of the major challenge to be solved successfully during the development of such storage devices (Capurso et al., 2016). Additionally, the filling of this type of storage tank is quite slow and the thermal management is complex and demanding.

5.3.1.3 Markets and Perspectives

There are already some applications for which metal hydrides are used as hydrogen storage.

One popular example of the use of metal hydrides in hydrogen storage is in submarines, in which a fuel cell is integrated into the propulsion system. To release hydrogen from storage, it is heated with the cooling fluid of the fuel cell. Under these very special frame conditions metal hydrides have some advantages, which are not applicable for other mobile applications. Additionally, due to the high weight, the use of extra weight in the form of lead can be reduced within the submarines; lead is typically used to stabilize a submarine. Also the high volumetric energy density is advantageous, because space is scarce in submarines (Emonts and Stolten, 2016a).

Another example is a small delivery scooter, which has been designed for the French post. This car was tested in the "MobyPost project" and it used a fuel cell and metal hydride storage for the propulsion system. The metal hydride storage tank inside the car could store about 0.9 wt% of hydrogen (MAHYTEC, 2015).

Furthermore, there are commercially available scooters equipped with a hydrogen tank called "H2Tank2Go" that are based on metal hydrides. A stored hydrogen amount of 1.8 wt% in the material is claimed for this tank (Zoz GmbH, 2011). Another related storage tank contains a reactive hydride composite (RHC) made of $LiBH_4$ and MgH_2; this allows a storage capacity of around 8 wt% (Zoz GmbH, 2016).

These examples show that metal hydrides for hydrogen storage are very promising and have a realistic chance to be implemented in the market at some point. Another advantage is that metal hydride storage vehicles may be refueled at gas stations for compressed hydrogen; that is, they are compatible with the infrastructure for hydrogen cars currently under development. Nevertheless, considerable research and development effort is still needed to provide a cheap, efficient, and reliable storage system with a long technical lifetime.

5.3.2 Liquid Organic Hydrogen Carrier (LOHC)

A liquid organic hydrogen carrier (LOHC) is a fluid that can be loaded and unloaded more or less easily with hydrogen. In such LOHCs, hydrogen can be stored essentially without any loss for a long time at ambient conditions. Additionally, the liquid phase of the storage medium allows some considerable advantages during the storage operation. Thus, this options has been discussed more frequently recently (Müller et al., 2013).

5.3.2.1 *Principle*

LOHCs are organic fluids (e.g., N-ethyl-carbazole (NEC), dibenzyltoluene (MSH), benzyltoluene (MLH)) that are able to absorb and desorb hydrogen in a fluid state. Therefore, the storage principle is comparable with that of metal hydride storage, with the main difference being that metal hydrides are solid and LOHCs are liquid. This fact has a great influence on the storage as well as on the vehicle design. Exemplarily for the various LOHCs currently under investigation, the chemical reaction of NEC is shown in Eq. (5.2).

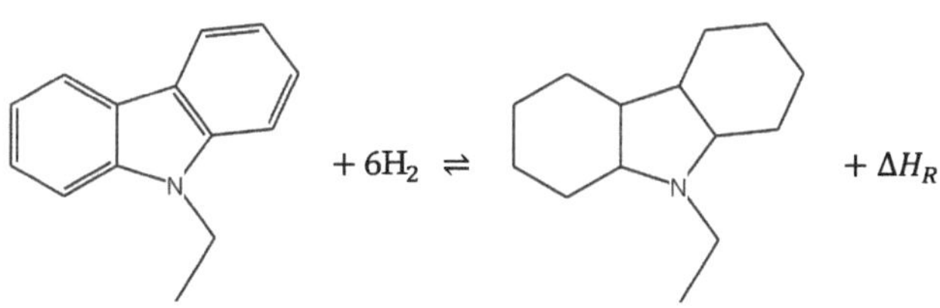

$$+ 6H_2 \rightleftharpoons \quad + \Delta H_R \qquad \Delta H_R > 0 \quad (5.2)$$

During a storage interval, first these fluids are hydrogenated. This is typically realized within a reactor. Here a chemical hydrogenation reaction takes place at pressures between 20 and 50 bar and at temperatures ranging from 150°C to 200°C, depending on the fluid. Typically, this chemical reaction is controlled by a catalyst (e.g., ruthenium or nickel-based).

During the withdrawal cycle, hydrogen is dehydrogenated from the fluid again within a reactor. Here the catalyst-controlled back-reaction typically takes place at a relatively low pressure level and at temperatures of 270–310°C depending on the storage fluid (LOHC). This reaction is controlled by a catalyst consisting, for example, of platinum on an aluminum oxide structure or of noble metals on carbon (Müller et al., 2016; Teichmann et al., 2011).

The liquid storage fluid contains between 5.8 and 7.2 wt% hydrogen. Thus, the efficiency of such LOHC storage is around 44%. The filling rate of LOHCs is comparable to the filling rate of fueling stations for liquid fuels based on fossil energy. Thus, a fill rate of 4 kg_{H2}/min might be possible.

5.3.2.2 Characterization and Status

Such storage systems can be used for mobile applications (e.g., within a car or a truck) following several different concepts (Teichmann et al., 2011). The loaded or unloaded LOHC can be transported and stored within an ordinary tank used typically for fossil fuels. But in contrast to standard tanks, the loaded as well as the unloaded fluid needs to be transported and stored separately. This might be possible by placing a membrane inside the tank, thus subdividing the tank into two different segments with varying volume. This would allow using only one tank instead of two.

The storage of the loaded and unloaded LOHCs might be realized based on the already available technology from tanks for diesel and gasoline. Additionally, a lot of experience is available from the storage of liquid fluids within the chemical industry. Thus, easy solutions can most likely be identified. In addition, a catalytic-controlled hydrogen discharge unit is needed to realize the release of hydrogen from the loaded organic fluid. This process needs heat, clearly above 100°C. But the fuel cell (PEM) used inside cars provides waste heat typically only at a temperature of 100°C. Thus, additional heat is needed, for example, by burning hydrogen; this will lower the overall storage efficiency (Teichmann et al., 2011; Müller et al., 2016). Additionally, a fuel cell needs highly purified hydrogen to avoid destruction of the electrodes. But most likely hydrogen coming from the discharge unit will not be completely pure because the gas might still contain some traces of LOHC or decomposition components of the LOHC. Thus, an additional cleaning step might be necessary.

LOHCs are liquid in a loaded and unloaded state. Thus, such liquids as are under discussion can be easily transported with the known and existing distribution technology used for fossil fuels or chemicals. Thus, it is not necessary to build completely new fueling stations, nor to develop new technologies for the refueling. When a car with a LOHC tank comes to a fueling station, two steps have to be performed. First, the old (dehydrogenated) LOHC has to be removed from the tank, and second, fresh (hydrogenated) LOHC has to be filled into the tank in succession or in parallel. While it is only required to change the fluid, the refueling process is very fast and might be comparable to the fueling times currently needed for gasoline or diesel.

5.3.2.3 Markets and Perspectives

One major advantage of LOHCs is the possibility to use the existing infrastructure with this kind of energy storage to a certain extent. This is especially true, because the LOHCs currently under discussion show more or less the same properties as gasoline at ambient conditions (Teichmann et al., 2011). Thus, it is easy to transport LOHCs in large quantities to the various fueling stations with typical trucks, as is done with other conventional fuels.

There are already some examples of LOHCs in stationary applications. These show that the technology could be used to store large amounts of hydrogen (Hydrogenious Technologies GmbH, 2016). Furthermore, there has been some effort to implement LOHCs as hydrogen storage for large-scale fuel cell ships or for powering large-scale combustion engines on ships. For this application, a high refueling rate is most important, because a very large amount of energy has to be refueled in a short time.

Concepts for this technology, especially for mobile applications, are under development. Most of them are dedicated to ships, because the machinery to load and unload the LOHC requires a large volume. That is also one of the reasons that this technology is only commercially available to bring hydrogen to industrial facilities which do not have hydrogen production on their own grounds.

For other applications, it is not expected that this storage technology will become market mature in the years to come. One reason is that the side aggregates are very expensive. Additionally, hydrogen is not yet fully clean and it might be necessary to add another cleaning step before hydrogen can be used, for example, within a fuel cell. Besides this, the loading and unloading is relatively energy intensive.

5.3.3 Metal Organic Framework (MOF)

A Metal Organic Framework (MOF) consists of coordinated polymers combined from metal ions, metal oxide cluster, or inorganic parts of larger dimensionality. The structure is composed of organic ligands, which coordinate the inorganic 1D chains or 2D layers that form in combination a regular structure able to absorb hydrogen. The hydrogen accumulates at the corner with the inorganic part of the materials (Hirscher, 2010).

5.3.3.1 Principle

Here hydrogen is stored with physisorption inside the porous media. Hydrogen is adsorbed and fixed with van der Waals forces at the surface of the inorganic part of the pores of the MOF. This is possible because the metal organic framework (MOF) is usually characterized by a highly porous structure resulting in a very big surface area per gram (approximately $1000\,m^2/g$). But this type of storage works only at relatively low temperature (below 100 K), allowing it to store 5–7.5 wt% hydrogen.

One of the first-used MOFs (MOF-5; Fig. 5.4) has an inorganic part (Zn_4O unit), that is connected with 1,4 benzene-dicarboxylate within a regular cubic structure. Thus, in comparison to activated carbon (Section 5.3.4), different possibilities exist to optimize the adsorption behavior for hydrogen. On one hand, the surface area could be increased by using shorter organic link molecules. On the other hand, the adsorption enthalpies could be optimized by using organic link molecules that have hydrogen or oxygen terminations and are not dominated by graphitic hexagonal groups (Klebanoff, 2013).

5.3.3.2 Characterization and Status

Another example of a MOF is the material MIL-101. Bimbo et al. (2016) have tested whether it is possible to increase the volumetric storage density, or to reduce the pressure with equal storage density, in comparison with 700-bar

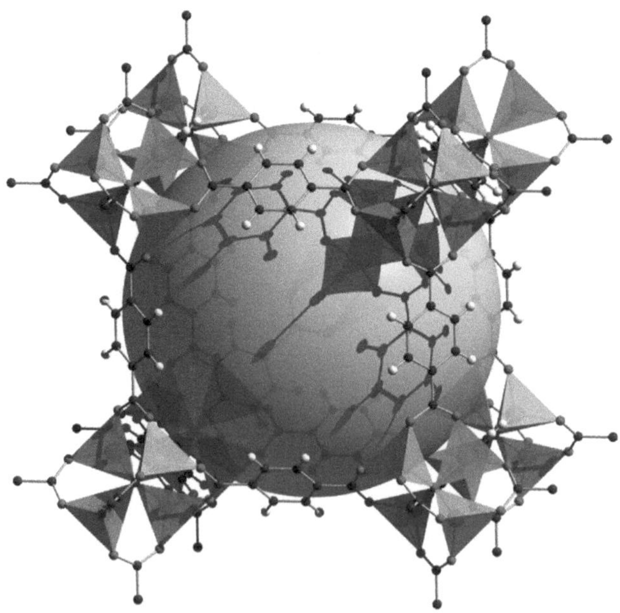

FIG. 5.4 MOF-5 Structure (the triangular parts are the inorganic component, the thin lines and small balls are the organic components, and the big ball represents the free pore).

pressure tanks based on several simulations and experiments. The main aim was to reduce the refilling pressure while filling the storage tank with hydrogen at a temperature of 70 K, and then to store it at ambient temperature. The idea was that no expensive insulation is required. The result was that the refilling pressure could be reduced significantly for an empty pressure tank, as well as for a tank filled with MOF MIL-101, if hydrogen was cooled down to 70 K. This temperature was chosen because it is the temperature of liquid nitrogen, which is technically easy to realize. The crucial factor here is the pressure that should be reached after the refilling process. The best results are achieved when the pressure in the tank should reach 350 bar at 298 K. At these conditions, in all analyzed options, the loading pressure is much lower than the final pressure. Therefore, the loading pressure for the tank without MOF is 72 bar, for the half-filled tank 22 bar, and for the tank fully loaded with MIL-101 3 bar. So, the compression work is reduced by about 69 bar in that case. The final storage pressure at 298 K is in all three cases nearly equal, ranging from 340 to 350 bar. With such a system, it is possible to reduce the effort of filling the storage tank even if the forces that keep hydrogen closer together only work during the filling process. Unfortunately, this concept only works up to a final pressure of 350 bar, because with higher final pressure the storage pressure of the half-filled and fully filled tank rise substantially above the aim of 700 bar; this concept creates then a new safety problem because of the high pressure (Bimbo et al., 2016).

5.3.3.3 Markets and Perspectives

So far only small test storage tanks, or experiments with a small amount of material, exist. Thus, it is not expected that this technology will become market ready in the years to come. One advantage of this concept is the possibility to combine this storage with other options to lower, for example, the refueling pressure of high-pressure storage tanks or to maximize the energy density in cryo-compressed storage tanks. In addition to the efforts to use different MOFs as hydrogen storage, MOFs can also be used for methane storage. Perhaps MOFs will be used in combination with cryo-compressed tanks. This option needs much more R&D before reliable statements can be made about possible market introduction.

5.3.4 Activated Carbon

Activated carbon is a highly porous material, widely used for many different industrial, chemical, medical, and other applications. This material is very interesting for hydrogen storage because hydrogen can be adsorbed on the large free surface of the pores. Thus, quite similar to the MOF storage, activated carbon accumulates hydrogen on the free surface.

5.3.4.1 Principle

In activated carbon, hydrogen is stored by adsorption at the surface of the carbon based on van der Waals forces. These forces are typically not stable enough to hold hydrogen at ambient temperature. Thus, hydrogen needs to be cooled to cryogenic temperature (approximately 70 K before storage). Therefore, oftentimes, activated carbon is combined with cryo-compressed storage to optimize the volumetric energy density of the vessel (Bimbo et al., 2016). Additionally, carbon is easy to handle, has a low cost of synthesis, and is light weight. Therefore, this material has been investigated by many different research groups (Burchell et al., 2017; Javaid, 2017; Klebanoff, 2013).

Although storage in activated carbon usually employs the powder form, there are also other possibilities to store hydrogen in carbon, such as graphite, carbon nanotubes, and activated carbon fibers; all these options are characterized by a high specific inner surface on which the hydrogen can accumulate. The main difference between the various modifications of porous carbon is the appearance.

- Graphite is usually formed in trigonal planer sheets leading to a relatively small surface of 20 m^2/g. This results in a very limited capacity for hydrogen storage. Therefore, graphite needs to be modified by implementing spaces between the different planes to allow for the accommodation of hydrogen molecules. These pores are called slit-pores, and they have an optimal size of twice the diameter of a hydrogen molecule; that is, both layers could hold one molecule. But under these conditions, hydrogen only absorbs at supercritical temperatures. Thus, further modifications are necessary, and this is a subject of intensive research. Because of these challenges, it cannot be foreseen whether or not this option can be more promising than storing pure hydrogen at the same conditions (Klebanoff, 2013).
- The hydrogen-adsorbing properties of carbon nanotubes are very similar to activated carbon (see below).
- Activated carbon fibers have a high micropore volume that is easily accessible from the surface. The good accessibility of the pore volume may be traced back directly to the fiber form, because there are many more open pores directly at the surface. But the carbon fibers used for hydrogen storage so far are in the same range of hydrogen storage conditions and storage parameters as the best activated carbon (like AX-21, Maxsorb). Therefore, activated carbon powders are mostly used in most of the ongoing research activities because the packaging of these materials is typically easier.

The best performance of granular activated carbon that has been reached at room temperature so far is close to 1 wt% at 100 bar and around 2 wt% at pressures up to 400 bar. If the material is loaded at low temperatures (77 K), it is possible to reach 5.5–6 wt% at 20 bar. Thus, this is the configuration that has been most used so far (Klebanoff, 2013). An amount of up to 8 wt% of hydrogen has been stored at cryogenic temperature (77 K) and high pressure (400 bar) (Burchell et al., 2017).

5.3.4.2 Characterization and Status

Most prototype tanks that store hydrogen in highly activated carbon (e.g., AX-21) work usually at a temperature of around 70 K; that is, mostly cryogenic storage is realized with physisorption, similar to MOF storage (Paggiaro et al., 2010a; Paggiaro et al., 2010b; Ahluwalia and Peng, 2009).

Recently a research project has been carried out on a cryo-compressed storage tank filled with activated carbon (Bimbo et al., 2016). Here the influence of the refueling conditions on the possible increase of the volumetric energy density has been tested. Therefore, the loading pressure as well as the temperature has been reduced. After loading, the temperature increases slowly to room temperature and in parallel the pressure rises. The goal is to reach the standard pressure of compressed hydrogen at room temperature and to reduce the energy for the hydrogen compression. The results are similar to the results described above for MOF storage (Section 5.3.3), so the reduction of the loading pressure is possible. However, at most of the values investigated, AX-21 has a tiny drawback in comparison to the MOF. As described above this is only possible up to a pressure of 350 bar, because when attempting to reach 700 bar, the pressure always increases up to 816 bar, leading to severe safety risks. However, there is still an advantage related to the necessary energy for the loading of the high-pressure storage. Also, it is possible to increase the volumetric hydrogen density of the storage with activated carbon to 13.1 kg_{H2}/m^3 instead of 0.3 kg_{H2}/m^3, without any active material inside the tank at the right temperature (Bimbo et al., 2016).

Sometimes activated carbon is tested at room temperature (298 K) (Schaefer et al., 2016). But under these conditions, the hydrogen stored inside amounts to only 0.59 wt%. Even though this is an easy and safe storage option, the question arises whether such an approach can be economically viable within a competitive market.

5.3.4.3 Markets and Perspectives

At the moment there is no market-ready hydrogen storage with activated carbon. However, compared to MOF, a lot of experience with this material is already available and it can be produced easily and cheaply. Thus, if hydrogen storage based on activated carbon is to be introduced, it is foreseeable that this storage will most likely be a combination of activated carbon and a cryo-compressed storage vessel. The reason concerns the weak forces that lead to the physisorption of hydrogen inside the pores of the activated carbon. This works best at cryogenic conditions. But it is not foreseeable that in the years to come such concepts will become market ready, because all projects are currently at the lab scale.

5.4 COMPARISON

Hydrogen storage concepts for mobile application have to fulfill numerous demands, and thus need to be compared using various factors, as it is not enough to meet the expectations and overarching goals in only one aspect. The figures used for characterization that are assessed here are defined in detail below. A distinction is made among them between technical values, economic figures, market aspects, and the research and development (R&D) status.

When possible, the values summarized below will be compared with the DOE targets. These targets, published by the US DOE, represent the internationally accepted technical and economic development targets for hydrogen storage systems for the time horizon 2020 as well as the overall long-term goals (so-called DOE ultimate targets). They are shown in Table 5.1 and will be compared with the single values in all figures, where such targets exist.

- Technical values. Under this category, the technical aspects of the different technologies are summarized.
  - Gravimetric energy density. This value describes the amount of hydrogen or energy that can be stored in 1 kg of storage material; it is expressed in wt% of hydrogen inside the storage material.
  - Volumetric energy density. This figure depicts the amount of hydrogen or energy that can be stored in 1 L of hydrogen storage material; it is expressed mostly in kg_{H2}/L or kWh/L of hydrogen inside the storage material.
  - Overall energy density. This figure combines the gravimetric and the volumetric energy density. Thus, the ideal mobile hydrogen storage material or system is characterized by a high volumetric and a high gravimetric energy density.
  - Well-to-Fuel Cell (WtFC) efficiency. This characteristic value describes the efficiency of the overall hydrogen storage system. It includes all treatment processes of hydrogen after production necessary to reach the right status for the storage within the respective storage system (e.g., liquefaction, compression, cooling) as well as redelivery to the fuel cell at the required conditions (e.g., regasification, pressure reduction, purification).
  - System fill rate. This figure describes the amount of hydrogen or storage material that can be refueled in 1 min; it is expressed mostly in kg_{H2}/min.
  - Storage conditions at idle state. This value depicts the pressure and temperature conditions needed for storage when there is no hydrogen demand; these conditions are expressed in bar and °C.
- Economic figures. In addition to the technical values, the performance of a hydrogen storage system or concept is also characterized by economic figures. Here the following characteristic values are investigated.
  - Storage system costs. This figure describes the specific investment for one kWh stored hydrogen within an overall hydrogen storage system; it is expressed in US$/kWh.
  - Infrastructure costs. This parameter describes the expenditures needed for the infrastructure required by the overall hydrogen storage system.

TABLE 5.1 Comparison of the Values of Different Hydrogen Storage Technologies (Lozano Martinez, 2010; Zoz GmbH, 2016; Nonobe, 2017; Brunner et al., 2016; Toyota, 2016; Long and Head-Gordon, 2015; Durr et al., 2017; Franzen, 2009; Kunowsky et al., 2014; Emonts and Stolten, 2016b; DOE, 2015; US Department of Energy, 2015)

Storage Technology	Gravimetric Energy Density in kg_{H2}/kg	Volumetric Energy Density in kg_{H2}/L	System Fill Rate in kg_{H2}/min	WtFC Efficiency in %	Storage Conditions at Idle State		Storage System Cost in $/kWh	Investment Cost	Market Proximity Today	Market Proximity 2025–30	Compatibility With the Existing Infrastructure	Research and Development Status
					p in bar	*T in °C*						
Pressure (350 bar)	0.055	0.017	1.5–2	56	350	Ambient	13	Low	Market ready	Established	Yes	Optimization of existing technology
Pressure (700 bar)	0.057	0.04 [a]	1.5–2	54	700	Ambient	20	Low	Market ready	Established	Yes	Optimization of existing technology
Liquid	0.075	0.062[a]	1.5–2	22	1–70	−253 to −244	x	Very high	Demonstrators available	Market ready, for commercial vehicle	No	Basic and applied research
Cryo-Compressed	0.055	0.072[a]	1.5–2	41	50–350	−223 to −203	12–30	High	Demonstrators available	Market ready, for commercial vehicle	With some modifications	Basic and applied research
Metal hydride (LiB + MgH_2)	0.08[a]	0.08[a]	slow	x	Ambient	Ambient	x	Medium	Demonstrators available	Market ready for special applications	With some modifications	Basic and applied research
Metal hydride ($NaAlH_4$)	0.035[a]	0.07[a]	0.5	x	Ambient	Ambient	x	Medium	Demonstrators available	Market ready for special applications	With some modifications	Basic and applied research

Continued

TABLE 5.1 Comparison of the Values of Different Hydrogen Storage Technologies (Lozano Martinez, 2010; Zoz GmbH, 2016; Nonobe, 2017; Brunner et al., 2016; Toyota, 2016; Long and Head-Gordon, 2015; Durr et al., 2017; Franzen, 2009; Kunowsky et al., 2014; Emonts and Stolten, 2016b; DOE, 2015; US Department of Energy, 2015)—cont'd

Storage Technology	Gravimetric Energy Density in kg_{H2}/kg	Volumetric Energy Density in kg_{H2}/L	System Fill Rate in kg_{H2}/min	WtFC Efficiency in %	Storage Conditions at Idle State		Storage System Cost in $/kWh	Investment Cost	Market Proximity Today	Market Proximity 2025–30	Compatibility With the Existing Infrastructure	Research and Development Status
					p in bar	*T in °C*						
Liquid organic hydrogen carrier (LOHC)	0.065[a]	0.056[a]	4	44	Ambient	Ambient	2–20[a]	Medium (conversion of fossil filling station)	Far, for the usage in vehicle	Market ready for hydrogen delivery	No	Basic and applied research
Metal organic framework (MOF-05)	0.05[a]	0.013[a]	1.5 - 2	41	3–350	−203 to Ambient	18–25	High	Far away	Demonstrators inside cryo-compressed storages	With some modifications	Basic research
Activated Carbon	0.054[a]	0.016[a]	1.5–2	39	3–350	−203 to Ambient	x	High	Far away	Demonstrators inside cryo-compressed storages	With some modifications	Basic research
DOE 2020 Targets	0.055	0.04	1.5	60	x	x	10	x	x	x	x	x
DOE Ultimate Targets	0.075	0.07	2	60	x	x	8	x	x	x	x	x

[a]*Data for the storage material without the system.*

This includes the process steps needed to be performed before hydrogen is in the right state for the refueling (e.g., compression, cooling, liquefaction); this includes the cost for the overall required infrastructure. Because of the strongly scattered values, the respective overall investments are here compared to a hydrogen refueling system based on 700-bar tanks, as these are currently the most widely used technology. Thus, this economic parameter is rated as (0) if the investment costs are comparable to this storage system. If they are most likely lower, the system is rated with a (+). If they are higher, it is assessed with a (-) and with a (- -) for significantly higher investments compared to the 700-bar option.

- Market aspects. Under this category, the market-related aspects are considered. This category covers how far the technologies are distributed already and which developments are expected.
  - Market proximity today. This aspect depicts the current development status of a hydrogen storage system. It is rated with a (+) for a market-ready technology, (0) for technology that has been already successfully demonstrated in a vehicle, and (-) for technologies that have been demonstrated on an industrial scale, but not yet in vehicles, and (- -) for technologies that are only available at the lab scale so far.
  - Market proximity 2025/30. This figure describes the expected development status of the hydrogen storage systems for the time horizon 2025/30. Here, the results will be represented with (+ +) for a technology most likely established by 2025/30, (+) for technologies that are market ready for commercially used vehicles, (0) for technologies that can be used for some special applications, and (-) for technologies that are expected to still be in a demonstration or lab status. Always the reference point is the time horizon 2025/30.
  - Compatibility with the existing infrastructure. This figure describes the ability of a storage technology to provide refueling with the already existing infrastructure. The assessment is shown with the following symbols: (+) hydrogen storage system is more or less fully compatible with the existing infrastructure, (0) compatibility with some modifications, and (-) for an incompatible technology.
- R&D status. This figure describes the research and development (R&D) status of a hydrogen storage system. If the technology has already been implemented in cars available on the market (in theory), a (+) is given. If there exits already a demonstrator, it is rated with (0). If the storage technology is only available at lab scale, it is rated with (-).

Based on these characterizing figures, the various hydrogen storage options outlined in Sections 5.2 and 5.3 are discussed in detail below. In addition, Table 5.1 provides an overview of the comparison of the different figures.

5.4.1 Technical Values

Below the technical data on the storage options that have been described are compared to each other. The data are presented according the various characterizing figures defined above. In addition to Table 5.1, Figs. 5.5–5.7 also provide a sound overview of the technical data of hydrogen storages options.

5.4.1.1 *Gravimetric Energy Density*

The highest value occurs in complex metal hydride storage materials (e.g., LiB $+MgH_2$ show a value of 0.08 kg_{H2}/kg). Liquid hydrogen storage is characterized by an energy density of 0.075 kg_{H2}/kg; this is the second highest value of all storage options discussed here. Therefore, liquid hydrogen is being widely discussed for vehicles that need high gravimetric energy density. Most hydrogen storage options (pressure, cryo-compressed, LOHC, MOF, activated carbons) are characterized by a gravimetric energy density of roughly 0.055 kg_{H2}/kg. The lowest values are found in most metal hydrides (e.g., $NaAlH_4$ with a value of 0.035 kg_{H2}/kg). But these values are only comparable to a limited extent because the values for the material-based storage systems do not include the required tank shell. So, it is possible that liquid hydrogen has a higher value in reality, taking the overall storage system into consideration.

In conclusion, to compare these values with the DOE targets, the values for pure storage material (metal hydride ($LiB+MgH_2$)) even exceed the DOE ultimate target of 0.075 kg_{H2}/kg. But there is one important difference between these values; the goal refers to the whole storage system while the figure for

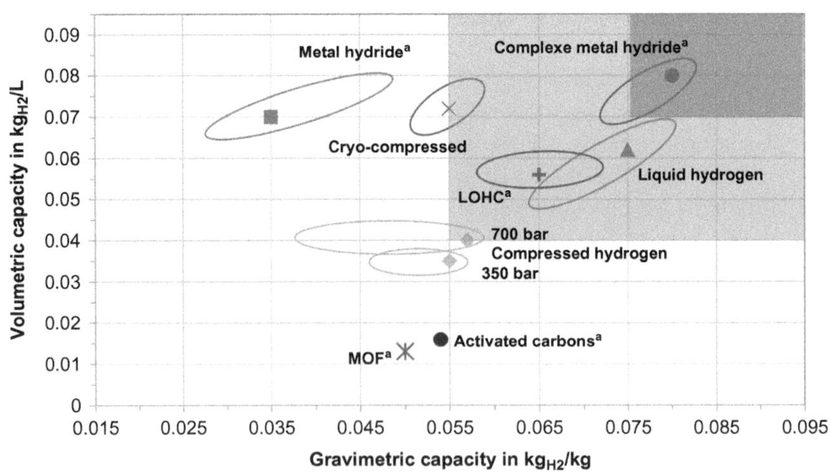

FIG. 5.5 Overview of the capacities of the explained hydrogen storage technologies (*bright gray*: DOE 2020 targets; *dark gray*: DOE ultimate targets) ([a] data for the storage material without the overall storage system).

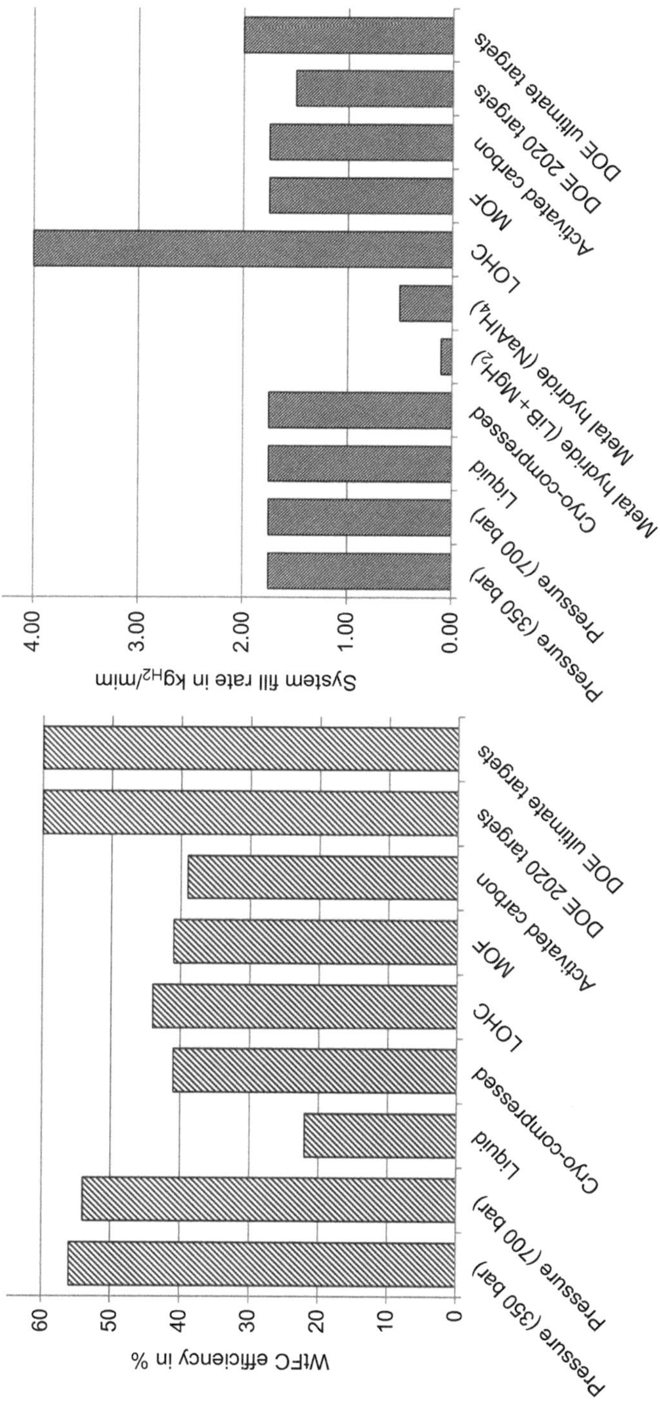

FIG. 5.6 Comparison of the system fill rate and the efficiency of the single storage systems.

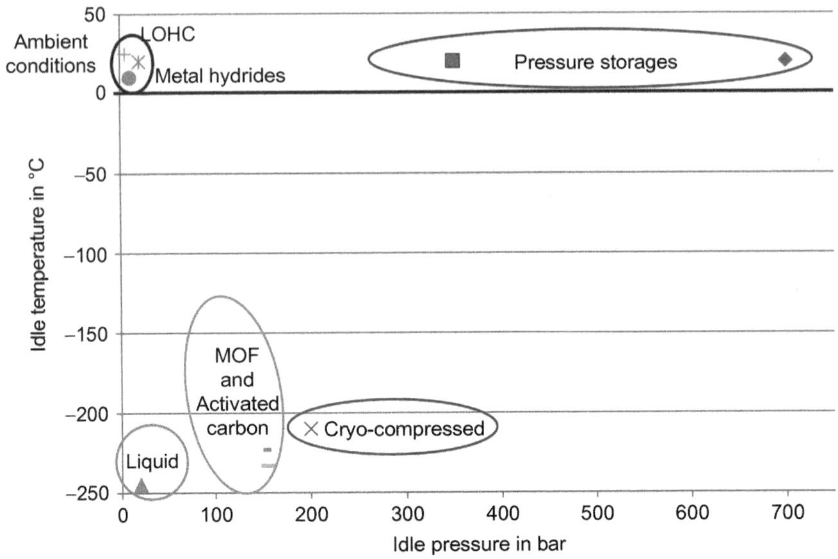

FIG. 5.7 Idle storage conditions of different technologies.

metal hydride takes only the material into consideration. This leads to lower values for the overall system. Therefore, the only system that comes near the DOE ultimate target is liquid hydrogen storage. The DOE 2020 target of 0.055 kg_{H2}/kg_{System} is met by the most advanced high-pressure storage systems and it is possible that some of the material-based storage systems will also meet the target.

5.4.1.2 Volumetric Energy Density

The distribution of the characteristic values for the volumetric energy density within the various analyzed storage options is slightly different compared to the gravimetric energy density (Fig. 5.5). Metal hydride (LiB + MgH_2) again shows high characteristic values, but here all metal hydrides are characterized by high energy densities. Other storage technologies that allow a high volumetric energy density are cryo-compressed hydrogen (0.072 kg_{H2}/L), liquid hydrogen (0.062 kg_{H2}/L), and LOHC (0.056 kg_{H2}/L). Compared to these possibilities, pressure tanks are characterized by a volumetric energy density of 0.04 kg_{H2}/L. The lowest values are found for MOF and activated carbon.

A comparison of these material values with the DOE ultimate targets leads to a similar result as for the gravimetric energy density; the DOE goals are mainly met by the storage materials assessed here, but not by the overall storage system. The only storage system that attains the DOE 2020 targets is liquid hydrogen.

5.4.1.3 Overall Energy Density

The overall energy density (gravimetric and volumetric) describes the characteristics of hydrogen storage systems that focus on compact design and light weight (Fig. 5.5). Metal hydride ($LiB+MgH_2$) shows the best performance on the development goal (a high volumetric as well as a high gravimetric energy density), in general. But liquid hydrogen is also a more promising solution in this respect, because lightweight metal hydride only has high values without the surrounding system. The other options mostly have either high volumetric energy densities or high gravimetric energy densities. Therefore, cryo-compressed hydrogen as well as metal hydrides (e.g., $NaAlH_4$) only have high volumetric values. In comparison, compressed hydrogen, activated carbon, and MOFs are characterized by higher gravimetric values.

5.4.1.4 Well-to-Fuel Cell (WtFC) Efficiency

An overview and comparison of the well-to-fuel cell efficiency of all technologies assessed here is shown in Fig. 5.6. Therefore, both high-pressure storage systems show the highest efficiency values (approx. 55%). The efficiency of most other storage options is roughly 15%-points lower (approx. 40%). Liquid hydrogen storage has the lowest value; this is due to the fact that a significant amount of energy is needed for the liquefaction and re-gasification. Thus, options characterized by less cooling, less compressing, and/or fewer transformation steps show higher values on average (they are more efficient on average). Moreover, the efficiency of the pressure storage systems nearly meets the DOE ultimate target of 60% already today.

5.4.1.5 System Fill Rate

The fastest filling rates can be realized with LOHC. Nearly all other storage technologies discussed here have system fill rates on the same order of magnitude, namely 1.5 up to $2\,kg_{H2}$/min. These values already meet the DOE 2020 targets. This is the reason that the filling rate is not the main focus of the globally ongoing R&D activities at this time. Only hydrogen storage systems based on metal hydrides are slower so far (Fig. 5.6).

5.4.1.6 Storage Conditions at Idle State

This parameter is the more advantageous the closer the pressure and/or temperature conditions are to ambient values (Fig. 5.7). This is true because especially high or very low storage temperatures require an expensive technological effort to keep these challenging temperature and/or pressure conditions throughout the overall storage time; it is challenging from a technological point of view to achieve storage of hydrogen under these conditions for a long time. Metal hydride and LOHC storage have values that are the closest to ambient conditions. The pressure storage options follow, because they are operated at ambient temperature and the high pressure does not generate any leakage over time as would be the case, for example, for liquid or cryo-compressed hydrogen storage systems.

5.4.2 Economic Figures

Economic figures are important to assess market readiness. Thus, the storage system costs, as well as the infrastructure costs, are assessed below. The latter includes the costs of the pretreatment of hydrogen before the refueling takes place and the necessary infrastructure. An overview of the values for each technology assessed here is given in Fig. 5.8.

5.4.2.1 Storage System Costs

A scientifically sound comparison of the storage costs is challenging because most of the options for hydrogen storage systems assessed here are not produced at a technical scale that allows a prediction of the cost after the technology has entered the market and the option is market mature. Thus, the figures discussed below represent only rough first estimations and there are only data available for a few storage options.

High-pressure storage with a pressure level of 350 bar has storage system costs of 13 US$/kWh. This cost is the lowest one at the moment for mobile hydrogen storage systems. All other systems assessed here have estimated costs roughly on the same order of magnitude. This includes high-pressure storage systems with 700 bar, characterized by approximately 20 US$/kWh; this is a commercial value found on the current (limited) market.

The values for the other options are only rough estimates.

- The technology with the widest range is LOHC. Here, the investment costs for the material are expected to be between 2 and 20 US$/kWh. In addition, the storage shell, as well as the hydrogenation and the dehydrogenation reactor, have to be paid for and operated. Thus, currently, in total the costs are expected to be equal or higher than these of the other options.
- The other two technologies for which the possible costs are available are cryo-compressed hydrogen storage systems, with a cost of 12–30 US$/kWh, and MOFs, with an expected cost of 18–25 US$/kWh.

In comparison with the DOE goals for 2020, all of these storage options are still too expensive, and none of them reach the goal of 10 US$/kWh for the time being.

5.4.2.2 Infrastructure Costs

For this assessment criterion, all technologies are compared to the infrastructure investments, including a refueling station, for high-pressure storage systems (350 and 700 bar) as these systems are market mature. Thus, the high-pressure

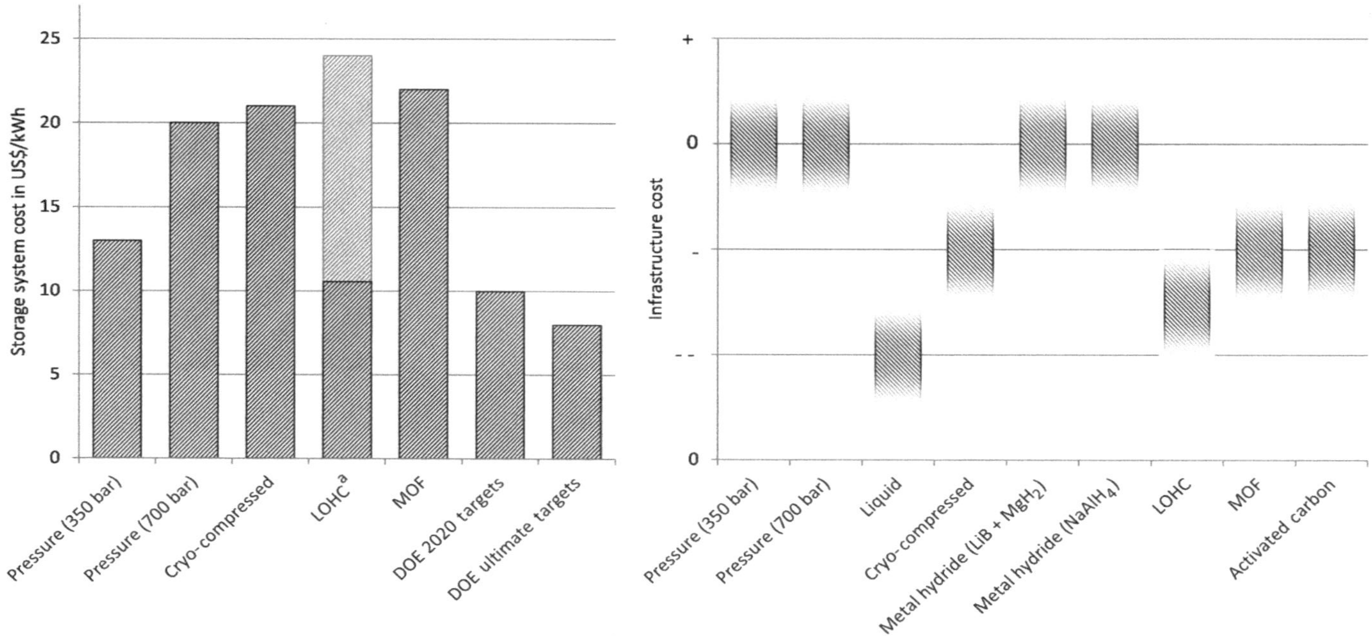

FIG. 5.8 Comparison of the economic figures (Infrastructure cost: 0 comparable costs; - higher costs; - - significantly higher costs); ([a] Data for the storage material without the system, light gray part: expected storage system cost).

storage tank is rated by definition as (0). Additionally, during the assessment, it is taken into consideration whether the infrastructure currently under development can be used or adapted to the other storage systems.

Metal hydride storage systems need, in comparison to high-pressure storage systems, a lower pressure and sometimes higher temperatures. Additionally, some metal hydride storage systems also require cooling during recharging, but the necessary system elements can be installed within the vehicle. Thus, metal hydrides are rated (0) because they can use the high-pressure infrastructure.

In comparison, LOHC needs a completely different infrastructure than that used for pressure storage. However, for LOHC a filling station similar to the gas stations in operation today can, in general, be used. The only difference is that it is necessary to implement a system to pump the discharged LOHC out of the vehicle and store it for recharging. This leads to some technical effort in converting an existing gas station before it can be used for LOHC refueling. Thus, LOHC is rated (0 to -) because there are some changes necessary, but part of the existing infrastructure and know-how can be used.

Activated carbon and MOF require a lower temperature, and depending on the respective storage concept, a lower pressure. The lower temperature especially will increase the effort needed for the necessary infrastructure. Here, they are both rated (-).

The cryo-compressed hydrogen systems have the advantage that it is possible to refuel them with compressed, liquid, or cryo-compressed hydrogen. But if they are to have an advantage in comparison to high-pressure storage, a refueling infrastructure with a very low temperature and a high pressure is needed (comparable to the required infrastructure for activated carbon and MOF). Therefore, this technology is also rated (-).

The last option to consider is liquid hydrogen storage. This technology requires a completely different infrastructure compared to the other options, one that allows for refueling the vehicle with liquid hydrogen. This makes the technological needs high and the investment for the refueling station quite expensive; special material and safety requirements have to be considered because of the extremely low temperatures. Thus, the infrastructure throughout the overall provision chain for the hydrogen pretreatment is very costly. For these reasons, liquid hydrogen storage systems are rated (- -).

There are no currently available technologies that are expected to have lower infrastructure costs than high-pressure refueling. For this reason, no technology is rated with (+).

5.4.3 Market Aspects

Market aspects include all figures that are important for market integration of the hydrogen storage systems for today and in the years to come. These are discussed below (Fig. 5.9).

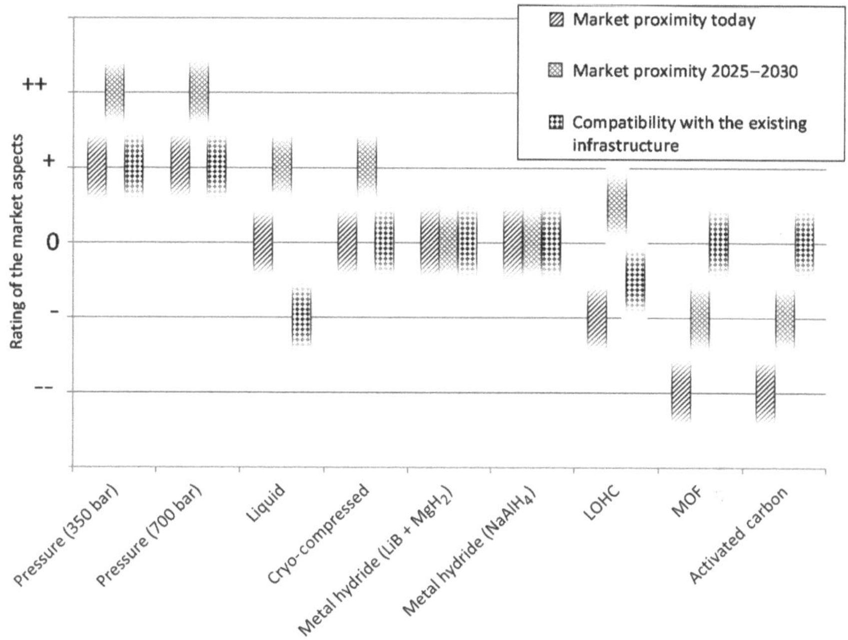

FIG. 5.9 Comparison of the market aspects (+ + very high value; + high value; 0 medium value; - low value; - - very low value).

5.4.3.1 *Market Proximity Today*

An important point for the use of a hydrogen storage technology is the possibility for an upscale to mass production of the storage as well as of the needed material and other auxiliary devices.

At the moment, only high-pressure storage tanks are made in small-series production, due to the fact that hydrogen can be stored in such devices at ambient temperature. Additionally, there are no storage losses during long parking times as would be the case with, for example, liquid or cryo-compressed storage systems. In addition, high-pressure storage tanks show the best efficiency and can be produced based on technologies that have already been developed for other technical applications; this advantage can contribute to cost reduction in the manufacture of such storage tanks. Thus, it is likely that in the coming years, the state of the art storage technology for hydrogen in vehicles will be 700-bar high-pressure storage for private cars and 350-bar storage for urban busses. These systems have the advantage that all available infrastructure for hydrogen mobility that has been built so far delivers high-pressure hydrogen at these pressure levels. Therefore, for other hydrogen storage technologies, it will be very difficult to achieve the same status in the market and/or to compete with high-pressure storage. This observation is supported by all commercially available hydrogen cars that are currently available (e.g., Toyota Mirai,

Hyundai IX35, Honda Clarity, Mercedes GLC F-Cell), as all of them have 700-bar pressure tank systems. All these points lead to a (+) rating.

For some other hydrogen storage technologies, a few demonstrators are available for different vehicles. These include liquid, cryo-compressed, and metal hydride demonstrators. Thus, these technologies are rated (0).

For LOHC, the first commercial systems are available. But these systems are all intended for industrial processes; no vehicular demonstrator exists for the time being. Therefore, LOHC is rated here with (-).

The two technologies that are far away from market readiness are active carbon and MOF. Because these options are still in an early research stage, with only some existing lab-scale applications, the rating for these is very low (- -).

5.4.3.2 Market Proximity 2025/30

In 2025/30 the overall market for mobile hydrogen storage systems will also still be dominated by high-pressure storage systems. They are characterized by an advanced tank shell structure and maybe even a pressure exceeding the 700-bar level. Thus, this storage system is rated here with (+ +).

In the years to come, the use of liquid or cryo-compressed hydrogen storage systems is not expected for cars or other private vehicles. The high temperature difference to ambient temperatures and the resulting boil-off gas are only one important reason. The expensive infrastructure that would have to be developed is another one. Therefore, this kind of storage is perhaps expected for public transportation or heavy-duty vehicles in the next 10 years. Thus, there are limited market opportunities for the time horizon 2025/30; therefore, the rating for these technologies is (+).

LOHCs are characterized by a very fast refueling rate, making this option very attractive for vehicles to be recharged quickly. The same is true in a figurative sense for big consumers (e.g., large ships), which need a large amount of fuel within a relatively short time frame. Additionally, the energy density is comparable to pressure storage, which is the state of the art technology at the moment. The most important drawback of the LOHC storage concept is the hydrogenation and de-hydrogenation device needed to get hydrogen in and out of the LOHC. Therefore, on a medium time horizon, it is expected that this technology will be used to deliver hydrogen to the refueling station only and not for direct use within a fuel cell car. Therefore, the rating is (+ to 0).

Metal hydride storage might become market mature in 2025/30 on a very small scale. Thus, it is rated (0). This technological option can be used for special applications, depending on the respective storage material as well as the given storage demands. An example of a possible application is a high-end sports cars, for which light weight is a very important criterion and the price is not as important as for other commodity markets. Other niche markets might arise for applications in which space is very limited, but weight is not so important (e.g., use in submarines).

By 2025/30, pressure storage systems combined with some of the materials used for material-based storage (e.g., MOF or activated carbon) might gain more maturity in the marketplace. Such combinations allow an optimization of the storage energy density and would most likely avoid the disadvantage of boil-off gas. Thus, there is a chance that such systems might become a realistic option with a time horizon of 10–15 years. Nevertheless, due to numerous uncertainties and the very early R&D-status of such systems, they are rated (-).

5.4.3.3 Compatibility With the Existing Infrastructure

One important point is to use the existing refueling infrastructure for fossil energy or biomass-based transportation fuels. This is true because it is not expected that a completely new hydrogen refueling infrastructure, except for the already partially built high-pressure hydrogen infrastructure, will be built in the next 10 or 20 years in the public domain on account of the economic constraints. Therefore, high-pressure storage is rated with (+).

Most of the other technologies can, in theory, use the high-pressure refueling stations because the 700-bar hydrogen is usually delivered at −70°C. Thus, it is possible to use this already existing hydrogen refueling infrastructure, at least partly, for cryo-compressed storage systems, for MOF, and/or for activated carbon. But to get a very high energy density, hydrogen has to be delivered at even lower temperatures. Thus, these options are rated with (0).

A similar problem is expected for the use of high-pressure refueling stations with metal hydrides, because concepts based on such materials need much lower pressures and also mostly warmer hydrogen. Thus, this option is rated with (0).

Compared to other hydrogen storages options, LOHC can be integrated into conventional fueling stations due to similar characteristics; there might be a chance for integration of this technology within the existing infrastructure. However, some modification of the infrastructure would be required. Thus, this technology is rated (0 to -).

Under the preconditions described, it is not expected that liquid hydrogen can be used for private cars. Thus, this option is rated with a (-). For this technology to be used, a completely new refueling station would be necessary at each location where such vehicles should be refueled.

Concerning infrastructure, there is no big chance for technologies that need their own infrastructure (e.g., liquid hydrogen) in the coming years. This is true because, at the moment, the high-pressure infrastructure has just started to be built and therefore it will take some time before the infrastructure currently under development will be further developed and adjusted to the ongoing developments and latest findings. At the least, the money to be spent has to bring the desired return on investment.

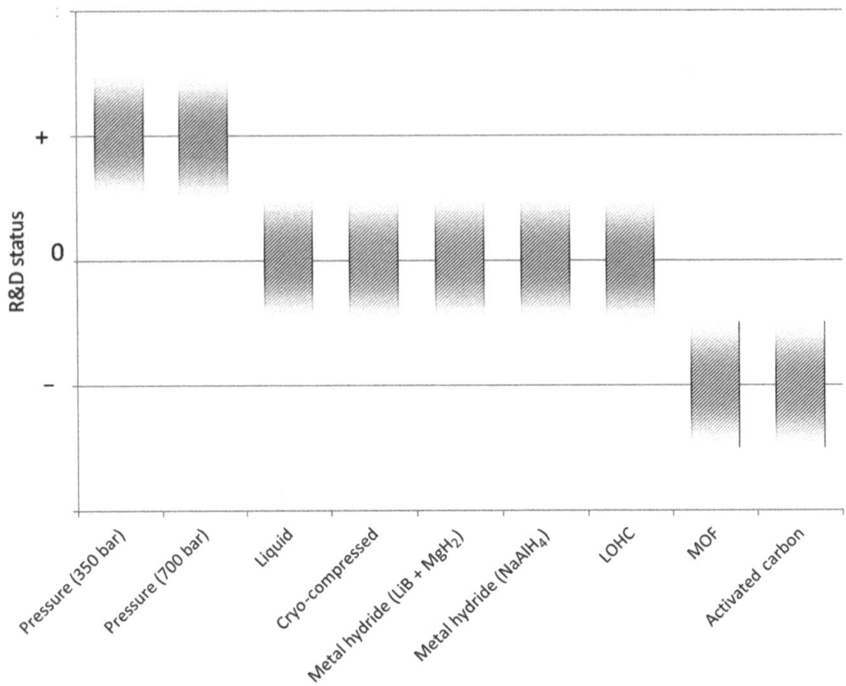

FIG. 5.10 R&Dstatus of the different storage technologies (+: advanced R&D status; 0: applied R&D status; -: basic R&D status).

5.4.4 R&D Status

In addition to the technical, economic, and market related aspects, the actual and foreseen R&D status is another important assessment metric for hydrogen storage systems (Fig. 5.10).

For all pure storage systems (storage of pure hydrogen), at least a prototype that could be used inside a vehicle exists. Compared to this, most options based on material-based storage are still in a basic research status. Even if some prototypes already exist for special materials (e.g., different metal hydrides), so far, the overall concepts have not been much developed. Additionally, it is not clear for the time being which material or material combination has a chance to be developed to a market-ready storage system or that would have the possibility of being scaled up for use in commercial products at a later stage.

Within the single storage options, high-pressure storage is already quite advanced, and therefore the focus is on the further optimization of the shell structure (+).

By comparison, the development of cryo-compressed and liquid storage systems is less advanced. But many companies and research institutions conduct R&D work on these two storage systems because of the high energy densities. Therefore, these options are rated with (0). More or less the same status is found for metal hydride systems; promising results have been published in recent years showing great potential to be exploited in the years to come. Thus, this option is also rated with (0).

LOHCs are different. They are also rated with (0), but here the focus of the research is not on the development of a storage system to be used for mobility of cars. The main focus of the R&D-activities ongoing within the field of LOHCs is currently on the implementation of the technology for transportation of hydrogen from the place of production to industrial facilities lacking their own hydrogen production; the same is true in a figurative sense for the transport of LOHCs from outside Europe to a destination inside the EU. Thus, all four technologies marked with (0) are still in an applied research phase.

The remaining two storage options, MOF and activated carbon, are still in a very basic research stage. Additionally, there is less ongoing R&D activity compared with the other technologies. Therefore, these are rated with (-).

In conclusion, still more R&D effort is needed for all storage options to meet the DOE targets in all categories. But it is expected that the DOE targets for 2020 will be reached by 2025 in nearly all values. The most challenging point is the cost reduction, because there is a large gap between the goal and the actual manufacturing costs.

5.5 FINAL CONSIDERATIONS

After describing and assessing the possible methods of mobile hydrogen storage, it is obvious that the overarching goal of storing hydrogen for mobile applications in a cheap, energy-efficient, and compact way has not been fully reached yet. For the time being, the best and most market mature compromise is storage at a high-pressure level of around 700 bar. This technology represents a good mix between a high energy density, good reliability, easy integration into existing refueling infrastructure, and the necessary cost efficiency. Thus, this option is the only storage technology that is available for mobile application on a commercial scale so far.

However, there are also other technologies that have a chance to become market mature in the years to come, for example, metal hydrides or LOHC. But these options have the highest potential in some special applications or niche markets, such as heavy-duty applications in which weight or temperature are not so crucial.

The decision on which storage solution to use does not only depend on the storage technology inside the car. It is also necessary to provide a compatible provision system (the refueling stations). Also in this respect the compressed-hydrogen storage has a clear advantage for the time being, because all stations, both existing and those under construction, are designed for the refueling of high-pressure storage tanks with either 350 or 700 bar. Thus, for every new hydrogen storage technology implemented on the market in the years to come, it is a precondition that they are based on the infrastructure currently under development (or on the already existing refueling stations as might be the case for LOHC).

Nevertheless, even with the quite promising R&D results in recent years, and with significant progress also in the years to come, it is most likely that high-pressure tanks will stay on the market. It is even more likely that they will gain greater market importance. This will be true even if the ongoing R&D activities are extraordinarily successful.

REFERENCES

Aceves, S.M., Espinosa-Loza, F., Ledesma-Orozco, E., Ross, T.O., Weisberg, A.H., Brunner, T.C., Kircher, O., 2010. High-density automotive hydrogen storage with cryogenic capable pressure vessels. Int. J. Hydrog. Energy 35 (3), 1219–1226.

Ahluwalia, R.K., Hua, T.Q., Peng, J.K., Kumar, R., 2010. System Level Analysis of Hydrogen Storage Options. DOE Hydrogen Program Review.

Ahluwalia, R.K., Peng, J.K., 2009. Automotive hydrogen storage system using cryo-adsorption on activated carbon. Int. J. Hydrog. Energy 34 (13), 5476–5487.

Bimbo, N., Xu, W., Sharpe, J.E., Ting, V.P., Mays, T.J., 2016. High-pressure adsorptive storage of hydrogen in mil-101 (Cr) and Ax-21 for mobile applications. Cryocharging and cryokinetics. Mater Des 89, 1086–1094.

Blagojević, V.A., Minić, D.M., Minić, D.G., Novaković, J.G., 2012. Hydrogen economy: modern concepts, challenges and perspectives. In: Minic, D. (Ed.), Hydrogen Energy—Challenges and Perspectives. Intech Open Access Publisher, London.

Brunner, T., Kampitsch, M., Kircher, O., 2016. Cryo-compressed hydrogen storage. In: Stolten, P.D., Samsun, D.R.C., Garland, D.N., Stolten, D., Garland, N., Samsun, R.C. (Eds.), Fuel Cells: Data, Facts, and Figures. Wiley-VCH, Weinheim, pp. 162–174.

Burchell, T.D., Contescu, C.I., Gallego, N.C., 2017. Activated carbon fibers for gas storage. Elsevier, Activated Carbon Fiber and Textiles, Washington, DC, pp. 305–335.

Capurso, G., Schiavo, B., Jepsen, J., Lozano, G., Metz, O., Saccone, A., De Negri, S., Bellosta Von Colbe, J.M., Klassen, T., Dornheim, M., 2016. Development of a modular room-temperature hydride storage system for vehicular applications. Appl. Phys. A. 122(3).

DOE (Ed.), 2015. Annual Progress Report IV. Hydrogen Storage.

Durr, S., Muller, M., Jorschick, H., Helmin, M., Bosmann, A., Palkovits, R., Wasserscheid, P., 2017. Carbon dioxide-free hydrogen production with integrated hydrogen separation and storage. ChemSusChem 10 (1), 42–47.

Emonts, B., Stolten, D. (Eds.), 2016a. Hydrogen Science and Engineering: Materials, Processes, Systems and Technology. Wiley-VCH, Weinheim.

Emonts, B., Stolten, D. (Eds.), 2016b. Hydrogen Science and Engineering: Materials, Processes, Systems and Technology. Wiley-VCH, Weinheim.

Franzen, J., 2009. Modellierung Und Simulation Eines Wasserstoffspeichers Auf Der Basis Von Natriumalanat. Fortschritt-Berichte Vdi Reihe 6, Energietechnik, Vol. 583. Als Ms. Gedr, VDI-Verl, Düsseldorf. Zugl.: Hamburg-Harburg, Techn. Univ., Institut Für Apparatebau, Diss.

Hirscher, M. (Ed.), 2010. Handbook of Hydrogen Storage: New Materials for Future Energy Storage. Wiley-VCH, Weinheim.

Hydrogenious Technologies Gmbh, 2016. Hydrogenious Produktlinien. Available at: http://www.hydrogenious.net/de/produkte/ (Accessed 21 December 2016).

Javaid, A., 2017. Activated carbon fiber for energy storage. In: Activated Carbon Fiber and Textiles. Elsevier, Oxford, pp. 281–303.

Klebanoff, L. (Ed.), 2013. Hydrogen Storage Technology: Materials and Applications. CRC Press, Boca Raton, FL.

Kunowsky, M., Marco-Lozar, J.P., Linares-Solano, Á., 2014. Activated carbon fibre monoliths for hydrogen storage. Adv. Sci. Technol. 93, 102–111.

Long, J.R., and Head-Gordon, M. Hydrogen storage in metal-organic frameworks, In: Doe, Doe Annual Progress Report, Hydrogen Storage, 2015, pp. Iv-63–Iv-69.

Lozano Martinez, G.A., 2010. Development of Hydrogen Storage Systems Using Sodium Alanate. Tuhh Universitätsbibliothek, Germany.

MAHYTEC, 2015. Solid and Compressed Hydrogen Storage Solutions.

Mori, D., Hirose, K., 2009. Recent challenges of hydrogen storage technologies for fuel cell vehicles. Int. J. Hydrog. Energy 34 (10), 4569–4574.

Müller, K., Aslam, R., Fischer, A., Stark, K., Wasserscheid, P., Arlt, W., 2016. Experimental assessment of the degree of hydrogen loading for the dibenzyl toluene based Lohc system. Int. J. Hydrog. Energy 41 (47), 22097–22103.

Müller, K., Völkl, J., Arlt, W., 2013. Thermodynamic evaluation of potential organic hydrogen carriers. Energy Technol. 1 (1), 20–24.

Nonobe, Y., 2017. Development of the fuel cell vehicle mirai. IEEJ Trans. Electr. Electron. Eng. 12 (1), 5–9.

Paggiaro, R., Bénard, P., Polifke, W., 2010a. Cryo-adsorptive hydrogen storage on activated carbon. I. Thermodynamic analysis of adsorption vessels and comparison with liquid and compressed gas hydrogen storage. Int. J. Hydrog. Energy 35 (2), 638–647.

Paggiaro, R., Michl, F., Bénard, P., Polifke, W., 2010b. Cryo-adsorptive hydrogen storage on activated carbon. II. Investigation of the thermal effects during filling at cryogenic temperatures. Int. J. Hydrog. Energy 35 (2), 648–659.

Schaefer, S., Fierro, V., Izquierdo, M.T., Celzard, A., 2016. Assessment of hydrogen storage in activated carbons produced from hydrothermally treated organic materials. Int. J. Hydrog. Energy 41 (28), 12146–12156.

Schlapbach, L., Zuttel, A., 2001. Hydrogen-storage materials for mobile applications. Nature 414 (6861), 353–358.

Simonovski, I., Baraldi, D., Melideo, D., Acosta-Iborra, B., 2015. Thermal simulations of a hydrogen storage tank during fast filling. Int. J. Hydrog. Energy 40 (36), 12560–12571.

Stolten, P.D., Samsun, D.R.C., Garland, D.N., Stolten, D., Garland, N., Samsun, R.C. (Eds.), 2016. Fuel Cells: Data, Facts, and Figures. Wiley-VCH, Weinheim.

Teichmann, D., Arlt, W., Wasserscheid, P., Freymann, R., 2011. A future energy supply based on liquid organic hydrogen carriers (Lohc). Energy Environ. Sci. 4 (8), 2767.

Thiangviriya, S., Utke, R., 2016. Improvement of dehydrogenation kinetics of 2libh4–Mgh2 composite by doping with activated carbon nanofibers. Int. J. Hydrog. Energy 41 (4), 2797–2806.

Toyota, 2016. Mirai Product Information.

US Department of Energy, 2015. Fuel Cell Technologies Office Multi-Year Research, Development, and Demonstration Plan. Hydrogen Storage. Available at: https://energy.gov/sites/prod/files/2015/05/f22/fcto_myrdd_storage.pdf (Accessed 27 January 2017).

Wenger, D., 2009. Metallhydridspeicher Zur Wasserstoffversorgung Und Kühlung Von Brennstoffzellenfahrzeugen. Zugl.: München, Techn. Univ., Diss., 2009, Fortschritt-Berichte Vdi Reihe 12, Verkehrstechnik/Fahrzeugtechnik, Vol. 699 Als Ms. Gedr, VDI-Verl, Düsseldorf.

Zoz Gmbh, 2011. H2tank2go® – Hydrogen storage cartridge. Fast, reliable and cost effective based on room temperature metal hydride hydrolium®. Available at: http://gmbh.zoz.de/wp-content/uploads/h2tank2go-e.pdf (Accessed 19 December 2016).

Zoz Gmbh, 2016. Komplexe Metallhydride - Wasserstoffspeicher Der Zukunft: Bor4store B4s-Sm/Mm Feststofftank. In: Erster Semi-Kommerzieller Borhydrid Feststoff-Wasserstoffspeicher Fertiggestellt. Wenden, Olpe.

FURTHER READING

DOE, n.d. Doe Annual Progress Report.

Chapter 6

Lowering Energy Spending Together With Compression, Storage, and Transportation Costs for Hydrogen Distribution in the Early Market

Didier Grouset* and Cyrille Ridart†
*Université de Toulouse, IMT Mines Albi, UMR CNRS 5302, Centre RAPSODEE, Campus Jarlard, Albi, France, †HERA—ALBHYON, Technopole Innoprod, rue Pierre Gilles de Gennes, Albi, France

6.1 INTRODUCTION

6.1.1 Hydrogen Supply Chain and Energy Requirements

Future developments in hydrogen fuel cell vehicles promise an important ***decrease in the final spent energy*** and in greenhouse gas and pollutant emissions from transportation. This is mainly due to the qualities of hydrogen fuel cells: high efficiency for chemical to electric conversion, no other product than water, no pollutant emission, and no noise. Nevertheless, to quantify the benefits of hydrogen as a new energy carrier, the ***overall energy chain*** has to be considered, from the primary energy used for hydrogen production to the final step of the useful energy spent for the vehicle movement. Indeed, great care has to be taken so that the decrease in the final spent energy does not induce too large an increase in the energy spent in the intermediate steps of hydrogen ***production, compression, storage, transportation***, and ***distribution***.

Nowadays, everybody would agree that the primary energy for the future large-scale ***production*** of hydrogen energy has to be renewable, through electrolysis of renewable electricity (photovoltaic, wind, hydraulic, etc.), through reforming of renewable hydrocarbons (e.g., biogas), through biomass gasification, and possibly others.

An advantage of hydrogen is that sources for its production are numerous and widespread over the world, so that its production is possible nearly everywhere and close to its valorization location. ***Transportation*** can then be avoided

Hydrogen Supply Chain. https://doi.org/10.1016/B978-0-12-811197-0.00006-3

or reduced to short distances and the costs and energy spending can be saved for hydrogen energy. This is not the case for fossil fuels, which will have to be transported over longer and longer distances of up to several thousand kilometers between their production sites and their distribution sites, resulting in losses of energy and CO_2 emissions; for example, a loss of about 20% of the energy content of natural gas for its pipeline transportation over 5000 km! Currently, with few hydrogen production sites and truck transportation over distances of several hundred kilometers, similar energy spending and CO_2 emissions can be encountered with industrial hydrogen or with hydrogen for energy in the very early market; but solutions under development will eliminate this problem, as shown in this chapter.

Compression of hydrogen for its storage has also to be considered with care; indeed, it is a highly energy consuming operation, as pointed out by Klell et al (2007). Hydrogen is a very light and bulky gas and its compression requires a lot of energy as the green curve in Fig. 6.1 shows. The isothermal compression work from 0.1 to 100 MPa represents >7% of the hydrogen energy content (7.2% of LHV_{H2}, hydrogen low heating value, which is 120 MJ/kg). One should keep in mind that this only represents the mechanical energy transferred to hydrogen in the ideal isothermal compression and that efficiency losses have to be added; any warming of hydrogen during compression, friction in the compressor, or inefficiency of the electric motor will increase the spent energy. Moreover, the efficiency for electric energy production from chemical energy should also be considered (50% in the best cases). Thus, if not processed with care, the compression step could be responsible for more than a 20% loss of the

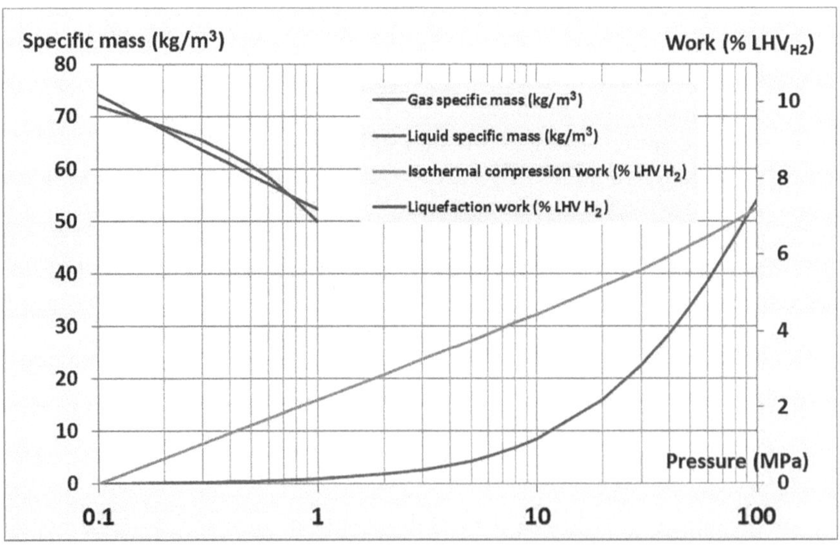

FIG. 6.1 Specific mass and minimum work for compression and liquefaction of hydrogen as a function of pressure, as in Klell et al. (2007).

hydrogen electric energy potential. However, compressing hydrogen to >70 MPa is a necessity to refill vehicle tanks at that pressure! On the other hand, using liquid hydrogen would not be a better solution, as the violet curve of the same figure shows even higher energy needs for the liquefaction of hydrogen. Hopefully, as shown in this chapter, some good practices can lower this energy spend.

The different steps of hydrogen production, compression, and transportation are linked together by the ***storage***, for which different technologies are now matured. Only high-pressure gaseous storage is considered here. Steel bottles or tubes (type 1 vessels) have been used for a very long time and are now challenged in cost, especially for very high pressure, by composite material bottles with an aluminum or plastic internal liner (type 3 or 4 vessels). This study quantifies the savings induced by the light weight of these storage bottles.

Finally, concerning the ***distribution***, many studies have been conducted during the last two decades concerning the most secure way to fill hydrogen car tanks. Some of them have been carried out by international teams that included researchers from different car companies, gas companies, hydrogen technology companies, and research centers, and have been reported, for example, at successive NHA (National Hydrogen Association) congresses, as in Schneider et al. (2005) or Maus et al. (2008). They resulted in the release in July 2014 of a last version of the SAE-J6201 standard (SAE International, 2014) for fueling protocols for gaseous hydrogen vehicles under 35 or 70 MPa pressures, on which are based all current and future refueling stations.

The contribution of compression and storage to the investment cost of a hydrogen refueling station (HRS) is known to be high; following Mintz et al. (2009), it is one-half to two-thirds the cost, according to the HRS size. Thus, it seems important to focus on these costs to understand how they contribute to the overall cost of the delivered hydrogen.

6.1.2 Refueling Principles: Current Practices

In the current practice for rapid filling (Parks et al., 2014; Rothuizen and Rokni, 2014), the vehicle tanks are filled at their nominal pressure simply by pouring available hydrogen from a cascade of buffers, as shown in Fig. 6.2. These buffers have been previously filled with hydrogen at a higher pressure by a compressor connected to the hydrogen production unit, or to a mass intermediate storage. A regulation valve controls the mass flow rate delivered to the tank and the pressure rise in the tank, according to the SAE-J2601 standard (SAE International, 2014).

The performance metrics of an HRS are: delivering pressure, time for filling a tank, and delivering capacity at the peak hour.

- Delivering pressure: two standards, 35 and 70 MPa, coexist for hydrogen vehicle tank pressure.

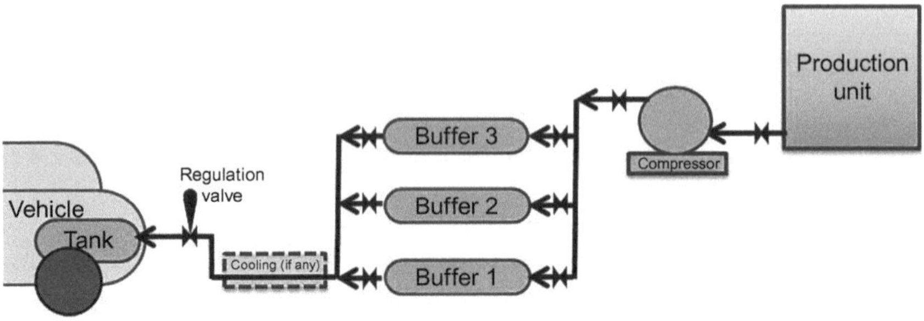

FIG. 6.2 Schematic view of the equipment used for filling hydrogen vehicle HP tanks.

- Time for refueling: the pressure difference between the very high pressure of the buffer storage (e.g., 45 or 90 MPa) and the tank ensures rapid filling. A cooling of the hydrogen is sometimes implemented: it is recommended by SAE-J2601 to allow filling within 3 min in nearly all conditions of ambient temperature and initial tank pressure. Small hydrogen refueling stations do not refrigerate hydrogen and cannot guarantee 3 min refueling.
- The peak hour performance relies mainly on the mass of hydrogen stored in the HP buffers and only incidentally on the mass flow rate of the compressor. In fact, this results from a cost optimization; compressors are expensive and it is cheaper to have a small compressor working nearly all day long to fill buffers than a large compressor with the possibility to fill the tank directly within a few minutes.

Bulk storage at an intermediate pressure is considered as a basic feature of an HRS when connected to a hydrogen pipeline (a 17.2 MPa bulk storage is considered in Parks et al. (2014)), when supplied by trucked trailers (Reddi et al., 2014), or with onsite production (between 20 and 35 MPa in Rothuizen and Rokni (2014)).

The HP buffer storage is usually made of several vessels that can be isolated and connected successively to the vehicle tank. They are always connected to the tank in the same order. The compressor fills the buffers from the bulk storage with a priority for the highest-pressure buffer (number 3 in Fig. 6.2), so that the buffers finally form a pressure cascade, from 35 to 90 MPa for a 70 MPa refueling as recommended in Rothuizen and Rokni (2014).

6.1.3 Content and Objectives

The present chapter is dedicated to the optimization of cost and of energy consumption of compression, transportation, and storage for hydrogen distribution in the current early hydrogen energy market.

Specific emphasis is put on energy needs; whereas costs have often been studied, quantitative information concerning energy consumption are in fact less available. For example, Rothuizen and Rokni (2014) writes as well that the energy consumption of HRSs are unknown; Wipke et al. (2012) reports variations by a factor of 10 for compression energy consumption; Gardiner (2009) gives more general figures for compression or liquefaction energies.

This chapter aims as well to present the equations necessary to calculate the basic design features of compression, storage, and transportation equipment, and to evaluate cost and energy. Simple models are formulated, together with their simplifying assumptions, so that it becomes possible to transpose the approach to specific cases other than those discussed here.

This chapter is concerned with current or near-future implementations of the emerging market for hydrogen energy, especially in France. So it only considers small HRSs (from 20 to 200kg/day), whereas most other papers are concerned with long-term perspectives for a developed hydrogen market with large HRSs; 850, 1000, and 1330kg/day in Parks et al. (2014), 250kg/day in Reddi et al. (2014). The costs generated here are for current implementations; they were issued from commercial consultations in 2015 and 2016 and no reduction factor has been applied for mass production, which will undoubtedly occur in the next decade, especially for composite pressure bottles, but also for compressors.

First, in Sections 6.2 and 6.3, general technical and economic data concerning compression and storage, the way they were obtained, and how to calculate them, is presented and discussed.

Then two different cases are studied. In Section 6.4, the case of a refueling station on the site of the hydrogen production is analyzed. In this case, the energy spent for the hydrogen distribution is linked to the compressor consumption, to the cooling of the compressor, and, if any, to the cooling of the hydrogen before being delivered to the vehicle tank.

In Section 6.5, the case of a production unit providing hydrogen to several distant refueling stations is considered. The different steps leading to hydrogen distribution are described both in the current practice and when using better practice, evaluating for each the energy consumption, CO_2 emission, investment cost, and operation cost in order to estimate globally for these steps (excluding the production step and the final distribution step), the total cost of ownership (TCO) in €/kg, the specific consumption in kWh/kgH_2, and the specific emission in t_{CO2}/t_{H2}.

6.2 TECHNICAL DATA FOR COMPRESSION AND STORAGE

This section presents the general technical data that will be used in following sections. There will also be a discussion of these data with respect to data used by other authors in similar papers, where appropriate.

6.2.1 Thermodynamic Data for Hydrogen

Hydrogen in not a perfect gas and correct thermodynamic data have to be used in order to get good estimations of heat and work exchanged during heating, cooling, or compression. More precisely, the evolutions of the specific mass, ρ, the specific internal energy, u, the enthalpy, h, the heat capacity, c_v, c_p, and the entropy, s, with temperature and pressure (p,T), are needed. Different sources can be used for this. Klell et al. (2007) gives interesting information and (T,s) diagrams at low temperature. SAE International (2014) gives, in its appendix, regressions for ρ, u, h, and c_v as functions of (p,T). Lemmon and Leachman (2008) from NIST derived a state equation for hydrogen in which the compressibility factor $Z(p,T)$ is calculated through a 9-term regression, each term needing three coefficients, according to the formula:

$$Z(p,T) = \frac{p}{\rho RT} = 1 + \sum_{i=1}^{9} a_i \left(\frac{100\,\mathrm{K}}{T}\right)^{b_i} \left(\frac{p}{1\,\mathrm{MPa}}\right)^{c_i} \tag{6.1}$$

The accuracy of this regression is very good; 0.15%, in a large range of p (up to 200 MPa) and T (150–1000 K). Then, knowing $Z(p,T)$ allows a calculation of u and h.

Calculations in this chapter also use hydrogen thermodynamic data available from the chemical data webbook published by NIST Chemical Webbook (n.d.). From the values of $\rho(p,T)$ read in the tables, $Z(p,T)$ can be calculated and polynomial regressions, simpler than Eq. (6.1), have been derived. The NIST webbook also gives the values for u and h, s, c_v,c_p used in this chapter. An example of the result for $Z(p,T)$ variation with p ranging from 0.1 to 90 MPa is given in Fig. 6.3; $Z(p,T)$ increases from 0.99 at ambient pressure to 1.63 at 90 MPa at $T=273$ K.

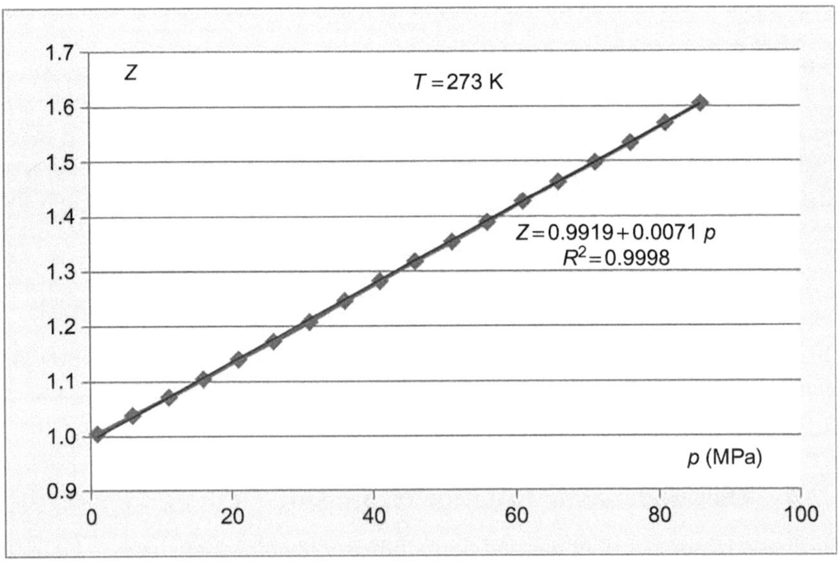

FIG. 6.3 Variation of hydrogen compressibility factor $Z(p,T)$ for $T=273$ K.

6.2.2 Compression Work, Isothermal or Adiabatic

The useful work developed by a compressor is equal to the work of the pressure forces during the volume decrease of the compressed hydrogen. Pressure and volume are linked through the gas state equation, so for a mass m or mole number n, the compression work is:

$$W_{useful} = -\int_{V=V_0}^{V_f} p * dV = -\int_{V=V_0}^{V_f} Z(p,T)\frac{nRT}{V} * dV = -m * {}^{R}\Big/_{M} \int_{V=V_0}^{V_f} Z(p,T) * \frac{T}{V} * dV \tag{6.2}$$

As hydrogen has the lowest molar mass M of all gases, the mass compression work will be the highest. Moreover, hydrogen is a nonperfect gas with $Z_{H2}(p,T) > 1$, which means that the compression work will be higher than for a perfect gas or for methane, which is an imperfect gas with $Z_{CH4}(p,T) < 1$. The high values of $Z(p,T)$ lead to a significant increase of the useful work with respect to a perfect gas.

In the case of an *isothermal* compression of a perfect gas, Eq. (6.2) can easily be integrated and leads to the *useful compression energy*, reference energy as the lowest possible:

$$W_{isothermal,\ perfect} = -m * {}^{R}\Big/_{M} * T_0 \int_{V=V_0}^{V_f} \frac{dV}{V} = m * {}^{R}\big/_{M} * T_0 * \ln\left({}^{p_f}/_{p_0}\right) \tag{6.3}$$

For hydrogen or any imperfect gas, the integral has to be calculated step by step, but the result can be expressed as:

$$W_{isothermal,\ H2} = \overline{Z_{H2}{}_0^f} * W_{isothermal,\ perfect} \tag{6.4}$$

where the coefficient $\overline{Z_{H2}{}_0^f}$ represents the real gas effect, comprised between $Z_{H2}(p_0,T_0)$ and $Z_{H2}(p_f,T_f)$. The result, divided by the hydrogen LHV, is plotted in Fig. 6.1 as a function of p_f for $p_0 = 1$ bar.

Furthermore, compressing a gas leads to an increase in its temperature, increasing its volume, and so increasing also the compression work. If the compression is *adiabatic*, temperature and pressure are linked, so that the final temperature and the *isentropic* compression work of a perfect gas can be expressed as:

$$W_{isentropic,\ perfect} = m * {}^{R}\Big/_{M} * T_0 * \frac{\gamma}{\gamma-1} * \left[\left({}^{p_f}/_{p_0}\right)^{\frac{\gamma-1}{\gamma}} - 1\right] \tag{6.5}$$

$$T_f = T_0 * \left({}^{p_f}/_{p_0}\right)^{\frac{\gamma-1}{\gamma}} \tag{6.6}$$

For hydrogen, an imperfect gas, the result can be written as:

$$W_{isentropic,\ H2} = \overline{Z'_{H2}{}_0^f} * W_{isentropic,\ perfect} \tag{6.7}$$

Eqs. (6.3) and (6.5) clearly show that the main parameter is the pressure ratio $r_p = {}^{p_f}/_{p_0}$ and also that the lower the initial temperature, T_0, the lower is the compression work. In the case of large compression ratios, the isentropic temperature increase can be large and the isentropic compression work will be much larger than the isothermal compression work. For example, for a compression from 2 to 45 MPa, the pressure ratio is 22.5 and

$$T_{f,\,isentropic} = T_0 * (22.5)^{\frac{\gamma-1}{\gamma}} = 2.43 * T_0, \quad \text{so}: W_{isentropic,\,perfect} = 1.61 * W_{isothermal,\,perfect}.$$

It is clear that cooling the gas and the compressor is necessary to lower the compression work. Splitting the compression into several successive intercooled stages is also beneficial.

6.2.3 Compression Efficiency

Moreover, friction and efficiency losses also increase the gas temperature and the electric power needed by the compressor. The question is how to estimate compressor efficiency? A lot of parameters influence the efficiency, and first of all is the compressor technology. Several mature technologies can be found for hydrogen compression, such as reciprocating compressors, diaphragm compressors, and pneumatic or hydraulic boosters.

It is not the purpose of this chapter to present details about these safe and mature technologies, but to focus on information concerning the energy needs in compression. The diaphragm compressor will have better energy efficiency, defined as the ratio between the energy transferred to the compressed hydrogen and the consumed electric energy:

- The friction of pistons in the booster cylinders generates higher losses than the deformation of the diaphragm. It can also be understood that friction will be relatively higher for smaller capacity compressors.
- Losses occur in the compression of the working fluid (oil or air) in boosters, and these losses will be more elevated for compressed air than for compressed oil; while the diaphragm compressor crank benefits from a direct electric drive.
- In case of low or medium charge of the compressor (e.g., when the storage pressure is only at 5 or 20 MPa for a nominal pressure of 45 or 90 MPa), the diaphragm compressor will adapt and will need less energy, whereas the booster always consumes the same energy; the oil has been compressed up to 20 MPa (or the air compressed up to 0.8 MPa) and the extra energy will be lost.

It is important to define the way to calculate the efficiency. Nexant Inc. et al. (2008) recommends defining efficiency with respect to isentropic (or adiabatic) work and reports isentropic efficiencies in the range of 86%–92% for large reciprocating compressors. But the compressors considered for hydrogen distribution in general, and specifically in the present study, are really smaller. Parks et al. (2014), Rothuizen and Rokni (2014), and Reddi et al. (2014) follow this recommendation and use isentropic efficiency. Rothuizen and Rokni (2014) uses a formula from chemical processes (Smith, 2005) for the variation of isentropic efficiency with the pressure ratio and assumes that all efficiency losses contribute to increase hydrogen enthalpy (no external losses), which again is acceptable for large uncooled compressors. Rothuizen and Rokni (2014) and Reddi et al. (2014) use a 65% isentropic efficiency. Parks et al. (2014) underlines the lack of experimental data, and a large dispersion, by a factor of 10, of the few compressor consumption data reported in Wipke et al. (2012) for hydrogen distribution, while in 2013 DOE estimated consumption from 2 to 4 kWh/kgH2 for 35 MPa refueling with an efficiency of about 65% and targets 80% in 2020.

Indeed, efficiency varies greatly with the compressor technology, its capacity, its nominal pressure ratio, and the current pressure ratio. It is also important to consider each single stage, as recommended in Gardiner (2009) and followed in Rothuizen and Rokni (2014), as intercooling considerably reduces the final temperature and isentropic power.

Yet, compressors always exchange heat with the ambient air; for small compressors this can be significant with respect to the necessary heat to cool the compressed hydrogen, while for medium or large capacities, boosters or diaphragm compressor heads are equipped with cooling jackets. Isothermal work is the reference as the lowest needed to compress a gas, and so its calculation relevant. Nevertheless, cooling is never totally efficient, leaving a residual heating of hydrogen during compression. Thus, it seems relevant to calculate the efficiency with respect to both isothermal and isentropic power. In this study, the average of isothermal and isentropic work for each stage is used (Eqs. 6.5 and 6.7). Thus, the electric power needed *for each stage* of a compressor is calculated as follows:

$$P_{eleccomp\,H2} = \frac{1}{h_{comp}} \left(\dot{W}_{isothermal,H2} + \dot{W}_{isentropic,H2} \right) / 2 \qquad (6.8)$$

Indeed, very large variations in efficiency are found. For small capacity pneumatic boosters, the efficiency can be very low. For example, for compressing 10 kg/day from 5 to 45 MPa an efficiency of 10%–15% has been calculated according to supplier data. Hydraulic boosters are slightly higher, and 15%–20% has been calculated. On the opposite end of the spectrum, for a large 2-stage cooled diaphragm compressor, compressing at its nominal point 850 kg/day from 1.2 to 25 MPa at 27°C and consuming 73.5 kW, an efficiency of 59% has been calculated following Eq. (6.8). With respect to isentropic power only, which is 48 kW with two stages, the isentropic efficiency is 65.2%, which is fully coherent with Parks et al. (2014) and Reddi et al. (2014). Forgetting the two stages and considering only one, would incorrectly lead to an isentropic power of 62 kW and an apparent efficiency of 84.5%.

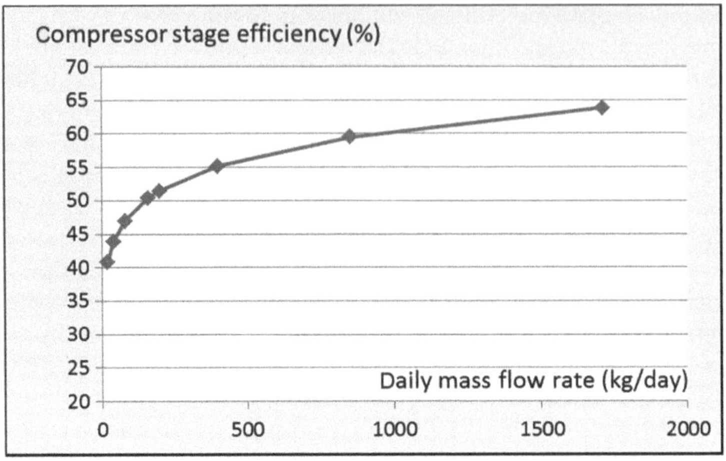

FIG. 6.4 Estimated effect of nominal capacity of a cooled diaphragm compressor on its stage efficiency.

Furthermore, it is known that large compressors have a better efficiency than small ones, and in this study considering cooled diaphragm compressors, while more data from suppliers are needed, the effect of capacity on efficiency will be calculated by a power law according to Eq. (6.9) and presented in Fig. 6.4.

$$h_{comp} = h_{comp,0}\left({}^{\dot{m}}/_{\dot{m}_0}\right)^p \quad h_{comp,\,0} = 60\% \quad \dot{m}_0 = 850^{\mathrm{kg}}/_{\mathrm{day}} \quad p = 0.1 \tag{6.9}$$

6.2.4 Cooling Needs

Compressor heads have to be cooled to keep the compression as isothermal as possible; hydrogen has to be cooled at the exit of each compression stage as well. The necessary cooling power can easily be estimated according to the thermodynamic law, which teaches that during a fluid transformation, the enthalpy variation is equal to the sum of the heat and work exchanged with the outside of the system:

$$\Delta H = W + Q \tag{6.10}$$

In the case of a perfect gas, the enthalpy only depends on temperature; as the gas recovers its initial temperature at the end of the compression + cooling process, then:

$$T_f = T_0 \quad \Delta H_{perfect} = 0 \ \text{ and } \ Q_{cooling} = -W_{comp} \tag{6.11}$$

This equation shows that a heat equivalent to the whole compression work provided to the gas has to be extracted from the system. Moreover, all losses in the compressor, drive, and electric motor convert into heat and also have to be extracted. Then, it is a cooling power equivalent to the electric power that has to be provided, through exchange with a cooling fluid and through natural convection with the ambient air.

Now, hydrogen is not a perfect gas and the variation of its enthalpy with temperature and pressure can be calculated using data from NIST Chemical Webbook (n.d.). According to Eq. (6.10), the cooling needs will slightly decrease due to the variation of enthalpy with pressure:

$$Q_{cooling,H2} = -W_{comp,H2} + m\left(h_{H2}(T_0, p_f) - h_{H2}(T_0, p_0)\right) \tag{6.12}$$

Cooling is assumed to be provided through a frigorific machine with a performance coefficient $COP_{cooling}$ of 3 and thus the electrical power of this machine is:

$$P_{eleccoolingH2} = \frac{1}{COP_{cooling}}\left(P_{eleccomp\,H2} - \dot{m}\left(h_{H2}(T_0, p_f) - h_{H2}(T_0, p_0)\right)\right) \tag{6.13}$$

6.3 ECONOMIC DATA FOR COMPRESSION AND STORAGE

This section presents economic data used in the following sections and discusses them with respect to data from other authors in similar papers, where appropriate.

6.3.1 Compressor Investment Cost

Some spots of information can be found in a number of papers, and especially in Parks et al. (2014), Reddi et al. (2014), and Nexant Inc. et al. (2008) concerning the investment cost of compressors, but few correlations or models are suggested. For example, Reddi et al. (2014) gives a correlation of a parabolic form for the investment cost as a function of capacity (kg/h). Nexant Inc. et al. (2008) recommends a linear increase of price with flow rate, but the range of flow rates is higher than that considered in this paper, mainly above 500 Nm^3/h.

Three compressor manufacturers were queried during 2014 and 2015, covering large ranges for normal flow rate, Q_v, from 5 to 460 Nm^3/h, for inlet pressure, p_0, from 0.7 MPa to 2.5 MPa and for pressure ratios, r_p, from 18 to 64. Quotations were analyzed in order to derive a model for estimating the investment cost of compressors. The size of the cylinders or diaphragms of a compressor depends on the real volume flow rate at the inlet, which is proportional to the normal flow rate, Q_v, divided by the inlet pressure. Indeed, the parameter, Q_v/p_0, appears to be the most relevant and a correlation of the form of Eq. (6.14) has been built:

$$Cost_{comp} = Cost_{comp,0} * a * e * \left\{\left({}^{Q_v}/_{p_0}\right) \Big/ \left({}^{Q_{v,0}}/_{p_{0,0}}\right)\right\}^d \tag{6.14}$$

with $Cost_{comp,0} = 130\text{k\$}, a = 1.04, Q_{v,0} = 40\text{Nm}^3/\text{h}, p_{0,0} = 2\text{MPa}$ and $d = 0.31$.

A shift appeared between manufacturers, represented by the factor e, called the trademark effect, and varied between 0.78 and 1.0 according to the manufacturer. When corrected by this e effect, the costs compare well between two manufacturers, as shown in Fig. 6.5.

It was not possible to find a clear effect of the inlet pressure, nor of the pressure ratio; attempts led to nonrelevant effects and bad correlation coefficients. However, increasing these parameters should have an effect on the design of the compressor, increasing its cost, so a light effect is suggested and introduced in Eq. (6.15):

$$Cost_{comp} = Cost_{comp,0} * a * e * \left\{ \left({}^{Q_v}/_{p_0} \right) / \left({}^{Q_{v,0}}/_{p_{0,0}} \right) \right\}^d * \left({}^{p_0}/_{p_{0,0}} \right)^b * \left({}^{r_p}/_{r_{p,0}} \right)^c \tag{6.15}$$

with $b = 0.1, c = 0.1, r_{p,0} = 22.5$.

- When the correlation from Reddi et al. (2014) is used, the cost obtained for a 3.6 kg/h compressor with an output pressure of 97 MPa would be 172 k\$, whereas our correlation gives 145 k\$, 15% less, but the quotation seems to date from 2007 in Reddi et al. (2014).
- In Parks et al. (2014) the cost given is also higher. A 35 kg/h 2-stage diaphragm compressor with a 2 MPa inlet pressure and a 45 MPa outlet is 376 k\$ (2013), whereas the correlation leads to 280 k\$, 26% lower, but Parks et al. (2014) estimated the price would drop by 25% by 2020 with high production.

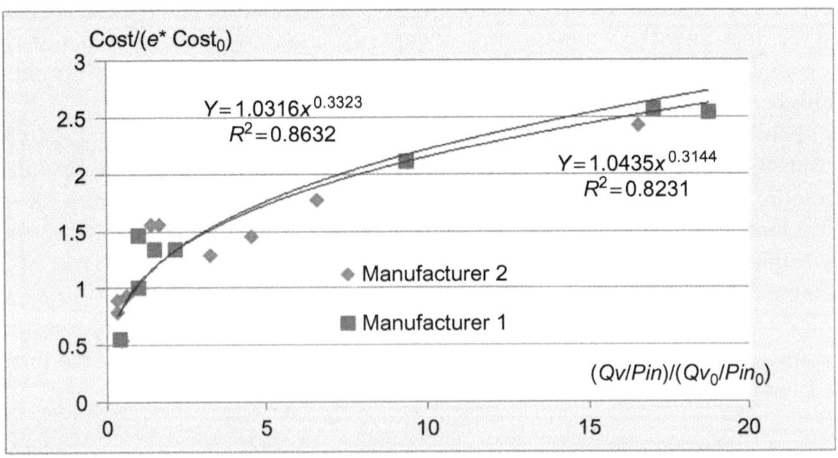

FIG. 6.5 Comparison of the power law for the cost of compressors for two manufacturers, after the trademark effect has been removed.

- On the other hand, when compared with data from Nexant Inc. et al. (2008) (2007), the present correlation gives a higher cost for a 50 kg/h 20–400 MPa 2-stage diaphragm compressor, 308 k$ instead of 170 k$ for the quotation. However, when the flow rate is doubled, the correlation gives the same price as the quotation: 385 k$.

6.3.2 Cost of Pressure Vessels

As for compressor investment cost, some information can be found about high-pressure vessel cost in a number of papers, for example, Parks et al. (2014), Reddi et al. (2014), Nexant Inc. et al. (2008), but no model is available.

- Nexant Inc. et al. (2008) considers large steel vessels of 21 kg H_2 each under 43 MPa for a price of 843 $/kg (2007, uninstalled).
- Reddi et al. (2014) considers 100 MPa steel bottles with a capacity of 12 kg for a rather high cost of 1475 $/kg (2013, +30% for installation).
- Parks et al. (2014) reports previous figures from Nexant Inc. et al. (2008) and adds others, with lower costs and for higher pressure. For 61 kg at 25 MPa, a 5 kg type 4 bottle container is reported at a cost of 450 $/kg (2007) and at 95 MPa, 12 kg type 4 bottles are selected at a cost of 911 $/kg.

It is interesting to note that composite bottles have reached lower cost than steel vessels and will continue to drop in cost with mass production, whereas steel vessels will drop less in cost as the technology is has been mature for a long time.

Three manufacturers of composite type 3 and type 4 pressure bottles were queried in 2015 and 2016. They covered a range of bottles with nominal pressures from 20 to 52.5 MPa and volumes from 75 to 500 L. It was then possible to show that the costs of the composite bottles could fit a correlation based on the nominal pressure and the mass capacity of the bottles of the following form:

$$Cost_{bottle} = c * (p_{nom\ bottle})^p * (capa_{bottle})^m \tag{6.16}$$

with $p_{nombottle}$ in MPa, $capa_{bottle} = \rho_{H2nombottle} * V_{bottle}$ in kg, $p = 0.27$, $m = 0.875$, and $c = 331€$. When applied to the bottle characteristics of Parks et al. (2014), the correlation gives a price of 635 €/kg = 730 $/kg at 25 MPa and 840 €/kg = 965 $/kg at 95 MPa. Thus, the order of magnitude is correct while the effect of pressure seems to be underestimated.

In fact, in (Parks et al., 2014) at 25 MPa, the bottles were assembled in a container for a larger capacity and the cost should be compared to costs collected by enquiries for containers varying from 160 to 850 kg. The cost of a container made of elementary bottles can be expressed as:

$$Cost_{container} = n^q * c' * c * (p_{nom\ bottle})^p * (capa_{bottle})^m \tag{6.17}$$

where c' is a coefficient relative to the cost of building the frame of the container, the supports of vessels and their connection, with a value of 1.3–1.5 (from quotation analysis, relevant with Nexant Inc. et al., 2008), n is the number of vessels necessary for the container capacity requirement, and the exponent q reveals the number effect, from 0.90 to 0.94 (from quotations analysis from different suppliers). When applied to the 616kg container of Parks et al. (2014), the correlation gives a cost of 680 €/kg for the assembled container instead of 635 €/kg for a single 5kg bottle; the number effect has nearly compensated the assembly cost.

These correlations are used in Sections 6.4 and 6.5 without taking into account the cost drop with large quantity production, which will indubitably occur in the coming years.

6.3.3 Preliminary Considerations and Recommendations

At this stage, good practices have already appeared concerning hydrogen compression:

- Use high p_0: as far as possible, produce the hydrogen at the highest pressure, using a high-pressure electrolyzer or a high-pressure reformer and PSA. It is less costly to compress the water feeding the electrolyzer or the biogas at the inlet of the reformer than the hydrogen at the exit. In the VABHYOGAZ project, the reformer operates at $p_0 = 1.5$ MPa.
- Use low T_0: cool the hydrogen before compression, as far as the compressor accepts it. Calculations (see later) show that with a $COP_{cooling}$ of 3, the hydrogen refrigeration will require less electric energy than can be spared when compressing hydrogen at lower temperature.
- Cool the compressor heads effectively so that the compression process is as near as possible to an isothermal process. Consider a multistage compressor with intercooling to lower the compression ratio of each stage and isentropic heating.
- Choose best compression technology: diaphragm compressors are more effective than boosters.
- Prefer large-scale units: it is difficult to reach good compressor efficiency in small production or distribution units.
- Choose an effective cooling system to cool compressor heads and hydrogen.

6.4 CASE OF H_2 DISTRIBUTION ON THE PRODUCTION SITE

When hydrogen production and distribution are located on the same site, there is no transportation to be considered and the energy requirements are mainly from the compression and associated cooling of the hydrogen.

6.4.1 Current Practices for Refueling: Energy Costs for Reference Cases

In the current practice, referring to Fig. 6.2, during the filling of successive vehicle tanks, the buffers numbered 1, 2, and 3 are always connected in the same order to the vehicle and they are then refilled by the compressor in the opposite order, with a priority to the highest-pressure buffer, number 3. Then, when buffer number 3 has recovered its nominal pressure, it is the turn of buffer number 2 to be refilled by the compressor up to its nominal pressure. And finally, it is the turn of buffer number 1 to be refilled. Even though Rothuizen and Rokni (2014) writes that the compressor refills the buffers in the order they are filling the tanks, i.e., number 1 first, this practice seems not to be implemented.

Indeed, refilling the buffer number 3 first is useful for the peak hour performance; this highest-pressure buffer will recover its nominal pressure (45 MPa or 90 MPa) within the shortest time so the probability of the refueling station to have a buffer able to achieve the next tank refueling at the nominal pressure is the highest. Then, a cascade naturally appears in the buffer pressures when cars follow each other at peak hour, but after these peak hours, or during the night, all of the buffers are refilled at the highest pressure.

Using the previous section, and specifically Eqs. (6.8), (6.9), and (6.12), it is now easy to calculate the electric consumption of the compressor and associated cooling of any refueling station. The necessary data are: daily delivering capacity: 80 kg/day, buffers pressure: 45 MPa, production unit working pressure: 1.5 MPa, number of stages of the diaphragm compressor: 2, inlet temperature: 20°C, and finally, cost of electricity: 80 €/MWh. An example of detailed results for this refueling station at nominal load is given in Table 6.1.

TABLE 6.1 Consumption and Cost for Compression and Cooling of 80 kg/Day From a Production Pressure of 1.5 MPa to a Buffer Pressure of 45 MPa

Useful isothermal compression power, real gas (kW)	4.07
Isentropic compression power, real gas (kW)	5.14
Compressor electric consumption ($kW_{élec}$)	8.82
Cooling consumption (kW_{cool})	8.48
Total electric power ($kW_{élec}$)	11.65
Annual electricity consumption (MWh/an)	96
Annual electricity cost (k€/an)	7.69
Specific consumption (kWh/kgH_2)	3.50
Ratio of specific consumption to LHV (% LHV)	10.6
Specific cost (€/kgH_2)	0.266

TABLE 6.2 Effect of Capacity and Distribution Pressure of Refueling Station (With OnSite Production) on Electricity Consumption and Cost for Cooled Compression at Nominal Load

Daily Capacity of the Refueling Station (With On-Site Production)	20 kg/Day		80 kg/Day		200 kg/Day	
Distribution pressure (MPa)	35	70	35	70	35	70
Electricity consumption for cooled compression (MWh/an)	28	34.6	96	119	222	275
Specific consumption (% LHV)	12.4	15.2	10.6	13.1	9.8	12.1
Electricity cost (k€/an)	2.24	2.77	7.69	9.53	17.8	22.0

The results show rather high specific energy and specific cost for this reference case, >10% of the LHV. They will be even higher for smaller distribution units, but hopefully lower for larger ones, as given in Table 6.2, which shows also the effect of the delivering pressure: 35 MPa or 70 MPa (with buffers at 45 MPa or 90 MPa).

6.4.2 Minimization of the Compression Energy

6.4.2.1 The Geometric Progression Pressure Cascade

In fact, it is not necessary to use a buffer at the highest pressure during the first moments of the refueling process and it generates a waste of energy; indeed, all the hydrogen has been compressed to the highest pressure, whereas an intermediate pressure would have been sufficient during these first moments.

It has been suggested that the buffer nominal pressures be staged and that the buffers never be refilled at a higher pressure than their respective staged pressures. If vehicles usually have to be refilled when their tank pressure, $p_{0,tank}$, is equal to a fraction, v_{tank}, of their nominal tank pressure, $p_{f,tank}$, with:

$$p_{0,tank} = v_{tank} * p_{f,tank} \tag{6.18}$$

in a refueling station equipped with n buffers, the tank pressure will increase in a ratio $R_p = {}^{1}/_{v_{tank}}$ through n steps, and each step will contribute for a smaller pressure ratio increase r_p :

$$r_p = (R_p)^{1/n} \tag{6.19}$$

TABLE 6.3 Geometric Progression Staged Buffer Pressures for Different Numbers of Stages for a Refueling at 70 MPa and $v_{tank}=4\%$

Number of Stages n	Stage Pressure Ratio r_p	Pressure Cascades: Geometric Progression of the Staged Buffer Pressures (MPa)			
$n=1$	$R_p={}^1/_v=25$	90			
$n=2$	$(25)^{1/2}=5.0$	18.0	90		
$n=3$	$(25)^{1/3}=2.92$	10.53	30.78	90	
$n=4$	$(25)^{1/4}=2.236$	8.05	18.0	40.25	90

For a refueling at 35 MPa instead of 70 MPa, just divide the values of pressures by 2.

Then, the buffers form a pressure cascade staged in a geometric progression. For example, for a refueling at 70 MPa, with the highest pressure buffer at 90 MPa and $v_{tank}=4\%$, the pressure cascades are given in Table 6.3.

It can be noted that for a given mass of perfect gas, n compression steps with the same pressure ratio, r_p, require equal compression energy; according to Eqs. (6.3) and (6.5); the pressure cascade is isoenergy. For a real gas, such as hydrogen, the higher compression steps will require slightly higher compression energy, even with the same r_p.

6.4.2.2 *Highlighting of Energy Savings*

The effect of a pressure cascade on the energy needs is depicted in Fig. 6.6, which presents the progressive filling of a 70 MPa tank initially empty at $v_{tank}=4\%$ (corresponding to a residual mass $w_{tank}=5.8\%$ in the tank):

- With only 1 buffer at 90 MPa, or with all buffers at 90 MPa, according to Eq. (6.8), all hydrogen requires a specific compression energy equal to 3.308 kWh/kg before being transferred to the tank (red curve).
- If 2 staged buffers at 18 and 90 MPa are used, the first 22% of mass transferred to the tank comes from the first buffer at 18 MPa and only requires a specific compression energy equal to 1.525 kWh/kg, while the following 72.2% comes from the second buffer at 90 MPa and requires 3.308 kWh/kg (blue curve).
- If 4 staged buffers are used, at 8, 18, 40.2, and 90 MPa, the first 7% transferred mass comes from the first buffer at 8 MPa and only requires a specific compression energy equal to 0.967 kWh/kg; the following 15% transferred mass comes from the second buffer at 18 MPa and requires an energy equal to 1.525 kWh/kg; the following 28% transferred mass comes from the third buffer at 40.2 MPa and requires an energy equal to 2.236 kWh/kg; while the last 44.2% comes from the fourth buffer at 90 MPa and requires 3.308 kWh/kg (orange curve).

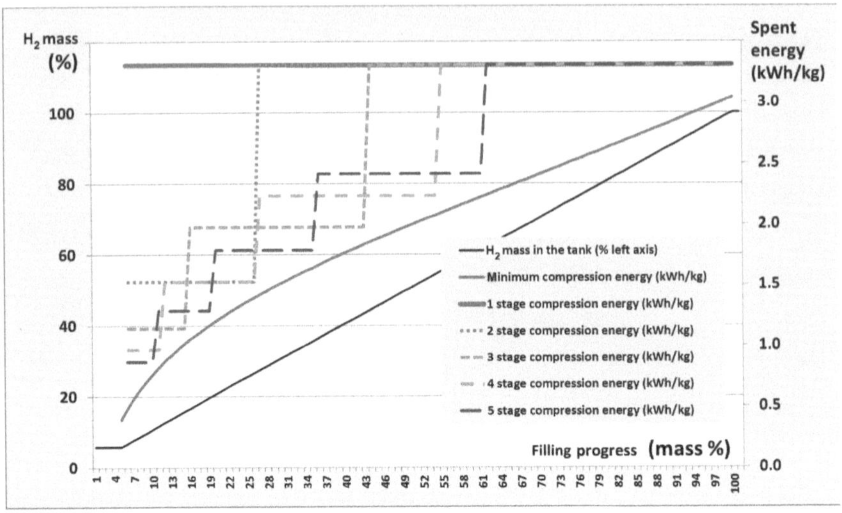

FIG. 6.6 Specific compression energies (kWh/kg) involved in the filling of a 70 MPa hydrogen tank when using buffers arranged in geometric pressure cascades with 2–5 stages, compared to the case of 1 unique very-high-pressure buffer (VHPB) and to the case of an infinite number of staged buffers (minimum energy).

- On this plot, the areas under the stair-shaped lines represent the energies spent for the compression of the hydrogen to be transferred. So, the area between the highest, red, straight line, and the other stair-shaped lines represent the energy saved thanks to the use of pressure cascades with 2, 3, 4, or 5 stages.
- On the opposite side, the green curve shows the minimum compression energy that would be spent with an infinite number of stages and buffers at increasing pressures equal to that of the tank. It is also the energy that would be spent by a compressor connected to the tank and filling it directly. The areas between the stair-shaped lines and that green curve represent the compression energy lost in the process of filling a tank by transfer from higher pressure buffers.

6.4.2.3 *Energy Savings as a Function of Number of Stages and Tank Pressure*

The areas under the stair-shaped lines have been calculated to obtain the numerical values of the energies shown in Table 6.4 for the filling of a 70 MPa hydrogen tank:

The same calculations have been made for the filling of a 35 MPa hydrogen tank (Table 6.5):

TABLE 6.4 **Specific Compression Energies (kWh/kg) Involved for Filling of a 70 MPa Hydrogen Tank, Gains Generated When Using Buffers Arranged in Geometric Pressure Cascades With 2–5 Stages in Comparison With the Case of a Unique Very-High-Pressure Buffer (VHPB) and to the Case of an Infinite Number of Staged Buffers (Minimum Energy)**

Number of Stages	1	2	3	4	5	Infinite
Specific compression energy spent (kWh/kg)	3.33	2.94	2.71	2.58	2.51	2.01
Specific compression energy spent/LHV (% LHV)	10.1	8.9	8.2	7.8	7.6	6.1
Losses with respect to minimum energy (%)	66	46	35	29	25	0
Gains with respect to unique pressure buffer (%)	0	11.9	18.8	22.5	24.9	39.8

TABLE 6.5 **Same as Table 6.4 but for the Filling of a 35 MPa Hydrogen Tank**

Number of Stages	1	2	3	4	5	Infinite
Specific compression energy spent (kWh/kg)	2.65	2.38	2.21	2.12	2.06	1.64
Specific compression energy spent/LHV (% LHV)	8.0	7.2	6.7	6.4	6.2	5.0
Losses with respect to minimum energy (%)	61	45	35	29	25	0.0
Gains with respect to unique pressure buffer (%)	0.0	10.4	16.6	20.2	22.4	38.0

Fig. 6.7 shows the energy saved when using buffers arranged in 2–5 stage geometric pressure cascades in comparison with the current practice in which all buffers are filled at very high pressure.

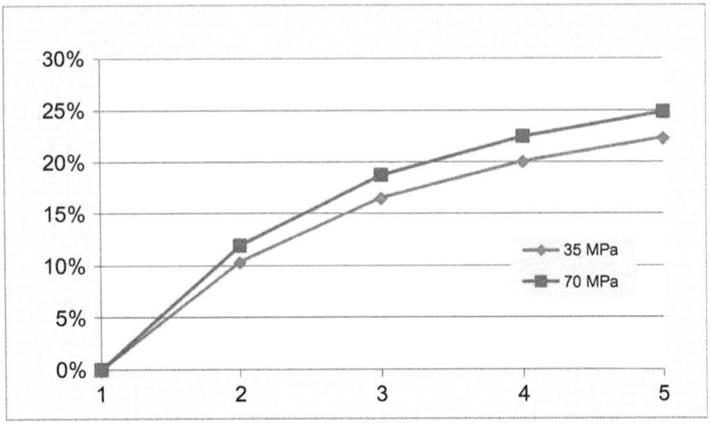

FIG. 6.7 Saved energy when using buffers arranged in geometric pressure cascades with 2–5 stages in comparison with the current practice of all buffers filled at very high pressure (45 or 90 MPa).

6.4.2.4 *Effect of the Shape of the Pressure Cascade on the Energy Savings*

The previous results have been obtained with pressure cascades organized in geometric progressions. For a perfect gas, these cascades are isoenergy and require isovolume buffers at each cascade stage (see next section). But nothing says that this arrangement will lead to the lowest compression energy.

Thus, the same calculations have been performed for other cascade shapes; arithmetic progressions have been considered, together with combinations of arithmetic and geometric progressions (Table 6.6; Fig. 6.8).

It appears in Fig. 6.8 that an arithmetic progression gives better results than a geometric one and the best results are obtained with a linear combination of 0.3

TABLE 6.6 Arithmetic Progression Staged Buffer Pressures as Function of Number of Stages for a Refueling at 70 MPa

Number of Stages *n*	Pressure Increase (MPa)	Pressure Cascade: Arithmetic Progression of the Staged Buffer Pressures (MPa)			
n = 1	90	90			
n = 2	90/2 = 45	45	90		
n = 3	90/3 = 30	30	60	90	
n = 4	90/4 = 22.5	22.5	45	67.5	90

For a refueling at 35 MPa instead of 70 MPa, just divide the values of pressures by 2.

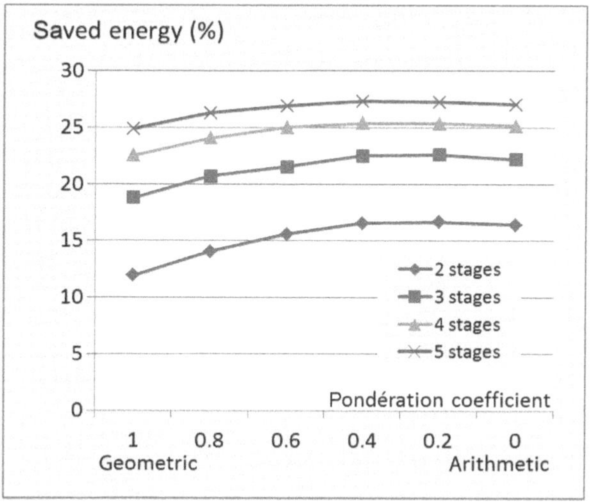

FIG. 6.8 Influence of the shape of the pressure cascade, from a geometric progression to an arithmetic progression, with 2–5 stages, on the saved energy for a 70 MPa distribution.

geometric +0.7 arithmetic progression, but the variations are rather flat, leaving a large possibility for energy savings higher than 25% in comparison with the current practice where all buffers are filled at the highest pressure (90 MPa).

An additional point of interest of lower pressure buffers for the early stages of refueling is to minimize the pressure ratio of the Joule-Thomson expansion occurring in the control valve of the refueling line and to minimize the temperature rise. With 10 MPa in place of 90 MPa, the Joule-Thomson temperature rise is 40 K lower, thus reducing the cooling need!

At this stage, a comparison with similar previous approaches is interesting. Rothuizen and Rokni (2014) is dedicated to the optimization of the energy consumption in cascade refueling. Its detailed modeling reports an energy consumption (compression + cooling) of 5.97 kWh for refueling 5 kg in a 70 MPa tank (3.6% of LVH_{H2}) with 1 stage and a decrease to 5 kWh with 2 stages (−16.2%) and to 4 kWh with 5 stages (−33%). Similar to the present study, the largest saving comes from the lower output pressure for the compressor; similar to Fig. 6.7, the shape of the curves shows that the first stages bring most of the savings while the 5th and higher stages bring smaller and smaller contributions. But it is difficult to go further in the comparison because Rothuizen and Rokni (2014) considers a refueling station supplied from a bulk storage at a higher pressure than the present 1.5 MPa (more comparable with cases treated in Section 6.5.3) and this explains the low energy demand, 3.6% of LHV_{H2}, not including the compression on the production site. Then, our saving, 24.9% of 12.1% of LHV_{H2} (Tables 6.2 and 6.4), is 3% of LHV_{H2} and is much more than the saving of 33% of 3.6% of LHV_{H2} (=1.2% of LHV_{H2}) calculated in Rothuizen and Rokni (2014).

6.4.3 Effect of Precooling on the Compression Energy

Considering Eqs. (6.3) or (6.4), another possibility to decrease the compression work is to decrease T_0 by cooling the hydrogen before compression. Calculations show that both the compression work and the compression cooling energy decrease, respectively, by −11 Wh/kg°C and −7 Wh/kg°C. Indeed, it is necessary to take into account also the initial precooling (+7.4 Wh/kg/°C), which cancels the gain of the final cooling. But globally, with a COP of 3, an overall saving of 10.6 Wh/kg/°C can be reached; thus, 0.22 kWh/kg can be saved for a cooling of 20°C at the entrance of the compressor, and this is an extra 5% energy savings, as shown in Table 6.7.

Thus, it is recommended to run the compressors with precooled hydrogen, at the lowest temperature possible, according to the compressor requirements.

6.4.4 Necessary Volume of the Buffers

The delivery capacity of a refueling station at the peak hour mainly relies on the mass of hydrogen stored in the HP buffers. Now, if the buffers are not all at the highest pressure, but are organized in a pressure cascade, their volume has to

TABLE 6.7 Effect of Hydrogen Precooling on Consumption for Compression and Cooling of 80 kg/Day from a Production Pressure of 1.5 MPa to a Buffer Pressure of 90 MPa

Compressor Inlet Temperature (°C)	20	0	−20	Variation
Compressor electric consumption ($kW_{élec}$)	11.03	10.30	9.57	
Specific variation (Wh/kg°C)				−11
Inter and final cooling needs (kW_{cool})	10.16	9.69	9.22	
Specific variation (Wh/kg°C)				−7.0
Initial cooling needs (kW_{cool})	0	0.53	1.06	
Specific variation (Wh/kg°C)				+7.4
Total electric power ($kW_{élec}$)	14.42	13.71	12.99	
Specific variation (Wh/kg°C)				10,6
Specific energy (kWh/kgH_2)	4.33	4.11	3.90	
Specific energy to LHV ratio (% PCI)	13.1	12.5	11.8	
Energy saved (%)	0	5	10	

be increased or else the peak hour performance will be reduced. Then, the investment cost of the staged buffers could be higher than that of smaller HP buffers. This section calculates the volume of the necessary buffers and estimates their investment cost in order to compare overinvestment and energy savings and to quantify the return on investment (ROI). But first the peak demand is defined.

6.4.4.1 Peak Hour Demand and Buffer Capacity

The demand at a refueling station is not constant. It varies with the day and with the week. The result of a statistical treatment of 385 US refueling stations is presented in Nexant Inc. et al. (2008), from which the data of Fig. 6.9A and B are extracted and analyzed here:

- Friday is the busiest day of the week, with a demand 7.5% above the average, while Monday is the quietest day.
- 3 p.m. is the busiest hour in the day with, on Friday, 7.8% of the total of the day, or 1.87 times the average of the day.
- 7 a.m. and 7 p.m. are on the Friday daily average, while the 12 h period in between is above the average; thus, it appears there is no peak hour but a large peak period.
- The area above the average line represents 27.5% of the total area (Fig. 6.9).

The minimum capacity for the HRS compressor would correspond to the average flowrate, with a 24 h/day operation. Then it will be necessary for the buffer to contain an extra mass of hydrogen, deliverable at the desired pressure, equal to 27.5% of the daily dispensed hydrogen. During the period 7 a.m. to 7 p.m., destocking hydrogen will occur from the buffer and its pressure will decrease down to the minimum acceptable pressure. After 7 pm, the hydrogen dispensed to the vehicle tanks is less than the average and the compressor delivers more than what is dispensed to the vehicles, thus stocking in the buffer will occur and its pressure will increase. The compressor will refill the buffer at its nominal pressure until 7 a.m., with an extra mass equal to 27.5% of the daily hydrogen dispensed.

In fact, it is recommended to oversize the compressor to be able to cope with some extra affluence, invisible in the average statistics. With a higher capacity compressor, the buffer will recover its nominal pressure sooner and then the compressor will stop, so that it will operate <24 h/day.

Also, when dealing with small capacity HRSs, dedicated to small captive fleets, the refueling behavior may be different and the delivery profile of the station has to be defined carefully in order to adjust the peak hour performance.

The following calculations consider a *compressor oversized by a factor of 2* and a mass of hydrogen that can be dispensed at the nominal delivery pressure during the peak hour or period, m_{peak}, *equal to 25% of its daily capacity* (i.e., m_{peak}= 20 kg for an 80 kg/day HRS). Thus, the compressor only runs 12 h/day

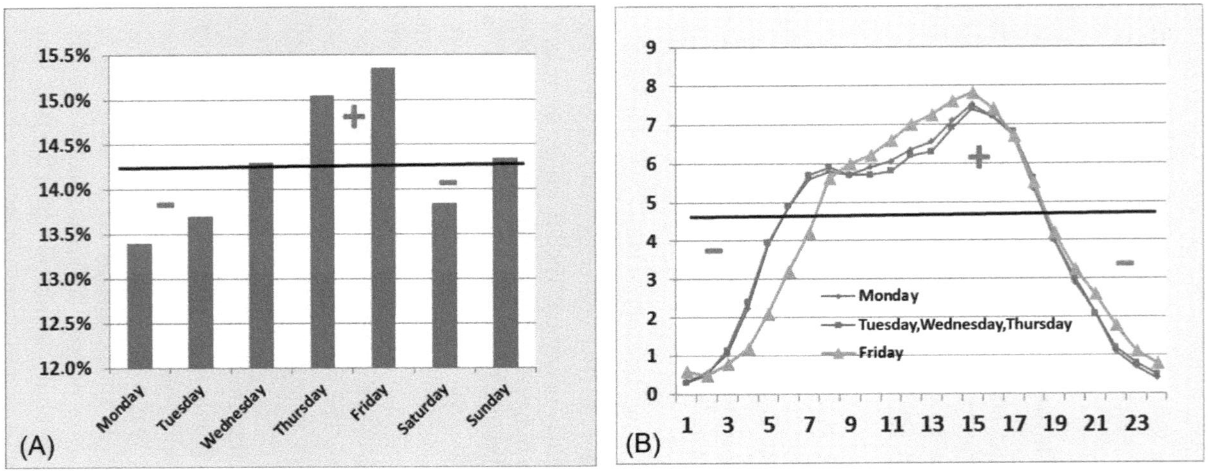

FIG. 6.9 Average hourly (A) and daily (B) gasoline distribution profile. *((Data from Nexant Inc., Air Liquide, Argonne National Laboratory, Chevron Technology Venture, Gas Technology Institute, National Renewable Energy Laboratory, Pacific Northwest National Laboratory, TIAX LLC, May 2008. H2A: Hydrogen delivery infrastructure analysis models and conventional pathway options analysis results. Report DE-FG36-05GO15032, DOE. www1.eere.energy.gov/hydrogenandfuelcells/pdfs/nexant_h2a.pdf.))*

on average and with such a value of m_{peak}, it is possible to face 3 h of peak demand at 8% of the daily average without the help of the compressor. With the help of the compressor, capable of delivering 8.33% of the daily capacity per hour, the buffer will have recovered its nominal pressure at the end of the most demanding hour. In fact, the *maximum capacity of the refueling station is twice its nominal capacity* and the vehicle fleet can increase by a factor of 2 before the station is overwhelmed.

6.4.4.2 Analytical Formulation of Buffer Volumes

The mass of hydrogen, m_{peak}, that can be delivered at the nominal pressure (35 or 70 MPa) is now defined and this mass corresponds to a volume V_{peak}:

$$V_{peak} = m_{peak} / \rho_{H2,tank} \tag{6.20}$$

Tanks of volume V_{peak} have to be refilled from $p_{tank,0}$ to $p_{tank,f}$ with m_{peak} while the buffer pressure decreases from $p_{buff,0}$ to $p_{tank,f}$. Besides R_p, the tank filling pressure ratio, a new pressure ratio is introduced, S_p, the overpressure of the initial full buffer with respect to the objective full tank pressure:

$$R_p = p_{tank,f} / p_{tank,0} = 1 / \nu_{tank} \tag{6.21}$$

$$S_p = p_{buff,0} / p_{tank,f} \tag{6.22}$$

Considering the mass conservation during the balancing of pressures between buffer and tanks and a perfect gas, the volume of the buffer, V_{buff}, necessary to fill the tank volume, V_{peak}, in a pressure ratio, R_p, with a buffer overpressure, S_p, can be calculated as:

$$V_{buff} / V_{peak} = (R_p - 1) / \left[R_p * (S_p - 1)\right] \tag{6.23}$$

It should be noted when writing Eq. (6.17), that the balancing of pressure is supposed to be isothermal, whereas compression occurs in the tank and heats hydrogen and expansion occurs in the buffer. Thus Eq. (6.17) is only valid after tank and buffer have recovered their initial temperature.

Now the flow of hydrogen is not conducted until the exact isothermal balancing of pressure, because it would be long. In fact, taking into account the temperature rise in the tank during the refueling, it is necessary to fill the tank with an overpressure so that, after the tank has cooled down to ambient temperature, the pressure has decreased to the desired $p_{tank,f}$.

So, it is considered that the tanks must be refilled while the buffer pressure decreases to $p_{buff,f,min}$, keeping a minimum overpressure $s_{p,min}$:

$$p_{buff,f,\min} / p_{tank,f} = S_{p,\min} \tag{6.24}$$

Then, the volume of the necessary buffer is calculated as:

$$V_{buff}/V_{peak} = (R_p - 1) \Big/ \left[R_p * \left(S_p - s_{p,\min}\right)\right] \tag{6.25}$$

Moreover, hydrogen is not a perfect gas and a correction has to be introduced as a ratio of compressibility factors:

$$V_{buff}/V_{peak} = \left[Z_{p_{buff},0}/Z_{p_{tank,f}}\right] * (R_p - 1) \Big/ \left[R_p * \left(S_p - s_{p,\min}\right)\right] \tag{6.26}$$

Finally, Eq. (6.26) allows the volume of the necessary buffer for the peak hour demand to be estimated easily, without detailed modeling of the unsteady filling of the tank, including hydrogen heating and heat transfer to the tank walls, as done in Rothuizen and Rokni (2014) and Reddi et al. (2014). In fact, all the thermal behavior of hydrogen and tank is represented by the factor $s_{p,min}$. According to the SAE-J2601 standard (SAE International, 2014), to compensate the heating of the hydrogen and tank, it is possible to overfill the tank (over its nominal pressure), so that after the natural cooling, its pressure decreases to its nominal pressure. Target pressures with a 1.10 overpressure factor are current when fueling at high ambient temperature (e.g., a target of 77.9 MPa for the refueling end for a nominal 70 MPa after natural cooling). Thus, using a $s_{p,min} = 1.1$ in Eq. (6.26) assumes that the pressure can increase by a factor 1.1 in the tank due to the hydrogen temperature increase in the same ratio, an increase from 290 to 320 K, which is rather low, and would mean that hydrogen has been cooled before entering the tank. Without precooling, the hydrogen temperature would be higher (but under 85°C), and either $s_{p,min}$ should be chosen higher than 1.10 to keep good peak hour performance, or the flow rate should be decreased in order to decrease the hydrogen temperature.

Thus, as long as $s_{p,min}$ is correctly estimated, there is no necessity of detailed thermal modeling, nor to consider filling rate, tank filling duration, or time between two successive vehicles…

Now, to estimate correctly $s_{p,min}$, information can be gained from detailed transient heat transfer modeling inside the tank and Bourgeois et al. (2015, 2017), Lei et al. (2010), Woodfield et al. (2007), Monde et al. (2007), and Lee et al. (2009) will be useful, especially Bourgeois et al. (2017) for the model description, and Lei et al. (2010) for results concerning the temperature rise as a function of filling rate, initial tank pressure, and ambient temperature.

6.4.4.3 Buffer Volume With Only 1 Very-High-Pressure Buffer (VHPB) at the Highest Pressure 45 MPa or 90 MPa (Reference Case)

For an 80 kg/day refueling station, $m_{peak} = 20$ kg; with $\rho_{H2,tank} = 24\,\text{kg/m}^3$ at 35 MPa or $40.24\,\text{kg/m}^3$ at 70 MPa, V_{peak} is obtained: $V_{peak,\ 35MPa} = 0.833\text{m}^3$ and $V_{peak,70MPa} = 0.497\text{m}^3$. $R_p = 25$ for $v = 4\%$ and $S_p = 1{,}285$ for $p_{buff,0} = 90$ MPa and $p_{tank,f} = 70$ MPa or for $p_{buff,\ 0} = 45$ MPa and $p_{tank,f} = 35$ MPa, thus, following Eq. (6.26): $V_{buff}/V_{peak} = 5.48$ and finally: $V_{buff,45MPa} = 4.56\,m^3$, $V_{buff,90MPa} = 2.72\,m^3$.

TABLE 6.8 Volume of Necessary Buffers in Case of Geometric Progression Staged Buffer Pressures

Number of Stages n	Stage Pressure Ratio r_p	V_{buff}/V_{peak} for Each Stage
$n=1$	$r_p=R_p=25$	5.48
$n=2$	$r_p=(25)^{1/2}=5.0$	4.57
$n=3$	$r_p=(25)^{1/3}=2.92$	3.76
$n=4$	$r_p=(25)^{1/4}=2.236$	3.16
$n=5$	$r_p=(25)^{1/5}=1.903$	2.71

6.4.4.4 Buffer Volumes With Staged Pressure

In the case of staged pressure buffers, the approach is formally the same and the only difference in the formulas is the substitution of the overall large pressure ratio, R_p, by the smaller pressure ratio, r_p, of each stage.

If the buffer pressures are staged in a geometric progression, all r_p are equal and then the buffer volumes are equal at each stage; the volumes are given in Table 6.8. It is clear that each stage requires a smaller volume of buffer, but globally, the total volume of buffers is larger.

6.4.5 Cost of the Storage Buffers

If all buffers are designed for the highest pressure, as in Rothuizen and Rokni (2014) or Reddi et al. (2014), the investment cost of the buffers will be higher. But as the lower pressure buffers will never experience the highest pressure, they can be designed for lower pressure; the overinvestment could be small.

Yet, composite bottles are only available for a few capacities and a few nominal pressures. The commercially available nominal pressures do not respect a geometrical progression (but they could lead to better energy savings as shown in Fig. 6.6). Adequate staged volumes have to be calculated as a function of each pressure ratio according to Eq. (6.20); they do not have equal values and furthermore will not correspond to the commercially available volumes. Thus, some buffers will have to be oversized, inducing additional cost.

It is considered here that bottles are available only with a unit volume of 300 L and for nominal pressures of 10, 20, 30, 52.5, and 90 MPa. The number of necessary 300 L bottles is calculated in Table 6.9 for each stage of the storage of an 80 kg/day 70 MPa refueling station. The cost of the storage is then obtained, using Eq. (6.10). The saving percentage is obtained from Fig. 6.8.

It can be seen that:

- In the reference case of current practice, with one only very-high-pressure stage (90 MPa), 6 bottles of 300 L would be needed (exact theoretical total

TABLE 6.9 Effect of Number of Stages on Number of Bottles of 300 L Needed at Each Stage, Cost of Storage, Energy Savings and Return on Investment for an 80 kg/Day 70 MPa HRS

Number of Stages	1	2	3	4	5
10 MPa buffer number					3
20 MPa buffer number				4	3
30 MPa buffer number		6	5	3	3
52.5 MPa buffer number			3	3	3
90 MPa buffer number	6	4	3	3	3
Storage cost (k€)	63.0	66.1	72.6	75.5	76.8
Annual energy savings (%)	Reference	16.0	22.4	25.2	26.3
Annual energy savings (k€)	Reference	1.52	2.13	2.40	2.51
Return on investment (years)	Reference	2.0	4.5	5.2	5.5

volume needed is 1816 L) for a total cost of 63 k€; the annual operation reference cost of energy is 9.53 k€/year, according to Table 6.2.

- With 2 stages, when introducing an intermediate 30 MPa stage, 4,300 L 90 MPa bottles would be enough (theoretical volume needed is 1263 L) together with 6,300 L 30 MPa bottles (theoretical volume needed 1663 L); the total cost is 66.1 k€ and the energy saving is 16% with respect to the reference case; thus, the return on investment is only 2 years.
- With 4 stages, introducing intermediate stages at 14.5 MPa, 30 MPa, and 52.5 MPa, 3,300 L 90 MPa bottles would be used, (even if too large, the theoretical volume needed is 793 L) together with 3,300 L 52.5 MPa bottles (theoretical volume 809 L), 3,300 L 30 MPa bottles (theoretical volume 979 L) and 4,300 L 20 MPa bottles (theoretical volume 1479 L); the total cost is 75.5 k€ and the energy saving is 25.2%, thus the return on investment is 5.2 years.
- The use of staged pressure buffers increases the total storage volume and the investment cost. The overinvestment payback time increases with the number of stages but remains acceptable.

6.4.6 Conclusion for Hydrogen Distribution on the Production Site

Compressing hydrogen is inevitable when fuel cell cars have to be refueled at high pressure. The energy cost of compressing and cooling hydrogen is high. In the case of hydrogen dispensed on the production site, it can reach *3.50 or 4.4 kWh per kg* of hydrogen transferred to the car tank at *35 or 70 MPa*, in a reference case corresponding to best current practice. Recommendations have been made in order to avoid spending even more energy, which is current in small refueling stations.

The study shows that this energy need can be reduced by *22%*, *25%*, or even *27%* when judiciously using *3*, *4*, or *5* stages of buffers organized in a pressure cascade for the filling of the tank. Whereas the total volume of the staged pressure buffers is higher than the volume of a unique very-high-pressure buffer, the extra cost is acceptable and the energy saving results in an acceptable payback time for the overinvestment of *4.5 to 5.5 years*.

Precooling the hydrogen before the compression would also lead to energy savings; an extra *5% to 10%* can be gained, and compressor technology could be improved to admit cooled hydrogen.

6.5 CASE OF A PRODUCTION UNIT SUPPLYING SEVERAL DISTANT REFUELING STATIONS

Currently, most of the hydrogen dispensed is supplied from large and distant production units: only a few HRSs have onsite production.

Hydrogen is usually supplied in steel bottle bundles or trailer tubes, which are trucked to the station. These steel containers have a low specific content with a bundle of 12 steel 50 L 20 MPa bottles weighing 1010 kg for a content of 9 kg H_2 only, or a payload of only 0.9%! A trailer with 14 steel 1535 L 20 MPa tubes weighs 31 t for a content of 320 kg H_2, a 1.04% payload!

Transportation energy costs are high as distances between production units and refueling stations can be long and the tractor and its trailer are heavy and greedy. A 38 t hauler can need 35 L of diesel fuel per 100 km. Then, for 320 kg_{H2} of a tube trailer, with 9.85 kWh/L_{Diesel}, and 0.270 kg_{CO2}/kWh_{Diesel}, it is 2.15 kWh/kg_{H2}, 6.5% of LHV_{H2}, and 0.6 t_{CO2}/t_{H2} for each 100 km of distance between production and distribution! And even much more for bottle bundles!

6.5.1 Potential for Reducing Energy Demand

When the production unit supplies hydrogen to several distant refueling stations, energy for transportation by truck and for recompression on the distribution site must be considered, but the distribution unit works in the same way as when on the production site (Fig. 6.2). Potentials explored in this paper to lower the energy spending *of actual or near-future small refueling stations* are:

- reduce the distances between production units and distribution sites to a 50 km average.
- use small transportable containers of high-pressure light composite bottles at 30, 50 MPa or even more, when available.
- use these transportable containers in place of intermediate pressure buffers on the distribution site (Fig. 6.10).

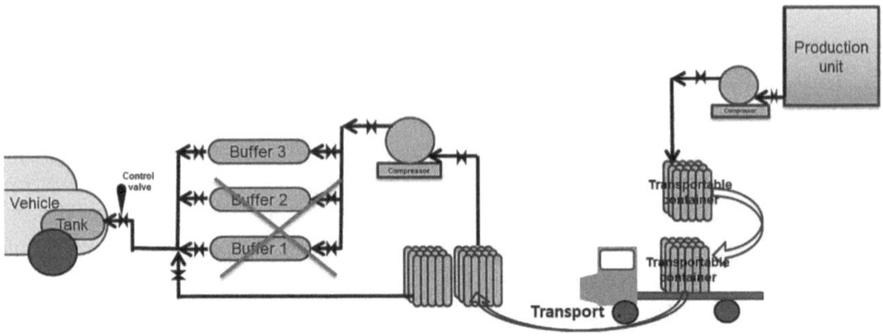

FIG. 6.10 Schematic view of the equipment for filling the HP tank of a hydrogen vehicle in a refueling station supplied by truck from a distant hydrogen production unit.

6.5.1.1 *Distributed Hydrogen Production to Reduce Transportation Distance*

Reducing the hydrogen transportation distance to about 50km (always <100km) is a specific possible advantage of distributed hydrogen production, promoted by the VABHYOGAZ3 project, which considers hydrogen production from biogas. As biogas can be produced from many kinds of waste and in lots of places, hydrogen refueling stations will never be far from a hydrogen source. This is also the case for hydrogen production from an electrolyzer. To the contrary, it is not the case in the current practice as hydrogen is now mainly produced by steam reforming in large units located, for example, on refinery sites, and hydrogen transportation distances can then be very long, sometimes >500km, with 10 times more energy spent and CO_2 emissions for transportation!

6.5.1.2 *Small High-Pressure Light Composite Bottle Transportable Containers*

The use of high-pressure composite bottle containers for hydrogen transportation, in order to reduce transportation cost and energy, has been suggested for a long time. Nexant Inc. et al. (2008) considers 7000psi (48MPa) trailers with a capacity of 1000kg H_2, as does the NREL report (Wipke et al., 2012) which considers a tube trailer at 35MPa with a capacity of about 800kg. Commercial offers can be obtained for containers at 25 or 30MPa (as reported in Section 6.3.2), but these containers are currently only used for natural gas transportation. European Commission programs have promoted developments in this field, but only a few prototypes have been realized, and the current practice still involves 20MPa steel tube trailers with a capacity around 300kg, or bundles of steel bottles for smaller quantities. This paper, in the frame of the VABHYOGAZ3 project, *considers the near future (as soon as 2018)*, with smaller production units and smaller refueling stations, and so it considers also smaller containers from 20 to 200 kg H_2 made of composite bottles with a working pressure of 30 or 52.5 MPa, which can be made available in the near future. The objective is to multiply by >3 the payload in transportation.

6.5.1.3 Optimized Use of the Transportable Containers to Fill Vehicle Tanks

The use of a trailer as the first stage of the pressure cascade for vehicle tank filling (by bypassing the compressor) has already been considered. Reddi et al. (2014) simulates the tube trailers as the first stage of refueling as long as the pressure is over 5 MPa and shows a decrease in the compressor flow rate and storage cost for large stations (250 kg/day). But this does not seem to be the current approach. And if the NREL report (Wipke et al., 2012) finally recommends the use of "low-pressure" bulk storage as a first stage in the cascade filling, this concerns in fact a fixed storage for a production site distribution or for a pipeline supplied distribution.

This paper differs from Reddi et al. (2014) as it considers smaller stations (20–200 kg/day) and the use of small transportable containers for several stages of the pressure cascade, so that just a small very-high-pressure buffer (VHPB) is to be kept and only a small part of the hydrogen is to be compressed to this VHPB.

The question is the level of investment in composite transportable bottles and the resulting overall cost for compression, storage, and transportation of the delivered hydrogen. This section estimates that overall cost, together with the overall energy expenditure, the energy savings and the green gas emissions, and makes comparisons to current practices.

6.5.2 Compression on the Production Site

On the production site where the transportable storage containers are filled, the nominal power for compression and associated cooling can be calculated using Eqs. (6.4), (6.7), (6.8), and (6.13), as in Tables 6.1 and 6.2, taking into account the hydrogen production pressure $p_{prod}=1.5$ MPa and the storage pressure $p_{sto,0}=20$, 30, or 52.5 MPa. The production unit is assumed to be running at nominal charge for 8120 h/year, and the compressor does also. But it does not always work at its nominal power. Beginning with empty storage containers at a return pressure $p_{sto,f}=2$ MPa, the lower pressure ratios induce energy savings, as already noticed in Rothuizen and Rokni (2014). These energy savings are estimated according to Appendix 1 and a reduction coefficient is introduced in Table 6.10. The investment cost of the compressor is estimated according to Eq. (6.15) and a global cost for compression can be calculated assuming a lifetime for the compressor of 8 years, together with an annual maintenance cost equal to 8% of the investment cost.

TABLE 6.10 Energy Spent and Compression Cost for Cooled Compression on Production Site, as Function of the Production Unit Capacity and of the Transportable Storage Pressure (Hydrogen Production Pressure 1.5 MPa, Initial Storage Pressure 2 MPa)

Production Unit Capacity (kg/Day)	100			200			400		
Storage pressure (MPa)	20	30	52.5	20	30	52.5	20	30	52.5
Compressor nominal electric power ($kW_{él}$)	9.1	10.8	13.5	16.9	20.2	25.2	31.6	37.7	47.0
Cooling nominal power (kW_{cool})	8.9	10.6	13.0	16.6	19.7	24.1	31.0	36.6	44.9
Reduction factor for progressive filling	0.71	0.73	0.75	0.71	0.73	0.75	0.71	0.73	0.75
Electricity consumption for cooled compression (MWh/an)	70	85	145	130	159	204	243	296	380
Electricity cost (k€/an)	5.6	6.8	8.7	10.4	12.7	16.3	19.4	23.7	30.4
Specific consumption for cooled compression (kWh/kg_{H2})	1.91	2.33	2.99	1.78	2.18	2.79	1.66	2.03	2.60
Specific consumption for cooled compression (% PCI)	5.7	7.0	9.0	5.3	6.5	8.4	5.0	6.1	7.8
Compressor investment cost (k€)	140	146	155	173	180	190	213	222	235
Global cost for compression and cooling (€/kg H_2)	0.90	0.96	1.05	0.59	0.64	0.70	0.40	0.43	0.49

Compression on the production site benefits from the scale effect; the higher the flow rate, the better the efficiency of the compressor and the lower the specific investment cost of it. Thus, the compression specific cost for hydrogen decreases significantly when increasing the production unit size, −55% for 400 kg/day with respect to 100 kg/day.

6.5.3 Compression on the Distribution Site

On the distribution site, the compressor power can be calculated if all the daily distributed hydrogen is assumed to be recompressed from the transportable storage pressure $p_{sto,0}=20$, 30, or 52.5 MPa, to the high-pressure buffers at $p_{buff}=45$ or 90 MPa, as it is usual in the current practice (Reddi et al. 2014).

Again, the compressor will start at lower pressure ratios, which induces energy savings estimated according to Appendix 2, and a reduction coefficient is applied to the electric energy demand.

Moreover, the use of the transportable containers as first stages for direct filling of the vehicle tank reduces the quantity of hydrogen to be compressed to the higher buffer pressure. Thus, the nominal flow rate of the compressor is reduced by a "bypass coefficient" (with respect to the daily distribution capacity) to be estimated as a function of the utilization strategy of the transportable containers, detailed in the following sections.

Table 6.11 gives the reference values of the current practices, namely, storage pressure of 20 MPa, with the first reduction factor, but without the second, as no bypass of the compressor is considered yet.

It can be seen that for small HRSs, with the current practice, the cost of compression can be very high, higher that the cost of compression in the production unit, even if the compressor power is lower.

The next sections discuss how improved container utilization scenarios can lower compressor investment cost and energy demand when using 30 MPa and 52.5 MPa storage containers.

6.5.4 Scenarios for Transportable Container Utilization

The idea is to place several transportable containers on the distribution site and to use them, much as they are, as first stages for filling vehicle tanks, while the compressor takes hydrogen in the lowest-pressure container, compresses it, and fills a higher-pressure buffer. This unique buffer allows the vehicle tank filling to be completed to its nominal pressure.

In the case in which the transportable containers have a nominal pressure higher than the vehicle tank nominal pressure (i.e., $p_{sto}=52.5$ MPa for $p_{tank}=35$ MPa), the compressor in no longer necessary, nor is the highest-pressure buffer. This case results in a drastic simplification of the refueling station with a considerably reduced investment and an energy consumption that is also drastically reduced. This case is considered first.

TABLE 6.11 Reference Values for Current Practices: Energy Spent for Cooled Compression on the Distribution Site, for Initial Storage Pressure of 20 MPa, Final Storage Pressure 2 MPa, and Compression of all the Hydrogen to the Buffer Pressure

Refueling Station Capacity (kg/Day)	20		80		200	
Buffer pressure (MPa)	45	90	45	90	45	90
Compressor nominal electric power (kW_{el})	4.75	6.25	16.5	21.8	37.7	49.6
Cooling nominal power (kW_{cool})	4.58	5.85	15.9	20.2	36.1	45.6
Reduction factor for progressive emptying	0.50	0.59	0.50	0.59	0.50	0.59
Electric consumption for cooled compression (MWh/an)	13.7	21.2	48	74	109	168
Electricity cost (k€/an)	1.1	1.7	3.8	5.9	8.7	13.4
Specific consumption for cooled compression (kWh/kg_{H2})	1.88	2.9	1.63	2.52	1.49	2.30
Specific consumption for cooled compression (% LHV)	5.6	8.7	4.9	7.5	4.45	6.87
Compressor investment cost (k€)	104	111	157	168	207	222
Global cost for compression and cooling (€/kg H_2)	3.07	3.36	1.24	1.39	0.70	0.81

6.5.4.1 *Refueling at $p_{tank} = 35\,MPa$ With Storage Containers at $p_{sto} = 52.5\,MPa$*

With storage at a higher pressure than the tank to fill, the compressor is of no use, but it is necessary to dispose of several storage containers on the distribution site, and these containers will have to be returned to the production unit with a nonnegligible residual hydrogen pressure.

An analytical formulation has been developed to calculate this residual pressure and it is presented in Appendix 3. The result depends on the number of storage containers present on the distribution site. It is easy to understand that with only one available storage container, this one will not be able to ensure a complete 35 MPa tank filling as soon as its pressure is lower than 35 MPa, or even lower than 38.5 MPa to take into account the heating of hydrogen during filling ($s_{p,min} = 1.10$). So, it should be substituted by another one and returned to the production unit with a residual pressure of $p_{sto,f,1} = 38.5$ MPa, i.e., $v_{sto,1} = 73.3\%$. With a second storage container available, this first one can be further used as a first stage for tank filling and its residual hydrogen pressure will then depend on the filling state of the vehicle tank v_{tank}.

The results are given in Table 6.12 as a function of the number of containers on the distribution site, which is the number of stages in the refueling, for $v_{tank} = 5\%$ ($p_{tank,0} = 1.75$ MPa). Table 6.12 also gives the contribution of each stage to refueling (percentage of tank filling). In the case in which the average v_{tank} is higher, the values of $v_{sto,n}$ would be higher.

Even with 4 stages, the residue returned to the production unit is high, $v_{sto,4} = 16.7\%$, but the advantages are that 100% of the energy mentioned in Table 6.11 is saved and the compressor and buffer investment costs are also saved.

6.5.4.2 *Refueling at* p_{tank} *=35 MPa With Storage Containers at* p_{sto} *=30 MPa*

With storage containers at $p_{sto} = 30$ MPa, it is necessary to use a compressor, which will increase the pressure of a fraction of the hydrogen to fill a buffer at $p_{buff} = 45$ MPa. The analytical equations to calculate this fraction, together with the energy saving, have been formulated and are presented in Appendix 4. The results depend again on the filling state of the vehicle tanks, $v_{tank} = 5\%$, but the residual hydrogen pressure with which the containers are returned to the production unit can now be chosen and are here $v_{sto,f,n} = 6.67\%$, or $p_{sto,f,n} = 2$ MPa.

In Table 6.13 column 2, the fraction of hydrogen to be compressed is shown (on a total of $(1 - v_{tank}) = 95\%$) and this number allows sizing a smaller compressor for the distribution unit; its capacity can be reduced by a factor of 2.5–3 with 3 or 4 stages. The columns 3, 4, and 5 give the contributions of the previous stages to the refueling and the pressure at which the storage is shifted to the lower stage. The last column gives the reduction factor in the energy demand with respect to the reference case in which 20 MPa storage containers would be completely transferred to the high-pressure buffer. It can be noticed that even with 1 available storage container only, it is possible to organize a direct connection for a prefilling of the tank before completing the refueling from the buffer.

Other calculations have been made for $p_{tank} = 70$ MPa and $p_{sto} = 52.5$ and 30 MPa. Because of the greater difference between tank and storage pressures, the contribution of the compressor is higher and the energy saving lower. But the compressor capacity can still be reduced, by a factor of 1.5–2.5.

TABLE 6.12 Effect of the Number of Available 52.5 MPa Storage Containers: Contribution of Each Stage to the Refueling at p_{tank} = 35 MPa, Starting With v_{tank} = 5%; Residual Pressure for Each Stage and Residual Fraction Returned to the Production Unit

Number of Storages n	Stage n $p_{sto,f,1}$ $\Delta v_{tank,1}$	Stage $n-1$ $p_{sto,f,2}$ $\Delta v_{tank,2}$	Stage $n-2$ $p_{sto,f,3}$ $\Delta v_{tank,31}$	Stage $n-3$ $p_{sto,f,4}$ $\Delta v_{tank,4}$	Residual Fraction Returned to Production $v_{sto,n}$ (%)
1	38.5 MPa 95%				73.3
2	38.5 MPa 43.3%	21.8 MPa 51.7%			41.5
3	38.5 MPa 33.9%	25.4 MPa 31.6%	13.3 MPa 29.5%		25.3
4	38,5 MPa 30.4%	26.7 MPa 25.5%	17.0 MPa 21.3%	8.8 MPa 17.8%	16.7

6.5.5 Detailed Characteristics and Costs for 20 kg/Day Distribution Units

A detailed example is given; named p_100_d_20, it considers five small distribution units of 20 kg/day within a distance of 50 km of their supplying production unit. The technical details and costs are analyzed and compared in three cases in Table 6.14:

(a) steel 20 MPa storage, 1 stage, no compressor bypass (reference case);
(b) 4 stages 30 MPa composite storage;
(c) 4 stages 52.5 MPa composite storage.

6.5.5.1 Cooled Compression

It appears that for a distribution at 35 MPa, the global cost for cooled compression is reduced by 28% and spent energy and CO_2 emissions are reduced by 55% in case b (30 MPa storage containers) compared to case a (20 MPa reference). They are reduced to 0 in case c (52.5 MPa storage containers). In case b, the nominal power of the compressor is much lower as a large fraction of hydrogen can bypass it, and so are the consumption and CO_2 emissions (green-house gas emissive power of French electricity is taken at 60 g_{CO2}/kWh_{el}).

TABLE 6.13 Effect of the Number of Available 30 MPa Storage Containers; Stage *n* is connected to the compressor; $p_{sto,f,n} = 2$ MPa; contribution of each stage to the refueling at $p_{tank} = 35$ MPa, starting with $v_{tank} = 5\%$; residual pressure for each stage and reduction factor for the compressor energy demand

	Stage *n* (Compressor)	Stage *n* − 1	Stage *n* − 2	Stage *n* − 3	
Number of Storages *n*	$p_{sto,f,1}$ $\Delta v_{tank,1}$	$p_{sto,f,2}$ $\Delta v_{tank,2}$	$p_{sto,f,3}$ $\Delta v_{tank,31}$	$p_{sto,f,4}$ $\Delta v_{tank,4}$	Energy Demand Reduction With Respect to Reference Case
1	2 MPa 75%	2 MPa 20%			0.75
2	2 MPa 53,7%	17.8 MPa 41.3%			0.591
3	2 MPa 40,3%	23.3 MPa 23.7%	13.9 MPa 31.0%		0.480
4	2 MPa 34%	25.4 MPa 15.6%	19.4 MPa 20.4%	11.5 MPa 24.9%	0.359

Nevertheless, for such small distribution units, the cooled compression cost is high, between 2.2 and 3.36 €/kg. Having no compression on the distribution site (case c) is a very large advantage.

6.5.5.2 *High-Pressure Buffer*

Using higher pressure storage containers induces also a beneficial effect on the necessary volume for high-pressure buffers in the distribution station. However, this cost remains low, being between 0 and 0.26 €/kg.

6.5.5.3 *Transportable Storage Containers*

Even if the transportable storage containers might rather be used on a rental basis, investment costs and specific costs are estimated (costs without margin). The calculations have shown that the best is having storage units with a capacity of about 1 day of distribution (thus, here 20kg).

(a) in current practice, 20MPa storage containers consist of bundles with 12 50L steel bottles for an overall mass of 1010kg (0.9% of hydrogen content only).
(b) 30MPa composite storage containers would be made from 2 bottles of 350L containing 14.7kg H_2 for a mass of 480kg (3.2% hydrogen content).
(c) 52.5MPa composite storage containers will be made of 2 bottles of 300L containing 19.8kg H_2 for a mass of 660kg (3.0% hydrogen content).

These scenarios take the following into account:

- v_{sto}, the residual hydrogen returned to the production unit when a storage container is considered as empty,
- a minimum net autonomy of 2days on the distribution site,
- a container number at least equal to the stage number (4 for b and c),
- a rotation every 2 or 3 days for the delivery of the bundles or containers,
- at least 1 container left on the production site for being filled.

This gives the total minimum number of containers, the overall autonomy, and the investment cost for the storage. The autonomy is then between 6.1 and 8.6days. The absence of a compressor induces a high residual pressure in the returned containers and increases by 1 unit the number of necessary containers for 35MPa distribution. With 52.5MPa storage containers, due to the larger H_2 mass of the containers, the autonomy of the station exceeds 3days. Thanks to their long lifetime, the depreciation of the storage investment cost is taken over 20years, together with a 1.5% annual maintenance cost. Finally, the specific storage cost remains high, from 0.63 to 1.86 €/kg, but the increase for cases b and c with respect to case a is lower than the savings on the specific compression cost.

6.5.5.4 Transportation Material

The transportation material consists of a tractor and a flatbed trailer equipped with a handling crane; the calculation model shows:

(a) 5 bundles have to be delivered every 2 days to each of the 5 stations; the total mass is >25 tons and would necessitate a 38 t tractor and trailer with a high investment and a specific cost of 0.55 €/kg (including maintenance cost).
(b) 3 containers have to be delivered every 2 days to each HRS; the load is only 7.2 t and will require a lighter tractor and trailer, with a specific cost divided by 2, 0.26 €/kg
(c) the station autonomy is larger than 3 days and the supply of 2 or 3 stations every 3 days is planned, with the same material as for case b. The trailer load will be 8 t for 35 MPa distribution or 6 t for 70 MPa and with specific cost of the same order.

The round-trip delivery for 3 or 5 stations is 6 or 9 h long for 200 or 300 km, as the distance between stations is assumed to be 50 km, and with 30 min for loading and unloading in each place. The total mileage is 54,000 km/year for cases a and b, and 40,500 km/year for case c, thus the 3-day delivery period gives an advantage. The diesel consumption (at 1.04 €/L, VAT excluded) for transportation is highly affected by the mass of the hauler. It is 2.5 times higher for case a than for case b, and by the total distance, case c is even 20% and 35% less than b. Thus, the light composite containers reduce drastically the specific consumption and CO_2 emission to an acceptable level of 0.36 to 0.56 t_{CO2}/t_{H2}, instead of 1.46 t_{CO2}/t_{H2} for case a!

At this stage, it must be noticed that tractor and trailer are not used at full charge. They could make at least 3 times as many deliveries, decreasing the investment specific cost by a factor of 3 if this material were used in common for 3 production units.

6.5.5.5 Labor Cost

In all three cases, the deliveries can be made with one full time equivalent worker (1640 h/year) for a cost estimated to be 36 k€/year, which gives a high specific cost of 0.99 €/kg H2.

Table 6.14 also shows the premium costs of the 70 MPa distribution option with respect to the basic 35 MPa distribution, which is high at 0.74 €/kg with 30 MPa containers and 2.39 €/kg with 52.5 MPa! The availability of very-high-pressure transportable containers (e.g., 105 MPa) would reduce this gap for 70 MPa dispensing.

The cost for compression, storage, and transportation of hydrogen, for the purpose of the distribution of small quantities, was known to be high. Finally, this model confirms and quantifies these costs with details. It appears that the use of composite light bottle containers adapted to the distribution unit capacity

TABLE 6.14 Effect of Transportable Storage Choice on the Detailed Characteristics and Costs for Small 20 kg/Day HRSs Supplied by a 100 kg/Day Production Unit (Distance 50 km)

Refueling Stations: Number and Capacity (kg/Day)	5 HRSs of 20 kg/day Supplied From 1 Production Unit at 50 km					
Storage: case, type, pressure (MPa)	a, Steel, 20		b, Composite, 30		c, Composite, 52.5	
Distribution pressure (MPa)	35	70	35	70	35	70
Buffer pressure (MPa)	45	90	45	52.5 and 90	None	52.5 and 90
Fraction of compressed hydrogen (%)	100	100	35.8	63.9	0	41.2
Compressor investment cost (k€)	104	111	76	97	0	85
Compressor nominal electric power (kW_{el})	4.75	6.25	1.9	4.2	0	2.8
Reduction factor for progressive storage emptying	0.50	0.59	0.49	0.57	0	0.46
Electric consumption for cooled compression (MWh/an)	13.7	21.2	6.2	11.9	0	7.4
Electricity cost (k€/an)	1.1	1.7	0.5	0.9	0	0.6
Specific consumption for cooled compression (kWh/kg_{H2})	1.88	2.9	0.85	1.62	0	1.01
Specific consumption for cooled compression ($\%LHV_{H2}$)	5.6	8.7	2.5	4.9	0	3.0
Specific GHG emission (t_{CO2}/t_{H2})	0.11	0.17	0.05	0.10	0	0.06
Global cost for compression and cooling ($€/kg_{H2}$)	3.07	3.36	2.21	2.86	0	2.48
High pressure buffer volume (L)	750	450	270	210+150	0	60+150
Buffer investment costs (k€)	14.5	15	6.6	11.6	0	8.0
Buffer specific cost ($€/kg_{H2}$)	0.28	0.29	0.13	0.22	0	0.15
Unit storage hydrogen content (kg)	9.0		14.7		19.8	
Number of storages on distribution site	6		5		5	5

TABLE 6.14 Effect of Transportable Storage Choice on the Detailed Characteristics and Costs for Small 20 kg/Day HRSs Supplied by a 100 kg/Day Production Unit (Distance 50 km)—cont'd

Refueling Stations: Number and Capacity (kg/Day)	5 HRSs of 20 kg/day Supplied From 1 Production Unit at 50 km			
Number of storages in transit	5	3	3	2
Total number of storages	15	9	10	9
Overall autonomy (days)	6.1	6.2	8.2	8.6
Investment cost of storages (k€)	71	123	209	188
Storage specific cost (€/kg_{H2})	0.63	1.10	1.86	1.67
Mass of containers in transit (t)	5*5.05	5*1.44	3*2.65	2*2.65
Mass of tractor and trailer (t)	38	12	13	11
Equipment specific cost (€/kg $_{H2}$)	0.55	0.26	0.28	0.23
Periodicity of deliveries (days)	2	2	3	3
Total annual mileage (km)	54,000	54,000	40,500	
Transport Diesel energy (MWh/yr)	192	76	61	49
Specific transport energy (kWh/kg_{H2})	5.25	2.07	1.66	1.34
Specific transport energy (% LHV_{H2})	15.7	6.2	5	4
Specific GHG emission (t_{CO2}/t_{H2})	1.42	0.56	0.45	0.36
Diesel specific cost (€/kg_{H2})	0.56	0.22	0.18	0.14
Labor cost: equivalent full time	1.0			
Specific labor cost (€/kg_{H2})	0.99			

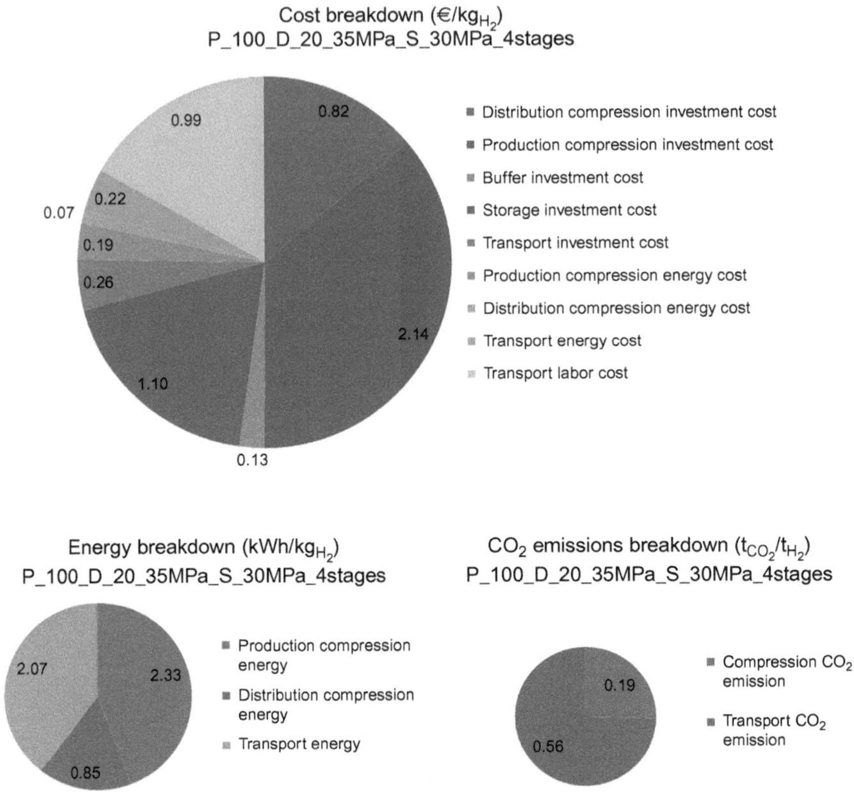

FIG. 6.11 Breakdown of cost, spent energy, and CO_2 emissions related to compression, storage, and transportation steps for a 20 kg/day 35 MPa refueling station supplied with 30 MPa hydrogen storage containers from a production unit at a distance of 50 km.

and a good distribution command-and-control strategy allow a significant decrease of this cost.

Moreover, one must keep in mind that these calculations have been made with the assumption that all refueling stations and production units are working at nominal load. A partial load would even increase the final specific cost. On the other hand, maximum load is twice the nominal load and working in between nominal and maximum load would decrease this cost.

The next section will show how the cost is drastically reduced for larger production units (200 and 400 kg/day) and larger HRSs (80 and 200 kg/day). These sizes are suitable for early hydrogen market deployment. For a mature market, HRSs of 400, 1000 kg/day, or even larger are considered, but it is probable that small capacity HRSs (80 and even 20 kg/day) will last and even expand in remote locations with low population densities.

Finally, the diagram of Fig. 6.11 illustrates the breakdown of cost, spent energy, and CO_2 emissions for this case, p_100_d_20, with 35 MPa distribution and 4-stage 30 MPa storage units. The investment payback represents a very large part of the global cost, 75% with a large contribution from the compressors, especially on a distribution site. The suppression of a distribution compressor thanks to the use of 52.5 MPa storage containers is a great advantage!

6.5.6 Estimation of Global Costs, Effect of Capacity and Stage Number

6.5.6.1 Comparison of Reference Case 20 MPa Steel Tubes and 30 MPa Composite Containers

Fig. 6.12 shows a comprehensive view of the global cost for compression, storage, and transport in the reference case (20 MPa steel tubes) in blue, and in cases of 30 MPa light transportable composite containers in green. The same methodology and assumptions as in the previous section are used. The different considered cases show the effect of the size or capacity of the units:

(a) Case p_100_d_20 (as in previous section): one 100 kg/day production unit supplying five 20 kg/day refueling stations,
(b) Case p_200_d_80: one 200 kg/day production unit supplying 80 kg/day refueling stations,
(c) Case p_400_d_80: one 400 kg/day production unit supplying 80 kg/day refueling stations,
(d) Case p_400_d_200: one 400 kg/day production unit supplying 200 kg/day refueling stations.

Results are given for 1, 2, 3, or 4 containers on the distribution site so that the effect of the number of stages can be seen. Despite the high investment cost of the composite containers, when amortized over 20 years, and provided their size is adapted to the distribution capacity, it appears clearly that:

- The composite solution is always better than the steel tube solution: with only 1 stage, the gain varies from *15% to 4.6%* according to the case (from 61 to 13 c€/kg H_2).
- Increasing the number of stages from 1 to 4 is beneficial to the global cost. The benefit is very high for small 30 MPa refueling stations and decreases with capacity and distribution pressure: at 35 MPa *12%* for p_100_d_20 and *9%* for p_400_d_200; at 70 MPa *5.5%* for p_100_d_20 and *4%* for p_400_d_200.
- The effect of capacity is large: the cost of p_400_d_200 is *2.31* €/kg_{H2}, a *60%* reduction of the cost of p_100_d_20, 5.91 €/kg_{H2}.
- The extra cost for the 70-MPa option is higher for small capacities: 76 c€ (12.5%) for p_100_d_20 and 25 c€ (10.5%) for p_400_d_200.

Concerning the energy spent, the gain of the composite solution is also clear: −37% for p_100_d_20 at 5.67 kWh/kg_{H2}, −32% for p_400_d_200 at

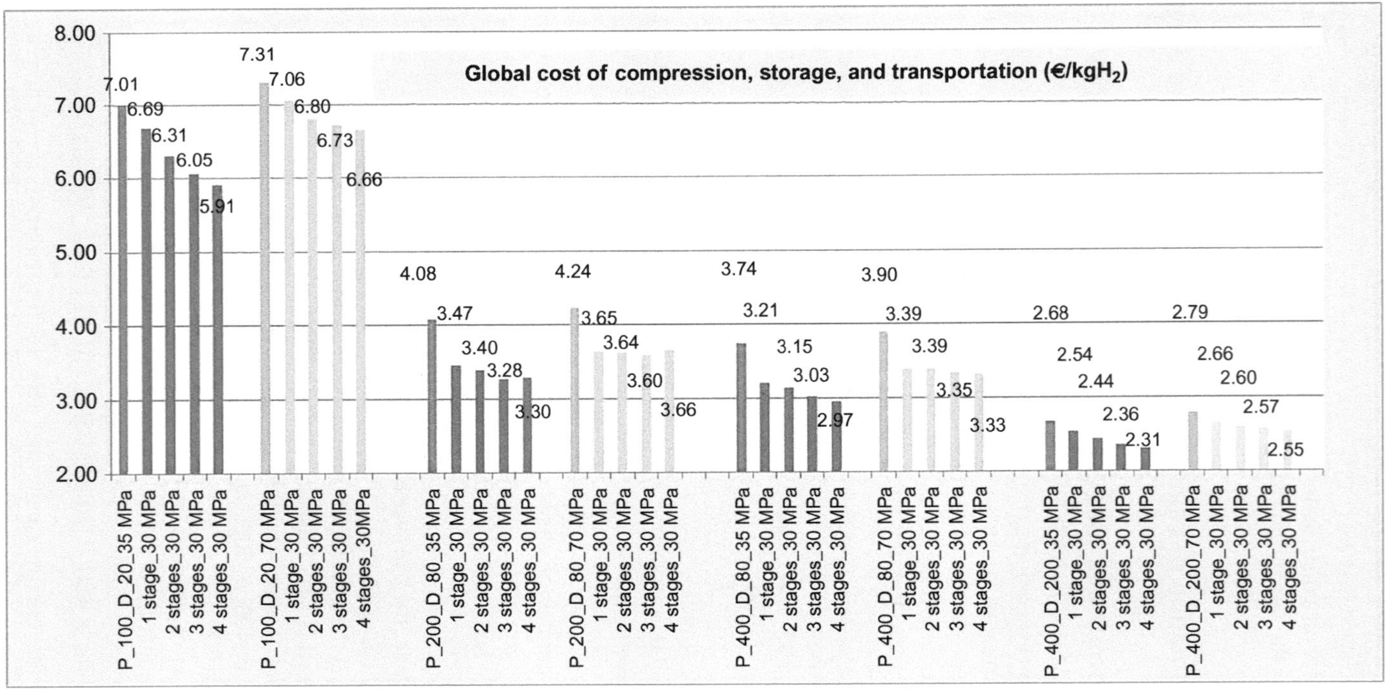

FIG. 6.12 Variation of global cost for compression, storage and transportation with the size of the units and the type of storage : reference steel 20 MPa (the *blue* and first bar of each series of 5) or composite 30 MPa with 1 to 4 stages (the 4 *green* following bars of each series of 5); 35 MPa distribution in *dark blue* or *green* (*dark grey* in print version) and 70 MPa distribution in *light blue* or *green* (*light grey* in print version).

3.52 kWh/kg $_{H2}$. Then, increasing the number of stages generates extra gains of −7.5% for p_100_d_20 and p_400_d_200. The 70-MPa refueling option costs 15 to 18% more energy than the 35-MPa option. The decay with the size is large: −35% from *5.25* kWh/kg_{H2} for p_100_d_20 to *3.26* kWh/kg_{H2} for p_400_d_200 (Fig. 6.13).

Concerning the CO_2 emissions, the results show similar trends. The large decrease for the reference steel storage between p_100_d_20 and p_200_d_80 is explained by the shift from bottle bundles to tube trailers with higher specific hydrogen content, inducing lower fuel consumption for transportation. The size effect reduces CO_2 emissions by 59% from *0.75* t_{CO2}/t_{H2} for p_100_d_20 to *0.31* t_{CO2}/t_{H2} for p_400_d_200 (Fig. 6.14).

6.5.6.2 Comparison Between 30-MPa and 52.5-MPa Composite Storage Containers

In Fig. 6.15, the best results for 30-MPa containers (with 4 stages) are compared to the results for 52.5-MPa containers with 1, 2, 3, and 4 stages for the same 4 cases as in the previous section in order to show the influence of increasing the pressure and the number of stages.

Of course, for 1-stage 35-MPa distribution without a compressor, because of the high level of hydrogen returned to production, 52.5-MPa storage generates a higher cost, but adding stages improves the results and finally generates appreciable savings with respect to the 30-MPa storage: *−29%*, i.e., −1.70 €/kg at 4.21 €/kg for p_100_d_20 and *−4%*, that is, −9 c€/kg at 2.22 €/kg for p_400_d_200.

For 70-MPa 80-kg/day distribution, the best costs are obtained with 3 stages and are very close (within 1% or 2%) of those obtained with 30-MPa storage; better for 200 kg/day production units and slightly higher for 400 kg/day production units.

Concerning the spent energy (Fig. 6.16), the best score is always obtained for four stages, with good improvements for p_100_d_20 at 35 MPa: −0.6 kWh/kg at 4.65 kWh/kg *(−13%)*, as at 70 MPa: −0.68 kWh/kg at 5.35 kWh/kg. At higher capacities, the advantages of the 52.5-MPa storage containers are less: the energy for their transportation is nearly unchanged and the compression energy saving at the distribution is nearly compensated by the extra compression energy at production. A specific energy as low as *3.11* kWh/kgH_2 can be reached.

For CO_2 emissions, the trends are the same and emissions as low as *0.29* t_{CO2}/t_{H2} can be reached (Fig. 6.17).

6.6 CONCLUSION

A methodology has been detailed to simply specify the components and evaluate the cost, the energy consumption, and the CO_2 emissions of the compression, storage, and transportation steps for hydrogen distribution. A numerical

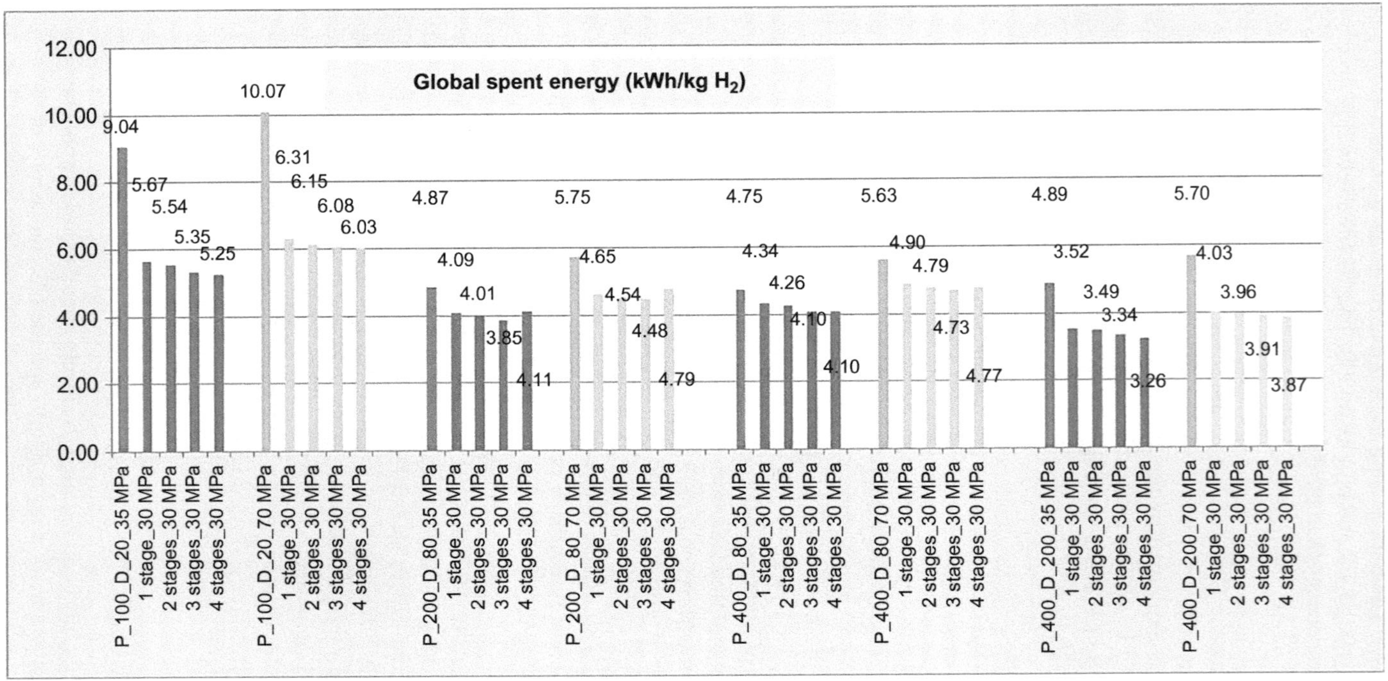

FIG. 6.13 Variation of global energy spent for compression, storage and transportation with the size of the units and the type of storage : reference steel 20 MPa (the *blue* and first bar of each series of 5) or composite 30 MPa with 1 to 4 stages (the 4 *green* following bars of each series of 5); 35 MPa distribution in *dark blue* or *green* (*dark grey* in print version) and 70 MPa distribution in *light blue* or *green* (*light grey* in print version).

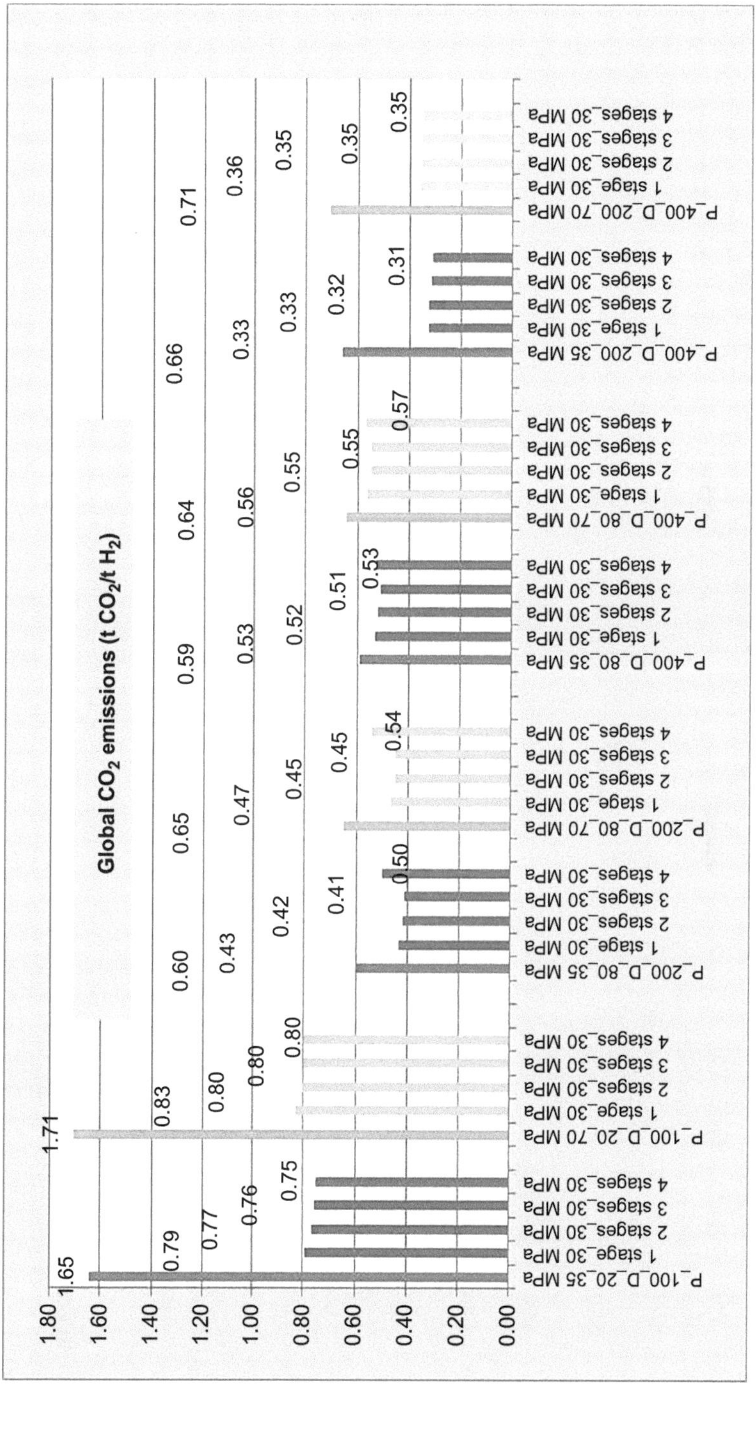

FIG. 6.14 Variation of the CO_2 emissions for compression, storage and transportation with the size of the units and the type of storage : reference steel 20 MPa (the *blue* and first bar of each series of 5) or composite 30 MPa with 1 to 4 stages (the 4 *green* following bars of each series of 5); 35 MPa distribution in *dark blue* or *green* (*dark grey* in print version) and 70 MPa distribution in *light blue* or *green* (*light grey* in print version).

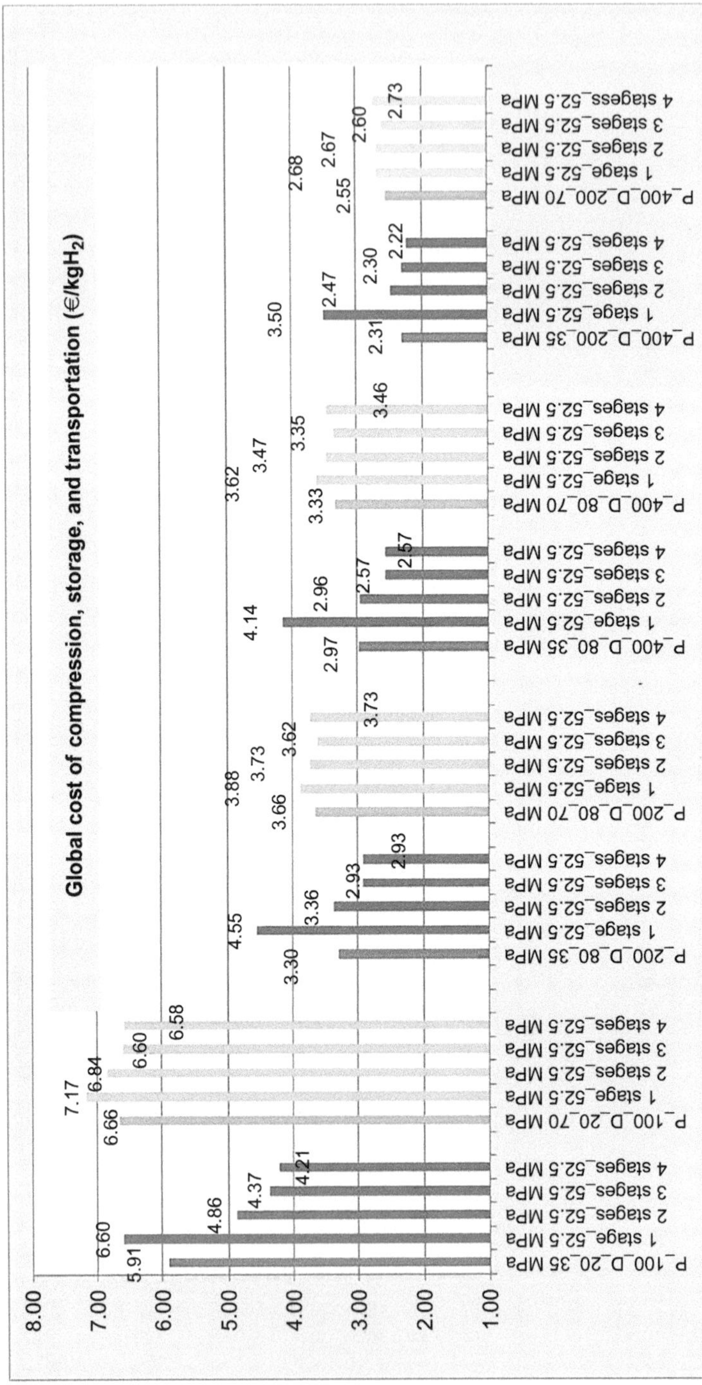

FIG. 6.15 Variation of global cost for compression, storage and transportation with the size of the units and the type of storage : reference steel 20 MPa (the *blue* and first bar of each series of 5) or composite 30 MPa with 1 to 4 stages (the 4 *green* following bars of each series of 5); 35 MPa distribution in *dark blue* or *green* (*dark grey* in print version) and 70 MPa distribution in *light blue* or *green* (*light grey* in print version).

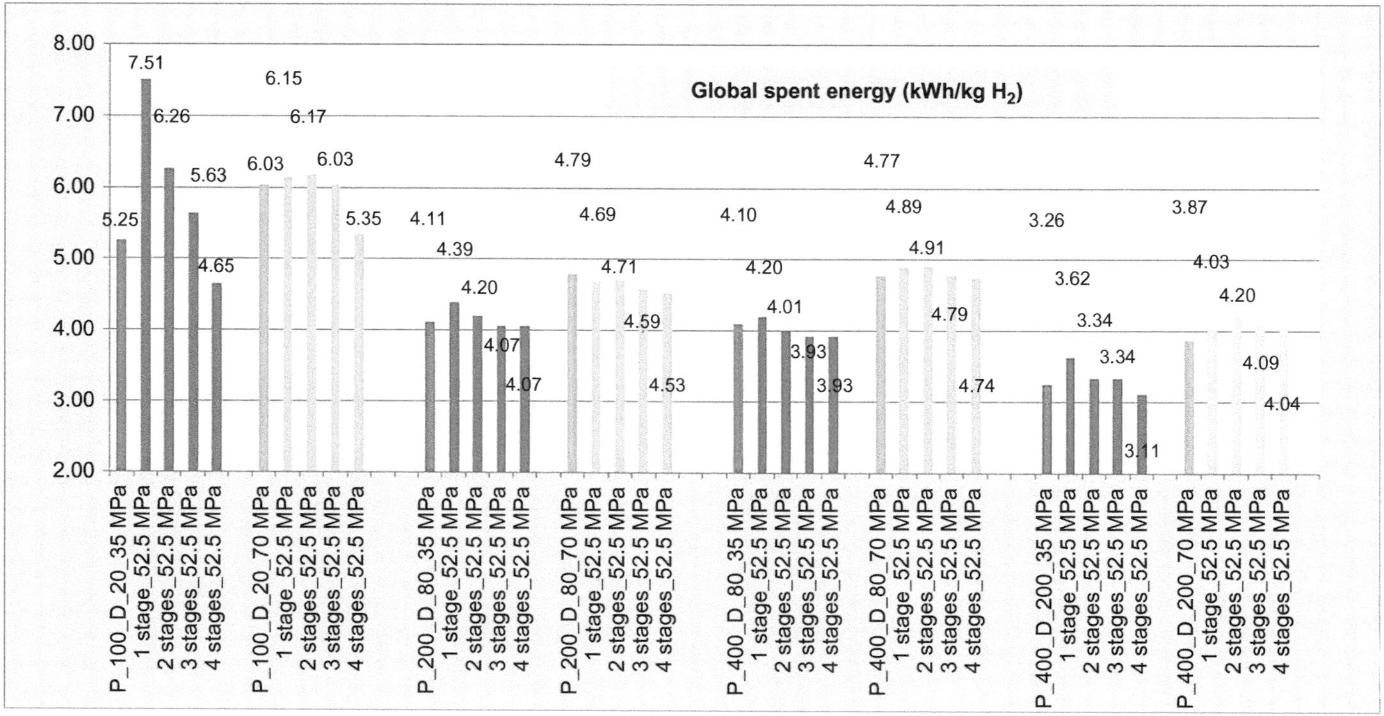

FIG. 6.16 Variation of global energy spent for compression, storage and transportation with the size of the units and the type of storage : reference steel 20 MPa (the blue and first bar of each series of 5) or composite 30 MPa with 1 to 4 stages (the 4 *green* following bars of each series of 5); 35 MPa distribution in *dark blue* or *green* (*dark grey* in print version) and 70 MPa distribution in *light blue* or *green* (*light grey* in print version).

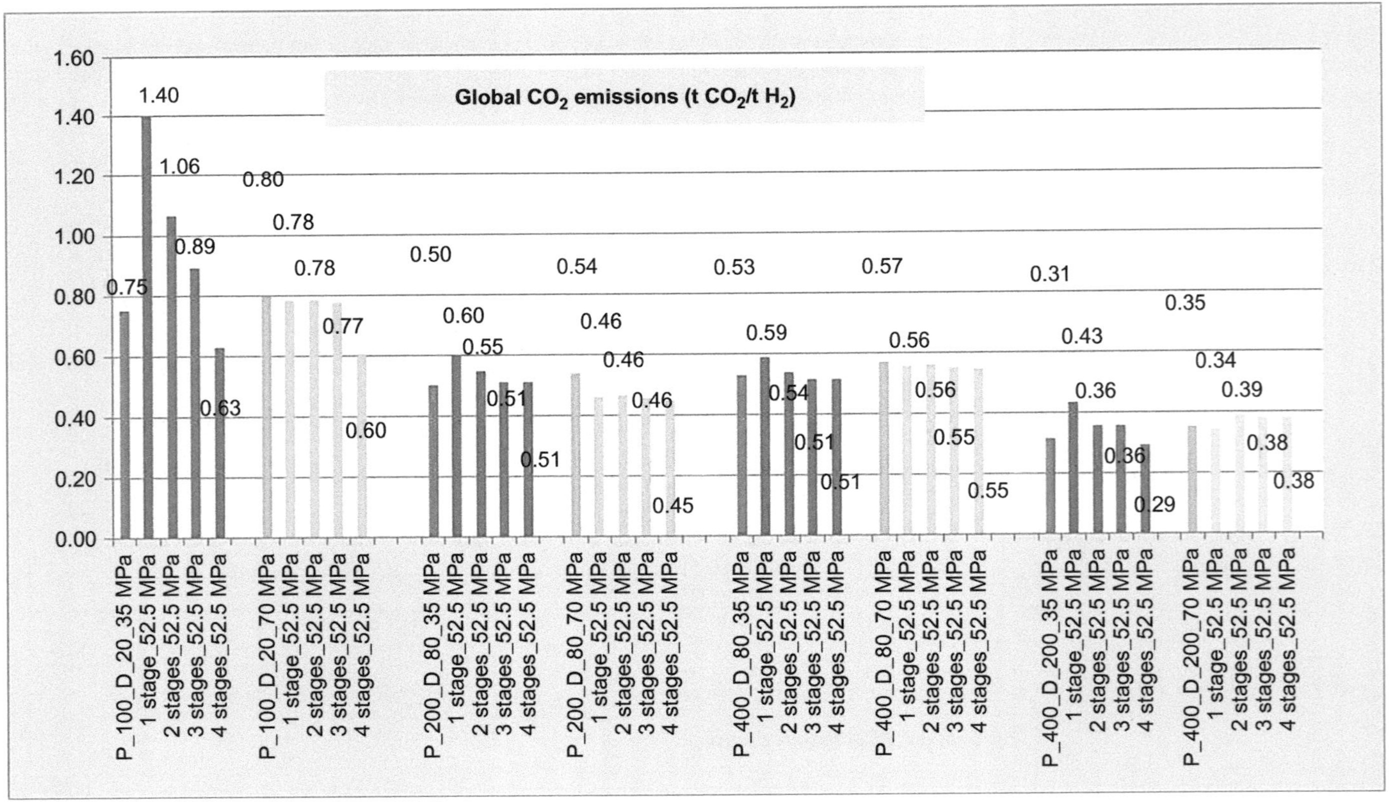

FIG. 6.17 Variation of the CO_2 emissions for compression, storage and transportation with the size of the units and the type of storage : reference steel 20 MPa (the *blue* and first bar of each series of 5) or composite 30 MPa with 1 to 4 stages (the 4 *green* following bars of each series of 5); 35 MPa distribution in *dark blue* or *green* (*dark grey* in print version) and 70 MPa distribution in *light blue* or *green* (*light grey* in print version).

tool has been developed on this basis. Results have been shown, together with the associated hypotheses, in some cases corresponding to small distribution units relevant to the early hydrogen market of the current and next years. The complete set of equations has been written to form a comprehensive model, which can be further used in other conditions or with other hypotheses by the reader.

Among the hypotheses of the model, the value of v_{tank}, the fraction of residual hydrogen in the tank as a vehicle comes to the refueling station, is a sensitive piece of data. The value used (2 MPa) can be considered low, thus inducing optimistic results. On the other hand, considering a load equal to the nominal load for the HRS is optimistic at the launch of the station, but as the maximum load equals twice the nominal load, with the further development of hydrogen vehicle fleets, this assumption will become pessimistic! And finally, it can be correct over the lifetime of the HRS.

For the case of *hydrogen distribution on the site of its production*, the energy need for cooled compression is evaluated at *3.50 or 4.4* kWh/kg_{H2} at *35 or 70 MPa, in the reference case*, corresponding to current practice. The study has shown that this energy need can be *reduced by >25%* when judiciously using *4 or 5 stages* of buffers organized in a pressure cascade for the filling of the tanks. Whereas these staged pressure buffers have higher volumes and costs than the usual very-high-pressure buffer (VHPB), the extra capital cost is acceptable and the energy saving results in an acceptable payback time of *4.5–5.5 years*.

In the case in which a production unit supplies hydrogen to *several distant distribution units*, energy for transportation by truck and for recompression on the distribution site must be added. The offsite distribution current practice considers the transportation of 20 MPa hydrogen steel bottle bundles or tube trailers and the recompression of all the hydrogen to the VHPB.

To lower the energy spending, small containers of 30 MPa light composite bottles can be used. Trucks can then transport loads with a three times larger hydrogen content. These containers can be used as well in the place of intermediate pressure buffers on the distribution site for prefilling of the tanks; just a small VHPB has to be kept and only a small part of the hydrogen has to be compressed to this VHPB.

The study shows that, even if the investment in composite bottle containers is high, the resulting overall cost is always lower. The use of up to 4 containers on the distribution site generates a global cost reduction of *27%–14%* for a distribution at 35 MPa, decreasing with the capacity (and 20.5–8.5% at 70 MPa). The energy savings are high, in a range of *42–14%* when shifting from steel 20 MPa to composite 30 MPa according to the different studied cases.

When higher pressure composite containers are available (52.5 MPa), then a compressor and VHPB are useless for distribution at 35 MPa and this results in *significant supplementary reductions* of investment cost, global cost *(−29%)*, and energy needs *(−13%)*, especially for small capacities, while for a distribution at 70 MPa, 52.5-MPa containers give very similar results to 30-MPa containers.

The overall energy expenditure can be *lower than 3.1* kWh/kg_{H2} for a 35 MPa distribution (*4.04* kWh/kg_{H2} for 70 MPa) and the CO_2 emissions lower than *0.30* t_{CO2}/t_{H2} for a 35 MPa distribution (*0.38* t_{CO2}/t_{H2} for 70 MPa).

These small, light, composite 52.5 MPa hydrogen containers generate interesting savings in operational costs and also drastic simplification for 35-MPa distributions. Thus, 20, 80, and 200 kg_{H2} containers are currently being developed in the framework of the VABHYOGAZ3 project by ALBHYON office of HERA, the company leader of the project. They will be certified for road transportation and implemented on the test sites of the project, and they will also be available as commercial products as a result of the project. Also being developed is the 35 MPa distribution unit adapted to these 52.5 MPa containers. It is very simple, it does not have any compressor nor any HP buffer, its cost will be low and it will make hydrogen available and affordable in many places. 70-MPa distribution units adapted to these 52.5-MPa container cascades will be another product resulting from the project, developed by ALBHYON, as they have the know-how for designing, manufacturing (with their supplier network), certifying, and implementing these solutions on those sites interested in small- or medium-scale hydrogen production and distribution.

Finally, it should be recalled that all costs presented in this paper are current costs as the results are to be applied now or in the next few years. No reduction for large quantity production is taken into account yet. However, these reductions will undoubtedly occur, as normally occurs in any domain when a new product progressively acquires a growing maturity and is produced in larger and larger quantities. This is known as "the learning curve" In a contiguous field of hydrogen energy, Staffell and Green (2013) shows how the prices of domestic fuel cells for combined heat and power generation have been decreasing in the past decade; it also recalls that this phenomenon has been noticed for other products from the energy sector, with a 15%–20% decay in the cost of a system noticed each time the produced quantity doubles. So, there is great hope that the costs announced in this paper are going to decrease significantly in the next decade.

APPENDIX 1 WORK FOR THE PROGRESSIVE FILLING OF A STORAGE CONTAINER

In some cases, the compressor does not always work at its nominal pressure ratio, for example when a trailer or a storage container comes back to the production site nearly empty, with a pressure $p_{sto,f}$ to be refilled up to the pressure $p_{sto,0}$. While the inlet pressure of the compressor is constant at p_{prod}, the filling begins with a low pressure ratio, $r_{p,f}$, and this pressure ratio increases progressively up to $r_{p,0}$ as the compressor fills the storage container. Thus the compressor starts at low power and ends at its nominal power.

$$r_{p,f} = {}^{p_{sto,f}}/_{p_{prod}} \quad r_{p,0} = {}^{p_{sto,0}}/_{p_{prod}}$$

It is easy to calculate the energy need along this progressive filling in the case of a perfect gas and of an isothermal compression.

For an isothermal compression of an increment of mass dm from p_{prod} up to p_{sto}, the useful energy need is:

$$dW_{filling} = k * dm * \ln\left({}^{p_{sto}}/_{p_{prod}}\right) \text{ with } k = \frac{R}{M} * T_0$$

The total energy need to fill the storage from $p_{sto,f}$ to $p_{sto,0}$ from a production unit at p_{prod} is equal to the integral:

$$W_{filling} = \int_{p_{sto}=p_{sto,f}}^{p_{sto,0}} k * dm * \ln\left({}^{p_{sto}}/_{p_{prod}}\right)$$

The pressure in the storage container in linked to the mass of gas present in the tank; for a perfect gas, an increment of mass dm induces an isothermal increment of pressure dp linked to the final pressure $p_{sto,0}$ and final mass $m_{sto,0}$ in the storage:

$$dm / {}_{m_{sto,0}} = {}^{+dp}/_{p_{sto,0}}$$

$$W_{filling} = +k * {}^{m_{sto,0}}/_{p_{sto,0}} * \int_{p=p_{sto,f}}^{p_{sto,0}} \ln\left({}^{p_{sto}}/_{p_{prod}}\right) * dp$$

Introducing the new variable:

$$r_p = {}^{p_{sto}}/_{p_{prod}}$$

$$W_{filling} = +k * m_{sto,0} * {}^{p_{prod}}/_{p_{sto,0}} * \int_{r_p=r_{p,f}=\frac{p_{sto,f}}{p_{prod}}}^{r_{p0}=\frac{p_{sto,0}}{p_{prod}}} \ln(r_p) * dr_p$$

The primitive function of $f(x) = \ln(x)$ is $F(x) = x*(\ln(x) - 1)$, thus it comes:

$$W_{filling} = +k * m_{sto,0}/r_{p,0} * \left[r_p(\ln(r_p) - 1)\right]_{r_{pf}}^{r_{p,0}}$$

$$W_{filling} = k * \frac{m_{sto,0}}{r_{p,0}} * \left\{r_{p,0} * (\ln(r_{p,0}) - 1) - r_{p,f} * (\ln(r_{p,f}) - 1)\right\}$$

TABLE A.1 Comparison of Progressive Filling of Storage (Increasing Pressure Ratio) With Constant Pressure Ratio Compression, Isothermal, Perfect Gas, $p_{prod} = 1.5\,\text{MPa}$, Storage Initial Pressure $p_{sto,f} = 2\,\text{MPa}$

Storage Pressure p_{sto} (MPa)	20	30	52.5
Initial pressure ratio $r_{p,\,f}$	1.33	1.33	1.33
Final pressure ratio $r_{p,0}$	13.3	20	35
Energy for progressive isothermal filling from $r_{p,\,f}$ to $r_{p,\,0}$	1.85*k	2.19*k	2.68*k
Energy for isothermal filling at constant $r_{p,\,0}$	2.59*k	3.00*k	3.56*k
Reduction factor for progressive filling	0.713	0.731	0.755

During this filling, the mass of the storage increases from $m_{sto,0} * v_{sto}$ to $m_{sto,0}$ with:

$$v_{sto} = {p_{sto,f}}/{p_{sto,0}} = {r_{p,f}}/{r_{p,0}}$$

Thus, the transferred mass is $m_{sto,0} * (1 - v_{sto})$ and finally the energy need per unit transferred mass is:

$$W_{filling}/m_{transf} = k * \frac{1}{r_{p,0} * (1 - v_{sto})} * \{r_{p,0} * (\ln(r_{p,0}) - 1) - r_{p,f} * (\ln(r_{p,f}) - 1)\}$$

Otherwise, this energy for filling a storage container with an increasing pressure ratio from $r_{p,\,f}$ to $r_{p,0}$ can be compared with the energy for a perfect gas isothermal compression of constant pressure ratio $r_{p,\,0}$, which is Eq. (6.3):

$$W_{isothermal,\,perfect}/m = k * \ln(r_{p,0})$$

Some numerical values are given in the Table A.1, which are used in Section 6.5:

It can be seen that the filling of a storage container with a progressively increasing pressure costs 25–30% less than filling at a constant pressure.

APPENDIX 2 WORK FOR EMPTYING A STORAGE CONTAINER TO A HIGHER-PRESSURE BUFFER

Sometimes, the opposite case happens and the compressor has always the same outlet pressure, but the inlet pressure varies and the compressor does not always work at its nominal pressure ratio. This happens, for example, when a trailer or a storage container is used to fill the buffer of a refueling station. While the pressure buffer is kept constant by the compressor at p_{buff}, at the beginning the trailer or the storage is at a high pressure, $p_{sto,0}$, and the buffer filling begins with a low pressure ratio $r_{p,0}$; then the pressure ratio increases progressively up to $r_{p,f}$ as the compressor empties the storage container down to $p_{sto,f}$. Again, the compressor starts at low power and ends at its nominal power. Again, it is easy to integrate the energy needs along this progressive emptying in the case of a perfect gas and of an isothermal compression.

The total energy need for an isothermal compression when emptying the storage container from $p_{sto,0}$ to $p_{sto,f}$ while keeping the buffer at p_{buff} is equal to the integral:

$$W_{emptying} = \int_{p_{sto}=p_{sto,0}}^{p_{sto,f}} k * dm * \ln\left({}^{p_{buff}}/_{p_{sto}} \right)$$

The pressure in the storage container is linked to the mass of gas present in the container; for a perfect gas:

$$dm/m_{sto,0} = -dp/p_{sto,0}$$

$$W_{emptying} = -k * {}^{m_{sto,0}}/_{p_{sto,0}} * \int_{p=p_{sto,0}}^{p_{sto,f}} \ln\left({}^{p_{buff}}/_{p_{sto}} \right) * dp$$

Introducing the new variable:

$$x = {}^{p_{sto}}/_{p_{buff}} = 1/r_p$$

$$W_{emptying} = +k * m_{sto,0} * {}^{p_{buff}}/_{p_{sto,0}} * \int_{x=x_0=\frac{p_{sto,0}}{p_{buff}}}^{x_f=\frac{p_{sto,f}}{p_{buff}}} \ln(x) * dx$$

$$W_{emptying} = +\ k * m_{sto,0} * r_{p,0} * [x(\ \ln(x) - 1)]_{x_0=\frac{1}{r_{p,0}}}^{x_f=1/r_{p,f}}$$

$$W_{emptying} = -k * \frac{m_{sto,0}}{r_{p,f}} * \left\{ r_{p,0} * \left(\ln\left(r_{p,f}\right) + 1 \right) - r_{p,f} * \left(\ln\left(r_{p,0}\right) + 1 \right) \right\}$$

During this emptying, the mass of the storage decreases from $m_{sto,0}$ to $m_{sto,0} * v_{sto}$. So, the transferred mass is $m_{sto,0} * (1 - v_{sto})$ and finally the energy need per unit transferred mass is:

$$W_{emptying}/m_{transf} = -k * \frac{1}{r_{p,f} * (1 - v_{sto})} * \left\{ r_{p,0} * \left(\ln\left(r_{p,f}\right) + 1 \right) - r_{p,f} * \left(\ln\left(r_{p,0}\right) + 1 \right) \right\}$$

TABLE A.2 Comparison of Progressive Emptying of Storage to Fill a Buffer (Increasing Pressure Ratio) With Constant Pressure Ratio Compression, Isothermal, Perfect Gas, p_{buff} = 45 or 90 MPa, Storage Final Pressure $p_{sto,f}$ = 2 MPa

Buffer Pressure p_{buff} (MPa)	45		90		
Nominal storage pressure $p_{sto,0}$ (MPa)	20	30	20	30	52.5
Initial pressure ratio $r_{p,0}$	2.25	1.5	4.5	3	1.71
Final pressure ratio $r_{p,f}$	22.5	22.5	45	45	45
Energy for progressive isothermal emptying from $r_{p,0}$ to $r_{p,f}$	1.56*k	1.21*k	2.25*k	1.90*k	1.41*k
Energy for isothermal emptying at constant $r_{p,0}$: $\ln(r_{p,f})$	3.11*k	3.11*k	3.81*k	3.81*k	3.81*k
Reduction factor for progressive emptying	0.499	0.389	0.591	0.500	0,370

Otherwise, this energy for keeping the buffer at its pressure while emptying a storage container with an increasing pressure ratio from $r_{p,0}$ to $r_{p,f}$ can be compared to the energy for a perfect gas isothermal compression in constant pressure ratio $r_{p,f}$, which is Eq. (6.3).

Some numerical values are given in the Table A.2, which are used in Section 6.5.

It can be seen that emptying a storage container with a progressively increasing pressure ratio in order to keep a buffer at its nominal pressure costs 41%–63% less than using the constant high-pressure ratio.

At this stage it is interesting to compare the compression energies in the cases of onsite and distant distributions. For distribution on the production site, only one compression is needed, at a constant high-pressure ratio. For a distant distribution, there are two compressions, but at increasing pressure ratio, with lower energy demand, as shown in Table A.3.

The last line of Table A.3 shows that the intermediate step of storage for transportation does not induce any increase in the compression energy demand, *provided the gas is perfect and the compression isothermal.* This result could be intuited, as this calculation is related to *only the useful work*, with no losses in heating the gas. It will no longer be the same in real cases with hydrogen real gas, imperfectly cooled compressions, and efficiency losses in the compressors.

TABLE A.3 Comparison of the Energies for Compression: On Production Site Distribution Versus Off Production Site Distribution; Isothermal, Perfect Gas, $p_{prod} = 1.5$ MPa, Storage Initial Pressure $p_{sto,f} = 2$ MPa

	OnSite Dispensing (35 or 70 MPa)		*OffSite* Dispensing (35 or 70 MPa)				
Buffer pressure p_{buff} (MPa)	45	90	45		90		
Nominal storage pressure $p_{sto,0}$ (MPa)	Without		20	30	20	30	52.5
Compression energy on the production site	3.40*k	4.09*k	1.85*k	2.19*k	1.85*k	2.19*k	2.68*k
Compression energy for dispensing on the distribution site	Included	Included	1.56*k	1.21*k	2.25*k	1.90*k	1.41*k
Total compression energy	3.40*k	4.09*k	3.40*k	3.40*k	4.09*k	4.09*k	4.09*k

APPENDIX 3 SCENARIO FOR REFUELING WITH SEVERAL STORAGE UNITS AT A HIGHER INITIAL PRESSURE THAN THE TANKS TO BE FILLED

In a hydrogen refueling station supplied with transportable containers at a nominal pressure higher than the nominal pressure of the tanks to be filled (e.g., $p_{sto}=52.5\,\text{MPa}$ for $p_{tank}=35\,\text{MPa}$), there is no need for a compressor, nor for high-pressure buffers. But it is necessary to dispose of several storage containers and these will be returned to the production unit with some residual hydrogen pressure.

This appendix formulates the analytical equations to calculate this residual pressure as a function of the number of storage containers available at the station. The equations are written for a perfect gas: at any temperature, pressure and specific mass in the vessel are proportional. This formulation is based on *a scenario in which all vehicles arrive at the station with the same filling state of their tank and are all filled in the same way from a cascade of storage containers.*

Consider the case of four storage containers available at the station numbered 1, 2, 3 and 4.

Consider a first vehicle with a tank of nominal contained mass m_{tank} at nominal pressure p_{tank} arriving at the station with a residual hydrogen pressure and mass $v_{tank,0}$:

$$p_{tank,0}=v_{tank,0}*p_{tank} \quad m_{tank,0}=v_{tank,0}*m_{tank}$$

Consider this vehicle has been filled to a fraction, $v_{tank,3}$, of its nominal mass from storage containers 4, 3, and 2. It contains a mass $v_{tank,3}*m_{tank}$. Then, the storage container 1 completes the filling with a mass $(1-v_{tank,3})*m_{tank}$ up to the pressure p_{tank}. The hydrogen mass of storage container 1, which was initially $m_{sto,0}$, decreases to $m_{sto,0}-(1-v_{tank,3})*m_{tank}$ while its pressure, initially $p_{sto,0}$, decreases down to

$$p_{sto,1}=p_{sto,0}*\left(1-(1-v_{tank,3})^{m_{tank}}/_{m_{sto,0}}\right)=p_{sto,0}*(1-(1-v_{tank,3})*a)$$

$$\text{with } a={}^{m_{tank}}/_{m_{sto,0}}$$

After having filled n such vehicles, the pressure and the mass in the storage container 1 are:

$$m_{sto,1}=m_{sto,0}*(1-n*(1-v_{tank,3})*a) \quad p_{sto,1}=p_{sto,0}*(1-n*(1-v_{tank,3})*a)$$

After N vehicles, storage container 1 pressure reaches its minimum value and is connected in the place of storage container 2:

$$m_{sto,1,\min}=m_{sto,2,0}=m_{sto,0}*(1-N*a*(1-v_{tank,3}))$$

$$p_{sto,1,\min}=p_{sto,2,0}=p_{sto,0}*(1-N*a*(1-v_{tank,3}))$$

Another new storage unit, at full pressure $p_{sto,0}$, replaces the unit 1; while the unit 2 replaces the unit 3 and unit 3 replaces the unit 4.

For the first vehicle, within the mass $v_{tank,3} * m_{tank}$, the contribution from the storage containers 4 and 3 has been $v_{tank,2} * m_{tank}$ and that of storage container 2 is $(v_{tank,3} - v_{tank,2}) * m_{tank}$.

In the same way, after N other vehicles have been filled, the pressure and mass in storage container 2 are:

$$m_{sto,2,\min} = m_{sto,3,0} = m_{sto,0} * (1 - N * a * (1 - v_{tank,2}))$$

$$p_{sto,2,\min} = p_{sto,3,0} = p_{sto,0} * (1 - N * a * (1 - v_{tank,2}))$$

At that moment, a new full storage container at $p_{sto,0}$ is connected in the place of storage container 1, while storage container 1 is connected in the place of storage container 2, storage container 2 takes the place of storage container 3, and storage container 3 takes the place of storage container 4.

For these vehicles, within the mass $v_{tank,2} * m_{tank}$, the contribution from storage container 4 has been $v_{tank,1} * m_{tank}$ and that of storage container 3 is $(v_{tank,2} - v_{tank,1}) * m_{tank}$.

In the same way, after N other vehicles have been filled, the pressure and mass in storage container 3 are:

$$m_{sto,3,\min} = m_{sto,4,0} = m_{sto,0} * (1 - N * a * (1 - v_{tank,1}))$$

$$p_{sto,3,\min} = p_{sto,4,0} = p_{sto,0} * (1 - N * a * (1 - v_{tank,1}))$$

At that moment the 3rd permutation takes place, a new full storage container at $p_{sto,0}$ is connected in the place of storage container 1 while storage container 1 is connected in the place of storage container 2, storage container 2 takes the place of storage container 3, and storage container 3 the place of storage container 4.

Within the mass $v_{tank,1} * m_{tank}$, the contribution of storage container 4 is $(v_{tank,1} - v_{tank,0}) * m_{tank}$ as the vehicle arrived with a residual mass $v_{tank,0} * m_{tank}$.

A 4th series of N vehicles is being filled, the residual pressure and mass in storage container 4 are:

$$m_{sto,4,\min} = m_{sto,residu} = m_{sto,0} * (1 - N * a * (1 - v_{tank,0}))$$

$$p_{sto,4,\min} = p_{sto,residu} = p_{sto,0} * (1 - N * a * (1 - v_{tank,0}))$$

These are the mass and pressure at which the storage containers are considered empty and returned to the hydrogen production unit:

$$p_{sto,residu} = p_{sto,0} * v_{sto} \quad so \ (1 - v_{sto}) = N * a * (1 - v_{tank,0}))$$

$(1 - v_{sto})$ is the fraction of hydrogen of the storage that is effectively used. $4*N$ is the number of vehicles that can be served with 4 stages of containers. d is introduced:

$$d = N * a = {}^{(1-v_{sto})} \Big/ {}_{\left(1-v_{tank,0}\right)}$$

At this stage, it can be noticed that it is not necessary that all the vehicles have the same tank capacity and the same filling rate as they arrive at the station. The important quantity is the product $N*a$ or the sum of the a_i from which their average can be calculated and the average of their filling rate at their arrival to the station:

$$\overline{a} = \frac{1}{N}\sum a_i$$

$$\overline{v_{tank,0}} = \frac{1}{N}\sum v_{tank,0,\ i}$$

The values of the coefficients $v_{tank,1}$, $v_{tank,2}$, $v_{tank,3}$ are essential and they cannot be chosen, they must be calculated considering constraints on the storage. The storage pressure must be high enough to ensure the transfer from the storage container to the tank, even after the Nth vehicle. The overpressure Sp_{min} is introduced (Fig. A.1). Then the complete set of constraints to consider is:

$$p_{sto,1,0} = p_{sto,0}$$

$$p_{sto,1,\min} = p_{sto,2,0} = p_{sto,0} * (1 - d * (1 - v_{tank,3})) > p_{tank,4} * Sp_{min} = 1 * p_{tank} * Sp_{min}$$

$$p_{sto,2,\min} = p_{sto,3,0} = p_{sto,0} * \left(1 - d * \left(1 - v_{tank,2}\right)\right) > p_{tank,3} * Sp_{min} = v_{tank,3} * p_{tank} * Sp_{min}$$

$$p_{sto,3,\min} = p_{sto,4,0} = p_{sto,0} * \left(1 - d * \left(1 - v_{tank,1}\right)\right) > p_{tank,2} * Sp_{min} = v_{tank,2} * p_{tank} * Sp_{min}$$

$$p_{sto,4,\min} = p_{sto,0} * v_{sto} = p_{sto,0} * \left(1 - d * \left(1 - v_{tank,0}\right)\right) > p_{tank,1} * Sp_{min} = v_{tank,1} * p_{tank} * Sp_{min}$$

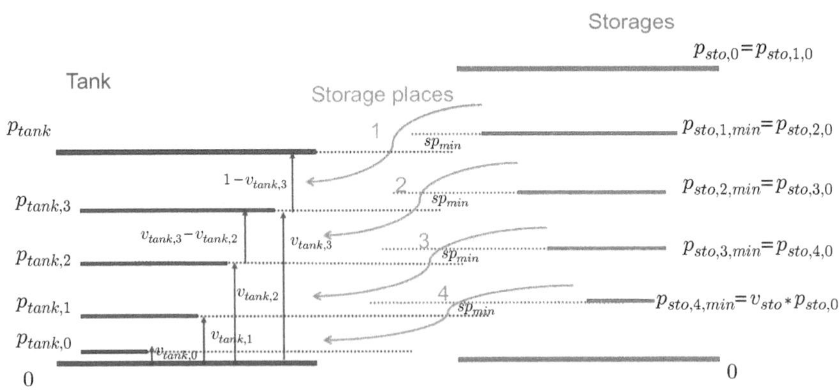

FIG. A.1 Schematic diagram for the filling of a tank from a cascade of storage containers.

When replacing the signs > by signs =, four equations are obtained, the solution of which allows calculating the minimum values of the four quantities $v_{tank,1}$,$v_{tank,2}$,$v_{tank,3}$, and $v_{sto,0}$. This means that the residual pressure with which storage containers are returned to the production unit cannot be chosen. It is imposed and highly sensitive to the quantity $v_{tank,0}$, the average filling rate at which the vehicles arrive at the station.

Another scenario may be considered for refueling. The quantity to transfer to the tank from a given storage container could be not maintained at the same level as the storage pressure decreases. It could be the maximum possible quantity transferable at any time. Then, the first vehicles of the series of the N vehicles would receive a higher contribution from the storage containers 4 and 3 and a lower contribution from the storage containers 2 and 1; but the opposite would happen for the last vehicles of the series; the contribution of storage containers 4 and 3 would be less as their pressure has decreased and storage containers 2 and 1 would have to complement with a larger contribution. Globally the equation for the definition of d shows that the same number of vehicles, $4*N$, could be served from the storage containers.

In the case in which only three storage containers are available at the station, the same series of equations can be used. Just do not calculate $v_{tank,3}$ and consider $v_{tank,3}=v_{tank,2}$ and $p_{sto,3,min}=p_{sto,4,min}$.

In the case in which only two storage containers are available at the station, the same series of equations can be used. Just do not calculate $v_{tank,\ 3}$ and $v_{tank,2}$ and consider: $v_{tank,3}=v_{tank,2}=v_{tank,1}$, $p_{sto,2,min}=p_{sto,3,min}=p_{sto,4,min}$.

APPENDIX 4 SCENARIO FOR REFUELING WITH A COMPRESSOR, A BUFFER AND SEVERAL STORAGE UNITS AT A LOWER INITIAL PRESSURE THAN THE TANKS TO BE FILLED

In a hydrogen refueling station supplied with transportable containers at a nominal pressure lower than the nominal pressure of the tanks to be filled, a compressor is mandatory and a very-high-pressure buffer as well (e.g., $p_{sto}=30$ MPa and $p_{tank}=35$ MPa with $p_{buff}=45$ MPa). This very-high-pressure buffer is filled by the compressor sucking hydrogen from one of the storage containers, and completes the filling of the tanks as the last stage of the cascade.

The analytical equations have been formulated for the compressor sucking from any storage stage; results have shown that the contribution of the compressor (the fraction of hydrogen to be compressed) is the lowest when the compressor sucks from the lowest pressure storage (as in Fig. A.2). Equations are shown in this case. The formulation is based on the same scenario as in Appendix 3 and the equations are very similar. The only differences are:

- The initial pressure of the 2nd storage (for the before last refueling stage) is equal to the nominal pressure of the full storage $p_{sto,0}$,
- 1 is substituted by $v_{tank,3}$ in the equations:

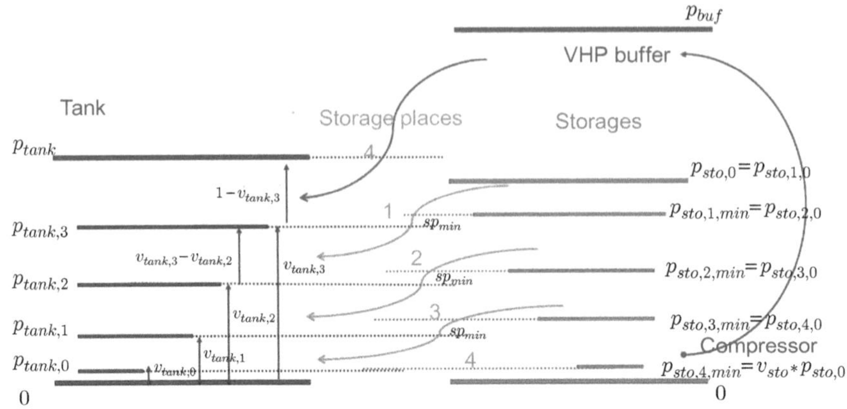

FIG. A.2 Schematic diagram for the filling of a tank from a cascade of storage containers, a compressor, and a very-high-pressure buffer.

$$p_{sto,2,0} = p_{sto,0}$$

$$p_{sto,2,\min} = p_{sto,3,0} = p_{sto,0} * \left(1 - d * \left(v_{tank,3} - v_{tank,2}\right)\right) > p_{tank,3} * Sp_{min} = v_{tank,3} * p_{tank} * Sp_{min}$$

$$p_{sto,3,\min} = p_{sto,4,0} = p_{sto,0} * \left(1 - d * \left(v_{tank,3} - v_{tank,1}\right)\right) > p_{tank,2} * Sp_{min} = v_{tank,2} * p_{tank} * Sp_{min}$$

$$p_{sto,4,\min} = p_{sto,1,0} = p_{sto,0} * \left(1 - d * \left(v_{tank,3} - v_{tank,0}\right)\right) > p_{tank,1} * Sp_{min} = v_{tank,1} * p_{tank} * Sp_{min}$$

$p_{sto,1,min} = p_{sto,residu} = p_{sto,0} * v_{sto}$ without any constraint as this storage is feeding the compressor.

When replacing the signs > by signs =, three equations are obtained the solution of which allows calculating the minimum values of the three quantities $v_{tank,1}$,$v_{tank,2}$, $v_{tank,3}$. The value of $v_{sto,0}$, giving the residual pressure of the storage containers when returned to the production unit, can be chosen. It allows calculating the pressure ratio of which the compressor must be capable. The value $(1 - v_{tank,3})$ is the fraction of the hydrogen to be compressed (on a total of $(1 - v_{tank,0})$).

ACKNOWLEDGMENT

This work is a part of the *VABHYOGAZ3* project supported by the French "Programme d'Investissements d'Avenir" under supervision of *ADEME*, the French Energy and Environment Agency. The project is conducted by *HERA France* office *ALBHYON*, in partnership with *TRIFYL, HP Systems, EMTA* (*VEOLIA* group) and the *IMT Mines-Albi RAPSODEE Research Center*. The VABHYOGAZ3 project considers hydrogen production from biogas with production units ranging from 100 to 800 kg/day to deliver hydrogen to several distribution units of 20–200 kg/day located within a distance <100 km from the production unit. The authors are grateful to the ADEME for its support to this project.

REFERENCES

Bourgeois, T., Ammouri, F., Weber, M., Knapik, C., 2015. Evaluating the temperature inside a tank during a filling with highly pressurized gas. Int. J. Hydrog. Energy 40, 11748–11755.

Bourgeois, T., Brachmann, T., Barth, F., Ammouri, F., Baraldi, D., Melideo, D., Acosta-Iborra, B., Zaepffel, D., Saury, D., Lemonnier, D., 2017. Optimization of hydrogen vehicle refuelling requirements. Int. J. Hydrog. Energy 1–21. https://doi.org/10.1016/j.ijhydene.2017.01.165.

Gardiner, M., July 2009. Energy Requirements for Hydrogen Gas Compression and Liquefaction as Related to Vehicle Storage Needs. U.S. Department of Energy Hydrogen and Fuel Cells Program Record No. 9013, http://www.hydrogen.energy.gov/pdfs/9013_energy_requirements_for_hydrogen_gas_compression.pdf.

Klell, M., Kindermann, H., Christian, J., 2007. Thermodynamics of gaseous and liquid hydrogen storage. In: Proceedings of IHEC 2007 the International Hydrogen Energy Congress and Exhibition, Istanbul, Turkey, 13–15 July. http://www.hycenta.tugraz.at/Image/Thermodynamics%20of%20gaseous%20and%20liquid%20hydrogen%20storage.pdf.

Lee, S., Kim, Y., Kim, S., Yoon, K., 2009. Temperature change of a type IV cylinder during hydrogen fueling process. In: Proceedings of the International Conference on Hydrogen Safety, Ajaccio. 2009.

Lei, Z., Yanlei, L., Jian, Y., Yongzhi, Z., Jinyang, Z., Haiyan, B., Xianxin, L., 2010. Numerical simulation of temperature rise within hydrogen vehicle cylinder during refueling. Int. J. Hydrog. Energy 35, 8092–8100. https://doi.org/10.1016/j.ijhydene.2010.01.027.

Lemmon, E.W., Leachman, J.W., 2008. Revised standardized equation for hydrogen densities for fuel consumption application. J. Res. Natl. Inst. Stand. Technol. 113, 341–350.

Maus, S., Hapke, J., Na Ranong, C., Wüchner, E., Friedlmeier, G., Wenger, D., 2008. Filling procedure for vehicles with compressed hydrogen tanks. Int. J. Hydrog. Energy 33 (17), 4612–4621.

Mintz, M., Elgowainy, A., Gardiner, M., 2009. Rethinking hydrogen fueling: insights from delivery modeling. J. Transp. Res. Board 2139, 46.

Monde, M., Mitsutake, Y., Woodfield, P., Maruyama, S., 2007. Characteristics of heat transfer and temperature rise of hydrogen during rapid hydrogen filling at high pressure. Heat Tran. Asian Res. 36 (1), 13–27.

Nexant Inc., Air Liquide, Argonne National Laboratory, Chevron Technology Venture, Gas Technology Institute, National Renewable Energy Laboratory, Pacific Northwest National Laboratory, TIAX LLC, May 2008. H2A: hydrogen delivery infrastructure analysis models and conventional pathway options analysis results. Report DE-FG36-05GO15032DOE. http://www1.eere.energy.gov/hydrogenandfuelcells/pdfs/nexant_h2a.pdf.

NIST Chemical Webbook: n.d. http://webbook.nist.gov/chemistry/fluid, last consulted December 2017.

Parks, G., Boyd, R., Cornish, J., Remick, R., May 2014. Hydrogen Station Compression, Storage, and Dispensing Technical Status and Costs. NREL Technical Monitor NREL/BK-6A10-58564, https://www.nrel.gov/docs/fy14osti/58564.pdf.

Reddi, K., Elgowainy, A., Sutherland, E., 2014. Hydrogen refueling station compression and storage optimization with tube-trailer deliveries. Int. J. Hydrog. Energy 39, 19169–19181. https://doi.org/10.1016/j.ijhydene.2014.09.099.

Rothuizen, E., Rokni, M., 2014. Optimization of the overall energy consumption in cascade fueling stations for hydrogen vehicles. Int. J. Hydrog. Energy 39, 582–592. https://doi.org/10.1016/j.ijhydene.2013.10.066.

SAE International, July 2014. Surface Vehicle Standard. J2601 Fueling Protocols for Light Duty Gaseous Hydrogen Surface Vehicles. http://www.sae.org/technical/standards/J2601_201407.

Schneider, J., Suckow, T., Lynch, F., Ward, J., Caldwell, M., Tillman, J., Mathison, S., et al., 2005. Optimizing hydrogen vehicle fueling. NHA 2005, http://static1.squarespace.com/static/53ab1feee4b0bef0179a1563/t/543e7eeae4b0983640ec4b93/1413381866875/NHA+2005+-+Optimizing+Hydrogen+Vehicle+Refueling.pdf.

Smith, R. (Ed.), 2005. Chemical Process. John Wiley & Sons, New York, p. 273.

Staffell, I., Green, R., 2013. The cost of domestic fuel cell micro-CHP systems. Imperial College Business Scholl of London, UK. Int. J. Hydrog. Energy 38 (2), 1088–1102. https://doi.org/10.1016/j.ijhydene.2012.10.090.

Wipke, K., Sprik, S., Kurtz, J., Ramsden, T., Ainscough, C., Saur, G., 2012. National Fuel Cell Electric Vehicle Learning Demonstration Final Report. National Renewable Energy Laboratory. http://www.nrel.gov/hydrogen/pdfs/54860.pdf.

Woodfield, P.L., Monde, M., Mitsutake, Y., 2007. Measurement of averaged heat transfer coefficients in high-pressure vessel during charging with hydrogen, nitrogen or argon gas. J. Therm. Sci. Technol 2 (2), 180–191.

Chapter 7

Hydrogen Applications: Overview of the Key Economic Issues and Perspectives

Christine Mansilla*, Cyril Bourasseau†, Camille Cany*,‡, Benjamin Guinot†, Alain Le Duigou* and Paul Lucchese§

*I-tésé, CEA, DAS, Université Paris Saclay, Gif-sur-Yvette, France, †LITEN, CEA, DRT, Grenoble Cedex 9, France, ‡Laboratoire Genie Industriel, CentraleSupélec, Université Paris-Saclay, Gif-sur-Yvette, France, §CEA and Capenergies, Domaine du Petit Arbois, Aix-en-Provence Cedex 4, France

7.1 OVERVIEW OF HYDROGEN APPLICATIONS

Despite the fact that hydrogen is mainly advertised as a promising energy carrier, especially for mobile applications, hydrogen can address a wide range of applications. This section aims at clarifying the diverse uses at stake and proposing an assessment of the related volumes, today and in the future.

7.1.1 Hydrogen: A Chemical Product and an Energy Carrier

What is hydrogen? From the market perspective, hydrogen is both an energy carrier that needs to be produced, even if "natural" hydrogen is a rising topic (Prinzhofer and Deville, 2015), and a chemical product involved in a range of industrial processes. To offer a relevant overview of hydrogen applications, both standpoints must be considered. This is what is intended the following text. Overall, as depicted by Fig. 7.1, hydrogen can be produced from any energy source, whether fossil, nuclear, or renewable, and involved in any use (or almost any).

7.1.1.1 Industry Applications

Hydrogen is required as a chemical input in a variety of industry chemical processes. The first application, in quantitative terms, is *the refining industry*. It uses 26.4 million tons of hydrogen per year worldwide (Alphéa Hydrogène, 2014).

Hydrogen Supply Chain. https://doi.org/10.1016/B978-0-12-811197-0.00007-5

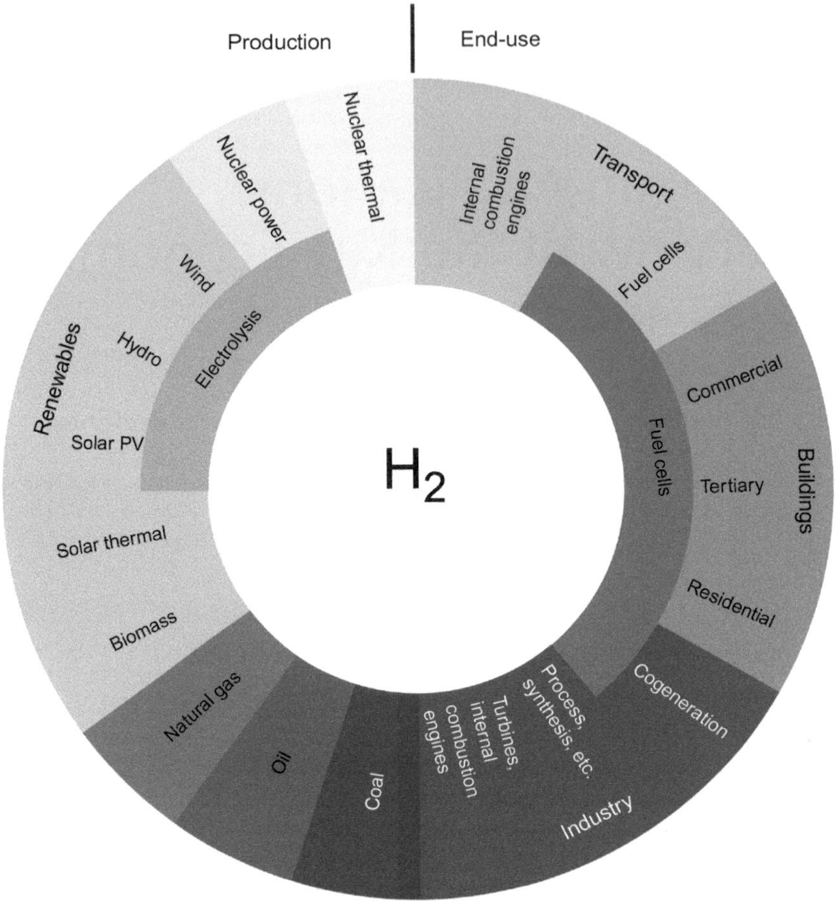

FIG. 7.1 Hydrogen production and uses *Adapted from 2011 connaissancedesenergies.com, from AFHYPAC—UE.*

Refining is a very complex process that involves a series of chemical reactions. Some require hydrogen as an input (e.g., hydrocracking, hydrotreating, desulphurization); others generate hydrogen as a byproduct (e.g., catalytic or thermal cracking, catalytic reforming). The global balance between these steps is negative; hydrogen must be produced to feed the refining process. Lately, the refining industry faces an increasing need for hydrogen due to two main factors: (1) fuel specifications have become more demanding (to be saleable, products must meet the standards), and (2) resources have become heavier.

The second major sector today is the *fertilizer industry* (ammonia production, NH_3), with demands similar to those of the refinery industry: 22.8 million tons per year.

Other applications also exist in a range of sectors more marginal in terms of hydrogen demand, and very diverse in terms of applications: methanol production, metallurgy, glass industry, semiconductor industry, food industry (for the well-known hydrogenated fats), and fuel for space rockets. Together, they represent 18% of the current hydrogen needs worldwide.

Finally, another category of industry applications is the production of *"new" fuels* that require hydrogen as an input. They include synthetic fuels and biofuels. We categorize these applications as a demand for industry because hydrogen is not directly used as an energy carrier but as a chemical input to generate energy carriers. These "new" fuels could be an attractive step toward a more sustainable energy system because they can be used directly (or almost directly) in classical internal combustion engines (drop-in fuels for either road or aviation, without major engine modification) (Imbach, 2012). The hydrogen input is also very valuable to generate these fuels because it helps enhance biomass use (Imbach, 2012), and production of fuels from carbon dioxide (Ademe, 2010). An overview of carbon dioxide hydrogenation routes is provided in Fig. 7.2. A variety of products could be generated from carbon dioxide emissions, together with a hydrogen input.

With regard to biofuels, second-generation biodiesel (also called BtL, for "biomass-to-liquid") can be produced without competing with food resources, as they are generated from lignocellulosic biomass (non-staple crops), primary resources of agriculture and forests including dedicated crops, or secondary resources (waste products and byproducts). The process requires adjusting the hydrogen-to-carbon-monoxide ratio from one to two, thereby necessitating an additional hydrogen feed. If supplied by another process (such as water electrolysis), the mass efficiency can be improved from 15% to 45%, resulting in a better use of the limited biomass feedstock.

7.1.1.2 "Green" Gas Applications

Hydrogen is also contemplated as an energy carrier, to be substituted—at least partially—for natural gas. This can be achieved through two different routes. The first option is to directly inject hydrogen into the natural gas network. A few percent in volume do not raise any difficulty, and diverse projects examine the possibility of injecting up to 20% (Florisson, 2010). Another option is to go from hydrogen to synthetic methane, through a methanation step (see Fig. 7.3). This would overcome any barrier in terms of injection limits, and also provide an additional sink for carbon dioxide emissions (provided that the carbon release is not delayed).

These options are often called "power-to-gas." However, they are very rarely properly and precisely defined. As a matter of fact, power-to-gas can refer to power to hydrogen, or to power to synthetic methane, for injection in the natural gas network, or for a range of different applications (including mobility, for instance). Besides, power-to-gas can rely on renewable power (exclusively),

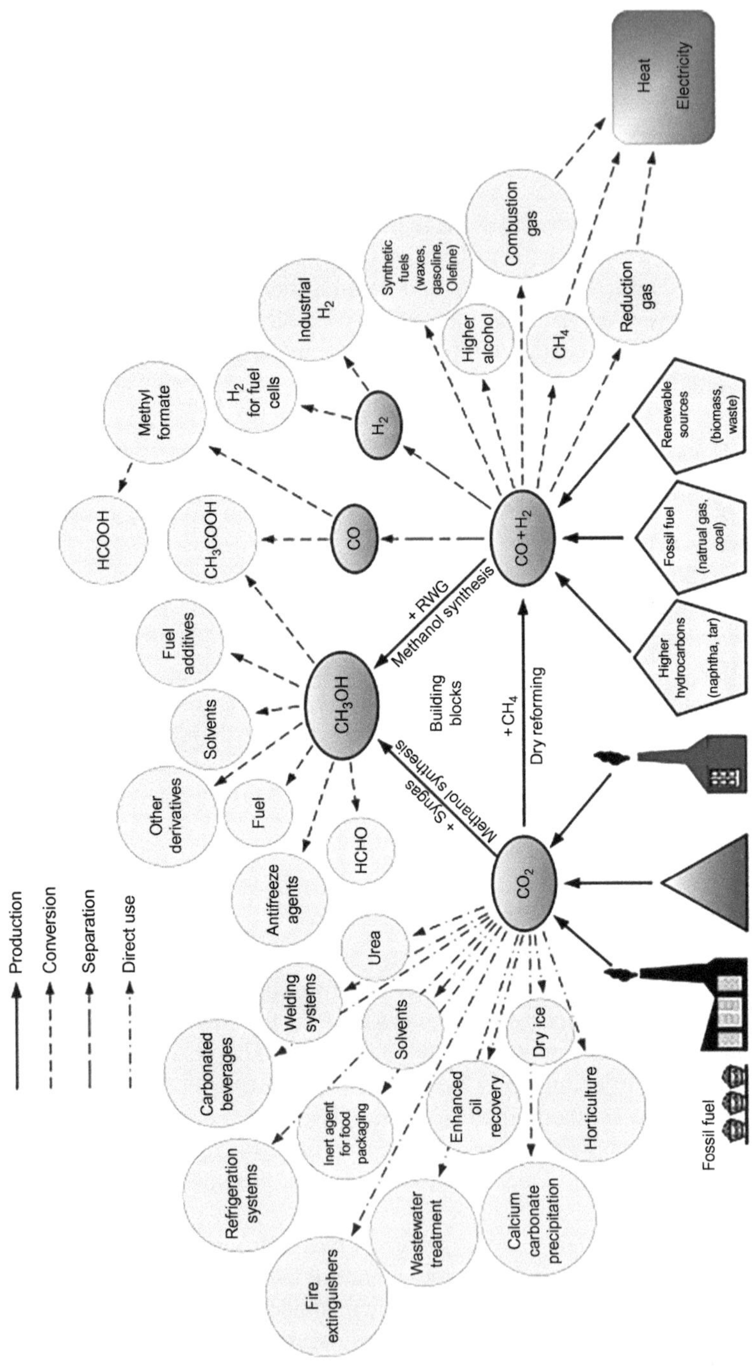

FIG. 7.2 CO_2 hydrogenation routes (Ademe, 2010). *From DiaMilani, R.K., Gholamreza, Z., Ali, A., 2015. A model-based analysis of CO_2 utilization in methanol synthesis plant. J. CO_2 Utilization 10, 12–22.*

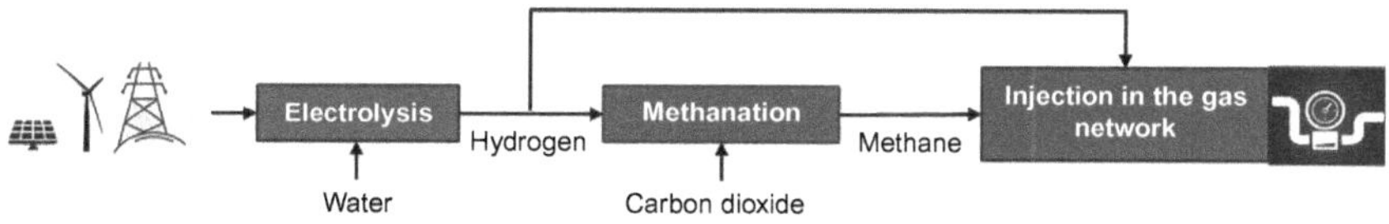

FIG. 7.3 "Green gas" pathways.

TABLE 7.1 Global Warming Potential of Methane vs. Carbon Dioxide (Climate Change, 1995)

	Global Warming Potential (Time Horizon)		
	20 Years	*100 Years*	*500 Years*
Carbon dioxide	1	1	1
Methane	56	21	6.5

sometimes only "excess" or "surplus" power, or simply power, disregarding the way it is produced (also trying to keep it low-carbon).

Overall, to promote this application as "green," several issues should be carefully examined:

- What is the life-cycle environmental footprint of the power-to-gas system?
- What is the alternative it will replace?

Indeed, such systems could improve the environmental footprint of pre-existing natural gas applications, but they could damage the global system environmental footprint if they lead to create brand new applications for natural gas (not replacing already existing ones, which are more carbon-intensive), as natural gas has a much greater global-warming potential than carbon dioxide (cf. Table 7.1). Moreover, losses during production and transportation steps due to unintentional methane equipment or pipeline leaks accentuate the impact (Robinson et al., 2003 in Camuzeaux, 2012). According to the French gas operator, losses amount to 0.6% (GRDF, 2009).

As a result, today, almost 3% of greenhouse gas emissions (all emissions included) are due to leakages and venting in natural gas and oil systems (Camuzeaux, 2012).

7.1.1.3 Mobility Applications

Hydrogen mobility is often understood as full-hydrogen mobility for passenger cars. However, it covers a range of diverse vehicles in land, maritime, and aeronautic mobility: light duty vehicles (LDV) such as cars and vans, commercial vehicles, buses, and trucks, two- and three-wheeled vehicles such as scooters and bicycles (cf. Fig. 7.4), forklift trucks, trains and trams, ferries and smaller

FIG. 7.4 Hydrogen bicycle. *Used with permission from Pragma-Industries*

boats. For aeronautics, applications include manned light aircraft, unmanned aerial vehicles (UAV), and unmanned underwater vehicles (UUV), for example, for recognition purposes. Aeronautics companies are also developing alternative power sources for non-propulsive aircraft systems such as a galley in a commercial aircraft, or to be used as a secondary power source onboard business jets, as an auxiliary power unit or last-resort turbine.

The market maturity is, of course, different depending on the market segments, from demonstration to products available on the market. For example, buses in California have been on the road for over 20,000h now (White, 2016). One spectacular example of market development is forklift trucks. As a matter of fact, they demonstrated their qualities in real conditions, in terms of charging time, autonomy, and costs. A number of companies implemented hydrogen forklift trucks in their warehouses. Among them are global brands like Walmart, FedEx, Sysco, IKEA, and Coca Cola (Denhoff, 2016).

Hydrogen vehicles have begun entering the market. Toyota, Hyundai, Honda, and BMW include hydrogen cars in their product portfolios. As an example, Toyota released the "MIRAI" in December 2014, at the price of 7.23 million yen (i.e., approximately 50,000 euros). They plan to sell 30,000 vehicles by 2020 (Ohira, 2016) (cf. Fig. 7.5).

With regard to passenger vehicles, hydrogen mobility may appear competitive with electric mobility. However, hydrogen mobility can also be realized via range-extenders, thus combining the advantages of hydrogen and electric mobility, with autonomy and recharge times similar to classical vehicles, smaller hydrogen storage and fuel cell capacity requirements as compared to full-hydrogen vehicles, and a smaller battery capacity requirement than a fully-electric vehicle.

As a result, hydrogen mobility should take part in the sustainable mobility panel, offering alternatives to electric vehicles and biofuels. In this respect, the way hydrogen is produced is of paramount importance, so that the life-cycle balance of hydrogen vehicles may be improved beyond that of diesel and

FIG. 7.5 Hydrogen car—example of the Toyota Mirai.

gasoline vehicles. As shown in, beyond comparison with gasoline or hybrid vehicles, it should be highlighted that the way hydrogen is produced dramatically impacts the environmental footprint of hydrogen vehicles.

Overall, the interest for hydrogen mobility also arises from the fact that, with regard to electric mobility, significant local benefits are realized in terms of air pollution and noise. In this respect, diffuse CO_2 emissions are avoided, whatever the hydrogen production process.

7.1.1.4 Stationary Applications

Stationary applications refer to the use of fuel cells to generate electricity from hydrogen. Fuel cells may be fed directly by natural gas, though (Hashimoto, 2015). Hydrogen is used as an energy carrier, and the market segment is clearly energy storage. It may also be named re-electrification, if hydrogen is initially produced from electricity. This kind of application is being developed for off-grid sites, for back-up power (complementary to renewable sources, for instance, and other kinds of storage such as batteries, instead of diesel gensets), and recently, for microgrids.

Stationary applications also refer to the coproduction of heat and electricity (i.e., CHP, or combined heat and power) that is useful for households, or buildings in general. In Japan, there is a massive development program of micro-CHP for households called Ene-Farm. The systems implemented are fueled by natural gas (Hashimoto, 2015). Over 154,000 units have already been installed, and 1.4 million units are planned by 2020, with an additional 5.3 million by 2030 (Ohira, 2016).

7.1.2 The Hydrogen Demand: Today and Tomorrow

7.1.2.1 Today

From the current hydrogen markets that were listed in the previous section, the global balance can be assessed. Today, the world's hydrogen consumption is

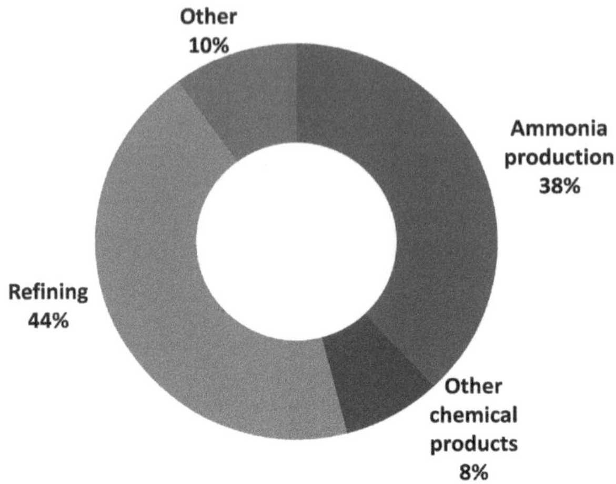

FIG. 7.6 Today world hydrogen demand. *(Data from Alphéa Hydrogène, 2014. Actualité Chimique n°347. In: AFHYPAC, 2016, Mémento de l'Hydrogène - FICHE 1.3 (in French).)*

60 million tons per year. As it is pointed out in next figure, this demand is driven by industrial needs (cf. Fig. 7.6).

7.1.2.2 Tomorrow

In the years to come, hydrogen applications are expected to grow:

- Oil refining will require additional hydrogen to meet the standards.
- Hydrogen mobility (through synthetic fuels, biofuels, full-hydrogen cars or range extenders) will develop.
- Interconnection between energy carriers (through power-to-gas) is also a promising route.

Overall, the global need for hydrogen will increase, as depicted in Fig. 7.7, which provides an outlook of a selection of hydrogen markets by 2050. Of course, such a development will be tightly linked to exogenous factors, among which the fossil prices and the carbon one. From the figure, it can be pointed out that the greatest volumes are expected from the mobility market.

With a similar perspective, the US Department of Energy also published a study called "Hydrogen at Scale," in which it is foreseen that hydrogen will represent more than 10% of the US energy flow by 2050, compared to 2% today (H2 at scale, 2016).

Hydrogen is expected to play a significant role in the energy system in the years to come, even if the dynamics of its development are still to be clarified. Two major drivers are, of course, political will (and policies implemented as a result) and the economic competitiveness of hydrogen solutions for different market segments, with the first influencing the latter. This is what is discussed in next section.

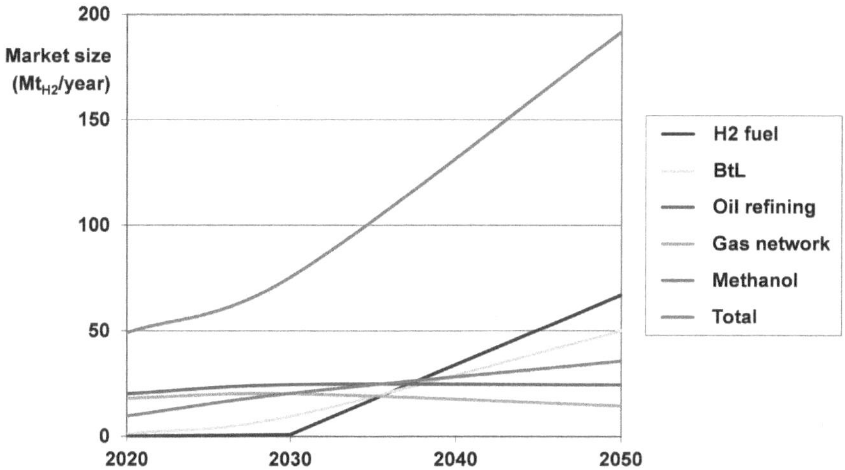

FIG. 7.7 Prospective assessment of world hydrogen demand, according to the market (Mansilla et al., 2012a).

7.2 THE HYDROGEN MARKETS: WHAT IS THE ECONOMIC EQUATION? WHAT IS THE POTENTIAL?

As previously mentioned, the economic competitiveness of hydrogen solutions will be key. As one key opens one lock, economic competitiveness depends on the competitors. As a result, the economic equation will not be unique, but will depend on the market segment. Depending upon the targeted market, the target price (and therefore, production cost) will be different, as will the drivers.

7.2.1 The Key Drivers

7.2.1.1 For Industry

In the industrial markets, hydrogen is needed as a chemical product. Therefore, hydrogen needs are related to the development of the industrial market.

Depending on the industry, hydrogen production can be either captive (which means that the hydrogen production process is installed on-site), or merchant (hydrogen is then produced by a gas company and sold to the client).

In both cases, sustainable hydrogen production will compete with the benchmark process, which is steam methane reforming.

The key drivers in this market will be the key variables of steam methane reforming: the natural gas and carbon dioxide prices, as depicted in Fig. 7.8.

If water electrolysis is considered the competitor (a sustainable one only if low-carbon power feeds the process), power price becomes a critical factor for economic competitiveness. This is shown by Fig. 7.9.

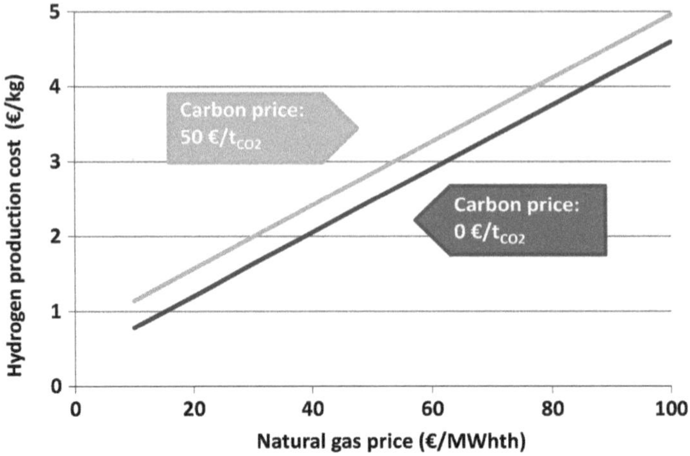

FIG. 7.8 Hydrogen production cost by steam methane reforming. *(Data from Afhypac, 2014. Production d'hydrogène à partir des combustibles fossiles. Mémento de l'Hydrogène - FICHE 3.1.1 (in French).)*

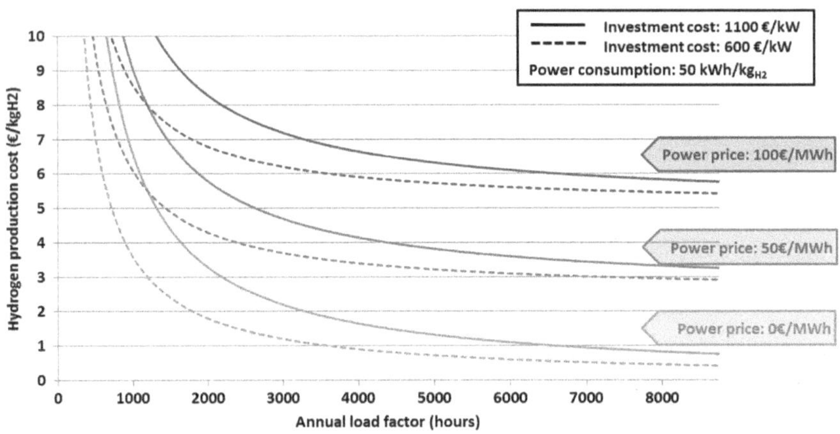

FIG. 7.9 Hydrogen production cost by water electrolysis (own calculation). *(Adapted from Caumon, P., Lopez-Botet Zulueta, M., Louyrette, J., Albou, S., Bourasseau, C., Mansilla, C., 2015. Flexible hydrogen production implementation in the French power system: expected impacts at the French and European levels. Energy 81, 556–562.)*

7.2.1.2 For "Green" Gas

In the "green" gas application, hydrogen is used as an energy carrier to be substituted (at least partially) for natural gas.

The straightforward competitor is then natural gas price on an energy basis. Indeed, the customer pays for an energy service, not a volume of gas. Comparing low heating values on a volume basis (cf. Table 7.2) shows that hydrogen's

TABLE 7.2 Hydrogen Physico-Chemical Properties and Comparison With Natural Gas

	Hydrogen	Natural Gas
Density	0.08988 kg/Nm^3	0.6512 kg/Nm^3
Low heating value	120 MJ/kg	50 MJ/kg
	10.8 MJ/Nm^3	32.6 MJ/Nm^3

performance is three times lower. So, depending on the amount of hydrogen blended with natural gas, the customer must buy more cubic meters for the same service (+7% for 10%$_{vol}$ hydrogen, +13% for 20%$_{vol}$ hydrogen).

As a result, competition is tough in this market segment because of the low price of natural gas.

To ease competition, the regulatory framework (carbon tax, or generally, the fiscal environmental framework), can propose measures to be leveraged.

7.2.1.3 For Mobility

Hydrogen as an energy carrier for mobility applications competes with classical fuels, such as diesel and gasoline. In the mid- to long term, if the hydrogen vehicle prices reach those of internal combustion engine vehicles (diesel or gasoline, as foreseen by McKinsey, 2010), the differing total costs of ownership will rise from fuel consumption.

On a purely economic basis, the customer will not want to spend more to drive 100 km with a hydrogen car, compared to a classical internal combustion engine vehicle (figures will be provided in Section 7.2.2.1 of this chapter). So, fuel consumption should not be more expensive, taking into account the improved performance expected for diesel or gasoline engine specifications over time. A 25% gain is foreseen by 2030 (Ademe, 2013), and a 30% gain by 2050 (McKinsey, 2010).

Other factors will matter, such as the charging time at the station, or the range of the car. These factors that may affect electric mobility are assets for hydrogen mobility.

So, from the customer's viewpoint, oil prices and tax policies will make the difference in deciding whether to switch to hydrogen mobility. Tax incentives could help make the decision. However, the applicable taxes represent substantial income for the government. In the United States, 15%–20% of gasoline's retail price is related to federal and state taxes. This percentage can exceed 50% in some countries such as France. However, macro-economically speaking, one could argue that this loss in income could be offset by the reduction of fossil fuel imports, thereby reducing the energy bill responsible for a large fraction of commercial deficits. Other benefits of developing hydrogen mobility could rise from the development of local employment, if the industry is developed in the country.

7.2.1.4 For Stationary Applications

When focusing on hydrogen use for stationary application, the local context is of paramount importance. First, the distinction should be made between on-grid and off-grid applications.

As a matter of fact, current power market prices rule out hydrogen from its competitors. With the massive penetration of variable renewables in the electric system (such as photovoltaic and wind power), power prices dropped. Indeed, variable renewables were first included in the electric system, but not in the market, being paid via feed-in tariffs, which contributed to the decrease of the prices on the day-ahead market. Variable renewables have low variable costs, which make them the first power plants to operate in merit order.

For off-grid applications, the picture looks quite different. As a matter of fact, energy prices are higher for remote areas. In such contexts, hydrogen systems can be a good alternative, especially if the environmental benefit is accounted for, and especially when considered complementarily with batteries (this will be discussed at the end of this chapter). Intermediate cases (such as micro-grids connected to the main power grid) should also be investigated, taking into account the externalities.

For stationary applications that involve heat generation (combined heat and power systems), in addition to the power price, the gas price is the other economic driver, but more importantly, this is the price spread between gas and power price, which generates the economic competitiveness of hydrogen solutions.

7.2.2 The Economic Target

After review of the drivers for economic competitiveness of hydrogen systems, we propose a few figures.

7.2.2.1 What Target Prices?

To achieve economic competitiveness on the markets we listed, we need to make assumptions on the economic drivers. Let us focus on the major ones: fossil fuel prices (oil and gas), and policy incentives. Here, we choose to focus on carbon prices. Of course, results strongly depend on the assumptions, so one should not concentrate on the precise numbers, but rather focus on the general trends.

Table 7.3 presents the assumptions, and Table 7.4 the resulting target prices necessary for hydrogen systems to be competitive on the selected market segments. The target price is the maximum price that hydrogen demand while still

TABLE 7.3 Assumptions: Fossil Fuel and Carbon Prices (Cany et al., 2017)

	2030	2050
Oil price ($\$_{2014}$/boe)	130	150
Gas price ($\$_{2014}$/MBtu)	13	14
CO_2 price (€$_{2014}$/tonne CO_2)	100	200

TABLE 7.4 Target Prices for Hydrogen Market Segments (€/kg_{H2}) (Cany et al., 2017)

	2030		2050	
€/kg_{H2}	*Without CO_2 Price*	*With CO_2 Price*	*Without CO_2 Price*	*With CO_2 Price*
H_2 as a fuel for mobility	2.8	4.6	3.0	6.4
H_2 to produce biodiesel	0.4	1.8	1.4	4.1
H_2 blending into gas network	1.1	1.9	1.2	2.8
H_2 for industries	1.7	2.6	1.9	3.6

remaining cost-competitive with its competitors on the selected market. Results are presented for two cases: one with the economic incentive of carbon price, and one without.

The higher the target price, the easier it is for hydrogen systems to enter the market. It appears that the best perspectives are for the mobility market. This is especially true because it is also from this market that the greatest volumes are expected. The target price assessed here focuses on car fuel consumption (in other words, by a timeframe when the hydrogen vehicle price will have converged with the diesel or gasoline ones; McKinsey, 2010). In accordance with the durability criteria defined by the European Union, hydrogen fuel should be competitive in the long term, without subsidies, with alternative fuels (European Parliament, 2009). Diesel/gasoline and hydrogen fuels were thus considered to be subject to the same amount of tax.

It must also be underlined that environmental incentives can really make a difference. With high carbon prices, hydrogen target prices can be increased anywhere from 50% to over 100%.

With regard to stationary applications, economic interest is highly dependent on the local context. To illustrate this, a comparison is made between three systems, in the context of the Gulf of Guinea (Guinot et al., 2015a). To replace the genset which is required as a back-up of a photovoltaic system (cf. Fig. 7.10), two alternatives are studied: a set of batteries (cf. Fig. 7.11) and a hybrid system made of batteries and a hydrogen solution (cf. Fig. 7.12).

As shown in Fig. 7.13, implementing a hydrogen solution (PV-Batt-H_2) makes it possible to reduce battery capacity by 45%, compared to the battery-only case (PV-Batt). The levelized cost of electricity is 15% less costly, making the electricity delivered by the hybrid system only 20% more expensive than by diesel. The hydrogen part of the system only represents 8% of the levelized cost of electricity produced by the hybrid system.

The competing options will be greatly preferred until environmental policies are implemented. Once again, the legal framework (regulatory and fiscal) is a key driver, especially for emerging markets. Demand-pull measures should be promoted (combined with accurate technology-push measures), to create the

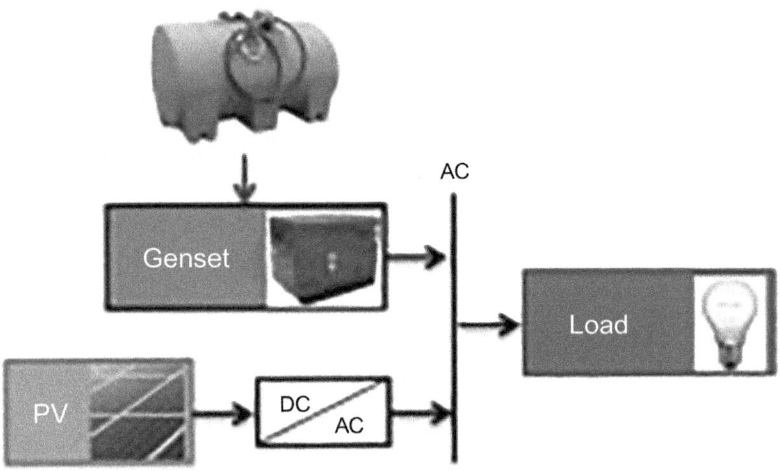

FIG. 7.10 System to be replaced (Guinot et al., 2015a).

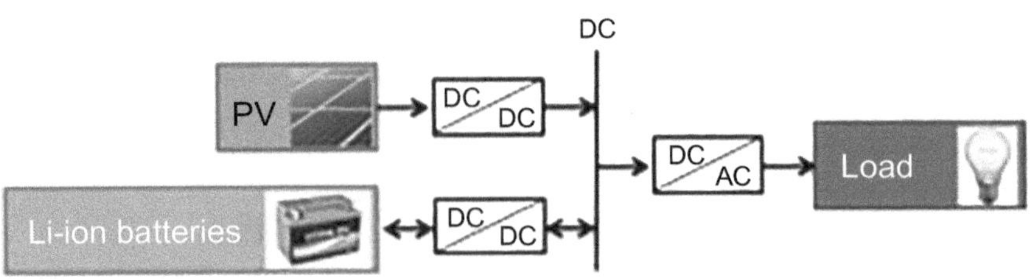

FIG. 7.11 Alternative 1: PV-Batt system (Guinot et al., 2015a).

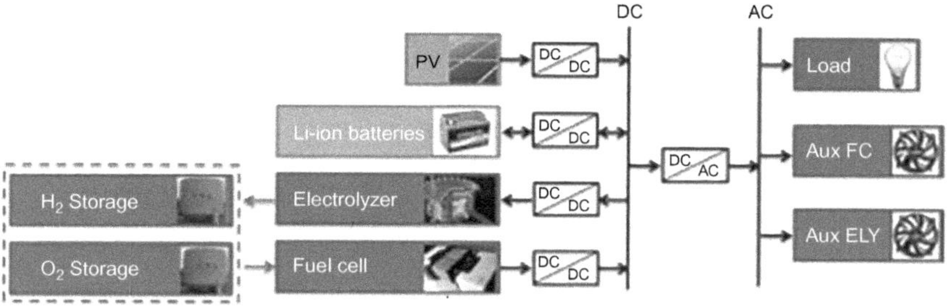

FIG. 7.12 Alternative 2: PV-Batt-H_2 system (Guinot et al., 2015a). FC, fuel cell; ELY, electrolyser.

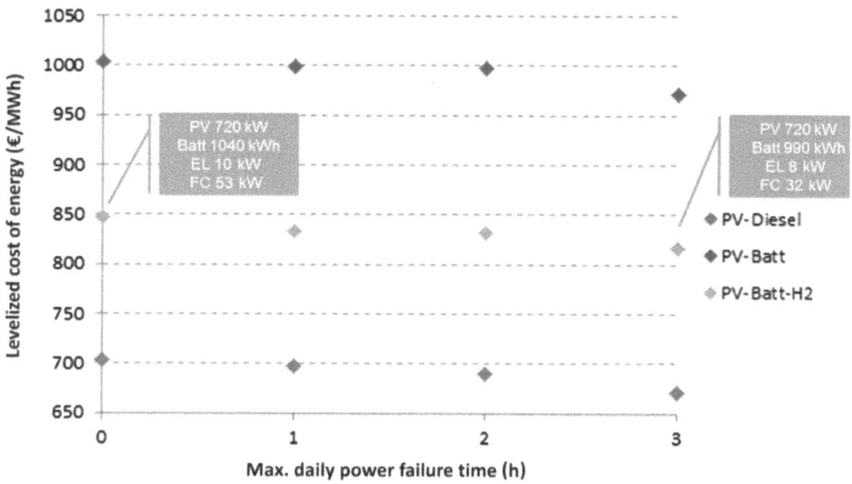

FIG. 7.13 Levelized cost of electricity according to the selected option, and the maximum power failure duration allowed each day (Guinot et al., 2015a). FC, fuel cell; EL, electrolyzer.

markets and the economic conditions for system deployment (Avril et al., 2012).

7.2.2.2 *What Target Costs?*

Target prices represent the upper boundary for target costs. The gap between the production cost and the target price will leave room for having more competitive systems, meaning systems that will be able to gain market shares. The gap is also required for profiting from hydrogen system sales.

From the comparison between current hydrogen production costs and the target prices, objectives in terms of investment costs (and load factor for non-steady state applications) and energy consumption can be established. Based on the current values, the economic equation can be tricky. Business models need to be examined on a case-by-case basis, and by considering

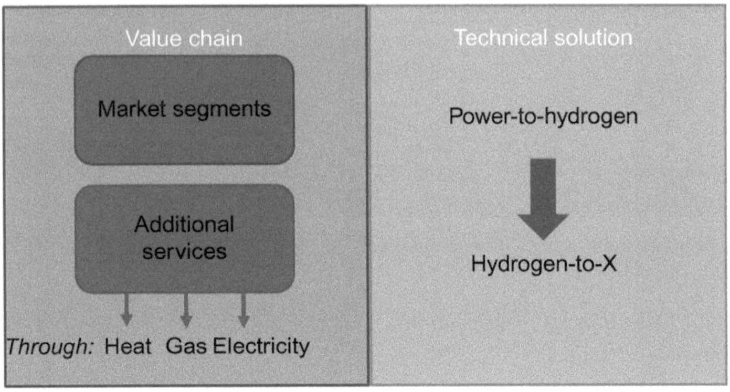

FIG. 7.14 Provision of services by hydrogen systems.

additional services. The idea is not only to valorize the hydrogen produced by the equipment (i.e., the electrolyzers), but also the equipment or the system itself, through additional services, as illustrated by Fig. 7.14 for power-to-hydrogen systems (i.e., hydrogen produced from power, via electrolysis).

In this way, the value chain can be expanded through the provision of services. The next section will address the topic of possible services provided to the electric system.

7.3 HYDROGEN SYSTEMS: NOT A SINGLE PRODUCT, BUT A PROVISION OF SERVICES

7.3.1 Services for the Electric System

The electric system is like a scale: balance must be achieved between the supply and demand, at all times, to maintain frequency (Fig. 7.15). On the supply side, we find all the "injections" into the electric grid, from power plants, imports, and storage systems, and also "negative" injection, thanks to demand-side management. On the demand-side, all the customers "withdraw" power from the

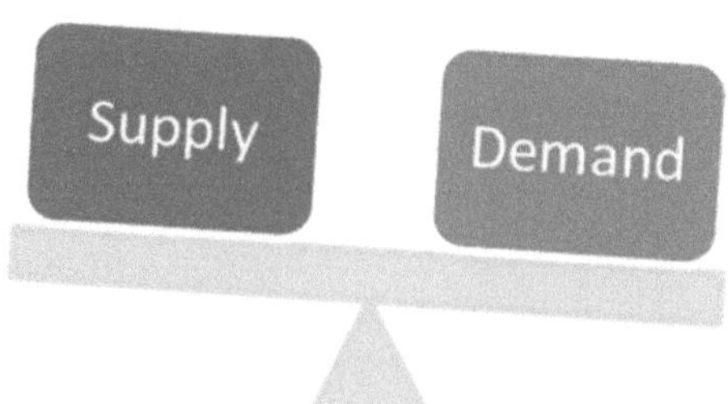

FIG. 7.15 Balance of the electric system.

electric grid, be they industrial customers or households, as well as exports and storage systems.

The supply and demand balance can be achieved:

- by adjusting supply though power modulation or curtailment;
- by adjusting demand through demand-side management or load shedding;
- by adjusting both supply and demand through storage or interconnections.

The characteristics of hydrogen production through electrolysis make it possible to quickly adjust power consumption, thus providing a new means to contribute to electric system management, either via local hydrogen production or central production.

7.3.1.1 Local Production

The first option is to connect the hydrogen production system directly to a renewable energy source, the whole system being connected to the grid (Guinot and Mansilla, 2016). As illustrated in Fig. 7.16, electrolysis consumption can be adjusted to smoothe the local production of a renewable energy source. The expected benefits will be to reduce the impact of intermittence, the forecasting needs, the needs for transmission capacity, and possibly enable the increase of renewable shares as a result.

The second option is to consider the hydrogen production system jointly to another power consumer, the whole system still being connected to the grid. Electrolysis consumption can then be adjusted to the intermittent consumption of the site it is connected to, as shown in Fig. 7.17.

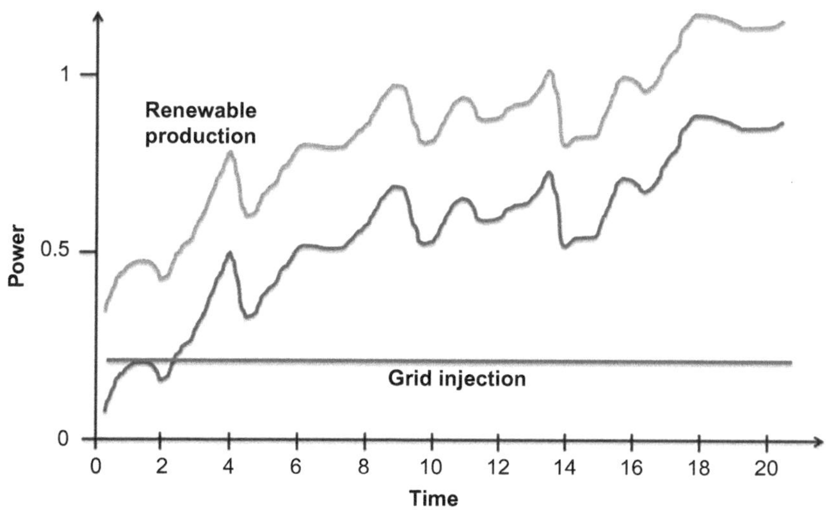

FIG. 7.16 Hydrogen local production to alleviate the renewable energy impact on the electric network (Guinot and Mansilla, 2016).

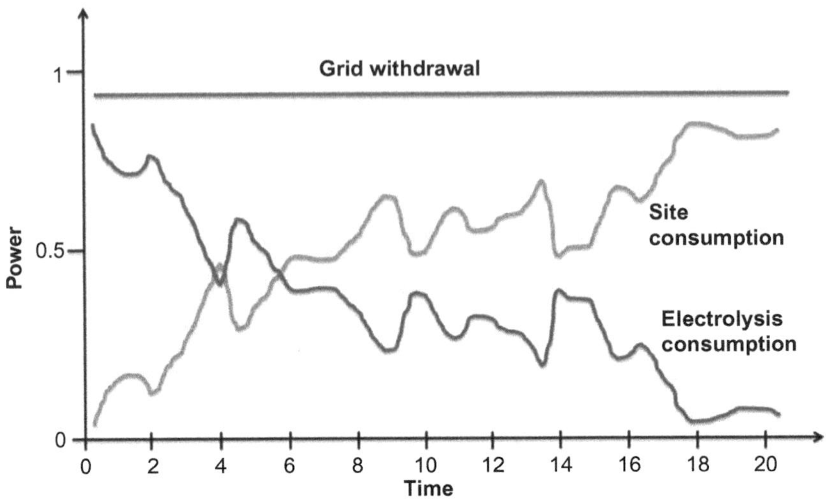

FIG. 7.17 Hydrogen local production to smooth the local demand (Guinot and Mansilla, 2016).

The benefits are similar to those in the previous case: reduction of the impact of variability (here of the demand-side), and forecasting needs.

7.3.1.2 Central Production

Hydrogen production can also be considered not as a power source or site consumption, but instead as a system (or a set of systems) simply connected to the grid (Guinot and Mansilla, 2016). From this perspective, electrolysis devices could participate in the balancing mechanisms or markets that are set up by the transmission system operators, in order to secure the balance between injections and withdrawals into the electric grid, to stabilize the grid frequency at any time. These mechanisms or markets are often named reserves, primary, secondary, and tertiary, according to their corresponding timescale (cf. Fig. 7.18).

The primary reserve (or frequency containment reserve) is used to restore the balance between supply and demand. Then, the secondary reserve (or automatic frequency restoration reserve) is harnessed to restore the frequency at its nominal level. Finally, the tertiary reserve (or manual reserve) is called upon to complete the frequency restoration if required, and reconstitute the reserves.

Several studies examined the economic relevance for hydrogen systems to participate in such mechanisms (Guinot et al., 2015b; Mansilla et al., 2012b). It was ascertained that, in the considered case, contributing to the electric system managements through these mechanisms is interesting from a purely economic viewpoint. The major uncertainty in this respect is the competition to provide

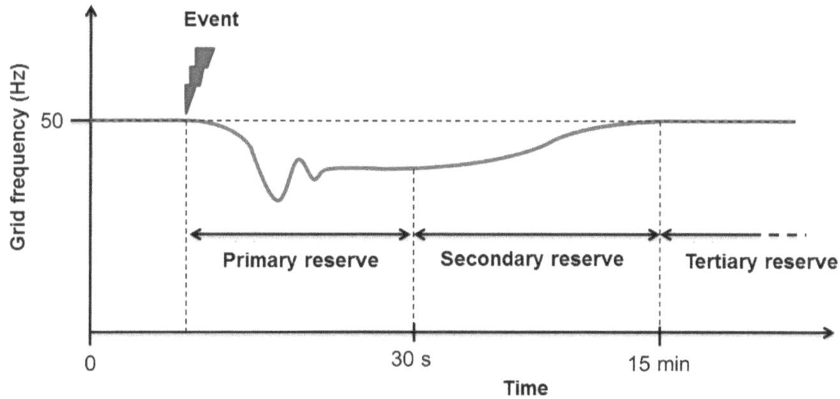

FIG. 7.18 Ancillary services to be provided to the electric grid (Guinot and Mansilla, 2016).

these services. As a matter of fact, a number of other technologies could supply similar services: storage systems, electric vehicles, or (more generally) any kind of flexible demand for power. The market for electric services will be fought for by these systems, according to the extent of their massive development and their relative price demand for providing these services.

7.3.2 Hydrogen to Foster Synergies

In conclusion, let us put things into perspective. Beyond enabling decarbonization of several sectors such as transport or industry, hydrogen could be a key ingredient to foster synergies:

- Between low-carbon energy sources, and
- Between energy networks.

In the current context of massive development of variable renewables, there exists an opportunity of developing deeply decarbonized power mixes by combining non-dispatchable renewable energy sources with low-carbon dispatchable sources, such as nuclear. In the French context, for instance, nuclear flexibility could be an asset to develop such mixes. Instead of reducing the nuclear load factor, the newly available energy could be taken advantage of by redirecting the "surplus" power (from the electric system viewpoint) toward other applications such as hydrogen production. This is what is illustrated by Fig. 7.19.

In this way, the nuclear fleet is fully used, and so are renewable energy sources. Indeed, hydrogen production as a flexible demand for power can also be useful in order to avoid curtailment when there appears to be a renewable energy surplus. This can be realized since interconnections make it possible to take advantage of the diversity of power mixes between neighboring countries (Caumon et al., 2015).

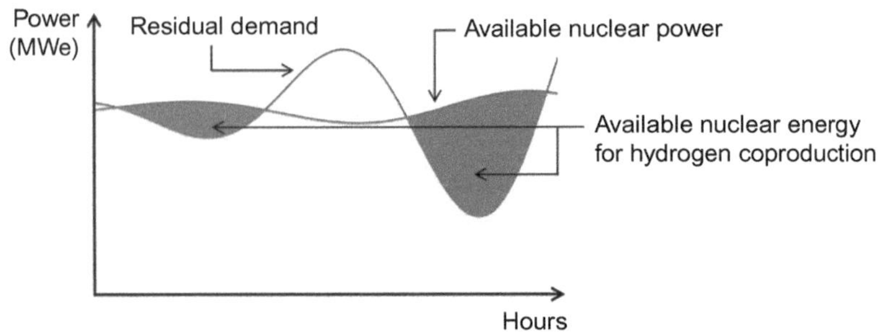

FIG. 7.19 Available nuclear energy assessment (Cany et al., 2017).

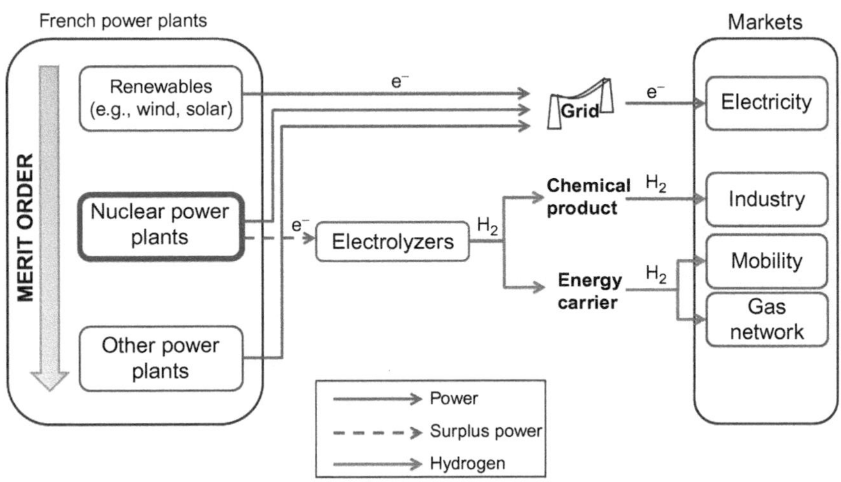

FIG. 7.20 Hydrogen fostering synergies between low-carbon sources and energy systems (Cany et al., 2017).

Another asset of the hydrogen system is its capacity to interconnect energy systems. The electric system was already mentioned; the gas network is a privileged channel for power-to-hydrogen systems (depending on the end-use for the hydrogen-natural gas mixture). Interconnections could also be done with heat networks. Thus, hydrogen can prove a privileged ingredient to foster synergies between low-carbon sources and energy systems, as depicted by Fig. 7.20.

Overall, system approaches are required to apprehend both hydrogen solutions in local contexts, and hydrogen systems' potential role in the energy system, as a provider of diverse energy services.

REFERENCES

Ademe, 2010. Joint seminar of BMBF & Siemens: "CO2 utilization potential", Bonn, September 22, 2009. ADEME, MEDDEM, ALCIMED, Panorama Des Voies de Valorisation du CO_2, Document final, June 2010(in French).

Ademe, 2013. L'exercice de prospective de l'ADEME «Vision 2030–2050». (in French).

Alphéa Hydrogène, 2014. Actualité Chimique n°347. In: AFHYPAC, 2016, Mémento de l'Hydrogène - FICHE 1.3. (in French).

Avril, S., Mansilla, C., Busson, M., Lemaire, T., 2012. Photovoltaic energy policy: financial estimation and performance comparison of the public support in five representative countries. Energ Policy 51, 244–258.

Camuzeaux, J., 2012. Mitigating Methane Emissions from Natural Gas and Oil Systems—The GNCS Factsheets. Columbia Climate Center, Earth Institute, Columbia University. Available online: http://climate.columbia.edu/files/2012/04/GNCS-Methane-from-Oil-Gas-Factsheet.pdf (Accessed 24 January 2017).

Cany, C., Mansilla, C., da Costa, P., Mathonnière, G., 2017. Adapting the French nuclear fleet to integrate variable renewable energies via the production of hydrogen: towards massive production of low carbon hydrogen? Int. J. Hydrog. Energy 42 (19), 13339–13356. https://doi.org/10.1016/j.ijhydene.2017.01.146.

Caumon, P., Lopez-Botet Zulueta, M., Louyrette, J., Albou, S., Bourasseau, C., Mansilla, C., 2015. Flexible hydrogen production implementation in the French power system: expected impacts at the French and European levels. Energy 81, 556–562.

Climate Change, 1995. The Science of Climate Change: Summary for Policymakers and Technical Summary of the Working Group I Report. page 22 http://unfccc.int/ghg_data/items/3825.php (Accessed 11 July 2016).

Denhoff, E., 2016. The state of Canada's fuel cell industry—successes and challenges.21st World Hydrogen Energy Conference (WHEC 2016), Zaragoza, Spain, June 13–16, 2016.

European Parliament, 2009. Directive 2009/28/EC of the European Parliament and of the Council of 23 April 2009 on the promotion of the use of energy from renewable sources and amending and subsequently repealing Directives 2001/77/EC and 2003/30/EC. 2009.

Florisson, O., 2010. NATURALHY: Assessing the Potential of the Existing Natural Gas Network for Hydrogen Delivery. GERG Academic Network Event, Brussels, Belgium. June 4, 2010. Available online: http://www.gerg.eu/public/uploads/files/publications/academic_network/2010/1b_Florisson.pdf (Accessed 24 January 2017).

GRDF, 2009. Le gaz naturel: une énergie qui répond aux enjeux du Grenelle de l'Environnement. Grenelle de l'Environnement, December 4, 2009. (In French).

Guinot, B., Champel, B., Montignac, F., Lemaire, E., Vannucci, D., Sailler, S., Bultel, Y., 2015a. Techno-economic study of a PV-hydrogen-battery hybrid system for off-grid power supply: impact of performances' ageing on optimal system sizing and competitiveness. Int. J. Hydrog. Energy 40 (1), 623–632.

Guinot, B., Montignac, F., Champel, B., Vannucci, D., 2015b. Profitability of an electrolysis based hydrogen production plant providing grid balancing services. Int. J. Hydrog. Energy 40 (29), 8778–8787.

Guinot, B., Mansilla, C., 2016. Hydrogen systems: a wide panel of services to help manage the electric systems.21st World Hydrogen Energy Conference (WHEC 2016), Zaragoza, Spain, June 13–16, 2016.

H2 at scale, 2016. H2 at scale: deeply decarbonizing our energy system.Touch Screen Presentation at AMR, June 6–10, 2016. Available at: https://www.hydrogen.energy.gov/pdfs/review16/2016_amr_h2_at_scale.pdf (Accessed 3 October 2016).

Hashimoto, M., 2015. Japan's Hydrogen Policy and Fuel Cells Development in NEDO. Available online: http://www.ieafuelcell.com/documents/excodocs/50/5_Japan_NEDO_Update.pdf (Accessed 24 January 2017).

Imbach, J., 2012. BtL industry: overview and outlook.Fuels of the Future conference, Berlin, Germany, January 23–24, 2012.

Mansilla, C., Avril, S., Imbach, J., Le Duigou, A., 2012a. CO2-free hydrogen as a substitute to fossil fuels: what are the targets? Prospective assessment of the hydrogen market attractiveness. Int. J. Hydrog. Energy 37 (12), 9451–9458. https://doi.org/10.1016/j.ijhydene.2012.03.149.

Mansilla, C., Louyrette, J., Albou, S., Barbieri, G., Collignon, N., Bourasseau, C., Salasc, B., Valentin, S., Dautremont, S., Martin, J., Thais, F., 2012b. Electric system management through hydrogen production—a market driven approach in the French context. Int. J. Hydrog. Energy 37 (15), 10986–10991.

McKinsey, 2010. McKinsey. A portfolio of power-trains for Europe: a fact-based analysis. The role of Battery Electric Vehicles, Plug-in Hybrids and Fuel Cell Electric Vehicles. McKinsey & Company, p. 2010.

Ohira, E., 2016. Current policy and R&D activity on hydrogen energy in Japan. 21st World Hydrogen Energy Conference (WHEC 2016), Zaragoza, Spain, June 13–16, 2016.

Prinzhofer, A., Deville, E., 2015. De l'hydrogène naturel sous nos pieds.Pour la Science n°456, October 2015.

Robinson, D.R., Fernandez, R., Kantamaneni, R.K., 2003. Methane Emissions Mitigation Options in the Global Oil and Natural Gas Industries. ICF Consulting & EPA. Available online: http://www.coalinfo.net.cn/coalbed/meeting/2203/papers/naturalgas/NG020.pdf (Accessed 24 January 2017).

White, C., 2016. Status and progress on FCEV market launch in California.21st World Hydrogen Energy Conference (WHEC 2016), Zaragoza, Spain, June 13–16, 2016.

FURTHER READING

Afhypac, 2014. Production d'hydrogène à partir des combustibles fossiles. In: Mémento de l'Hydrogène - FICHE 3.1.1.(in French).

Afhypac, 2017. AlterBike Pragma Industries. Available: http://www.afhypac.org/documentation/phototheque/deux-roues-pile-a-combustible/ (Accessed 19 January 2017).

Toyota, 2015. The Mirai Life Cycle Assessment Report. Available online:http://www.toyota-global.com/sustainability/environment/low_carbon/lca_and_eco_actions/pdf/life_cycle_assessment_report.pdf (Accessed 30 September 2016).

Chapter 8

Social Aspects of H_2 Supply Chains: Hydrogen Technologies Genesis and Development: The Case of Myrte Platform

Nicolas Simoncini*, Katy Cabaret*, Fabienne Picard*, Mohamed Becherif[†], Haithman Ramadan[†,‡] and Olivier Bethoux[§]

*UBFC-UTBM, FCLab FR CNRS 3539, FEMTO-ST/RECITS UMR CNRS 6174, Belfort Cedex, France, [†]UBFC-UTBM, FCLab FR CNRS 3539, FEMTO-ST UMR CNRS 6174, Belfort Cedex, France, [‡]Zagazig University, Faculty of Engineering, Electrical Power and Machines Dept., Zagazig, Egypt, [§]LGEP, University of Paris Sud 11, Gif sur Yvette Cedex, France

8.1 INTRODUCTION

The transition toward a sustainable energy system is situated at the intersection of technical innovations, the evolution of consumption, new economic models, and new forms of governance. Smart grids that aim to integrate all users' behaviors and actions in order to allow a direct management of offer and demand, the injection of renewable electricity into the network, and the increase of effectiveness of the whole system, fully contribute to this issue. Along with energy storage, they are recognized as one of the most important factors in the current energy transitions (ETPS, 2006, 2007, 2010, 2012). If current technologies show some limits, interesting perspectives arise around hydrogen (Becherif et al., 2015) and experimental projects have been developed in several European countries, such as Belgium (Don Quixote Project), Germany (Enertag Hybrid station), and France (Grhyd a Dunkirk urban community). Nevertheless, it still does not appear as an obvious solution, at least in France, as the recent energy transition law project makes little mention of hydrogen.[1] In this context, initiatives come from territorial regions and various stakeholders, as in Corsica where the Myrte platform (Mission hydrogen renewable for the integration to the

1. The first law project did not mention hydrogen. It was finally inserted thanks to amendments made during the votes of the Parliament at the end of 2014 (National Assembly) and at the beginning of 2015 (Senate) linked to renewable energy storage issues.

Hydrogen Supply Chain. https://doi.org/10.1016/B978-0-12-811197-0.00008-7

electrical network), which associates a solar central, an electrolyzer, and fuel cells, has been subjected to a complex arrangement between academics, politicians, and manufacturers. Moreover, this case refers to technical parameters as well as social processes.

Based on an ethnographic investigation carried out in 2015[2] on the development of this specific technological system, this paper aims to analyze the Corsican hydrogen storage emergence dynamics. Our investigation was based on seven semidirective interviews with various stakeholders of the platform (industrial companies, scientific research team), on the analysis of archives (reports of the Scientific Interest Group, activity reports), and on ethnographic observations of the site. More precisely, and thanks to the theoretical and methodological tools of the science studies, we would like to look into the sociotechnical conditions according to which this kind of project can expand, as well as the series of trials and controversies through which it passes. As we pay attention to the action "in the making," we make perceptible the states of doubt, oppositions of value, legitimizations, and the commitments that take part in the construction of the hydrogen economy, and its diffusion or its blockage. We will first clarify the particularities of the island context and their impact on the project. The second part will demonstrate how the shaping of the project depends on a treatment of the problem that closely links technical aspects to political obligations. We will next see that structural modifications are necessary to make the platform with the multiplication of stakeholders. Finally, we will question the stability of their relations to understand the mechanisms that determine the success and the failure of such an operation.

8.2 AN ISLAND CONTEXT PROMOTING INNOVATION

The European policy on energy has known different phases forming the pattern of a slow European energy system construction process (Cabaret and Picard, 2015). If energy was historically one of the first fields of expertise in the European communities, energy policy has only been part of the European treaties since the Treaty of Lisbon in 2007. The objectives retained in the field of electricity aim to develop a more distributed electricity network with ambitious goals in terms of the diffusion of renewable energy sources, especially through decentralized renewable energy production gathered around mini grids. On islands, energy issues are more important because there is no connection, or very few connections, with the continental network. Islanders then have to produce the energy they use and are more exposed to overconsumption risks and blackouts. In addition, this production is sometimes more expensive and eco-destructive. These territories then represent favorable places for experimentation, with identifiable market potential.

In addition to these specific characteristics in Corsica, the technical, economic, political, and social situation creates perfect conditions for innovation.

2. The authors would like to thank all the people who agreed to answer their questions. This research project was supported by GDR SEEDS in 2014.

The winter of 2005 forced EDF (French power supplier) to organize selective power cuts to avoid blackouts between February the 28th and March the 11th. This highlighted the imperfections of the regional electricity network, and new requirements in terms of territorial energy policy emerged for Corsica.[3] It was decided to diversify production and supply to reduce disruptions. A connection with Sardinia, the development of renewable energy on the island, and efforts toward energy efficiency completed the role that was essentially held by thermical plants until then. At the same time, a regional delegation from the CapEnergies center of excellence was created to galvanize employment and research in the energy sector. It is in this context that the Université of Corsica began to forge links with the CEA (the French Alternative Energies and Atomic Energy Commission) and Helion, an Areva subsidiary specializing in hydrogen technologies.[4] Philippe Poggy, energy university lecturer and researcher at the University of Corsica and Pierre Serre-Combe, and the head of the CEA Hydrogen and Fuel Cells Program, saw an opportunity to present a large-scale experimental project that associates a photovoltaic plant to a hydrogen network.

Myrte is generally seen as a life-sized test, which aims to demonstrate the feasibility of a solar energy storage solution overcoming the intermittency of renewable energy production. More precisely, the electricity produced by the photovoltaic panels can be directly injected into the network or stored during a certain amount of time in tanks after its transformation into hydrogen thanks to an electrolyzer (Darras et al., 2012)[5] When the power produced by the photovoltaic panels is not sufficient compared to needs, fuel cells take over and transform once more the stored hydrogen into electricity (Fig. 8.1).

8.3 WHEN TECHNOLOGY ENTERS POLITICS AND POLITICS ENTER TECHNOLOGY

However, the task is quite difficult. Philippe Poggi and Pierre Serre-Combe needed to convince academics, manufacturers, and politicians of the legitimacy of their project. As Christian Morel (2002) shows in his analysis of the steps that led to the explosion of the Challenger space shuttle in 1986, a problem is perceived differently by social stakeholders according to their rationality, their cognitive characteristics, their jobs, and their respective hierarchical positions.[6] Likewise, Picard and Rey (2012) reveal the importance of the contradictory visions that stakeholders have of them and others in the orientation of scientific

3. See: "Plan Energétique de la Corse", Délibération N 05-225 AC du 24 novembre 2005.

4. Helion was created in 2001 to bring Areva Technicatome's military technologies and applications (fuel cells and electrolytes used in submarines) into civil society.

5. An electrolyzer aims to break down water (H_2O) thanks to electric power to produce hydrogen (H_2) and oxygen (O_2).

6. With the analysis of the space shuttle explosion, the sociologist demonstrated that all individuals did not possess the same perception of risks: the seals distention problem was known. However, some engineers estimated the launch failure probability around 1%, whereas NASA managers evaluated it around 1 over 10,000 and even 1 over 100,000.

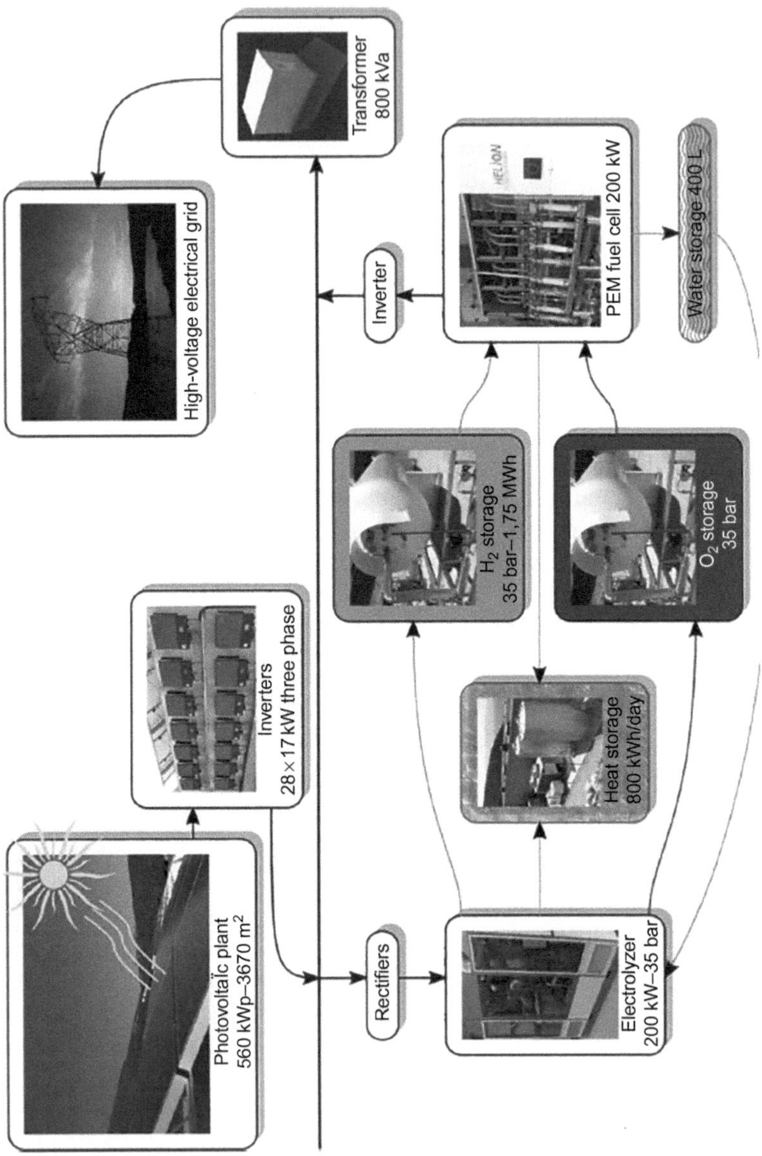

FIG. 8.1 Simplified illustration of the components and operations of the MYRTE platform (Darras et al., 2012).

research. How then to fill this gap between the logic of the researcher, the manufacturer, and the politician? How to gather varied representations and divergent interests around the same goal?

When they offered to develop this technology, Philippe Poggi and Pierre Serre-Combe did not only make a scientific hypothesis. To convince potential partners to join them, they formulated a real scenario for society; they put human desires into technology (Barthe, 2009), allocated skills to objects, territories and, institutions, and split the work among all the program stakeholders they designated. Thus, from the beginning, the Myrte project incorporated in its definition a new world in which it is believed to be built:

> *"Corsica has to deal with important stakes in the fields of energy supply, economic development and environment quality control. In our island particular context, there is no doubt that the renewable energies large-scale development will be a solution and a top opportunity at the same time. This project aims to test and developing life-size practical technical solutions for the Corsican network's electricity supply and to federate all different stakeholders from R&D to technology transfer, (initial, lifelong and professional) formation and the industrial economic world, the whole of them being a part of a strong territorial dynamics in terms of renewable energies, under the authority of the Corsican Economical Development Agency, the CapEnergies center of excellence's regional organisation. [...] This reference project for Corsica also answer to the objectives defined in the energy plan adopted in November 2005 by the Corsican Territorial Collectivity to satisfy the island needs by mobilizing all the energy power sources and in particular the new renewable energies."*
>
> (Christian Cristofari, Philippe Poggi et Jean-Pierre Roger, CapEnergies, 2007)

As the problem was formulated, the first speeches on Myrte involved a certain number of stakeholders and linked them and met their interests at the same time. As such, this socio-technical project formed a "device of interessement" (Akrich, 1989). Thanks to its technical attributes, all of the stakeholders should be satisfied. For the University of Corsica, Myrte is a scientific research instrument that lets the institution increase its skills and even legitimize its academic position after it obtained recently more autonomy in the framework of the LRU, the French law on the liberties and responsibilities of universities'.[7] For the CEA, it is a life-size test that could prove the validity of hydrogen technologies, while for Helion it is a demonstration of its equipment quality for the conquest of future markets. Raffalli, a local public works company, joined the team with the objective of developing the photovoltaic industry in the territory. Regional public institutions (the city of Ajaccio, the Territorial Collectivity, the Corsican Economic Development Agency) also see the project as a way to improve the

7. The French law on universities' liberties and responsibilities promulgated in august 2007 aims to reinforce the budgetary autonomy of French universities and transferring them skills in terms of human resources and property assets.

quality of energy supply and to preserve their environment. The French State and the European Union at last support this kind of initiative, which prefigures the energy transition.

Although Philippe Poggi had suffered failure in 2004 with a similar project, his aptitude to translate everyone's concerns into the conception of the Myrte platform permitted him to build it this time. Concretely, the researcher multiplied contacts with partners, especially the regional institutions (direction of the University, administrative centers, General Secretary of Corsican Affairs) and discursively transformed the technical characteristics (solar panels, fuel cells, hydrogen) into social and economic activities (benefits for the region, technology transfer, eco-friendly energy), which are supposed to influence all participants. In short, as he moved objects as well as utterances, Philippe Poggi made science step outside the laboratory and changed it into a homogeneous collective experience:

> *"In 2004, we were a small team of 4 or 5 researchers and we thought we could develop an experimental project on the site which would combine a small wind turbine, photovoltaic panels, fuel cells and an electrolyzer. We set up a scientific project and then we met with the Research and Technology Regional Delegate*[8] *and the Energy and Environment Protection Agency*[9] *vice-director who told us: 'your project, well, not bad but how will it benefit the region?' We are researchers so the region... (laughs) [...] [About the Myrte platform] There was the center of excellence which gathered institutions, politicians, a research and formation organism, a research and technology transfer organism, the CEA, Raffalli, the local company on which we could have operated the transfer. There was also a national firm which developed processes. It gained skills, it validated its products and made them progress thanks to us. So we had exactly what it takes." (interview with Philippe Poggi, 2015)*

The research made technology enter politics and politics enter technology (Akrich, 1987) in a circular movement, which gave his action some strength: a heterogeneous engineering work was produced (Law, 1989; Hubert and Vinck, 2014). Myrte became a boundary object that coordinates several points of view and carries in itself knowledge, practices, and representations from different social spaces. When the stakeholders are finally convinced and enrolled (Callon, 1986) through reasoning, discursive and shaping operations—not only about the project but also about the reality where it has to be inserted—a first trial is passed: a network is formed where functions and objectives of everyone have been assigned. Thus, technologies, politics, environment, and Corsica's specific characteristics involving geography and social acceptability of solar panels, are linked together. As a direct consequence of this successful

8. In French: DRRT (Délégué régional à la recherche et à la technologie).
9. In French: ADEME (Agence de l'environnement et de la maîtrise de l'énergie).

interessement, a Scientific Interest Group and a steering committee[10] were created in 2009 to fix, legally speaking, the collaboration and to guarantee the work proceedings in which all partners were asked to contribute to the project: the European Union through the European Regional Development Fund (8 millions of euros), the French State (3 millions of euros), the Regional Collectivity (4.5 millions of euros), the University of Corsica (2.1 millions of euros) and the industrial partners (3.4 millions of euros). However, during the construction of the Myrte platform, several other trials successively broke and then the relations established before had to be rebuilt.

8.4 HOW TO RENEGOCIATE THE SOCIO-TECHNICAL NETWORK AND TO PRODUCE HOMOGENEITY

A site located in Vignola near Ajaccio, used before as a concentrated solar power station, and built in the 1980s, was selected to build to the platform. Many dismantling and clean-up[11] procedures were carried out, as well as landscape impact studies. At the same time, a file was completed to classify the installation for the protection of the environment.[12] The Regional Direction of Industry, Research and Environment,[13] a regional representative of the industry ministry in charge of industrial development and environmental topics, validated it in the first place. But in 2009, a legislative measure from the Urbanism Code and the Environment Code fixed the rules about photovoltaic systems based on the ground.[14] Power stations that produce more than 250 kWc required new authorizations (building permit, impact study, public study).[15] Moreover, the impact study revealed that a protected species lived on the site, the Hermann's tortoise. A unique land-based tortoise in France, the *Testudo hermanni* is one of the world's most threatened reptiles and has been protected by a national plan since 2007. The construction was thus directly put in danger. The Environment, Planning and Housing Regional Agency, which replaced the former Regional Direction of Industry, Research and Environment,[16] wanted to stop the project.

We can see how definitions and categories, especially the ones concerning the Corsican environment, evolve according to network composition. When unexpected stakeholders are added, they disrupt the group, move frontiers,

10. In 2009, this committee was composed of representatives from CapEnergies, the University of Corsica, Raffalli, Helion, a law firm, and an energy expert assessment office.
11. For example, the old concentrated solar power station contained asbestos.
12. The installations likely to present risks are submitted to specific regulations.
13. In French: DRIRE (Direction régionale de l'industrie, de la recherche et de l'environnement).
14. Décret n° 2009-1414 du 19 novembre 2009.
15. Article R421-1 du code de l'urbanisme.
16. The DRIRE were nationally reorganized between 2009 and 2010. Their missions were given to the Environment, Planning and Housing Regional Agencies (DREAL in French) and to the Companies, Concurrency, Consumption, Work and Employment Regional Agencies (DIRECCTE in French).

and establish others (Simoncini, 2014). Before the environment experts, who became spokespersons and translators for the fauna and flora, transposed tortoises into a document that they presented against the platform (Latour, 2010), they did not exist. Philippe Poggi said it very well himself, "Tortoises appeared when there was the impact study." The perception of the project suddenly turned. From eco-friendly, it became eco-destructive. Philippe Poggi and his team then tried to renegotiate these conditions by concluding new alliances (Charvolin, 2012). Whereas a first step of the project consisted of putting the territory into the device, the device was now connected to the territory. First, a compromise was found with the Environment, Planning and Housing Regional Agency. An offer to dig holes into the fence to let tortoises, which have a large range, go around the site as they want, was made. During the building work, the team committed to catch and place the few present specimens into an adapted pen.[17]

Other socio-technical devices were implemented to interest and enroll traditional users of the local environment, including the Ajaccio Astronomy Club, which uses the old power station located on a small hill as a sky observation point, the close natural reserve,[18] and hunters who track prey around. They were not only subjected to communication strategies, but also to technical operations:

> *"The site was not enclosed enough, it was in the middle of the maquis and there are wild boars. So we asked ourselves a lot of questions about the tanks. I was not relaxed: Could the tanks resist to shotgun bullets? That was my problem. Areva gave us all the calculations that proved it could resist. But I was still not convinced. So we buried them [aimed to reduce landscape impact]." (interview with Philippe Laclau, Head of the University of Corsica's assets, 2015)*

Likewise, the area was redefined to welcome astronomers thanks to the construction of an appropriated building on the site, and all the structures were reconfigured to melt into the landscape (wood cover, ground turned green, trees plantation, etc.). As the network of its components became more complex after passing through a hybrid forum (Callon et al., 2001; Laborie et al., 2014), the platform became a mediator of relations between human and nonhuman stakeholders (tortoises, stars, wild boars, shotgun bullets). Moreover, the research team integrated these incidents into its orders of justification to increase the legitimacy of its approach (Lafaye and Thevenot, 1993). By demonstrating

17. Electricité de France will come later to ask advice from the Myrte team to settle a similar problem elsewhere on the island.

18. The Sanguinaires Archipelago and the Punta di Parata are classified "Zone Natura" 2000 and "Zone naturelle d'intérêt écologique, faunistique et floristique".

in public space that it was capable of adapting technology to wild life, it changed again the definition of the project:

> *"We settle that with the regional administrative center and it transformed into a symbol for the operation since Europe wrote articles on it: we protect Hermann's tortoises. We managed to turn the constraint into an advantage. [...] The dialogue process with the media was very interesting, the way we dealt with it. We did not do it deliberately but when you put things into perspective, we did it the way it had to be done." (interview with Philippe Poggi, 2015)*

Resolving this controversy allowed the team to acquire new media competencies that it made the most of to enroll the general public and to establish the importance of its work in the long term. In particular, the team conducted campaigns to raise awareness in schools, organized site visits, and produced meticulous educational short films. In one of them, entitled "Bottling the sun,"[19] an eloquent equivalence is set up between hydrogen and our star:

> *"[...] there is no nuclear plant here, we are on an island, Corsica. And as much other islands, the electricity power production is reduced. [...] The wind does not blow constantly. Some days, there is less sun light. And during the night, there is no sun light at all. These are called intermittent energy sources. We are never sure how much electricity they will produce. Nevertheless when the sun shines, it produced a lot of electricity. But you cannot bottle the sun. Although... Do you see these big demijohns? Well in a way they hold the energy of the sun."*

This "cosmogram" (Latour, 2010) aimed to produce homogeneity. By arranging and linking different beings and modes of existence together into practical life forms, it inserted the platform in the territory and allowed it an essential role: when the sun disappears, Myrte replaces it. Thanks to multiple operations, delicate intervals between visibility and invisibility were performed here. On one hand, stabilization and smoothing devices aligned all the stakeholders in the network (Houdart, 2015) and made their interactions more fluid. On the other hand, some translations were employed to make the technology step outside the laboratory and to connect it to the world (Pictures 8.1 and 8.2).

8.5 BACK TO THE LAB: THE DISINTEGRATION OF RELATIONSHIPS

Nevertheless, in December 2010, an unexpected development arose. Just when negotiations with relevant institutions about tortoises came to an end, a new regulation came into effect that suspended the obligation to buy electricity

19. In French: "Du soleil en bouteilles". This is a play on words that already existed in the 19th century when the first solar energy power devices were developed (Jarrigue, 2010).

PICTURE 8.1 Roads for turtles in fences (authors' pictures, April 2015).

PICTURE 8.2 Burials of hydrogen and oxygen tanks (authors' pictures, April 2015).

produced by photovoltaic installations of more than 3 kW.[20] In this context, a network connection application had to be submitted to ErDF (a power energy company) and new studies had to be done. This was another substantial problem for the University and Raffalli, which planned to sell the electricity generated. Therefore, negotiations were initiated with the Energy Control Commission in charge of power regulations and EDF SEI (EDF in Corsica and overseas departments) to fix a purchase price that could satisfy everyone. But the Energy Control Commission gave a negative decision, arguing that its role is not to fund research activities. This interessement failure had consequences. Indeed, Raffalli, which was supposed to install the electric nodes (lightening study, high-voltage network, low-voltage distribution boards, etc.), decided to withdraw from the project. Without any short term prospect of economic profits, the company could not hope for an amortization of its investment in the Myrte construction. To reduce the impact of this incident on the work planning, the University of Corsica took over the responsibility for its missions, which were entrusted to its assets department. This last trial clearly demonstrated the instability of relations between the stakeholders. Everything can still be called into question; nothing is definitively placed in a black box (Callon and Latour, 1981).

The platform was finally inaugurated in January 2012 after 1 year of work. Nonetheless it was still subject to controversies. Myrte especially concentrated struggles around its identity and its objectives. In the same time period, the principal industrial partner suffered important changes in its structure and its commercial and industrial strategy. Helion disappeared under the pressure of the market and was replaced by Areva ES (Energy Storage), which brought its hydrogen activities closer to the group research division. This phenomenon was a response to the lateness of the European storage energy commercial development in relation to the optimistic estimations of rising markets made at the end of the nineties and at the beginning of the 21st century. According to an Areva ES executive, this lateness can be explained by "the gas and oil prices drop, the absence of political pressure on carbon and the heterogeneity of storage applications" (2015). Despite the Greenergy Box production, Myrte did not represent a priority for the French industry firm:

> *"Current affairs imply to know what is going to be kept and how we can capitalize things. We do not want everything just to disappear. Unfortunately the logic we have now is declining. On the University's side it has been a strategic position tool, it has combined development works on electric networks, renewable energies and island issues. It has been a huge success in comparison to initial objectives.*

20. Décret n° 2010-1510 du 9 décembre 2010. After the State created support measures for the photovoltaic industry (especially with purchase prices around 50–60 cents/kWc), many companies invested in the market, which quickly reached saturation point. In December 2010, as the price of installations began to decrease—after the Chinese imports increased—the State decided to reduce its subsidies and temporarily suspended authorizations for new projects.

The question is: What will happen next? It does not have to be especially about Myrte future developments but we need to do something after, from an academic and an industrial point of view, for Corsica and the network. For more operational projects for the Corsican collectivity. On Helion's side, it was a very important step that allowed us to demonstrate the reliability of our technologies. It showed that it was possible to do things after, to start a more industrial stage. That already has been done with Myrte 2, with the container. The logic stage for Helion was to start standardization and sales. [...] Which has still not been done today." (interview with a former Helion executive, 2015)

On the other hand, the political support given to the platform started to decline. A specific event allows this to be understood. During the project proceedings, political turnover was important. The people who composed the national (Government, Parliament), regional (presidency of the University, prefecture, Territorial Collectivity, Corsican Economic Development Agency, Environment, Planning and Housing Regional Agency), and local institutions (the city of Ajaccio) were replaced. Then, the technology could not be appropriated. Alliances were broken too often and the Myrte team had to reengage new translations continually in the hope of stirring interest again in public authorities. Philippe Poggi explains:

"At one point elections, intern reorganizations and a change of majority happened. It took some time before everything was reorganized and messages were not heard very well. We were not supported by the Corsican Economic Development Agency anymore, etc. So the situation was confused at a local level. And after we lost time because of planning issues linked to regulatory aspects, we came too late in relation to political turn-over. That is why there was no technology appropriation by politicians and administrations. And that is what I miss today. They all consider that it is my project and they come to ask me what are the benefits for the region. But I do not have to answer this question. Politicians and administrations have to." (interview with Philippe Poggi, 2015)

Thus, the definition of Myrte seemed to be revised downward. After the platform stepped outside the laboratory, it was sent back. Whereas it had obtained a preindustrial experimental status, it became a scientific experiment again that could not be transferred to society anymore. This became visible when the Myrte transfer to a local stakeholder was aborted. As Ajaccio's hospital needed to be rebuilt in line with High Quality Environment Standards, the head of the project offered to start a partnership with the University of Corsica and Areva for the installation of a device just like Myrte. But the lack of public funds and the withdrawal of the manufacturer led to the cancellation of the operation. As a result of the division (Dodier, 1995), the network structure ended up contracting and organizing around the research team with a new Scientific Interest Group for 2015–20. Nevertheless, the research team had difficulties in highlighting the platform's scientific aspects through scientific papers and risks to be downgraded.

> *"When I hold conferences, I am the only one to show actual curves, not projects. The problem is we are judged on our papers, our conferences. We did some. But how do I increase the Myrte's value? It is complicated: it is an industrial-type platform, managed by a research and formation organism with its own indicators. This is not a paper [on a screen he show us the curve of a test carried out during the 20/03/2015 solar eclipse. During the eclipse fuel cells took over from photovoltaic panels to maintain continuous electricity supply] but for hydrogen industry it is very important." (interview with Philippe Poggi, 2015)*

If this is a typical example of the difficulties endured by public and private stakeholders in the framework of cooperative innovation situations (Laperche et al., 2013), then differences between them in terms of relations to time and future prospects are well-known factors of failure. But this kind of conflict also refers to classifications in action (Chauvin, 2006), which question the frame of reference (Flichy, 2003) stabilized before. With this in mind, opposition between the different functions given to Myrte constitute a mechanism of negotiation and correction of categories. As the platform does not mean the same thing for all stakeholders, its properties can always be discussed again and reinterpreted. Thus, these conflicts about the technology definition are also legitimization operations of identities from stakeholders, who describe in their technical choices their political and economic positions and in the modification of their commitments what are their new domains of competences and their new interests.

8.6 CONCLUSIONS

In this chapter, we tried to clarify a device that prefigures the energy transition by apprehending the political processes it contains. When discussing the technical details, we showed that the conception and the construction of the Myrte platform strongly depended on the ability of stakeholders to model its form to integrate within it social constraints from all of the different trials experienced. In return, the territory, the Corsican environment, and the stakeholders themselves also changed through their relation with the machine. Therefore, we can see how it is impossible to think of the platform outside the network in which it works: both are intimately interconnected. The movements of the first have an effect on the second and this situation leads to technical, political, economic, and social negotiations and conflicts. From tortoises to hunters, the sun, regional public authorities, and the general public, it is an entire universe that is aligned at the end as the technology unfolds. For this reason, the definition of Myrte never really stabilized: it continues to evolve according to the association of heterogeneous entities and beings who support it.

REFERENCES

Akrich, M., 1987. Comment décrire les objets techniques? Technol. Cult. (9), 49–64.

Akrich, M., 1989. La construction d'un système socio-technique. Esquisse pour une anthropologie des techniques. Anthropol. Soc. 13 (2), 31–51.

Barthe, Y., 2009. Les qualités politiques des technologies. Irréversibilité et réversibilité dans la gestion des déchets nucléaires. Tracés. Rev. Sci. Hum. (16). (mis en ligne le 20 mai 2011).
Becherif, M., Ramadan, H.S., Cabaret, K., Picard, F., Simoncini, N., Bethoux, O., 2015. Hydrogen energy storage: new techno-economic emergence solution analysis. Energy Procedia (74), 371–380.
Cabaret, K., Picard, F., 2015. Politique européenne de l'énergie et transition vers un système énergétique décarboné fondé sur les smart grids. Inno. Rev. Eco. Manag. Inno. 1 (46), 33–69.
Callon, M., 1986. Éléments pour une sociologie de la traduction. La domestication des coquilles Saint-Jacques dans la Baie de Saint-Brieuc. Ann. Soc. (36), 169–208.
Callon, M., Latour, B., 1981. Unscrewing the big leviathan: how actors macrostructure reality and how sociologists help them to do so. In: Knorr Cetina, K., Cicourel, A.V. (Eds.), Advances in Social Theory and Methodology: Toward an Integration of Micro- and Macro-Sociologies. Routledge and Kegan Paul, Boston, pp. 277–303.
Callon, M., Lascoumes, P., Barthe, Y., 2001. Agir dans un monde incertain. Essai sur la démocratie technique. Le Seuil, Paris.
Charvolin, F., 2012. L'affaire de la Vanoise et son analyste. Vingtième Siècle. Rev. Hist. 1 (113), 82–93.
Chauvin, P.-M., 2006. Les classifications en actions. De la sociologie durkheimienne des classifications aux hiérarchies des vins de Bordeaux. Tracés. Rev. Sci. Hum. (10). (mis en ligne le 11 févier 2008).
Darras, C., Muselli, M., Poggi, P., Voyant, C., Hoguet, J.-C., Montignac, F., 2012. PV output power fluctuations smoothing: the MYRTE platform experience. Int. J. Hydrog. Energy 37, 14015–14025.
Dodier, N., 1995. Les hommes et les machines. Métailié, Paris.
European Technology Platforms Smartgrids, 2006. Vision and Strategy for European Electricity Networks of the Future. Office for Official Publications of the European Communities, Luxembourg, p. 44.
European Technology Platforms Smartgrids, 2007. Strategic Research Agenda, 2007–2020. Office for Official Publications of the European Communities, Luxembourg.
European Technology Platforms Smartgrids, 2010. Strategic Deployment Document for Europe's Electricity Networks of the Future. Office for Official Publications of the European Communities, Luxembourg.
European Technology Platforms Smartgrids, 2012. Strategic Research Agenda, 2012–2035. Office for Official Publications of the European Communities, Luxembourg.
Flichy, P., 2003. L'innovation technique. Récents développements en sciences sociales, vers une nouvelle théorie de l'innovation. La découverte, Paris.
Houdart, S., 2015. Les Incommensurables. Zones Sensibles, Bruxelles.
Hubert, M., Vinck, D., 2014. Des pratiques d'ingénierie aux transitions socio-techniques. Rev. Anthro. Con. 2 (8), 361–389.
Jarrigue, F., 2010. Mettre le soleil en bouteille: les appareils de Mouchot et l'imaginaire solaire au début de la Troisième République. Romantisme 4 (150), 85–96.
Laborie, L., Mazoyer, H., Pinsolle, D., 2014. Les grands réseaux en démocratie technique? In: Lequin, Y.-C., Lamard, P. (Eds.), Eléments de démocratie technique. Université de technologie de Belfort-Montbéliard, Trevenans.
Lafaye, C., Thevenot, L., 1993. Une justification écologique? Conflits dans l'aménagement de la nature. Rev. Fr. Sociol. 4 (34), 495–524.

Laperche, B., Picard, F., Yacoub, N., 2013. Origine, fonctionnement et gouvernance de l'innovation multi-partenariale: une lecture néo-institutionnelle. In: Boutillier, S. et al., (Ed.), L'Innovation. Analyser, Anticiper, Agir. P.I.E Peter Lang, Bern, pp. 143–174.

Latour, B., 2010. Cogitamus. Six lettres sur les humanités scientifiques. La Découverte, Paris.

Law, J., 1989. Technology and heterogeneous engineering: the case of Portuguese expansion. In: Bijker, W., Hughes, T., Pinch, T. (Eds.), The social Construction of Technological Systems. New Directions in the Sociology and History of Technology. MIT Press, Cambridge, MA, pp. 111–134.

Morel, C., 2002. Les décisions absurdes: sociologie des erreurs radicales et persistantes. Gallimard, Paris.

Picard, F., Rey, B., 2012. Diffusion socio-technique de la pile à combustible en France – une première approche par les représentations des chercheurs. Premières Journées Internationales de Sociologie de l'Energie, Toulouse.

Simoncini, N., 2014. Comment les surfeurs ont inventé la pollution de l'océan. Ter. Trav. 2 (25), 161–177.

FURTHER READING

Latour, B., 2006. Les 'vues' de l'esprit. Une introduction à l'anthropologie des sciences et des techniques. In: Akrich, M., Callon, M., Latou, B. (Eds.), Sociologie de la traduction. Textes fondateurs. Ecole des Mines de Paris, Paris.

Star, S.L., Griesemer, J., 1989. Institutional ecology, 'translations' and boundary objects: amateurs and professionals on Berkeley's museum of vertebrate zoology. Soc. Stud. Sci. (19), 387–420.

Chapter 9

Power-to-Gas—Concepts, Demonstration, and Prospects

Christina Wulf, Jochen Linssen and Petra Zapp
Institute of Energy and Climate Research—Systems Analysis and Technology Evaluation (IEK-STE), Forschungszentrum Julich, Julich, Germany

9.1 INTRODUCTION

The Power-to-Gas concept (other terms used: power to gas, P2G, PtG) uses renewable or excess electricity to produce hydrogen (Power-to-Hydrogen) via water electrolysis. This hydrogen can be used directly as a final energy carrier or converted to methane, synthesis gas, electricity, liquid fuels, or chemicals, for example. The reasons for using PtG are diverse (Tichler et al., 2014). The main purpose is to store energy long term by converting it to other easily storable energy carries, and at the same time reducing the load of the electricity grid by controlled operation (flexible demand). Furthermore, the production of renewable fuels for transportation, households, or industry, as well as for chemical production, can be a main driver for PtG. Not only are the motivations for PtG diverse, but also the choice of technologies. Fig. 9.1 gives a schematic overview of the necessary PtG pathways and components that have been discussed. In the following sections, the different components needed for gas production are discussed in more detail regarding their special needs for PtG (for example, different electrolysis technologies) their process design (for example, methanation) and the requirements for the transport and distribution of gases in the natural gas grid. The uses of the produced gases that are unrelated to energy, for example, in the production of chemicals, is not in the scope of this chapter.

The actual existing PtG projects in Europe are analyzed in Section 9.3 regarding their purpose and the technologies used. In Section 9.4, an overview of scenario studies for the deployment of PtG considering the development of future energy systems gives an estimation of the expected installed capacities for selected regions or countries.

Hydrogen Supply Chain. https://doi.org/10.1016/B978-0-12-811197-0.00009-9

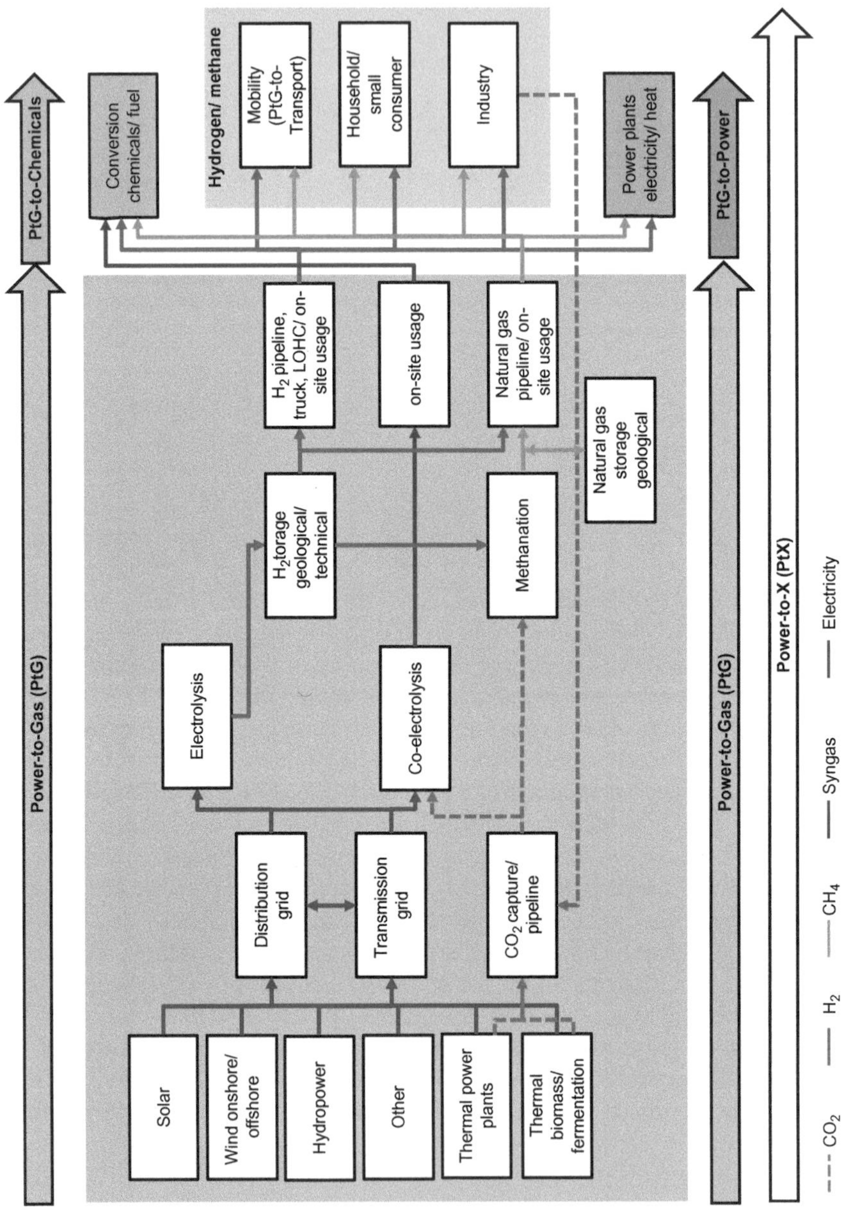

FIG. 9.1 Overview of Power-to-X concepts.

9.2 TECHNOLOGIES FOR POWER-TO-GAS

The main component of PtG systems is the gas production. This includes hydrogen production by electrolysis and its further processing to methane. Also, the electrolytic production of synthesis gas is an option. Transport and distribution of these gases is another important topic discussed here.

9.2.1 Hydrogen Production

Several concepts for electrolytic hydrogen production are already commercially available today. One option is alkaline water electrolysis. As PEM (polymer electrolyte membrane) electrolysis has a better part load behavior, this type of electrolysis became the first choice for PtG (Stenzel et al., 2016). However, for larger capacities, it has not reached a Technology Readiness Level of 9 yet, as the alkaline electrolyzer has done. Furthermore, a new type of electrolyzer, SOEC (solid oxide electrolyzer cell), which works under high-temperature conditions (700–800°C), is being explored for PtG applications. As hydrogen production by electrolysis is explained in more detail in another chapter, here only aspects are mentioned that are connected to the special requirements of PtG.

One important goal of PtG is to integrate fluctuating electricity generation from renewable energies (REN) into regional or national energy systems. Therefore, the main challenge for electrolysis within PtG is intermittent/dynamic operation. All three electrolyzers considered, PEM, alkaline electrolysis, and SOEC, are able to handle load changes. As alkaline and PEM electrolysis are more developed than SOEC, they are discussed here in more detail.

In general, PEM electrolysis can handle load changes better than alkaline electrolysis. Its minimum load lies between 0% and 10%, while alkaline electrolyzers only achieve 20%–40% (Smolinka et al., 2011). These values are defined due to the gas purities and subsequent safety concerns. Gas crossover takes places in every electrolyzer. While hydrogen occurrence at the anode, where oxygen is produced, causes no problems, oxygen at the cathode might be dangerous. An oxygen/hydrogen gas mixture with an oxygen content of >4 vol% is spontaneously combustible. Therefore, in technical applications, the maximum value of oxygen in hydrogen is set to 2 vol%. In alkaline electrolyzers hydrogen reaches the cathode by diffusion through the diaphragm or because it is dissolved in potassium hydroxide, which is mixed constantly. In PEM electrolyzers hydrogen diffuses through the membrane or is transported there dissolved in water. As these effects occur with a constant rate, independent from the actual hydrogen production, impurities increase with lower loads. Both electrolyzers are theoretically able to operate under overload conditions also. Since the performance range of the PEM is wider than for the alkaline electrolyzer, overload operation is normally only done with PEM electrolyzers. However, operating in overload leads to efficiency losses and stress for the materials (Tjarks et al., 2016).

Load changes affect temperature and pressure levels in the electrolyzer, and subsequently, its efficiency. Each electrolyzer has an optimal operation temperature. With lower loads, less heat is produced and the optimal temperature may not be reached. In contrast, under overload conditions, additional cooling is needed. A change of pressure has contradictory effects on the efficiency. A higher pressure level lowers the stack efficiency of the electrolyzer while it otherwise lowers the energy demand for gas drying and compression after the hydrogen production. In particular, for pressurized alkaline electrolyzers, lower pressure leads to lower system efficiencies. In those electrolyzers inherent to the system, a compression of the hydrogen up to 30 bar is done, which is more efficient than an external compressor. These electrolyzers are, however, more affected by stops of the system. If not enough electricity for the minimum load is provided by the fluctuating energy sources, the electrolyzer needs to be shut down completely after a short time, including a depressurization. Then the electrolyzer has to be flushed to prevent unwanted reactions of the remaining hydrogen. In conclusion, a stop leads to hydrogen losses as well as a loss in the pressure level, and subsequently to efficiency losses (Bourasseau and Guinot, 2016; Tjarks et al., 2016).

For small loads, additionally, the specific efficiency decreases significantly. The balance of the plant, for example, the pumps and cooling, has a constant energy demand independent of the hydrogen production. For loads lower than 30%, this leads to a significantly higher energy demand per kg hydrogen (Bourasseau and Guinot, 2016). In general, not only the electrolyzer itself, but also the process control, power rectifier, and the balance of the plant need to be designed for load changes.

Next to efficiency losses, a faster degradation of the materials used for the electrolyzer can be anticipated. At the moment, no studies are available that quantify the degradation and explain the reasons. However, with fluctuations of electricity input, the electrical as well as the mechanical parts of an electrolyzer system come under higher stress in comparison to constant load operation (Bourasseau and Guinot, 2016).

The amount of experience with the degradation of SOEC is even less than for the other electrolyzers. However, SOEC has been tested to check that it can work under fluctuating load conditions (Bourasseau and Guinot, 2016). Like the other electrolyzers, the SOEC is affected by fluctuating conditions with temperature and pressure changes. As the maintenance of a high temperature is of utmost importance, a cooldown due to a cessation of operation would lead to a long downtime and high energy losses. In total, SOEC has reached a TRL level of 5 at the moment.

The specific investments (per kW electrical input) for electrolyzers vary considerably. Values for alkaline and PEM electrolyzers range between 900 and 4200 €/kW for the year 2010–12 and are expected to decrease to between 250 and 900 €/kW in 2030 (Bertuccioli et al., 2014; Kasten et al., 2016). This big difference may be explained by the different technologies analyzed. At the moment, PEM electrolyzers are still more expensive than alkaline electrolyzers. For the year 2030, however, both might be in the same range (Bertuccioli et al., 2014). Other factors for the uncertainty are upscaling and learning effects. Even though the electrolysis cells cost the same in small and big electrolyzers, the balance of the plant (e.g., power supply, system control, gas drying) achieves cost reductions due to upscaling. The future costs therefore depend on the scenario chosen for the electrolyzer sizes and the absolute installed capacity. Specific gas production costs not only depend on the investments, but also strongly on the purchase costs for the needed electricity and the full load hours of the system. This affects also the other parts of PtG systems, such as methanation.

9.2.2 Methane Production

Besides hydrogen production, another special process for PtG is the methanation described by the Sabatier reaction (Eq. 9.1). If methanation is part of the PtG system, carbon dioxide has to be separated from industry processes, power plants, or biogas plants if they are to be carbon neutral.

$$CO_2 + 4H_2 \rightarrow CH_4 + 2H_2O \quad (9.1)$$

Two different concepts are available, so far. Most common is the catalytic methanation at higher temperatures (starting at 250–300°C) (Otten et al., 2016). Additionally, biological methanation is an option.

9.2.2.1 Catalytic

The above-mentioned Sabatier reaction is most commonly conducted as a thermochemical catalytic reaction. The highest conversion rate of this reaction is reached at low temperatures. However, to also ensure a reasonable reaction rate, a catalyst is needed. Today, mainly nickel catalysts (other metals like ruthenium are possible (Younas et al., 2016; Meylan et al., 2016)) with aluminum oxide as carrier material (also silica, zeolites, or metals under research (Lehner et al., 2014)) are applied. They require at least 300°C to work (Grond and Holstein, 2014). This reaction is highly exothermic ($\Delta H = -252.9\,\text{kJ/mol}$). The catalyst, however, starts sintering above 550°C (Younas et al., 2016). Thus, the main challenge of reactor design is temperature control. The pressure level varies between 1 and 200 bar (Benjaminsson et al., 2013; Götz et al., 2014; Leonzio, 2016). Next to temperature control, maximization of the conversion rate to synthetic natural gas (SNG) is the main goal for reactor design.

Therefore, different types of reactors are used or are under research. Fixed-bed reactors are state of the art for methanation (Götz et al., 2014). To control the temperature, several reactors (between two and five) are connected in series with cooling of the flow in between (adiabatic design) (TRL 7). As an alternative, directly cooled fixed-bed reactors, that is, tube bundle and plate reactors, are being tested as prototypes for an isothermal design (TRL 6) (Rönsch et al.,

2016). These reactors reach a sufficient conversion to SNG in just one reactor stage. However, they are more expensive than adiabatic reactors and the waste heat reaches lower temperatures, which makes further usage difficult (Specht et al., 2016; Benjaminsson et al., 2013). Currently, fluidized bed reactors are at the prototype stage, too (TRL 6) (Rönsch et al., 2016). They have a simple design and enable good heat removal. On the other hand, they also show faster attrition and incomplete carbon dioxide conversion caused by bubbling (Götz et al., 2016). Another possibility is three-phase reactors, either as fluidized bed or bubble columns, which are being investigated at lab scale at the moment (TRL 4) (Rönsch et al., 2016; Götz et al., 2014). An even less explored design is structured reactors with honeycombs or microchannels as well as sorption enhanced reaction (TRL 3) (Götz et al., 2016).

At the moment, the efficiency of fixed-bed reactors lies between 70% and 85% and is highly dependent on the use of the waste heat (Grond et al., 2013; Götz et al., 2014). Depending on the system integration of catalytic methanation, the heat might be used for biogas upgrading by an amine scrubber (Specht et al., 2016). Due to the sensitivity of the catalyst to impurities, the carbon dioxide needs to be cleaned from sulfur compounds, especially when coming from biogas upgrading. In theory, the use of desulfurized biogas is also possible. Because of a larger volume flow, this leads to a higher reactor volume and thus to higher investments. The produced SNG has a purity of approximately 92%. Depending on the requirements of the local natural gas grid, membrane processing of the SNG might be necessary before feeding it into the grid (Specht et al., 2016).

The dynamic and part load behavior of catalytic methanation is another important aspect in the design of PtG systems. The minimum load of adiabatic fixed-bed reactors is approximately 40% of the nominal load, while isothermal reactors reach lower values. More advanced systems, such as three-phase methanation even achieve values between 10% and 20% (Götz et al., 2016). Fast load changes and subsequent temperature changes in the methanation are no problem for the catalyst (Specht et al., 2016). However, state of the art catalysts have only a maximum lifetime of 24,000h (Grond and Holstein, 2014). Furthermore, after a period of downtime, the reactor needs several days for restart (Grond et al., 2013), and before shutting the reactor down it needs to be flushed with hydrogen or some other inert gas (Götz et al., 2016).

The specific investments (per kW methane capacity[1]) for catalytic methanation are also hard to define. For 2030, the expected range lies between 200 and 600 €/kW (Kasten et al., 2016).

According to Götz et al. (2016), for average size reactors, three-phase or fluidized-bed reactors, and for large-scale plants (>100MW), fixed-bed reactors are most favorable from a technical point of view.

1. Based on the lower heating value.

9.2.2.2 Biologic

Instead of the thermochemical catalytic process described above, the Sabatier reaction can also be conducted by metabolic processes of methanogenic archaea, chemoautotrophic microorganisms. This is a common process that is part of the production of biogas in anaerobic digesters. Therefore, different methods of implementing this reaction into biological reactors are possible. First, into a biogas digester or the post digester hydrogen is added to increase the methane yield. In this way, the methane concentration in the biogas can be elevated from 50% to 75% (Benjaminsson et al., 2013; Götz et al., 2014; Krassowski and Jochum, 2016). A higher hydrogen partial pressure in the digester, however, inhibits the degradation of biomass. Thus, the addition of hydrogen is limited (Benjaminsson et al., 2013).

As a second option, the methanation can also take place in a separate reactor with microorganisms, into which pure carbon dioxide or raw biogas is fed. Similar to the biogas digester, the microorganisms need thermophilic conditions (around 65°C). Archaea, which work under mesophilic (33–45°C) or hyperthermophilic (65–85°C) conditions, are used (Krassowski and Jochum, 2016). At higher temperatures, the solubility of hydrogen in water decreases, though (Benjaminsson et al., 2013). The gases (hydrogen and carbon dioxide) need to be solved into water to be processed by microorganisms. The pH value must stay between 6.8 and 7.2, which needs to be monitored closely as the carbon dioxide decreases it. An effective method for pH value control still needs to be developed (Götz et al., 2016). Furthermore, strictly anaerobic conditions are necessary as oxygen kills the microorganisms. Because the microorganisms are more resistant to impurities than the nickel catalyst in thermochemical methanation, the biological methanation can be fed with raw biogas and hydrogen without any pretreatment. Before SNG can be fed into the gas grid, however, it needs to be cleaned, for example, of sulfur components. In test reactors, higher methane yields were achieved with pure carbon dioxide in comparison to biogas (Krassowski and Jochum, 2016). Furthermore, the methanation reactor can be smaller if pure carbon dioxide is used. After methanation, only 85%–95% of the produced gas is methane (Götz et al., 2014; Krassowski and Jochum, 2016). The rest consists of hydrogen and carbon dioxide, which is sent back to the methanation process after cleaning. Due to unfinished reactions of the starting materials, the conversion efficiency of hydrogen to methane is lowered. Furthermore, 18% of the added hydrogen is transferred into heat and not methane (Benjaminsson et al., 2013). For this process, different types of reactors are under development. Most common for this application is a continuously stirred tank reactor. This type, however, needs additional energy to mix everything. At a certain size of the reactor, a homogeneous mixing becomes complicated (Götz et al., 2014). Alternative reactor designs are a trickle-bed reactor and reactors with hollow fiber membranes (Benjaminsson et al., 2013). Both designs avoid stirring by pumping the different media through the reactors.

The most important challenge at the moment for biological methanation is the supply of gaseous hydrogen to the microorganisms due to the low solubility of hydrogen in water (Götz et al., 2016). On the bright side, a first test showed that archaea are robust against sudden temperature, pressure, and load changes (Götz et al., 2014; Krassowski and Jochum, 2016). Even a start after a longer period of downtime is possible without problems (Grond and Holstein, 2014). Further research is needed to investigate the dynamic behavior. It can be assumed, however, that the process control system will be the limiting factor and not the process itself (Götz et al., 2016). Compared to thermochemical methanation, biological methanation happens at a lower rate due to the lower temperature, gas liquid mass transfer resistances, and backmixing (Götz et al., 2016). This leads to bigger reactor volumes when a similar volume flow should be reached. At the moment, biological methanation is still at the pilot and demonstration scale, leading to TRL between 6 and 7.

The investment required for biological methanation are, for smaller plants ($\sim$5 MW_{SNG}), in the same range as for catalytic methanation. Scaling effects, however, are clearly more distinct for catalytic methanation (Bär et al., 2015; Benjaminsson et al., 2013).

According to Götz et al. (2016), for small reactor sizes, biological methanation is technically most favorable.

9.2.3 Synthesis Gas Production

Besides hydrogen and methane, other synthesis gases with different H_2:CO ratios play an important role in PtG. As input to several subsequent production processes for fuels or chemicals they are the starting point for sector coupling, especially to the transport sector and the chemical industry. Today most of the syngas is produced from fossil fuels, either by steam reforming or partial oxidation with all their environmental constraints. The electrolytic production route consists of hydrogen production in an electrolyzer from water and subsequent reaction of parts of this hydrogen with carbon dioxide in a water-gas shift reactor to produce carbon monoxide. Hydrogen and carbon monoxide are then mixed in the desired ratio. For H_2:CO ratios lower than 3:1, the mature Fischer-Tropsch process converts syngas into liquid hydrocarbons.

The coupling of low-temperature water electrolysis and high-temperature gas-phase conversion, however, is challenging as both require different reaction conditions. In this context, the already mentioned solid oxide cells are gaining renewed interest due to steam and carbon dioxide reduction at high temperatures. In a so-called coelectrolysis process, a high-temperature CO_2-reduction reaction and steam electrolysis are combined in one simultaneous process step instead of subsequently coupling these two distinct reactions in separate reactors. As possible syngases, H_2:CO ratios from 1:1 to 3:1 are accessible (Foit et al., 2016), thus providing feed material for several fuels or chemical products, such as alcohols (1:1), diesel and kerosene (>1:1), ethanol and methanol (2:1), or even methane (3:1). At present, research focuses on the ability of the operating parameters, such as temperature, current density, and material, to steer the required distribution of the H_2:CO ratios to reach an extensive understanding of this complex process.

Conventional fossil fuel-based syngas production often comes with undesired degrading reactants, such as sulfuric components or soot (Speight, 2015). The coelectrolysis, however, starts with pure input flows of deionized water from cooling water and carbon dioxide of sufficiently high concentrations, for example, from industrial point sources, leading to almost pure products. The starting point of current research work is the expertise in well-known solid oxide fuel cells. The typical process temperatures from 750 to 1000°C, using an oxygen-ion conducting ceramic electrolyte, show thermodynamic advantages. Additionally, the kinetics of the electrode reactions is much faster, meaning there is no need for expensive noble metal catalysts. At the steam electrode, usually a Ni/8 yttrium oxide stabilized zirconium oxide (YSZ) cermet is used for the water and carbon dioxide reduction. 8YSZ is used for the electrolyte and an electronically conducting metal oxide is used as oxygen evolution electrode (Ebbesen and Mogensen, 2009). However, there is still a great need for further fundamental research, as especially under coelectrolysis conditions, at high current densities, considerable degradation rates have been reported. Possible reasons are impurities in the gas and air flows, delamination of the air electrode due to pores building at the grain boundary, or coarsening on nickel particles (Foit et al., 2016). The TRL is therefore not higher than 3. These challenges must be overcome first before coelectrolysis will play a major role in future PtG scenarios.

9.2.4 Blending of Hydrogen Into the Natural Gas Grid

Hydrogen produced by water electrolysis can be blended at a defined share directly to the natural gas grid, or, after the conversion to methane by methanation, be added to the natural gas system nearly without any restrictions. After adding hydrogen or methane to the natural gas grid, the full scale of the natural gas supply and application system can be used. Therefore, PtG enables the use of REN or excess electricity in gas-fired power plants, industrial applications, or final consumer technologies, such as heating systems or gas stoves. Worldwide there are demonstration projects ongoing to analyze the impact of adding hydrogen to the natural gas grid and to analyze the interaction of this cross-sectional solution for the use of renewable electricity and its storage (see Section 9.3).

The results of one of the most comprehensive studies on the restrictions of hydrogen blending to the natural gas grid was published in Müller-Syring et al. (2013). On the basis of a detailed analysis on the impacts of hydrogen blended to natural gas, the authors of the study summarize that the existing natural gas infrastructure is suitable for up to 10 vol% of hydrogen blending. This statement is only valid if the combustion parameters of the gas blend fulfill the prevailing

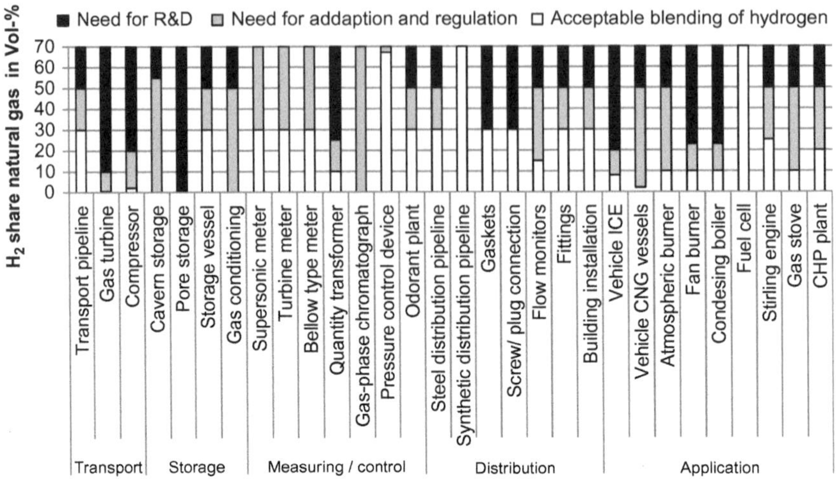

FIG. 9.2 Margin of hydrogen shares in natural blends for selected parts of the natural gas system. *(Source: Müller-Syring, G., Henel, M., Köppel, W., Mlaker, E., Sterner, M., Höcher, T., 2013. Entwicklung von modularen Konzepten zur Erzeugung, Speicherung und Einspeisung von Wasserstoff und Methan ins Erdgasnetz. DVGW-PRojekt G1-07-10. DVGW, Bonn. Own illustration.)*

rules and standards for gas quality. Fig. 9.2 points out the hydrogen compatibility for important components of natural gas transport and distribution, gas storage, measuring technologies, and final consumer technologies.

The authors of the study identified a need for research on natural gas and hydrogen blends in geological storage (especially pore-space storage), on gas turbines and changing combustion behavior of the blends and high-pressure vessels for CNG (Compressed Natural Gas) vehicles. Currently, the share of hydrogen in natural gas for the operation of gas turbines is limited to 1–5 vol %, depending on manufacture requirements. Higher hydrogen concentrations cause an increase in the rate of spread for the flame that may lead to thermal stress in the components of the combustor and therefore to damage. In addition, flame instabilities can occur, which can also cause damage depending on the hydrogen share of the blend. For pressure vessels in compressed natural gas (CNG) vehicles and the corresponding components of the filling stations, further research must be done for the assessment of the stability of the currently used steel tanks under alternating pressure conditions.

The suitability of geological pore-space storage for natural gas/hydrogen blends is still subject to fundamental research. Needs for additional knowledge are seen concerning the microbiological aspects (formation of iron sulfite through microorganisms) as well as the integrity of the geological overburden. In industrial natural gas applications, not the hydrogen share itself is the problem but the quality of the blend, for instance in ceramic or glass producing industries. A changing composition of the gas blend makes the application of additional measuring equipment necessary (e.g., Wobbe index measuring or process chromatographs). However, the authors of the study see fewer problems in final consumption technologies, such as heating appliances and gas engines.

The study of Müller-Syring et al. (2013) indicates that blending hydrogen to natural gas pipelines results in a reduction of the transport capacity. As one example calculation, for a pipeline diameter of DN 1000, a length of 250 km, and a capacity of 1 million Nm^3/h, blending 10 vol% of hydrogen to natural gas leads to a reduction in transport capacity of between 5% and 6% (according to the gas quality high- or low-grade gas). To reach the same transport capacities for a blended gas, the volume flow and the pressure level must be increased. The result is an increase of the compressor capacity of 25% when blending 10 vol% of hydrogen.

9.3 POWER-TO-GAS IN EUROPE

To integrate PtG into the energy system, in addition to the development of the single components, the system integration is of major importance. This can only be investigated by demonstration projects. In total in Europe 106 PtG projects have been conducted (January 2017). In the following, the projects are discussed in terms of their field of application, the processes used, and other important parameters (Table 9.1). While Bailera et al. (2017) focus only on projects that include methanation, here also Power-to-Hydrogen and PtG-to-Chemicals projects are included. The analysis is based on several previous review studies or projects (Bailera et al., 2017; Götz et al., 2016; Dena, 2017; Vartiainen, 2016; Iskov and Rasmussen, 2013; European Powertogas, 2017; Stenzel et al., 2016; DVGW, 2015). Furthermore, a great number of primary sources is used to fill knowledge gaps.

In total 15 different European countries have executed or are projecting PtG projects. The earliest projects started in 2003. For the following seven years, only a small number of projects were commissioned (between one and three, see also Fig. 9.6). Since 2011 the number of commissioned projects has increased rapidly and reached its maximum in 2015 with 19 projects. In 2017, eight projects will launch their plants. Also, for the years to come, several projects will start. The number of projected projects (seven), however, has to be regarded cautiously. Some of them sound very ambitious, for example, Power-to-Gas Hungary (Fründt and Pentz, 2016), while other notices of intent are already several years old, for example, Wijster (DNHK, 2014).

Even though Germany has realized the largest number of projects today and will continue to do so (47 have been executed or are projected, 44%), it had only one project in the early years (2003 – 10). In those days, Spain realized 28% of the projects (five projects). Also, Greece, Denmark, The Netherlands, and United Kingdom each realized two projects. Regarding the field of application of these projects, the distribution for Germany and 'other' countries correlate with the overall distribution, see Fig. 9.3. Spain, France, Denmark, and United

TABLE 9.1 Overview of Power-to-Gas projects in Europe

Acronym/ Location/Name of the Project	Type of Electrolyzer	Capacity kW	Commissioning	TRL	Methanation	Country	Source
Blending into natural gas grid							
Ameland	PEM	8.3	2008[a]	6		The Netherlands	(Kippers et al., 2011)
Hybrid power plant Prenzlau[b]	Alkaline	500	2011	7		Germany	(Enertrag, 2015)
Morbach	Alkaline	25	2011[a]	7		Germany	(Energielandschaft Morbach, 2015)
H2-Research centre BTU	Alkaline	145	2012	6		Germany	(Dena, 2012a)
Methanation at Eichhof							
1st	n.s.	25	2012[a]	6	Catalytic	Germany	(Krautkremer, 2017)
2nd	n.s.	50	2017	6	Catalytic	Germany	(Schröer and Krautkremer, 2016)
Power to Gas at Eucolino	n.s.	108	2012[a]	6	Catalytic	Germany	(Dena, 2012b)

Hybrid power plant Falkenhagen—STORE&GO Germany	Alkaline	2000	2013	7	Catalytic (u.c.)	Germany	(DVGW, 2016)
Audi e-gas	Alkaline	6000	2013	7	Catalytic	Germany	(Köbler, 2013)
Foulum	n.s.	250	2013[a]	6	Biological	Denmark	(Byman et al., 2013)
Viessmann microbial methanation	PEM	275	2013[a]	6	Biological	Germany	(Viessmann, 2015)
GRYHD[c]	PEM	n.s.	2014	6		France	(FCB, 2012)
Rozenburg	Alkaline	8.3	2014	6	Catalytic	The Netherlands	(Stedin Netbeheer, 2014; Vlap et al., 2015)
Thüga demonstration plant	PEM	300	2014	7		Germany	(Thüga, 2014)
RWE demonstration plant	PEM	150	2015	7		Germany	(RWE, 2015)
WindGas Hamburg	PEM	1000	2015	7		Germany	(Uniper, 2016)
Energiepark Mainz[d]	PEM	6000	2015	7		Germany	(Energiepark Mainz, 2016)
Energy storage—hydrogen injected into the gas grid via electrolysis field test	PEM	5.5	2015	7		Denmark	(Bruun et al., 2014)

Continued

TABLE 9.1 Overview of Power-to-Gas projects in Europe—cont'd

Acronym/ Location/Name of the Project	Type of Electrolyzer	Capacity kW	Commissioning	TRL	Methanation	Country	Source
Biogas upgrading	SOEC	~50	2016	5	Catalytic	Denmark	(Hansen, 2017)
DemoSNG	PEM	~60	2015	6	Catalytic	Sweden	(Graf, 2014)
BioPower2Gas	PEM	1200	2015	6	Biological	Germany	(IdE, 2015)
Power-to-Gas-pilot plant Allendorf	n.s.	1100	2015	6	Biological	Germany	(Viessmann, 2015)
Energy park Pirmasens-Winzeln	Alkaline	20,500	2016	7	Biological	Germany	(PFI Germany, 2017)
Hybrid power plant Aarmatt—STORE&GO Switzerland	PEM	700	2015	7	Biological	Switzerland	(DVGW, 2016)
BioCat Project/ POWERSTEP	Alkaline	1000	2016	7	Biological	Denmark	(Forstmeier, 2016)
bioCONNECT	PEM	n.s.	2016	5	Biological	Germany	(HS-OWL, 2015)
Ingrid—STORE&GO Italy	PEM	1.000	2016	7	Catalytic	Italy	(DVGW, 2016)
Renovagas	Alkaline SPE	15	2016	6	Biological	Spain	(Rubio et al., 2016)

Integrated High-Temperature Electrolysis and Methanation for Effective Power to Gas Conversion	SOEC	15	2017	5	Catalytic	Germany	(Founti, 2016)
Minerve	SOEC	n.s.	2017	5	n.s.	France	(AFUL, 2017)
CO2-SNG	n.s.	n.s.	2017	6	Catalytic	Poland	(BIOMA, 2015)
HyDeploy	PEM	500	2018	7		United Kingdom	(Markillie, 2016)
Jupiter 1000	Alkaline/ PEM	500+ 500	2018	7	Catalytic	France	(GRTgas, 2016)
Symbio	n.s.	n.s.	~2018	6	Biological	Denmark	(DTU, 2013)
SYNFUEL	SOEC	n.s.	~2019	5	Catalytic	Denmark	(Hendriksen, 2015)
Power-to-Gas Hungary	n.s.	10,000	projected	7–8	Biological	Hungary	(Fründt and Pentz, 2016)
Wijster	n.s.	n.s.	projected	7	Catalytic	The Netherlands	(DNHK, 2014)
Heat and power generation							
HARI	Alkaline	36	2004[a]	6		United Kingdom	(Gammon et al., 2006)
Utsira	Alkaline	50	2004[a]	6		Norway	(Statoil, 2004)
Vestenskov	n.s.	n.s.	2006[a]	6		Denmark	(DAC&CITIES, 2009)

Continued

TABLE 9.1 Overview of Power-to-Gas projects in Europe—cont'd

Acronym/Location/Name of the Project	Type of Electrolyzer	Capacity kW	Commissioning	TRL	Methanation	Country	Source
RES2H2 Pozo Izquierdo	Alkaline	100	2007[a]	6		Spain	(Argumosa and Cambreleng, 2009)
Hídrolica	PEM	30	2007[a]	6		Spain	(Andalusian Energy Agency, 2011)
HYRES	PEM	4.5	2008	6		Greece	(Voutetakis, 2015)
Abalone Energie	Alkaline	n.s.	2009	6		France	(Hydrogenics, 2010)
H_2KT	Alkaline	100	2010	6		Denmark	(Nielsen, 2010)
Myrthe	PEM	210	2013	6		France	(FCB, 2014)
RH2-WKA Grapzow	Alkaline	1.000	2013	7		Germany	(WIND-WASSERSTOFF-projekt, 2016)
La Croix Valmer	PEM	n.s.	2014	7		France	(Areva, 2015)
Spring Bank Farm[c]	Alkaline SPE	4.8	2014	6		United Kingdom	(Acta, 2016)
El Tubo	Alkaline SPE	2.4	2015	6		Spain	(Acta, 2016)

Exytron demonstration project	Alkaline	21	2015	7	Catalytic	Germany	(Dena, 2016a)
Zero-Emission-Wohnpark	Alkaline	62.5	2017	7–8	Catalytic	Germany	(Dena, 2016b)
Stromlückenfüller							(Rentzing, 2016)
Test	PEM	20	2015	6		Germany	
Pilot phase	PEM	200	projected	7		Germany	
Hybrid plant	PEM	1.000	projected	7		Germany	
Power-to-Gas Haßfurt	PEM	1.250	2016	7		Germany	(Stadwerk Haßfurt, 2015)
Smart Grid Solar	PEM	75	2016	6		Germany	(Gossens, 2016)
Power-to-Flex	n.s.	n.s.	2018	6–7	Biological	Germany/ The Netherlands	(Power to Flex, 2016)
HYPOS LocalHy	Alkaline	250	2018	7		Germany	(HYPOS, 2015a)
Fuels							
HyFLEET:CUTE Hamburg	Alkaline	400	2003[a]	6		Germany	(HyFLEET:CUTE, 2009)
HyFLEET:CUTE Amsterdam	Alkaline	400	2003[a]	6		The Netherlands	(HyFLEET:CUTE, 2009)
HyFLEET:CUTE Barcelona	Alkaline	400	2003[a]	6		Spain	(HyFLEET:CUTE, 2009)

Continued

TABLE 9.1 Overview of Power-to-Gas projects in Europe—cont'd

Acronym/ Location/Name of the Project	Type of Electrolyzer	Capacity kW	Commissioning	TRL	Methanation	Country	Source
HyFLEET:CUTE Stockholm	Alkaline	400	2003[a]	6		Sweden	(HyFLEET:CUTE, 2009)
ECTOS	Alkaline	300	2003[a]	6		Iceland	(Icelandic New Energy, 2013)
PURE	Alkaline	18	2005[a]	6		United Kingdom	(Lumsden, 2011)
RES2H2 Keratea	Alkaline	25	2005[a]	6		Greece	(Varkaraki, 2009)
Chic Aargau	Alkaline	300	2011[a]	7		Switzerland	(PostAuto, 2012)
George Olah Plant							
1.	Alkaline	~1700	2011	7	[h]	Iceland	(Steffansson, 2015)
2.	Alkaline	~5200	2014	7	[h]	Iceland	(Steffansson, 2015)
Hydrogen refueling station HafenCity	Alkaline	600	2012	7		Germany	(Wulf et al., 2011)
Solar hydrogen filling station Freiburg	PEM	30	2012	7		Germany	(Fraunhofer ISE, 2012)
Herten hydrogen center of excellence	Alkaline	280	2013	7		Germany	(Brautmeier, 2015)

Hydrogen filling station Stuttgart	Alkaline	400	2013	7		Germany	(EnBW, 2017)
Sunfire Research project	SOEC	10	2014[a]	5	Fischer-Tropsch	Germany	(BMBF, 2015)
Sunfire Power-to-Liquids	SOEC	150	2014	5	Fischer-Tropsch	Germany	(Jendrischik and Aldag, 2014)
Multienergy fueling station H_2BER[e]	Alkaline	500	2014	7		Germany	(Total, 2014)
Power to Gas Biogasbooster[f]	n.s.	10	2014	6	Biological	Germany	(Dena, 2015)
M1 Wind Hydrogen Refueling station	PEM	100	2015	7		United Kingdom	(Pearce, 2015)
Wind2Hydrogen	PEM	100	2015	6		Austria	(OMV, 2015)
Power-2-Hydrogen-Tankstelle	PEM	185	2015	7		Germany	(Falk, 2016)
H2 Aberdeen: Hydrogen Bus Project	Alkaline	1000	2015	7		United Kingdom	(Aberdeen City Promotions, 2016)
Rapperswil	Alkaline	25	2015	6	Catalytic	Switzerland	(IET, 2017)
Levenmouth Community Energy Project[g]	Alkaline/PEM	$2 \times 60+250$	2016	7		United Kingdom	(Todd, 2016)

Continued

TABLE 9.1 Overview of Power-to-Gas projects in Europe—cont'd

Acronym/ Location/Name of the Project	Type of Electrolyzer	Capacity kW	Commissioning	TRL	Methanation	Country	Source
HyFive London 1	PEM	100	2016	7		United Kingdom	(HyFive, 2016; Pearce, 2015)
HyFive London 2	PEM	100	2016	7		United Kingdom	
HyFive London 3	PEM	100	2017	7		United Kingdom	
Flagship project: Power-to-Gas Baden-Württemberg	Alkaline	1000+300	2017	7		Germany	(Del Regno and Vartmann, 2016)
MefCO2	PEM	1000	2018	6	[h]	Germany	(Spire, 2016; Steffansson, 2015)
Kitzingen	n.s.	4400–14,000	projected	7	n.s.	Germany	(Dieter, 2014)
Hydrogen for industry							
Sotavento	Alkaline	300	2007[a]	6		Spain	(Correas and Aso, 2010; Sotavento Galicia, 2017)
CO2RRECT	n.s.	300	2013[a]	7		Germany	(Rieks, 2011)
Osshy Pushy	Alkaline	60	2013	6		France	(Afhypac, 2014)
Lashy Pushy	Alkaline	65	n.s.	6		France	(Afhypac, 2014)

Hanau	PEM	35	2015	7		Germany	(Focht, 2015)
H_2Orizon[b]	PEM	1000	2017	7		Germany	(Andersen, 2016b, a)
GrInHy	SOEC	150	2018	6		Germany	(GrinHy, 2016)
Delfzijl	n.s.	12,000	projected	7		The Netherlands	(Port of Hamburg, 2016)
Usage of gas not specified							
ITHER	Alkaline/ PEM	63+7	2010[a]	6		Spain	(Correas and Aso, 2010)
SEE	PEM	6	2011[a]	6		Germany	(Fraunhofer ISE, 2014)
PtG 250	Alkaline	250	2012	6	Catalytic	Germany	(Zuberbühler, 2011)
Hydrogen Centre	Alkaline/ PEM	55+12+~1	2012	6		United Kingdom	(University of South Wales, 2017; Armata, 2013)
MeGa-stoRE							(Tornberg, 2015; Møller et al., 2015)
1	Alkaline	6	2014[a]	6	Biological	Denmark	
2	Alkaline	60	2016	6	Biological	Denmark	
ESI Platform	PEM	100	2016	6	Catalytic	Switzerland	(Büchi and Schmidt, 2015)

Continued

TABLE 9.1 Overview of Power-to-Gas projects in Europe—cont'd

Acronym/ Location/Name of the Project	Type of Electrolyzer	Capacity kW	Commissioning	TRL	Methanation	Country	Source
Underground Sun Storage[i]	Alkaline	600	2016	6		Austria	(Bauer, 2016)
HYPOS rSOC	SOEC	n.s.	projected	5		Germany	(HYPOS, 2015b)

Note: *TRL: Technology Readiness Level; u.c.: under construction; n.s.: not specified; PEM: polymer electrolyte membrane; SOEC: solid oxide electrolysis cell; SPE: solid polymeric electrolyte.*
[a]*Plant decommissioned.*
[b]*H_2 also for mobility and CHP.*
[c]*H_2 also for mobility.*
[d]*H_2 also for mobility and industry.*
[e]*H_2 also for CHP, industry, and natural gas grid.*
[f]*CH_4 also for natural gas grid.*
[g]*H_2 also for natural gas grid and re-electrification.*
[h]*Methanol production.*
[i]*Main objective: underground storage of H_2.*

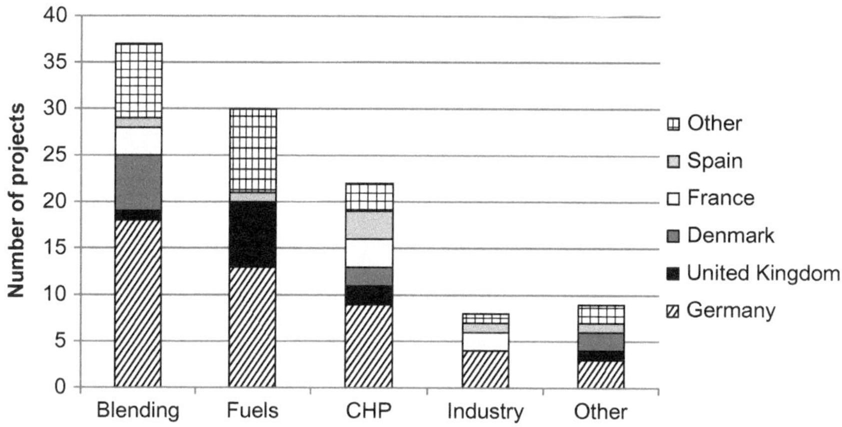

FIG. 9.3 Spatial distribution of Power-to-Gas technologies regarding field of application.

Kingdom, however, focus on certain fields of application. The United Kingdom, for example, conducts PtG mainly for fuel production ('Fuels'), while this field is not covered in Denmark and France. On the other hand, Denmark has a focus on blending hydrogen or methane into the natural gas grid (Blending), and France as well as Spain have a focus on electricity and heat production (CHP).

In the 106 projects, different types of electrolyzers are used. In addition to the alkaline, PEM, and SOEC electrolyzers described above (Section 9.2.1), a new technology has been developed, the so-called alkaline solid polymeric membrane (SPE) electrolyzer (Acta, 2016). It combines an alkaline with a PEM electrolyzer by using a solid membrane and an alkaline environment in one of the electrodes. By avoiding expensive metals for the electrodes, it should be cheaper than a PEM electrolyzer but better in part load behavior than an alkaline electrolyzer. In most of the projects, alkaline electrolyzers are used (41%) because the technology is more mature. The temporal development of the use of electrolyzers shows that the first projects were all operated with an alkaline electrolyzer. The first PEM electrolyzer was used in 2007 and other kinds of electrolyzers were not introduced before 2014. Even though, since 2014, the majority of the projects are no longer operated with alkaline electrolyzers, they are still an important technology for PtG.

The different characteristics of the electrolyzers influence the choice of electrolyzers for the different fields of application discussed. Fig. 9.4[2] shows that, at least for alkaline and PEM electrolysis, clear preferences are apparent. When the main goal is fuel production, alkaline electrolyzers are clearly favored. Fueled vehicles rely on a steady supply of fuel. For a mostly constant hydrogen production, alkaline electrolyzers are preferred still. When blending hydrogen or methane into the natural gas grid, the supply does not need to be that steady and more PEM than alkaline electrolyzers are used to buffer electricity from fluctuating renewable energy sources because PEM electrolyzers

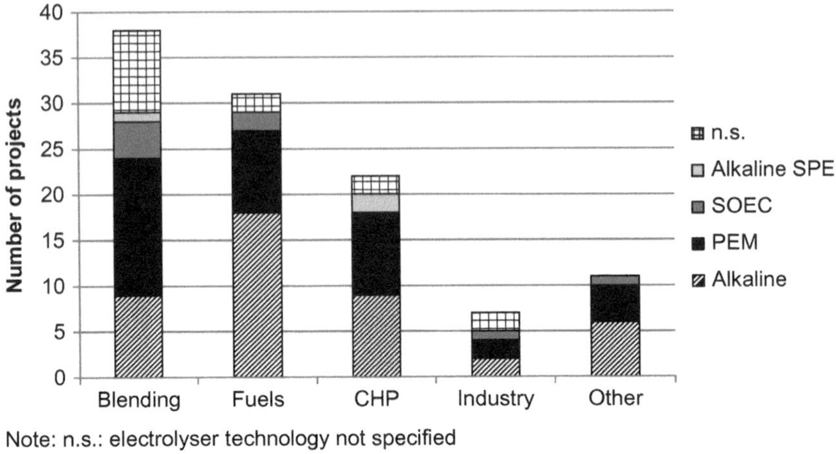

FIG. 9.4 Electrolyzer technologies in Power-to-Gas regarding field of application.

show a better dynamic behavior. This is also true when the hydrogen is used for electricity and heat generation. In those systems, often an alternative heat or electricity source is installed. Thus, the system can act more flexibly. Because the number of SOEC and alkaline SPE electrolyzers remains small, no clear trend can be identified yet.

Next to the field of application, the capacity (total input power) of the electrolyzer is an interesting parameter to categorize the projects. In this analysis, the size of the electrolyzers is grouped into five classes, see Fig. 9.5. Most of the electrolyzers have a capacity between 5 and 100 kW (36%) and are alkaline electrolyzers. In the smallest class (<5 kW), mostly PEM electrolyzers are used

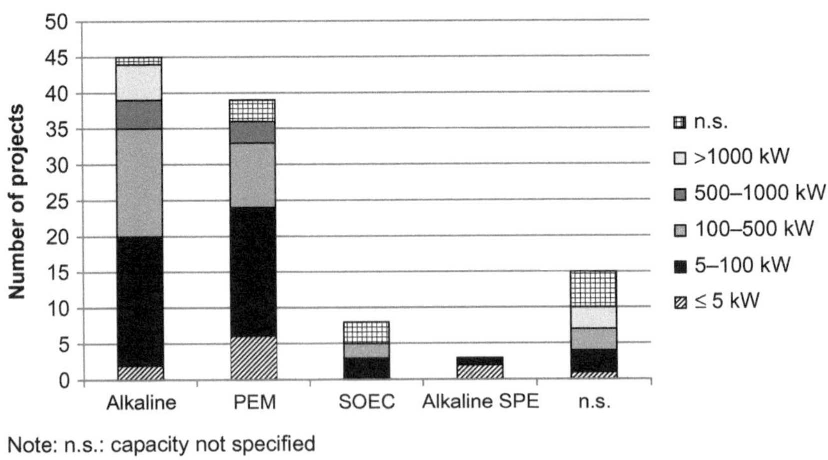

FIG. 9.5 Electrolyzer technologies in Power-to-Gas regarding their capacity.

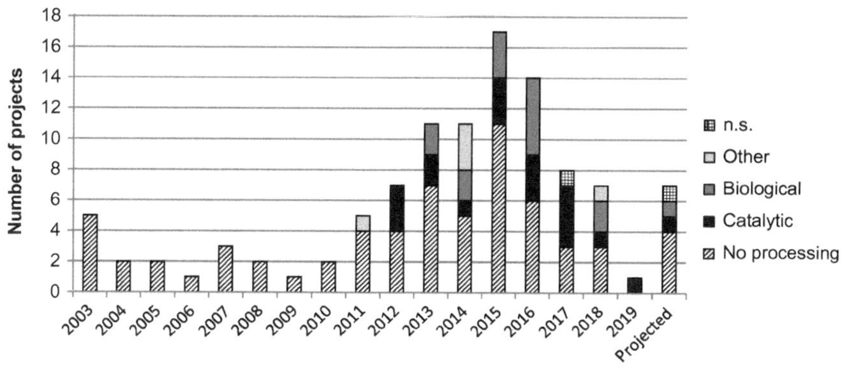

Note: n.s.: methanation technology not specified

FIG. 9.6 Temporal development of methanation in Power-to-Gas.

for more research-oriented projects. SOEC electrolyzers are not realized in such a small capacity. On the other extreme, the largest electrolyzers (>1000 kW) are only found among the alkaline electrolyzers and in some projects in which the technology is not specified.

Another important application within PtG is the processing of hydrogen to methane or other chemicals. In the 106 projects, hydrogen is directly used in 65 of them (61%). In 2011, the first processing of hydrogen to methanol (in Fig. 9.6 subsumed under 'other' together with the Fischer-Tropsch process) was conducted in the George Olah Plant in Iceland (Steffansson, 2015). In the subsequent years, processing of hydrogen, and mostly methanation, became more and more popular. Even though, catalytic methanation is more mature, almost as many projects with biological methanation (15 projects) are being operated as with catalytic methanation (19 projects). Both started roughly at the same time 2012/13. Since 2016 even more PtG projects are processing hydrogen further to other chemicals than using it directly. Sorting the kind of product by country, it becomes obvious that Denmark is very interested in methanation (70% of all projects include methanation). In particular, there is focus on biological methanation. Only Germany has more projects investigating biological methanation.

9.4 PtG IN ENERGY SUPPLY SCENARIOS

In current energy scenarios, storage options, such as PtG, are the subject of analysis. Nearly all scenarios with a focus on energy storage assume a future trend toward increasing storage capacity in conjunction with more REN in the electricity generation. The quantification of the required storage capacity in the scenarios depends on many assumptions and is subject to current research work.

A survey of energy scenarios for the world, the EU, and the national examples France, Sweden, and Germany investigates the range of calculated

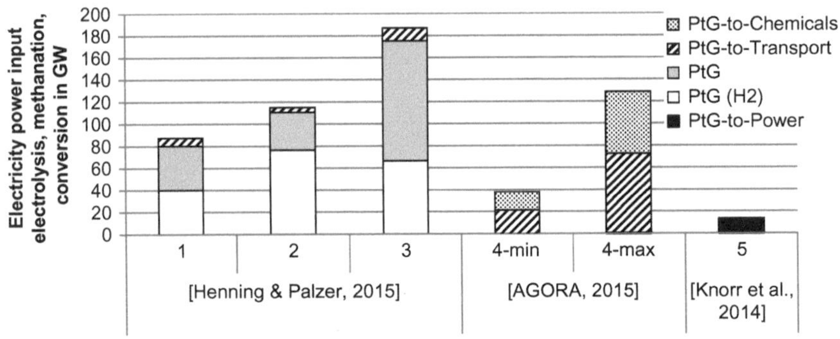

FIG. 9.7 Comparison of PtX capacities in different energy supply scenarios for Germany, year 2050, own assumptions (AGORA, 2015; Henning and Palzer, 2015; Knorr et al., 2014).

storage capacities with a special focus on the seasonal storage option by PtG. Important impacts on the results identified are in particular the share of REN in electricity generation, the composition of the conventional power plant mix, grid constraints, electricity export/import, storage costs, and carbon dioxide emission caps.

The IEA scenario study "Energy technology perspective (ETP)" (IEA, 2014) is accepted worldwide and is a reference for many national energy scenarios. The focus is on the application of established as well as new energy technologies and their future potential for climate protection. The 2014 version has a spotlight on electricity storage. For the 2DS scenario (ambitious scenario with a maximum global warming of 2°C worldwide resulting in high shares of REN worldwide and decarbonization in electricity generation in 2050), the trends for daily storage capacity are calculated. The installed capacity will reach over 300 GW in the regions of China, India, United States, and EU in 2050. This is more than four times the installed capacity of daily storage in 2011 for these regions. PtG is described as a technical option that can significantly increase the level of REN integration in the energy system by connecting the power system to the natural gas system, thus making use of the already existing storage potential of the natural gas infrastructure. However, the capacity of PtG applications in the scenarios is not quantified.

A European study supported by the Fuel Cell and Hydrogen Joint Undertaking (FCH JU) (FCH-JU, 2015) of the year 2015 analyzes the development of the European electric power system and the need for the installation of energy storage options. For daily storage (PtG-to-Power), the analysis provides scenario results for installed capacities in Germany, Spain, and Sweden. For PtG options (using excess electricity in the electricity, heat, and transport sectors) the installed electrolyzer capacity is calculated for Germany as an example. In the high-renewables scenario (85% share of REN in electricity generation in 2050), the calculated electrolyzer capacity is between 115 GW for no grid constraints and 170 GW for essential constraints in the transport and distribution grid. The European potential for electrolyzer capacity in 2050 of the high-renewables scenario is stated as hundreds of GWs for the full integration of excess electricity. The calculated capacities in this study are very high because a full use of REN excess electricity is assumed and no curtailment or flexibility of demand is taken into account. The demand for storage differs significantly among the analyzed countries. In contrast to Germany and Spain, Sweden has large reservoir hydro capacities, which can integrate REN and eliminate the need for daily and seasonal storage options. There is no economic demand for REN integration in Sweden, even in the high-REN scenario.

The French study (ADEME, 2015) investigates scenarios for the French electricity sector and its possible pathway to a 100% REN share in a time horizon to 2050. The applied model optimizes the generation mix and the design of the electric grid and takes into account regional effects on a one-hour time resolution for a whole year. In this study only PtG-to-Power is analyzed as seasonal storage. For a share of 80% REN in the generation mix (46.5% wind, 14.3% hydro, 12.5% PV and 6.7% biomass and others) the optimization results show no need for the installation of the seasonal storage option PtG. The installed capacity for hourly and daily storage options (mainly pumped hydro storage is assumed) is calculated to be 15 GW in the scenario year 2050 for France. Flexible generation, grid improvement, and flexibility in electricity demand reduce the need for additional storage options. For the scenario with a 95% and 100% share of REN in French electricity generation in 2050 (increased share of PV and wind onshore, wind offshore/offshore floating), the optimization results show a need for installed capacity of 9 and 17 GW, respectively, for PtG-to-Power options. The need for long-term seasonal storage options therefore is strongly connected to the share of REN generation in the electricity system.

For Germany, the largest electricity consumer in the EU with ambitious goals regarding REN electricity generation, a large set of scenario calculations exist with a focus on storage demand. Some of the scenarios take into account PtX for the integration of REN in the sectors of electricity, heat, or transport. A detailed look at three different studies is used for a comparison of the prospects of PtX applications in the German energy system. The selection of these scenarios is based on suitability for the comparison by detailed information about the share of REN electricity and mode of PtX applications.

(A) Henning and Palzer (2015)
The study analyzes the development of the current German energy system through 2050 to comply with the goals of the German "Energiewende." On the basis of nine scenarios (among other assumptions different shares of REN in electricity generation) cost-optimized transformation paths are indicated. The scenarios take into account PtX for electricity, heat, and transport.

(B) AGORA (2014)
The study investigates the need for daily and seasonal storage by means of cost optimization for the transformation of the current German energy systems toward a sustainable one in light of the "Energiewende" concept.

The different scenarios assume different REN shares in electricity generation for Germany and Europe. For the year 2050, a share of 60% (Europe) and 90% (Germany) could be reached.

(C) Knorr et al. (2014)
The study analyzes an electricity supply system in Germany that completely relies on REN. The question that this study seeks to answer is whether the security of the electricity supply can be guaranteed at any time of the year. The analysis has a strong focus on the electricity system but takes into account the demand for heat and transport with respect to its flexibilities.

The calculated PtG capacities in five scenarios of the three German studies are summarized in Table 9.2. The selected set of scenarios for the German energy system in the year 2050 indicates a large spectrum of results in terms of installed PtX capacities and their mode and coupling to different sectors.

In the three selected scenarios with 80%, 85%, and 90% carbon dioxide reduction, Henning and Palzer (2015) assume an installed PtG capacity between ~90 and 180 GW for the year 2050. The high capacities arise from the need for guaranteed energy supply with high shares of REN in conjunction with very low carbon dioxide emissions. In the most ambitious scenario, #3, PtG is the dominant option due to the substitution of fossil natural gas in domestic and transport applications. Power-to-Hydrogen delivers transportation fuel and additional blend to natural gas.

The study AGORA (2014) gives a spectrum of installed capacities due to different assumptions about flexible generation, curtailment, transportation, and distribution grid enhancement and flexible demand. The PtG-to-Transport option is seen in the study as more cost effective in comparison to PtG-to-Power. The minimum capacity for the latter option is at a very low level, below 1 GW installed capacity.

The study by Knorr et al. (2014), with its focus on the electricity system, calculates 13 GW for PtG-to-Power options for a guaranteed electricity supply by 100% REN generation. Seasonal storage by PtG-to-Power is seen to be in competition with fast responding decentralized biomass power plants and flexible demand in this scenario.

A more detailed look at the scenario assumptions reveals differences in the share of REN in electricity generation, REN installed capacities, and carbon dioxide reduction goals. Especially the costs and application of flexible generation and demand is one of the assumptions that determines the results.

High shares of REN electricity generation lead in all scenarios to a large amount of excess electricity. All of the analyzed scenarios show a potential for PtG in transport or chemical applications. The selected PtG option depends on the assumed cost structure of the entire energy conversion chain, as well as the costs of competitive options, such as flexibility of generation or demand.

TABLE 9.2 Comparison of Selected Scenario Assumptions (AGORA, 2015; Henning and Palzer, 2015; Knorr et al., 2014)

# (Referring Fig. 9.7)	1	2	3	4-min	4-max	5
Source	Henning and Palzer (2015)			AGORA (2015)		Knorr et al. (2014)
Assumed REN share in electricity generation	Not specified	Germany: 85%	Not specified	Germany: 90% Europe: 60%		Germany: 100% Europe: 100%
Assumed PV and wind generation capacity in GW	293	367	536	279		260
Application of PtX	Power, gas, fuel	Power, gas, liquid, transport	Power, gas, liquid, transport	Power, fuel, chemicals		Power
Other assumptions	80% CO_2 reduction; see scenario #2	85% CO_2 reduction, ambitious goals for energy efficiency in buildings, alternative drive trains in transportation, phase out of coal power plants till 2040	90% CO_2 reduction, high excess electricity, see scenario #2;	Minimum potential of flexible generation, flexible demand, curtailment, grid enhancement, resulting in low capacities of PtX	Maximum potential of flexible generation, flexible demand, curtailment, grid enhancement, resulting in high capacities of PtX	fast responding decentralized power plants and demand side management

The comparison of studies conducted here shows no clear trend on cost effectiveness because the scenario assumptions are not comparable and influence the results in an essential way.

9.5 OUTLOOK

In the previous sections it was revealed that the technologies for PtG, as well as the applications, are diversified. If production of fuels and chemicals is the main goal of PtG, then alkaline electrolyzers are preferred. If, in contrast, the use of excess electricity and ancillary services for the electricity grid are the focus of the project, then PEM electrolyzers are, despite their higher costs, used more often. Furthermore, new electrolyzers (SOEC, SPE) are under development to increase the efficiency. To avoid the limitations on hydrogen that can be fed into the natural gas grid, another focus is on the production of other gases, for example, by methanation and coelectrolysis. The technologies for that are quite different. However, up to now, all technologies have, under certain requirements, a right to exist.

Looking into the future and the integration of PtX into energy systems, the international energy scenario analyses reveal the potential application of PtX options in scenarios for energy systems with high (>80%–85%) REN penetration. The quantification of capacities, however, depends strongly on local requirements of the generation, grid, and demand. All scenarios highlight the use of PtG to significantly increase the interactions between the gas, power, and transport sectors, as well the option to support uses unrelated to energy, such as chemicals and other products.

REFERENCES

Aberdeen City Promotions, 2016. H2 Aberdeen: Hydrogen Bus Project [Online]. Aberdeen City Council, Aberdeen. Available: http://www.aberdeeninvestlivevisit.co.uk/Invest/Aberdeens-Economy/City-Projects/H2-Aberdeen/Hydrogen-Bus/Hydrogen-Bus-Project.aspx (Accessed 9 January 2017).

Acta, 2016. Powered by Nature [Online]. Crespina Acta Spa. Available: http://www.actaspa.com/ (Accessed 18 January 2017).

ADEME, 2015. Un mix électrique 100% renouvelable ? Analyses et optimisations. Final report, Agence de l'Environment et de la Maîtrise de l'Energie, Angers Cedex.

Afhypac, 2014. Rapport d'activités de la france sur l'hydrogène et les piles à combustible. Association française pour l'hydrogène el les piles à combustible (Afhypac), Paris.

AFUL, 2017. MINERVE: Un démonstrateur power-to-gas pour expérimenter l'avenir. Association Fonciére Urbaine libre AFUL Chantrerie, Nantes.

AGORA, 2014. Stromspeicher in der Energiewende: Untersuchung zum Bedarf an neuen Stromspeichern in Deutschland für den Erzeugungsausgleich, Systemdienstleistungen und im Verteilnetz. AGORA Energiewende, Berlin.

AGORA, 2015. Stromspeicher in der Energiewende. AGORA Energiewende, Berlin.

Andalusian Energy Agency, 2011. Status of Hydrogen and Fuel Cell Technologies in Andalusia. Andalusian Energy Agency, Regional Ministry of Economy, Innovation and Science, Seville.

Andersen, T., 2016a. H_2ORIZON – DLR investiert in Wasserstoff-Standort. HZwei 07 (2016), 16–17.

Andersen, T., 2016b. H_2ORIZON: Hydrogen at the DLR Site Lampoldshausen. Hardthausen, German Aerospace Center (DLR).

Areva, 2015. AREVA's Energy Storage Projects [Online]. Areva, Paris. Available: http://www.areva.com/EN/operations-6831/hydrogen-fuel-cells-and-energy-storage.html (Accessed 10 March 2017).

Argumosa, M.P., Cambreleng, T., 2009. RES_2H_2 Spanish Site. Instituto Tecnológico de Canarias, Las Palmas.

Armata, M., 2013. Hydrogenics Selected References: Grid Balancing, Power to Gas (PtG). Hydrogenics, Mississauga.

Bailera, M., Lisbona, P., Romeo, l.M., Espatolero, S., 2017. Power to gas projects review: Lab, pilot and demo plants for storing renewable energy and CO_2. Renew. Sust. Energ. Rev. 69, 292–312.

Bär, K., Mörs, F., Götz, M., Graf, F., 2015. Vergleich der biologischen und katalytischen Methanisierung für den Einsatz bei PtG-Konzepten. gwf - Gas|Erdgas 156, 466–473.

Bauer, S., 2016. Underground Sun Storage: Den Sonnenschein speichern. RAG Rohöl-Aufsuchungs Aktiengesellschaft, Vienna.

Benjaminsson, G., Benajminsson, J., Boogh Rudberg, R., 2013. Power-To-Gas—A Technical Review. Svenskt Gastekniskt Center AB, Malmö.

Bertuccioli, L., Chan, A., Hart, D., Lehner, F., Madden, B., Standen, E., 2014. Study on Development of Water Electrolysis in the EU Lausanne: Fuel Cells and Hydrogen Joint Undertaking FCH-JU.

BIOMA, 2015. TAURON: Nowatorska instalacja zagospodarowania CO2 w 2017 roku [Online]. BIOMA Odnawialne Źródła Energii, Jastrzębia. Available: http://odnawialnezrodlaenergii.pl/tech/item/2086-tauron-nowatorska-instalacja-zagospodarowania-co2-w-2017-roku (Accessed 24 January 2017).

BMBF, 2015. Projektbeschreibung Sunfire Geförderte Projekte im Rahmen der BMBF-Fördermaßnahme. In: Technologien für Nachhaltigkeit und Klimaschutz - Chemische Prozesse und stoffliche Nutzung von CO2. German Federal Ministry of Education and Research, Berlin.

Bourasseau, C., Guinot, B., 2016. Hydrogen: a storage means for renewable energies. In: Godula-Jopek, A. (Ed.), Hydrogen Production by Electrolysis. Wiley-VCH, Weinheim, pp. 311–382.

Brautmeier, P., 2015. Herten Pits on Hydrogen [Online]. HTVG, Herten. Available: http://www.wasserstoffstadt-herten.de/Wasserstoffstadt-Herten.99.0.html?&L=1 (Accessed 9 March 2017).

Bruun, J., Graf, T., Iskov, H., Koch, B., 2014. Energy Storage—Hydrogen Injected into the Gas Grid Via Electrolysis Field Test. Energinet.dk, Fredericia.

Büchi, F.N., Schmidt, T.J., 2015. ESIP: Energy Storage using Hydrogen. Energy System Integration (ESI) Platform. Villigen Paul Scherrer Institut.

Byman, K., Haraldsson, Jernelius, S., 2013. Power to Gas: Internationell utblick och potentialen i Sverige. ÅF-Infrastructure AB, Stockholm.

Correas, L., Aso, I., 2010. Stolten, D., Grube, T. (Eds.), Task 24: Wind energy and hydrogen integration. 18th World Hydrogen Energy Conference. Forschungszentrum Jülich GmbH, Zentralbibliothek, Verlag, Essen.

DAC&CITIES, 2009. Vestenskov: The world's first hydrogen community [Online]. Danish Architecture Centre, Copenhagen. Available: http://www.dac.dk/en/dac-cities/sustainable-cities/all-cases/energy/vestenskov-the-worlds-first-hydrogen-community/ (Accessed 5 January 2017).

Del Regno, A., Vartmann, A., 2016. Power-to-Gas-Leuchtturmprojekt: Betreiber und Standort gefunden [Online]. Zentrum für Sonnenenergie- und Wasserstoff-Forschung Baden-Württemberg (ZSW), Stuttgart.

Dena, 2012a. *H2-Forschungszentrum der BTU Cottbus* [Online]. Deutsche Energie-Agentur GmbH (dena), Berlin. Available: http://www.powertogas.info/power-to-gas/pilotprojekte-im-ueberblick/h2-forschungszentrum-der-btu-cottbus/ (Accessed 8 March 2017).

Dena, 2012b. Viessmann Power-to-Gas im Eucolino am Standort Schwandorf [Online]. Deutsche Energie-Agentur dena, Berlin. Available: http://www.powertogas.info/power-to-gas/pilotprojekte-im-ueberblick/viessmann-power-to-gas-im-eucolino/ (Accessed 8 March 2017).

Dena, 2015. Power to Gas Biogasbooster [Online]. Deutsche Energie-Agentur dena, Berlin. Available: http://www.powertogas.info/power-to-gas/pilotprojekte-im-ueberblick/power-to-gas-biogasbooster/ (Accessed 9 March 2017).

Dena, 2016a. EXYTRON Demonstrationsanlage [Online]. Deutsche Energie-Agendur dena, Berlin. Available: http://www.powertogas.info/power-to-gas/pilotprojekte-im-ueberblick/exytron-demonstrationsanlage/ (Accessed 10 March 2017).

Dena, 2016b. Exytron Zero-Emission-Wohnpark [Online]. Deutsche Energie-Agentur, Berlin. Available: http://www.powertogas.info/power-to-gas/pilotprojekte-im-ueberblick/extyron-zero-emission-wohnpark/ (Accessed 9 March 2017).

Dena, 2017. Pilotprojekte [Online]. Deutsche Energie-Agentur GmbH (dena), Berlin. Available: http://www.powertogas.info/power-to-gas/pilotprojekte-im-ueberblick/?no_cache=1 (Accessed 9 January 2017).

Dieter, R., 2014. Stadtrat bringt 15 Millionen-Projekt auf den Weg. Main Post 21 (02), 2014.

DNHK, 2014. Absichtserklärung für Power-to-Gas-Anlage in Wijster unterzeichnet [Online]. Den Haag, Deutsch-Niederländischen Handelskammer (DNHK). Available: http://www.dnhk.org/news/single-view/artikel/absichtserklaerung-fuer-power-to-gas-anlage-in-wijster-unterzeichnet/?cHash=09397eb74385a51812fd1ffabcdb8147 (Accessed 4 January 2017).

DTU, 2013. SYMBIO - Biogasupgrade [Online]. Technical University of Denmark, Lyngby. Available: http://www.biogasupgrade.dk/ (Accessed 16 January 2017).

DVGW, 2015. Wo aus Wind und Sonne Grünes Gas Wird. Deutscher Verein des Gas- und Wasserfaches e. V.; greenfacts, Bonn.

DVGW, 2016. STORE&GO [Online]. DVGW Deutscher Verein des Gas- und Wasserfaches e. V, Bonn. Available: http://www.storeandgo.info/ (Accessed 19 December 2016).

Ebbesen, S.D., Mogensen, M., 2009. Electrolysis of carbon dioxide in solid oxide electrolysis cells. J. Power Sources 193, 349–358.

EnBW, 2017. Wasserstofftankstelle: Wasserstoff – Energiespeicher und Kraftstoff fürs Auto [Online]. EnBW Energie Baden-Württemberg, Karlsruhe. Available: https://www.enbw.com/unternehmen/konzern/forschung/energiesystem/wasserstofftankstelle/index.html (Accessed 9 March 2017).

Energielandschaft Morbach, 2015. Stromspeicherung - Ökostrom zu Methangas [Online]. Morbach: Gemeindeverwaltung Morbach (Accessed 08.03.17).

Energiepark Mainz, 2016. Turning Wind into Gas [Online]. Stadtwerke Mainz, Mainz. Available: http://www.energiepark-mainz.de/en/ (Accessed 8 March 2017).

Enertrag, 2015. Hybrid Power Plant [Online]. Enertrag, Dauerthal. Available: https://www.enertrag.com/90_hybridkraftwerk.html?&L=1 (Accessed 8 March 2017).

European Powertogas, 2017. Power-to-gas (Demonstration) Projects in Europe [Online]. European Power to Gas Platform, Groningen. Available: http://www.europeanpowertogas.com/demonstrations (Accessed 9 January 2017).

Falk, H., 2016. Möglichkeiten der Industrieversorgung aus Erneuerbaren Energien. Niedersächsische Energietage, 1.-2.11.2016. Goslar. Energie-Forschungszentrum Niedersachsen.

FCB, 2012. GDF Suez, McPhy in French GRHYD project on methane, hydrogen. Fuel Cells Bull. 2012, 10.

FCB, 2014. MYRTE hydrogen energy storage test powers up in Corsica. Fuel Cells Bull. 2014, 8.

FCH-JU, 2015. Commercialisation of Energy Storage in Europe: A Factbased Analysis of the Implications of Projected Development of the European Electric Power System Towards 2030 and Beyond for the Role and Commercial Viability of Energy Storage. Fuel Cell and Hydrogen Joint Undertaking FCH-JU, Brussels.

Focht, P., 2015. Grüner Wasserstoff für Hanauer Industriepark Energie & Management. Energie & Management Verlagsgesellschaft mbH, Herrsching.

Foit, S., Eichel, R.-A., Vinke, I.C., De Haart, L.G.J., 2016. Power-to-Syngas—an enabling technology for the transition of the energy system? Production of tailored synfuels and chemicals using renewably generated electricity. Angew. Chem. Int. Ed. 56, 5402–5411.

Forstmeier, M., 2016. Erneuerbare Gase – Wohin geht die Reise? Erfa-Tag Biogas, Zürich.

Founti, M., 2016. Energy Storage Technologies: Focus on Power-To-Gas Technology. Workshop at NTUA National Technical University of Athens, Athens.

Fraunhofer ISE, 2012. Solare Wasserstofftankstelle Freiburg: H2Move - Informationen und Beschreibung. Fraunhofer Institute for Solar Energy Systems ISE, Freiburg.

Fraunhofer ISE, 2014. Methane Storage—Storage of Electric Energy from Renewable Sources in the Natural Gas Grid: H2O Electrolysis and Synthesis of Gas Components [Online]. Fraunhofer Institute for Solar Energy Systems ISE, Freiburg. Available: https://www.ise.fraunhofer.de/en/research-projects/methane-storage.html (Accessed 24 January 2017).

Fründt, R., Pentz, T.-Å.P., 2016. Münchner Cleantech-Startup Electrochaea und ungarischer Energieversorger MVM gründen Power-to-Gas-Joint Venture. Electrochaea, Munich.

Gammon, R., Roy, A., Barton, J., Little, M., 2006. Hydrogen and Renewables Integration (HARI). Centre for Renewable Energy Systems Technology, Loughborough University, Loughborough.

Gossens, S., 2016. Innovative Wasserstoff- Speichersysteme unter Nutzung der LOHC-Technologie. RheinMain Blue Cluster 24.05.2016 Höchst. Areva.

Götz, M., McDaniel Koch, A., Graf, F., 2014. State of the art and perspectives of CO_2 methanation process concepts for power-to-gas applications.International Gas Union Research Conference, Copenhagen.

Götz, M., Lefebvre, J., Mörs, F., Mcdaniel Koch, A., Graf, F., Bajohr, S., Reimert, R., Kolb, T., 2016. Renewable power-to-gas: a technological and economic review. Renew. Energy 85, 1371–1390.

Graf, F., 2014. Power to Gas—State of the Art and Perspectives. MARCOGAZ General Assembly: Workshop "New Developments" 04.06.2014, Prague.

GrinHy, 2016. Technology [Online]. Salzgitter Mannesmann Forschung GmbH, Salzgitter. Available: http://www.green-industrial-hydrogen.com/technology/technical-specifications/ (Accessed 4 January 2017).

Grond, L., Holstein, J., 2014. Power-to-gas: Climbing the technology readiness ladde. Gas for Energy 20–25.

Grond, L., Schulze, P., Holstein, J., 2013. Systems Analyses Power to Gas—Deliverable 1: Technology Review. KEMA Nederland B.V., Groningen.

GRTgas, 2016. The Project Jupiter 1000 [Online]. Fos-sur-Mer, GRTgas. Available: http://www.jupiter1000.com/en/projet.html (Accessed 17 January 2017).

Hansen, J.B., 2017. Biogas upgrading in a 50 kilowatt solid oxide cells based demonstration plant—design and operational experiences including strategies to counteract degradation.231st ECS Meeting, 28.5.-1.7.2017, New Orleans.

Hendriksen, P.V., 2015. SYNFUEL [Online]. Technical University of Denmark, Roskilde. Available: http://www.synfuel.dk/ (Accessed 23 January 2017).

Henning, H.-M., Palzer, A., 2015. Was kostet die Energiewende? Wege zur Transformation des deutschen Energiesystems bis 2050, Fraunhofer Institute for Solar Energy Systems ISE, Freiburg.

HS-OWL, 2015. Arbeitsgruppe Bioraffinerie/Workgroup Biorefinery. Ostwestfalen-Lippe University of Applied Sciences HS-OWL, Lemgo.

Hydrogenics, 2010. Hydrogen Energy Storage Applications. Canadian Institute: Energy Storage. 08.07.2010 Toronto. Hydrogenics.

HyFive, 2016. hyFIVE: Hydrogen for Innovative Vehicles [Online]. HyFIVE Project Coordination, London. Available: http://www.hyfive.eu/ (Accessed 9 March 2017).

HyFLEET:CUTE, 2009. Electrolysis [Online]. Fontaine Publishing Group, Fremantle. Available: http://www.global-hydrogen-bus-platform.com/www.global-hydrogen-bus-platform.com/Technology/HydrogenProduction/electrolysis.html (Accessed 9 January 2017).

HYPOS, 2015a. Hydrogen Power Storage & Solutions East Germany—Musterpräsentation. Hydrogen Power Storage & Solutions East Germany, Halle (Saale).

HYPOS, 2015b. Verbundprojekte Steckbriefe 1. Welle. Hydrogen Power Storage & Solutions East Germany, Halle (Saale).

Icelandic New Energy, 2013. Second NewsLetter. Icelandic New Energy, Reykjavik.

IdE, 2015. BioPower2Gas [Online]. Institut dezentrale Energietechnologien gemeinnützige GmbH. Available: http://www.biopower2gas.de (Accessed 26 January 2016).

IEA, 2014. Energy Technology Perspectives 2014—Harnessing Electricity's Potential. International Energy Agency (IEA), Paris.

IET, 2017. Pilot and Demonstration Plant Power-to-Methane [Online]. Hochschule für Technik Rapperswil, Rapperswil. Available: https://www.iet.hsr.ch/index.php?id=13510&L=4 (Accessed 9 March 2017).

Iskov, H., Rasmussen, N.B., 2013. Global Screening of Projects and Technologies for Power-To-Gas and Bio-SNG: A Refernce Report. Danish Gas Technology Centre, Hørsholm.

Jendrischik, M., Aldag, N., 2014. Sunfire präsentiert Power-to-Liquids. Sunfire, Dresden.

Kasten, P., Mottschall, M., Köppel, W., Degünther, C., Schmied, M., Wüthrich, P., 2016. Erarbeitung einer fachlichen Strategie zur Energieversorgung des Verkehrs bis zum Jahr 2050. Umweltbundesamt, Dessau-Roßlau.

Kippers, M.J., De Laat, J.C., Hermkens, R.J.M., Overdiep, J.J., Van Der Molen, A., Van Erp, W.C., Van Der Meer, A., 2011. Pilot project on hydrogen injection in natural gas on island of Ameland in the Netherlands.International Gas Union Research Conference, Seoul.

Knorr, K., Zimmermann, B., Kirchner, D., Speckmann, M., Spieckermann, R., Widdel, M., Wunderlich, M., Mackensen, R., Rohrig, K., 2014. Kombikraftwerk 2 - Abschlussbericht. Fraunhofer Institute for Wind Energy and Energy System Technology, Kassel.

Köbler, J., 2013. Audi Future Lab: Mobility. Audi, Ingoldstadt.

Krassowski, J., Jochum, O., 2016. Biological Methanation: Methane Production by Microbial Conversion of H_2 and CO_2. In: van Basshuysen, R. (Ed.), Natural Gas and Renewable Methane for Powertrains: Future Strategies for a Climate-Neutral Mobility. Springer International Publishing, Cham, pp. 167–181.

Krautkremer, B., 2017. Test Centers | Laboratories [Online]. Fraunhofer Institute for Wind Energy and Energy System Technology, Kassel. Available: https://www.energiesystemtechnik.iwes.fraunhofer.de/en/labore/hbfz.html (Accessed 8 March 2017).

Lehner, M., Tichler, R., Steinmüller, H., Koppe, M., 2014. Methanation. In: Power-to-Gas: Technology and Business Models. Springer International Publishing, Cham, pp. 41–61.

Leonzio, G., 2016. Process analysis of biological Sabatier reaction for bio-methane production. Chem. Eng. J. 290, 490–498.

Lumsden, M., 2011. Promoting Unst Renewable Energy (PURE) Project. Pure Energy Centre, Unst.

Markillie, R., 2016. National Grid HyDeploy Consortium wins £7m Ofgem Funding for UK Power-to-Gas [Online]. ITM Power, Sheffield. Available: http://www.itm-power.com/news-item/national-grid-hydeploy-consortium-wins-7m-ofgem-funding-for-uk-power-to-gas (Accessed 14 March 2017).

Meylan, F.D., Moreau, V., Erkman, S., 2016. Material constraints related to storage of future European renewable electricity surpluses with CO_2 methanation. Energy Policy 94, 366–376.

Møller, P., Yde, L., Bejder, K., Janulis, R., Jensen, J.K., Møller, J.C., 2015. The MeGa-stoRE Energy Project [Online]. Aarhus, methan.dk. Available: http://www.methan.dk/about.html (Accessed 9 March 2017).

Müller-Syring, G., Henel, M., Köppel, W., Mlaker, E., Sterner, M., Höcher, T., 2013. Entwicklung von modularen Konzepten zur Erzeugung, Speicherung und Einspeisung von Wasserstoff und Methan ins Erdgasnetz. DVGW-PRojekt G1-07-10. DVGW, Bonn.

Nielsen, J.H., 2010. Hydrogen Energy Storage in Greenland: Status H2KT Project & possible Next Steps. Nukissorfiit, Nuuk.

OMV, 2015. Factsheet Forschungsprojekt wind2hydrogen. OMV Aktiengesellschaft, Vienna.

Otten, R., Block, T., Hammer, E., 2016. Synthetic methane from renewable electric energy. In: Van Basshuysen, R. (Ed.), Natural Gas and Renewable Methane for Powertrains. Springer International Publishing, Cham, Heidelberg, New York, Dordrecht, London, pp. 131–180.

Pearce, A., 2015. Energy storage | clean fuel | clean air. In: Green Comission. ITM Power, Sheffield.

PFI Germany, 2017. PtG-Anlage in Pirmasens in Betrieb. BWK Das Energiefachmagazin 69, 47.

Port of Hamburg, 2016. WindEnergy Hamburg to Highlight Future Energy Solutions Utilising Smart Networks and Energy Storage Technology [Online]. Port of Hamburg, Hamburg. Available: https://www.hafen-hamburg.de/en/news/windenergy-hamburg-to-highlight-future-energy-solutions-utilising-smart-networks-and-energy-storage-technology—34750 (Accessed 4 January 2017).

PostAuto, 2012. Brennstoffzellenpostautos – Ein emissionsfreier Antrieb für unsere Umwelt. Forum Fribourg, Fribourg 26.04.2012.

Power to Flex, 2016. Warum Energie speichern? [Online]. Provincie Groningen, Groningen. Available: http://www.powertoflex.eu/de/ (Accessed 9 January 2017).

Rentzing, S., 2016. Power-to-Gas: Welche technische Fragen gibt es beim Hoffnungsträger der Energiewende noch zu lösen? [Online]. ee-news. Available: http://www.ee-news.ch/de/wind/politik/article/32912/power-to-gas-welche-technische-fragen-gibt-es-beim-hoffnungstrager-der-energiewende-noch-zu-losen (Accessed 9 March 2017).

Rieks, M., 2011. Verwertung von CO2 als Kohlenstoff-Baustein unter Verwendung überwiegend regenerativer Energie [Online]. INVITE, Leverkusen. Available: http://www.invite-research.com/en/contentseite/aktuelle-forschungsprojekte/co2rrect.html (Accessed 9 March 2017).

Rönsch, S., Schneider, J., Matthischke, S., Schlüter, M., Götz, M., Lefebvre, J., Prabhakaran, P., Bajohr, S., 2016. Review on methanation—from fundamentals to current projects. Fuel 166, 276–296.

Rubio, J., Cortés, P., Escudero, M.T., De Godos, I., Lana, J.A., Navarro, R., Pérez, S., Sánchez, M., 2016. RENOVAGAS: process for the production of renewable natural gas. 21st World Hydrogen Energy Conference, 13.-16.06.2016, Zaragoza, pp. 48–50.

RWE, 2015. Das RWE-Pilotprojekt zur Power to Gas-Technologie in Ibbenbüren. RWE Deutschland, Essen.

Schröer, R., Krautkremer, B., 2016. Direktmethanisierung von Biogas im Technikumsmaßstab: 50 kWel Power-to-Gas-Anlage [Online]. Fraunhofer Institute for Wind Energy and Energy

System Technology, Kassel. Available: https://www.iee.fraunhofer.de/de/projekte/suche/laufende/direktmethanisierung_ptg.html (Accessed 8 March 2017).

Smolinka, T., Günther, M., Garche, J., 2011. Stand und Entwicklungspotenzial der Wasserelektrolyse zur Herstellung von Wasserstoff aus regenerativen Energien: NOW-Studie. Fraunhofer ISE, FCBAT, Freiburg.

Sotavento Galicia, 2017. System of Generation and Energy Storage in the Form of Hydrogen [Online]. Sotavento Galicia, Santiago de Compostela. Available: http://www.sotaventogalicia.com/en/technical-area/renewable-facilities/hydrogen-plant (Accessed 4 January 2017).

Specht, M., Brellochs, J., Frick, V., Stürmer, B., Zuberbühler, U., 2016. Technical Realiziation of Power-To-Gas Technology (P2G®): Production of Substitute Natural Gas by Catalytic Methanation of H_2/CO_2. In: Van Basshuysen, R. (Ed.), Natural Gas and Renewable Methne for Powertrains: Future Strategies for a Climate-Neutral Mobility. Springer International Publishing, Cham, pp. 141–167.

Speight, J.G., 2015. Handbook of Petroleum Product Analysis. Wiley, Hoboken.

Spire, 2016. MefCO2 (Methanol Fuel From CO2)—Synthesis of Methanol From Captured Carbon Dioxide Using Surplus Electricity [Online]. Spire, Brussels. Available: https://www.spire2030.eu/mefco2 (Accessed 5 January 2017).

Stadwerk Haßfurt, 2015. Mehr als eine Mälzerei: Weyermann Malz in Haßfurt. tag&nacht das Magazin für Energiesparer 4, 2.

Statoil, 2004. Power From the Hydrogen Plant on Utsira [Online]. Statoil, Stavanger. Available: https://www.statoil.com/content/statoil/en/news/archive/2004/04/14/PowerFromTheHydrogenPlantOnUtsira.html (Accessed 8 March 2017).

Stedin Netbeheer, 2014. Power-to-Gas officieel geopend: elektriciteit wordt aardgas in Rozenburg [Online]. Amersfoort: TKI-gas. https://www.stedin.net/over-stedin/pers-en-media/persberichten/powertogas-officieel-geopend-elektriciteit-wordt-aardgas-in-rozenburg (accessed 09.01.17).

Steffansson, B., 2015. Power and CO_2 emissions to methanol. European Methanol Policy Forum. 13.10.2015 Carbon Recycling International, Brussels.

Stenzel, P., Hennings, W., Linssen, J., Wulf, C., 2016. Energiespeicher. BWK 68, 38–51.

Thüga, 2014. Strom zu Gas - Projektbeschreibung, Munich.

Tichler, R., Steinmüller, H., Reiter, G., 2014. Wirtschaftlichkeit und Systemanalyse von Power-to-Gas-Konzepten. In: Graf, F., Götz, M., Henel, M., Schaaf, T., Tichler, R. (Eds.), Technoökonomische Studie von Power-to-Gas-Konzepten. Deutscher Verein des Gas- und Wasserfaches e. V., Linz

Tjarks, G., Mergel, J., Stolten, D., 2016. Dynamic operation of electrolyzers—systems design and operating strategies. In: Stolten, D., Emonts, B. (Eds.), Hydrogen Science and Engineering: Materials, Processes, Systems and Technology. Wiley-VCH, Weinheim, pp. 309–330.

Todd, I., 2016. Levenmouth Community Energy Project: Project update.All Energy Conference and Exhibition, 05.05.2016, Glasgow.

Tornberg, J., 2015. El og brint sætter fut i produktionen af biogas. Energi 9, 8–9.

Total, 2014. Freie Fahrt für die Energiewende: Staatssekretärin Reiche eröffnet Wasserstoff-Großanlage in Berlin-Schönefeld - Pressemitteilung [Online]. Available: http://www.total.de/news/pressemitteilungen/Erstbetankung-Wasserstoff-BER.html (Accessed 28 January 2016).

Uniper, 2016. WindGas Hamburg [Online]. Uniper Innovation Energy Storage, Quickborn. Available: http://www.windgas-hamburg.com/startseite/ (Accessed 8 March 2017).

University of South Wales, 2017. Hydrogen Centre: The University of South Wales' Hydrogen Research and Demonstration Centre at Baglan [Online]. University of South Wales, Pontypridd. Available: http://www.h2wales.org.uk/pages/hydrogen-centre/hydrogen-centre.html (Accessed 9 March 2017).

Varkaraki, E., 2009. The RES_2H_2 Wind-Hydrogen Plant in Greece: Case Study Data. Centre for Renewable Energy Sources, Pikermi Attiki.

Vartiainen, V., 2016. Screening of Power to Gas Projects. Lappeenranta University of Technology, Master Thesis.

Viessmann, 2015. Power-to-Gas-Anlage in Betrieb genommen - Schlüsseltechnologie zum Gelingen der Energiewende [Online]. Viessmann Werke, Allendorf. Available: www.viessmann.de/content/dam/vi.../Pressetext-09032015-08.docx (Accessed 14 March 2017).

Vlap, H., van der Steen, A., Knijp, J., Holstein, J., Grond, L., 2015. An Overview of the Technical Assumptions and Results of the Power-To-Gas Demonstration Project in Rozenburg, The Netherlands. DNV GL Oil & Gas, Groningen.

Voutetakis, S., 2015. HYRES: Development of Integrated autonomous Electrical Power System Utilizing Renewable Energy Sources With Production and Storage of Hydrogen [Online]. Centre for Research and Technology—Hellas CERTH, Thessaloniki. Available: http://hyres.cperi.certh.gr/ (Accessed 10 March 2017).

WIND-WASSERSTOFF-projekt, 2016. Willkommen auf der Internetpräsenz des Demonstrations- und Innovationsprojekts RH2-WKA [Online]. WIND-WASSERSTOFF-projekt GmbH & Co. KG, Börgerende. Available: http://www.rh2-wka.de/ (Accessed 10 March 2017).

Wulf, C., Hustadt, D., Kaltschmitt, M., Weinmann, O., 2011. Hydrogen refueling station in Hamburg HafenCity: a status report.International Conference on Hydrogen Production, Thessaloniki.

Younas, M., Loong Kong, L., Bashir, M.J.K., Nadeem, H., Shehzad, A., Sethupathi, S., 2016. Recent advancements, fundamental challenges, and opportunities in catalytic methanation of CO_2. Energy Fuel 30, 8815–8831.

Zuberbühler, U., 2011. // PtG 250 (P2G®) [Online]. Zentrum für Sonnenenergie- und Wasserstoff-Forschung Baden-Württemberg (ZSW), Stuttgart. Available: https://www.zsw-bw.de/en/projects/regenerative-kraftstoffe/ptg-250-p2gr.html (Accessed 9 March 2017).

Part II

Exploring Methods and Tools for HSC design, Deployment and Operation

Chapter 10

Methods and Tools for Hydrogen Supply Chain Design

Jesus Ochoa Robles, Sofía De-León Almaraz and Catherine Azzaro-Pantel
Laboratoire de Génie Chimique, Université de Toulouse, CNRS, Toulouse, France

ACRONYMS

AHP	analytic hierarchy process
AML	algebraic modelling languages
ANP	analytic network process
AUGMECON	augmented ε-constraint method
CCS	carbon capture and storage
DEMATEL	decision making trial and evaluation laboratory
DP	dynamic programming
ELECTRE	elimination and choice translating reality
GA	genetic algorithms
GHG	greenhouse gas
GIS	geographic information systems
HFCV	hydrogen fuel cell vehicle
HSC	hydrogen supply chain
LCA	life cycle assessment
LCC	life cycle costing
LP	linear programming
MCDM	multicriteria decision making
MILP	mixed integer linear programming
MINLP	mixed integer nonlinear programming
M-TOPSIS	modified TOPSIS
NIS	negative ideal solution
NPV	net present value
NSGA-II	nondominated sorting genetic algorithm

Hydrogen Supply Chain. https://doi.org/10.1016/B978-0-12-811197-0.00010-5

PCA	principal component analysis
PEM	proton exchange membrane
RA	risk assessment
RHS	right-hand-side
SAA	sample average approximation
SC	supply chain
SCM	supply chain management
SCND	supply chain network design
SO	solid oxide
TOPSIS	Technique for Order Preference by Similarity to Ideal Solution

10.1 INTRODUCTION

This chapter presents the different approaches that can be used for the solution of hydrogen supply chain design problem. The reader must be aware that this chapter constitutes a companion paper to the following chapters of this book. The objective is to propose guidelines for the methodological choices that emerge as the best options for solution strategies. Some of the formulations are illustrated in the dedicated chapters of this book.

HSC analysis and design can be viewed as a multiscale and multiobjective problem, with several criteria related to cost, environmental impact, and safety, among others. Some of the decisions that must be made in HSC design are as follows: what are the best places to build hydrogen production facilities? How large should the facilities be? Where does each facility get its feedstock from? What kinds of energy sources and production technology constitute the best choices? Which demand centers are served by each production facility? Which mode of hydrogen delivery is used for each demand center? These questions must be answered by considering simultaneously the abovementioned criteria.

This chapter is divided into two sections. Section 10.2 first presents the description of some optimization frameworks according to the type of problem (e.g., linear, nonlinear) and some significant solution strategies that can be used. The HSC problem can be viewed as an optimization problem with both integer (number of production plants, storage facilities, and transport units) and continuous variables (e.g., hydrogen production and flow rates). This section also distinguishes the mono and multiobjective formulations. Section 10.3 focuses on multiobjective optimization methods because they are well suited to the HSC problem. Special attention is paid to the chosen techniques. Some approaches for decision support orientation based on multicriteria decision aid following the multiobjective optimization step are also reviewed. At the end of this section we examine how the HSC design optimization framework can be linked with a spatially detailed infrastructure model. Finally, this chapter ends with some guidelines that can be useful for the practitioner.

10.2 METHODOLOGICAL FRAMEWORKS FOR SUPPLY CHAIN DESIGN

In the design and management of a supply chain, the best global performance should ideally be sought, so as to achieve better performance of a single link of the chain. The decisions that must be made involve different levels of the supply chain and need to be supported by robust tools to evaluate the impact of various decisions prior to implementing them in the real environment. In this context, system modelling is used to predict the behaviors of the supply chain as variations of network configurations. Supply chain modelling aims at minimizing or maximizing an objective function through the identification of decisions and tradeoff solutions that satisfy conflicting objectives at the same time, so that optimization approaches, which are generally based on mathematical models, are largely used to design supply chains (Akgul et al., 2014; Kim et al., 2011; Liu and Papageorgiou, 2013; Pishvaee et al., 2011).

10.2.1 General Decision Levels in a Supply Chain

Several decision levels are classically considered in a supply chain related to different time horizons:

- *Strategic planning*: this level refers to a long-term horizon (several years) and has the objective of identifying strategic decisions for a production network and defining the optimal configuration of a supply chain: capacity sizing, technology selection, sourcing, facility location, production allocation, and others. Future demands resource and management for the entire supply chain must be anticipated.
- *Tactical planning*: this level refers to a mid-term horizon (around 1 year) and has the objective of fulfilling demand and managing material flows, with a strong focus on the tradeoff between the service level and cost reduction: production allocation, supply chain coordination, transportation policies, inventory policies, safety stock sizing.
- *Operational planning*: this level refers to a short-term period (1 day to 1 year) and has the objective of determining material/logistic requirement planning: allocation of customer demands, vehicle routing, and plant scheduling.

10.2.2 Methods for Supply Chain (SC) Management and Design

Different methods and tools have been used and reported in the supply chain management (SCM) and design literature (see Table 10.1 for some examples) and are not specific to the HSC case.

The literature review shows that the most common approach in designing and modelling supply chains is optimization through mathematical models. As opposed to simulation based approaches, these models utilize formal

TABLE 10.1 Different Ways to Optimize SCM

Technique	Reference	Application
Linear programming	Kim et al. (2011)	Optimal design of biomass SC network under uncertainty, in the South-eastern region of the United States.
	Liu and Papageorgiou (2013)	Global process SC optimization problem, considering cost, responsiveness, and customer service level simultaneously.
	Perea-López et al. (2003)	Predictive control strategy to find the optimal decision variables to maximize profit in SC with multiproduct batch plants.
	Pishvaee et al. (2011)	Robust optimization model for handling the inherent uncertainty of input data in a closed-loop SC network design problem in business environment.
	Soylu et al. (2006)	Systematic approach to identify the synergy among different energy systems.
	Tsiakis and Papageorgiou (2008)	Optimal configuration of a production and distribution network subject to operational and financial constraints.
	van Dyken et al. (2010)	Biomass supply chain with different types and the relationship between moisture and energy.
Nonlinear programming	Akgul et al. (2014)	Model of carbon negative energy generation in the UK to examine the potential for existing power generation assets.
	Shabani and Sowlati (2013)	SC configuration of a typical forest biomass power plant.
Dynamic programming (DP)	Buffett and Scott (2004)	Optimization of the inventory level and minimization of the total cost.
	Choi et al. (2006)	A multiproduct supply chain under demand uncertainty.
	Gigler et al. (2002)	DP model for an agricultural chain of willow biomass fuel to an energy plant.
Markov chains	Busse et al. (2012)	Price interdependencies between the German biodiesel and related agricultural and energy markets.
	Kurata and Liu (2007)	Determination of the frequency of the price discount, including or excluding a supplier's inventory decision.

TABLE 10.1 Different Ways to Optimize SCM—cont'd

Technique	Reference	Application
Analytical hierarchy (AH)	Haq and Kannan (2006)	Evaluation of vendor selection in a company in the southern part of India.
Analytical network process (ANP)	Agarwal et al. (2006)	SC encapsulating market sensitiveness, process integration, information driver and flexibility measurement.
	Tseng et al. (2009)	Novel hierarchical evaluation framework to assist the expert group for optimal supplier selection in SC management strategy.
Network equilibrium model	Nagurney and Toyasaki (2005)	Reverse supply chain management of electronic waste, including recycling.
Game theory	Bai et al. (2012)	SC design incorporating farmers' decisions on land use and market choice into the biofuel.
Fuzzy and neuro-fuzzy	Shaw et al. (2012)	Integrated approach for selecting the appropriate supplier in the supply chain, addressing the carbon emission issue.

optimization techniques to allow advanced decisions to be captured and to provide comprehensive integrated solutions (Hugo et al., 2005). The aim of such methods is to find optimal configurations according to some specific criteria (e.g., economic, safety, environmental aspects). One of the main advantages of this type of modelling is that mathematical models form a bridge to the use of high-powered mathematical techniques and computers to analyze the problems (Hillier and Lieberman, 2001).

The use of mathematical programming for designing a supply chain consists of three major steps as reported in Grossmann et al. (2000):

i. The representation of all possibilities from which the optimal solution is extracted by defining the so-called superstructure; a superstructure is defined as the set of all possible connections in a network.

ii. The formulation of a mathematical model includes generally discrete and continuous variables. The main components of a model are:

  (a) the optimization criteria, which are expressed as mathematical functions, and

  (b) the constraints, which can be either of the equality or inequality type.

iii. The resolution of the mathematical model to determine one or more optimal solutions.

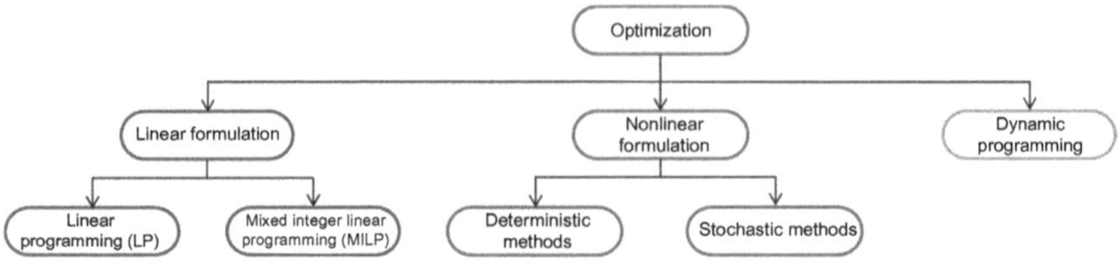

FIG. 10.1 Classification of the main methods of optimization *(Adapted from Collette, Y., Siarry, P., 2003. Multiobjective Optimization: Principles and Case Studies. Springer.)*

Traditionally, the main focus of the research studies dedicated to supply chains has been minimizing the overall cost or maximizing the total revenue as a single-objective optimization problem.

The most common optimization frameworks for capturing SC problems are summarized in Fig. 10.1. These can be classified as either *linear* or *nonlinear programming* or *dynamic programming*. Initially, the majority of these studies were based on a monoobjective formulation.

10.2.2.1 Linear Formulation

A linear formulation is used when the problem (objective functions and constraints) is linear (Hillier and Lieberman, 2001). Two methods can be used, linear programming (LP) and mixed integer linear programming (MILP).

- LP models are used for the efficient allocation of limited resources in known activities in order to meet the desired goals (for instance, maximizing profits or minimizing costs). Linear programming problems can involve decision variables that can take integer values. When integer variables are restricted to the binary variables (0–1), the corresponding problem is called the *binary* integer programming *problem.* An integer variable can be defined such that it determines whether a processing unit should be invested in or not.
- In the case of both integer and continuous variables, the problem is referred as a mixed-integer linear programming one. Because of its capability to naturally capture logical conditions, applications of MILP have been widespread in areas of investment planning, supply chain, and logistics management, energy industry planning, engineering design, and production scheduling (Hugo et al., 2005). MILP methods consist of maximizing or minimizing an objective function as a function of parameters, variables, and several constraints on these variables (Haeseldonckx and D'haeseleer, 2011).

The use of *integer variables* in general, and binary ones in particular, dramatically broadens the capabilities of linear programming modelling, enabling the disjunction of constraints, the logical implication and general restrictions to the model incorporating certain nonlinear behaviors of reality. Many practical optimization problems lead to the consideration of an extremely large number of feasible solutions, so that the problem can be viewed as a combinatorial one.

The solution of the system of linear equations that are involved in the problem formulation can be performed by using the Gauss-Jordan method. When problems become larger (more parameters, variables and constraints), the Gauss-Jordan method is generally combined with a branch-and-bound method in order to converge to an optimal solution as quickly as possible (Haeseldonckx and D'haeseleer, 2011).

Mathematically, the MILP problem can be expressed as follows:

$$\text{Min } cx + dy$$

subject to

$$Ax + By \geq b$$

$$L < x < U$$

$$y = \{0, 1, 2, \ldots\}$$

where x is a vector of variables that are continuous real numbers, and y is a vector composed of variables that can only take integer values. In this expression, $cx+dy$ is the objective function, and $Ax+By \geq b$ represents the set of constraints. L and U are vectors of lower and upper bounds on the continuous variables, and $y=\{0,1,2,\ldots\}$ represent the integer variables.

With regard to the solution of the MILP problems, several algebraic modeling languages (AML) were developed with the aim of allowing users to express LP and other optimization problems in a natural, algebraic form similar to the original mathematical expressions, such as AIMMS, AMPL, GAMS, etc. For instance, GAMS includes well-known algorithms for the solution of MILP (Geletu, 2008): Branch & Bound, Benders Decomposition, Cutting Plane (Gomory) algorithm and Branch & Cut. Usually these algorithms are used in combination with the simplex algorithm and/or the interior-point method. For instance, some of the solvers that can solve MILP problems are BARON, BDMLP, LINDO GLOBAL, MOSEK, OSL, XPRESS, and CPLEX (Mansini et al., 2015).

It must be highlighted that linear programming is the most used technique to optimize the SC. Several applications can be found, such as biomass supply chains (Kim et al., 2011; van Dyken et al., 2010), the optimization of the SC under financial constraints (Liu and Papageorgiou, 2013; Tsiakis and Papageorgiou, 2008), in energy systems (Soylu et al., 2006), in business environment (Pishvaee et al., 2011), or in multiproduct batch plants (Perea-López et al., 2003).

10.2.2.2 Nonlinear Formulation

The nonlinear formulation can be tackled by two main methods, either *deterministic* or *stochastic* algorithms procedures. In the nonlinear deterministic models, no randomness is associated. Then, given a particular input, a deterministic algorithm obviously produces the same type of output (Prawda, 2004).

- Mixed integer nonlinear programming (MINLP) refers to mathematical programming with, on the one hand, continuous and discrete variables, and, on the other hand, nonlinearities in the objective function(s) and constraints. The use of MINLP is a *deterministic* approach of formulating problems where it is necessary to simultaneously optimize the system structure (discrete variables) and its parameters (continuous variables). MINLP problems are difficult to solve because they combine all the difficulties of both of their subclasses: the combinatorial nature of mixed integer programs and the difficulty in solving nonconvex (and even convex) nonlinear programs (Bussiec and Pruessner, 2003).

  The general form of a MINLP is:

  $$\text{Min } f(x, y)$$

  subject to

  $$g(x, y) \leq 0$$
  $$x \in X$$
  $$y \in Y$$

  The function $f(x, y)$ is a nonlinear objective function and $g(x, y)$ a nonlinear constraint function. The variables x, y are the decision variables, where y is required to be an integer vector. X and Y are bounding-box-type restrictions on the variables. Nonlinear formulations with mathematical MINLP can be found in Akgul et al. (2014) and Shabani and Sowlati (2013).
- *Stochastic programming* is used when random-valued parameters and objective functions subject to statistical perturbations are part of the problem formulation (Coello et al., 2007). The stochastic models can incorporate uncertainty in parameters, such as demand, costs, potential sites, and distances, and then fall into probabilistic approaches and scenarios (Patay, 2008). Metaheuristics cannot guarantee that an optimum can be obtained. The stochastic methods are divided into neighborhood techniques, such as Simulated Annealing, Tabu Search, and evolutionary algorithms, and among others genetic algorithms, evolutionary strategies, and evolutionary programming (Tabkhi, 2007).

10.2.2.3 Dynamic Programming

Dynamic programming (DP) is an optimization approach that changes a complex problem into a less complex one by separation of the problem into simpler and smaller problems (Bellman and Dreyfus, 1962; Momoh, 2008). The method used by dynamic programming is recursive, which means that the method calls itself, adding information each time, until the conditions of stopping are met.

According to Chinneck (2006), the method steps are the following ones:

1. Dividing the problem into small problems and finding the optimum solution for each small problem.
2. Enlarging the small problem and finding the optimum solution to the next problem using the previously found optimum solution.
3. Continuing with the second step until the enlarged problem encompasses the entire original problem.
4. Tracking back the solution of the entire problem from the optimum solutions to the small problems solved along the way.

The requirements of this technique are (García and Moreno, 2000):

- The solution to the problem must be reached through a sequence of decisions, each one in each step.
- Such sequence of decisions must satisfy the optimum principle.

Several works using dynamic programming for supply chain problems have been reported in the dedicated literature. Williams (1983) develops a dynamic programming algorithm for simultaneously determining the production level and distribution batch sizes at each node within a supply chain network.

Buffett and Scott (2004) propose a technique for use in supply chain management that assists the decision making process for purchase of direct goods. Based on projections for future prices and demand, request-for-quotes (RFQs) are constructed and quotes are accepted that optimize the level of inventory each day, while minimizing total cost. The problem is modeled as a Markov decision process (MDP) and Dynamic programming is then used to determine the optimal quote requests and accepts at each state in the MDP.

A similar approach has been adopted by Choi et al. (2006) for multiproduct supply chains under uncertainty modelled through Markov chains. They use an approach based on stochastic dynamic programming (DP), which can generate a dynamic operating policy that incorporates information about the uncertainty in the problem at each time step.

Gigler et al. (2002) have also developed a methodology for optimization of agricultural chains using DP, taking into account quality development of a product as a function of the process conditions. The methodology optimizes the route of the chain that returns the minimum integral cost. DP has been firstly applied to a supply chain of bananas (four stages) and then to a chain of willow biomass fuel to an energy plant (7 stages).

Even if dynamic programming is a very elegant framework for analyzing supply chain systems, it is mostly used at a theoretical level to characterize the optimal policy. This approach is yet limited in its applicability; as the number of state variables increases, the state space size grows exponentially, a phenomenon known as the curse of dimensionality, rendering the standard dynamic programming approach impractical.

10.2.2.4 Other Methods for Supply Chain Modelling

In contrast to optimization methods, many researchers have developed an equilibrium model of competitive supply chain networks (Nagurney et al., 2002). The *equilibrium model* captures both the independent behavior of the various decision makers as well as the effect of their interactions. The equilibrium model is drawn from economics and, in particular, from network economics. Nagurney et al. (2002) developed a supply chain network equilibrium model for the case of consumers demand for the product that can be expressed as a deterministic function.

- *Queuing models*, such as *Markov chains*, have also been used to investigate supply chain problems for many years (e.g., Toktaş-Palut and Ülengin, 2010), in the determination of biodiesel prices (Busse et al., 2012) or discounted prices (Kurata and Liu, 2007).

  Markov chains can be described as follows. Let us denote S a set of states, $S=\{s_1, s_2, \ldots, s_r\}$. The process starts in one of these states and moves successively from one state to another. Each move is called a step. If the chain is currently in state s_1, then it moves to state s_j at the next step with a probability denoted by p_{ij}, and this probability does not depend upon the states the chain was in before the current state. In some SCs, the probabilities are used to model the times of some tasks.
- *Game theory* can be defined as "the study of mathematical models of conflict and cooperation between intelligent rational decision makers." It provides general mathematical techniques for analyzing situations in which two or more individuals make decisions that will influence one another's welfare (Myerson, 2013). In SC, for example, Majumder and Groenevelt (2001) present a two-period model of remanufacturing. An original equipment manufacturer competes with a local remanufacturer under many reverse logistics configurations. Another example is support in the decision between land use and market choice for biofuel (Bai et al., 2012).
- *Neural networks* and *fuzzy systems* combine the advantages of fuzzy systems (e.g., interpretability, use of vague or inexact data) with the learnability of neural networks, so that the parameters of fuzzy systems can be learnt by neural networks according to existing requirements. In Marx-Gomez et al. (2002), fuzzy logic is used to forecast the prognoses for the amount of time of returned product, and in Shaw et al. (2012) to select the correct supplier.

10.2.2.5 Multiobjective Formulation

Multiobjective optimization is now a popular approach to modelling supply chains, especially green supply chains (Srivastava, 2007), and in particular for sustainable design of distributed energy supply systems because it allows for the antagonistic objectives of economic and environmental performance

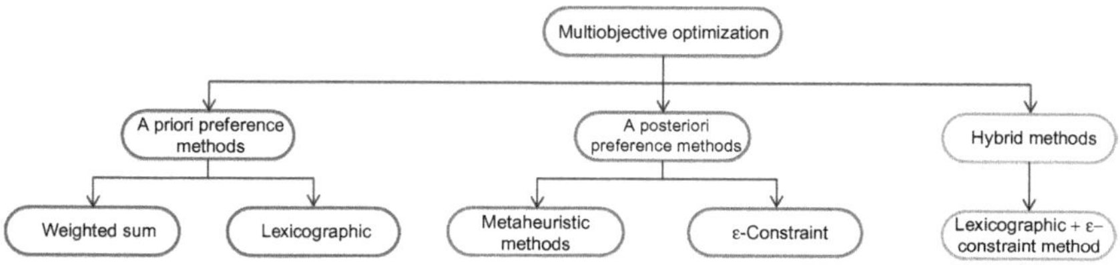

FIG. 10.2 Multiobjective optimization methods.

to be evaluated and optimized simultaneously. Tradeoff solutions are found through this approach giving the decision makers a way to incorporate many objectives and preferences in a single decision framework.

The general formulation of a multiobjective optimization problem is:

$$\text{Min } [f_1(x), f_2(x), \ldots, f_k(x)]$$

subject to

$$g(x) \leq 0$$
$$h(x) = 0$$
$$x \in S$$

where f_i (with $i=1$ to k) is a function of $R^{n1} \times [0,1]^{n2}$ in R, $g(x) \in R^m$, $h(x) \in R^p$ and x is an element of S. $R^{n1} \times [0,1]^{n2}$ in R.

Such an optimization scheme is implied when there is a conflict between two or more objectives, even if the most profitable infrastructure may not necessarily be the least environmentally damaging. Because of this tradeoff, there is no single solution to this class of problem, but rather a set of nondominated solutions called *Pareto front*. A solution belonging to the Pareto front is said to be Pareto-optimal if there are no other solutions that can better satisfy all of the objectives simultaneously and any improvement in one objective leads to the worsening of at least one other objective.

Several solution methods have been developed for multiobjective optimization problems and can be classified as the a priori, a posteriori, and *hybrid* methods (Collette and Siarry, 2003), including scalar, interactive, fuzzy, and meta-heuristic methods (see Fig. 10.2).

A Priori Preference Methods

With these methods, the decision maker defines the tradeoff to be applied (preferences) before running the optimization method. The aggregative methods belong to this family (in which the objective functions are gathered into one objective function). More precisely, the weighted sum, goal programming, and lexicographic methods (among others) can be mentioned (Collette and Siarry, 2003). The drawback is that the decision maker never sees the whole picture (the set of efficient solutions). Hence, the most preferred solution is "most preferred" in relation to what the decision maker has for comparison so far (Mavrotas, 2007, 2009).

- *Weighted sum*

  The goal of the weighted sum is to transform the problem so that it turns into a monoobjective optimization problem, for which various methods of solution exist. The simplest way to proceed is to take each objective function, associate a weight with the objective function, and then take a weighted sum of objective functions. Hence, a new, unique objective function is obtained. The weighting factors are assigned a priori, and are modified to obtain the Pareto front, with all nondominated solutions (or satisfactory solutions). The major problem with this method is the variation of the weighting factors, which often leads to Pareto fronts with a low density of solutions (Hernandez-Rodriguez, 2011). It can be used only when the feasible space of values of the objective function is convex. In the weighting method, the weighted sum of the objective functions is optimized. The problem is stated as follows:

$$\text{Min } \left(w_1 \times f_1(x) + w_2 \times f_2(x) + \ldots + w_p \times f_p(x)\right)$$

subject to

$$g(x) \leq 0$$
$$h(x) = 0$$
$$x \in S$$

  By varying the weights w_i it is possible to obtain different efficient solutions.

- *Lexicographic method*

  Lexicographic problems arise naturally when conflicting objectives exist in a decision problem but for reasons outside the control of the decision maker, the objectives have to be considered in a hierarchical manner (Khorram et al., 2010). This method can be viewed as an "a priori" approach with aggregation using constraints in a decoupled method. In the lexicographic ordering, the objectives are ranked according to the order of importance. The optimization process starts minimizing the most important objective and proceeds according to the assigned order of importance of the criteria. An alternative is to randomly select an objective when there is no more rank available. One disadvantage of this method is that it tends to favor certain objectives, making the Pareto front converge to a particular region. The main advantage is its simplicity and computational efficiency, making it competitive with other ideas, such as weighted sum of objectives

(Collette and Siarry, 2003). In general, the lexicographic problem can be expressed as follows:

$$\text{Lexmin } \{f_1(x), f_2(x), \ldots, f_r(x)\}$$

subject to

$$g(x) \le 0$$
$$h(x) = 0$$
$$x \in S$$

To solve the problem, the following procedure, known as the sequential method, is adopted. First, $f_1(x)$ is minimized, and an optimal solution x^* is determined ($f_1(x^*) = \beta_1$). The problem is then solved minimizing $f_2(x)$ subject to $f_1(x^*) = \beta_1$, and so on at the q iteration:

$$\text{Lexmin } \{f_q(x) : f_i(x) \le \beta_i, i = 1, \ldots, q-1\}$$

subject to

$$g(x) \le 0$$
$$h(x) = 0$$
$$x \in S$$

If either the last equation has a unique optimum or $q = r$, then its optimal solution is a preemptive optimum. Otherwise, one proceeds to iteration $q+1$ (Khorram et al., 2010).

A Posteriori Preference Methods

With these methods, the decision maker chooses the solution by examining solutions computed by the optimization model. Methods belonging to this family produce, at the end of the optimization, a tradeoff surface (Collette and Siarry, 2003). This kind of method produces many solutions, whereas only one will be chosen by the decision maker, and a lot of time may be invested to find the Pareto front. The value of using this kind of method within a multicriteria optimization framework is that it does not require the a priori articulation of preferences by the decision maker. Instead, the aim is to generate the full set of tradeoff solutions and not to present only one single "best" alternative. From the set of alternatives, the decision maker can then further investigate interesting tradeoffs and ultimately select a particular strategy that satisfies his/her willingness to compromise (Hugo et al., 2005). In the a posteriori method, the solutions of the problem are generated and then the decision maker is involved, in order to select among them, the most preferred one.

- *Metaheuristic methods*

  Metaheuristic methods can be used as a nonaggregative approach. They are particularly useful to treat problems known as "black box" ones in which no mathematical property of the problem is known (Boix, 2011). This category includes genetic algorithms, tabu search, simulated annealing, ant colonies, neural networks, etc. In the case of the HSC, this method has been rarely used (Nepal et al., 2011) because linear constraints and equality constraints (balance equations must be satisfied with a small tolerance, gap inferior to 0.001%) are involved in a major way in the formulation.
- *The ε-constraint method*

  In the ε-constraint method, introduced by Haimes et al. (1971), all but one of the objectives are converted into constraints by setting an upper or lower bound to each of them, and only one objective is to be optimized (Liu and Papageorgiou, 2013). By varying the numerical values of the upper bounds, a Pareto front can be obtained. The ε-constraint technique fits into the family of "a posteriori" approaches with aggregation using constraints in a decoupled method (Collette and Siarry, 2003). This method presents some advantages compared to the a priori methods, for example, for linear problems, the weighting method is applied to the original feasible region and results in a corner solution (extreme solution), thus generating only efficient extreme solutions. Yet, the ε-constraint method alters the original feasible region and can produce nonextreme efficient solutions. An additional advantage of the ε-constraint method is that the number of the generated efficient solutions can be controlled by properly adjusting the number of grid points in each one of the objective function ranges.

  In the ε-constraint method, one of the objective functions is optimized using the other objective functions as constraints, incorporating them in the constraint part of the model as shown below:

$$\text{Min}\, f_1(x)$$

subject to

$$f_2(x) \leq \varepsilon_2$$
$$f_3(x) \leq \varepsilon_3$$
$$f_p(x) \leq \varepsilon_p$$
$$x \in S$$

By parametrical variation in the right-hand-side (RHS) of the constrained objective functions (ε_i), the efficient solutions of the problem are obtained. This method is yet easy to implement even if, in some cases, an intensive computation time is required.

However, one of its key disadvantages is that the generated solution largely depends on the selected vector ε (Liu and Papageorgiou, 2013).

The main difficulty of this method lies in determining Nadir points (in which the criteria are their worst values). To tackle this problem, a hybrid method can be used as the augmented ε-constraint method (AUGMECON) proposed by Mavrotas (2007, 2009).

Hybrid Methods

The works reported in Mavrotas (2007, 2009) shed new light on determining Nadir points combining the ε-constraint method with the lexicographic one. According to Liu and Papageorgiou (2013), the decision makers may not have any preference for any objective, that is, all the objectives are equally important. In this case, it is crucial to generate a fair solution in which all normalized objective function values are as close to each other as possible. In order to generate such solutions, the lexicographic can easily be coupled to the ε-constraint method.

The AUGMECON method (Mavrotas, 2007, 2009) is an effort to effectively implement the ε-constraint method for producing efficient solutions. To determine Utopia and Nadir points in the classical ε-constraint method, the most common approach is to take upper and lower bounds from the payoff table (the table with the results from the individual optimization of the p objective functions). In a minimization problem, the Nadir value is usually approximated with the maximum of the corresponding column. However, even in this case, it must be sure that the obtained solutions from the individual optimization of the objective functions are efficient solutions. In order to overcome this limitation, the AUGMECON method proposes the use of lexicographic optimization for every objective function in order to construct the payoff table with only efficient solutions. A simple remedy to bypass the difficulty of estimating the Nadir values of the objective functions is to define reservation values for the objective functions. The reservation value acts like a lower (or upper for minimization objective functions) bound. Values worse than the reservation value are not allowed.

10.2.2.6 Multiple Criteria Decision-Making Approaches

Designing sustainable supply chains requires complex decision support models that must deal with multiple dimensions of sustainability while taking into account specific characteristics of products and their supply chain. When the decision space involves continuous variables, multiobjective optimization techniques, such as mathematical programming problems with multiple objective functions, can be used as abovementioned. Yet multiobjective optimization techniques lead to a set of alternatives, among which the decision maker has to choose a solution for implementation purposes. In that context, multicriteria decision making (MCDM) that deals with discrete decision spaces where the decision alternatives are predetermined is a useful approach to quantify tradeoffs between economic, social, and environmental criteria. The analysis of the dedicated literature shows that the use of MCDM approaches for designing SCs is a rather new, but emerging, research field (Banasik et al., 2016). Many of the MCDM methods share the concepts of alternatives and attributes. A set of finite alternatives represent different choices of action available to the decision maker. Alternatives need to be prioritized with respect to the multiple attributes with which the MCDM problems are associated. Attributes are also referred to as goals or decision criteria, and may be in conflict with each other, may not be easily represented in a quantitative way and may be stochastic or fuzzy. Without being exhaustive, some of the most used MCDM methods are the following ones:

AHP/ANP: Analytic hierarchy process (AHP) is a pairwise comparison-based method proposed by Saaty (1980). An MCDM problem is first formulated as a hierarchy including several levels. The first level represents the goal, the second level shows the main decision criteria, the next levels show the subcriteria, and the last level indicates the alternatives. The elements of each level are compared in a pairwise fashion forming a pairwise comparison matrix.

The *analytic network process* (ANP) developed by Saaty (1996) is a multistage decomposition method used to solve decision making problems involving more than one criterion. It is a comprehensive decision making technique that captures the outcome of the dependence and feedback within and between the clusters of elements. Analytical hierarchy process (AHP) (Saaty, 1980) serves as a starting point for ANP. ANP consists of two steps, the first is a control hierarchy or network of criteria controlling the interactions and the second is a network of influences among the elements and clusters. It can be said that ANP uses a network without levels, as it is used in AHP (Ravi et al., 2005). Typically, in AHP the top element of the hierarchy is the overall goal for the decision model. ANP can treat complex problems with strong dependencies among factors (Sarkis, 1999).

ELECTRE: The ELECTRE (elimination and choice translating reality) method was introduced by Roy (1968). The basic concept of the ELECTRE method is to deal with "outranking relations" by using pairwise comparisons among alternatives under each one of the criteria separately. The decision maker is requested to assign weights or importance factors in order to express their relative importance. The ELECTRE method elicits the so-called concordance index, defined as the amount of evidence to support the conclusion that alternative A_j outranks or dominates alternatives A_i, as well as the discordance index, the counterpart of the concordance index. The ELECTRE method is sometimes unable to identify the most preferred alternative. It only produces a core of leading alternatives. This method has a clearer view of alternatives by eliminating less favorable ones. This method is especially convenient when there are decision problems that involve a few criteria with a large number of alternatives as it saves much time.

TOPSIS/M-TOPSIS: TOPSIS (Technique for Order Preference by Similarly to Ideal Solution) was developed by Hwang and Yoon (1981) as an alternative to the ELECTRE method: the basic concept of this method is that the selected alternative should have the shortest distance from the Positive Ideal Solution (PIS) and the farthest distance from the Negative Ideal Solution (NIS) in a geometrical sense. Yet, one of the problems related to TOPSIS is that it can cause the phenomenon known as rank reversal (García-Cascales and Lamata, 2012), in which the alternative order of preference changes when an alternative is added to or removed from the decision problem.

Ren et al. (2007) presented a novel, modified TOPSIS (M-TOPSIS) method to evaluate the quality of the alternative and to deal the rank reversal problem. In M-TOPSIS, the positive ideal solution (D_i^+) and negative ideal solution (D_i^-) in finite planes are found (as in the TOPSIS Method) first and then the $D^+ D^-$ plane is constructed. D^+ is the x-axis and D^- is the y-axis. The point (D_i^+, D_i^-) represents each alternative ($i=1, 2..., n$). The point A (min(D_i^+), max(D_i^-)) is the "optimized ideal reference point." Finally, the relative distance from each evaluated alternative to the ideal reference point (A) is calculated to determine the ranking order of all alternatives.

10.2.2.7 Supply Chain Network Design Under Uncertainty

Uncertainty in the supply chain is an issue that is difficult to deal with, and that increases the complexity of a supply network. Supply chain uncertainty can be defined by the lack of information about the environment of the supply chain, about the processing capacities or the lack of prediction of the impact of some control actions (van der Vorst and Beulens, 2002).

Three sources of uncertainty can be identified as upstream (supply) uncertainty, internal (process) uncertainty, and downstream (demand) uncertainty (Davis, 1993). Among these three sources, the demand is seen as the most severe parameter due to its volatile nature and the consequences of an inaccurate forecast. In addition, because some of the relevant technologies are still in the process of maturing, many important parameters, such as processing costs and yields, are highly uncertain. There are also uncertainties regarding the future course of energy policies, such as carbon taxes.

Grossmann and Guillén-Gosálbez (2010) reviewed major contributions in process synthesis and supply chain management, including the handling of uncertainty and the multiobjective optimization of economic and environmental objectives, and highlighted the need to develop sophisticated optimization and decision support tools to help in exploring diverse system alternatives under uncertainty.

The majority of the approaches to managing these sources of uncertainty seek to reduce uncertainty at its source, and to cope with it, thereby minimizing its impact on performance (Simangunsong et al., 2012).

Table 10.2 shows the different techniques to treat the uncertainty of the demand. Three distinct methods are frequently mentioned for representing uncertainty (Chen and Lee, 2004). First, the distribution-based approach, in

TABLE 10.2 Different Ways to Model Demand Uncertainty

Technique	Reference
Scenario based approach	Tsiakis et al. (2001), Chen and Lee (2004), Almansoori and Shah (2012), Kim et al. (2008), Nunes et al. (2015)
Distribution based approach	Gupta and Maranas (2003), You and Grossmann (2008)
Demand generator	Jung et al. (2004)
Fuzzy based approach	Chen and Lee (2004), Peidro et al. (2009)
Spatially aggregated demand model	Dayhim et al. (2014)

which the normal distribution with specified mean and standard deviation is widely used for modelling uncertain demands and/or parameters. For example, You and Grossmann (2008) used normal distributions and triangular distributions to model the demand, transforming the disjunction of the triangular distribution into MINLP constraints. Second, in the fuzzy-based approach, the forecast parameters are considered as fuzzy numbers with accompanied membership functions. Third and finally, the scenario-based approach is a classical approach, in which several discrete scenarios with associated probability levels are used to describe the expected occurrence of particular outcomes. Yet, the creation of scenarios with their associated probabilities could be a problematic and cumbersome task, especially in real-life SC problems. Also, the use of an adequate number of scenarios could lead to a large-scale optimization problem that may be computationally time consuming.

The objective of any SCND under uncertainty is to achieve a configuration with a good performance even if uncertain parameters are involved. In general, the uncertainty sources include the existing uncertainty in parameters, such as supply, demand, and cost, that are inherently uncertain, and the uncertainty caused by natural or manmade disruptions (Govindan et al., 2017).

10.3 DESIGN OF HYDROGEN SUPPLY CHAINS

This section highlights the current trends for HSC design regarding the general context of supply chain modelling and the typical features of HSC.

Hydrogen-based energy systems have been widely studied and modelled (see Table 10.3). The dominant models used to describe them are supply chain models as opposed to equilibrium models and are quasi exclusively based on the MILP formulation.

TABLE 10.3 Taxonomy of HSC Studies

	Territorial Scale		Time Scale		Objective		Energy Source	Approach	Uncertain Parameters	Observations
	Region	*Country*	*Monoperiod*	*Multiperiod*	*Monoobjective*	*Multiobjective*				
Agnolucci et al. (2013)		Great Britain		9 (2020–2060)	Financial		Coal, Natural gas, Biomass (CCS), renewable	MILP		Development of a spatially-explicit MILP model, called SHIPMod (Spatial Hydrogen Infrastructure Mode)
Almansoori and Betancourt-Torcat (2016)		Germany	x		Financial		Natural gas, Coal (CCS), Biomass	MILP		
Almansoori and Shah (2006)		Great Britain	x		Financial		Natural gas, coal, biomass, other renewable sources	MILP		
Almansoori and Shah (2009)		Great Britain		5 (2005–2034)	Financial		Natural gas, coal, biomass, renewable	MILP		

Continued

TABLE 10.3 Taxonomy of HSC Studies—cont'd

	Territorial Scale		Time Scale		Objective		Energy Source	Approach	Uncertain Parameters	Observations
	Region	*Country*	*Monoperiod*	*Multiperiod*	*Monoobjective*	*Multiobjective*				
Almansoori and Shah (2012)		Great Britain		3 (2005–2022)	Financial		Natural gas, coal, biomass, other renewable sources	MILP	Demand	Demand uncertainty is modelled using scenario-based-approach
De-León Almaraz et al. (2013)		Great Britain	x			Cost, Ecological, Safety risk	Natural gas, coal, biomass	MILP		ε-constraint method for the multiperiod problem
De-León Almaraz et al. (2014)	Midi-Pyrénées	France		4 (2010–2050)		Cost, Ecological, Safety risk	Natural gas, photovoltaic, wind, hydro, nuclear	MILP		ε-constraint method for the multiperiod problem
Gondal and Sahir (2013)		Pakistan	x		Financial		Biomass	MINLP-GIS		
Guillén Gosálbez et al. (2010)		Great Britain		5 (5 years)		Cost, Ecological	Natural gas, coal, biomass	MILP		The Pareto front is obtained by the ε-constraint method

Han et al. (2013)		Korea	x			Financial, Ecological, Risk	Natural gas (CCS), renewable sources	Fuzzy multiple objective programming		
Hugo et al. (2005)	x			5 (2004–2038)		Financial, Ecological	Natural gas, coal, biomass, other renewable sources	MILP		The territorial scale is not specified, only defined as a "geographical region"
Kamarudin et al. (2009)		Malaysia	x		Cost		Natural gas, coal, biomass, water electrolysis	MILP		Two methods for demand determination: one based on the prediction of vehicle numbers and the other based on the supply of gasoline and diesel
Kim and Moon (2008)		Korea	x			Financial, Safety	Natural gas, renewable sources	MILP		The relative risk index proposed is based on the relative risks of individual components of hydrogen infrastructure

Continued

TABLE 10.3 Taxonomy of HSC Studies—cont'd

	Territorial Scale		Time Scale		Objective		Energy Source	Approach	Uncertain Parameters	Observations
	Region	*Country*	*Monoperiod*	*Multiperiod*	*Monoobjective*	*Multiobjective*				
Kim et al. (2008)		Korea	x		Financial		Natural gas, renewable sources	MILP	Demand	Demand uncertainty is modelled using scenario-based-approach
Li et al. (2008)		China		5 (2010–2034)		Financial, Ecological	Natural gas, coal, biomass, and other renewable sources	MILP		
Ochoa Robles et al. (2016)	Midi-Pyrénées	France		4 (2010–2050)		Cost, Ecological, Safety risk	Natural gas, renewable sources	Genetic Algorithms		
Samsatli et al. (2016)		Great Britain		4 (seasons)	Financial		Wind, renewable sources	MILP		
Sabio et al. (2010)		Spain		8		Financial Risk	Natural gas, coal (CCS), Biomass, other renewable sources	MILP	Fuel price	The uncertainty is associated to the operating costs

Woo et al. (2016)	Jeju Island	Korea		12 (months)	Financial		Biomass	MILP		A sensitivity analysis is conducted to provide insights into the efficient management of the B2H2 supply chain
Zhou et al. (2013)						Financial, Ecological		MINLP		The Pareto front is obtained by an adaptive weighted-sum method

A variety of potential hydrogen supply chain pathways can be found that make the problem original compared to more classical supply chain problems:

- the variety of feedstock and/or the major energy source from which the hydrogen is produced. These include fossil resources, such as natural gas and coal, as well as renewable resources, such as biomass with input from renewable energy sources (e.g., sunlight, wind, wave, or hydropower). The studied cases do not yet include all combinations of the factors. Intermittent technologies (wind, photovoltaics) can be used independently or in combination;
- the variety of technologies to produce (including chemical, biological, electrolytic, photolytic, and thermo-chemical) store and distribute hydrogen.
- the size of the facility at which the hydrogen is produced and the transportation requirements to deliver it to the customer;
- the state of the technology used, whether current or to be improved by future developments; most hydrogen and fuel cell technologies are still in the early stages of commercialization. The generation of hydrogen from fossil resources (such as natural gas), its transmission, distribution, and use within industry and the refining sector are based on mature technologies and applied on a large scale, and are not the main focus of this work, but meanwhile, they can help to build early markets and infrastructure. Major differences in the degree of maturity of some technologies must be highlighted; although alkaline electrolyzers are a mature and affordable technology, PEM and SO electrolyzers show a greater potential to reduce capital costs and to increase efficiency;
- differences also exist, for example, whether or not the carbon dioxide (CO_2) byproduct is sequestered when hydrogen is produced from fossil fuel;
- various markets with multiple uses (mobility, power, industry, buildings, and others);
- multiple stakeholders: policy and government decision makers, strategic investors, stakeholders of hydrogen technologies for production, distribution, and storage.
- integration of different geographical scales: regional and national levels to develop hydrogen solutions.

10.3.1 Problem Formulation for HSC Design

As presented in Table 10.3, several methods can be chosen to present the taxonomy of HSC problems. Following the guidelines of the general presentation on SC modelling, some typical features of HSC are highlighted in what follows.

10.3.1.1 Deterministic Optimization Approaches for HSC Design

Not surprisingly, following the general trends that have been previously observed for SC problems, the optimization formulation is the classical way to tackle HSC design with a specific focus on MILP, as can be observed in Table 10.3. The inputs of such models are constituted by a set of options for the production, storage, and transportation, while the outputs are relative to the type, numbers, location, and capacity of the production, storage, and transportation.

The network design problem can be characterized according to different levels of interest:

- type of problem: location, allocation, routing, location-allocation, location routing,
- planning level considering the strategic, tactical, or operational aspects,
- temporal dimension, for example, either static or dynamic,
- type of data (deterministic, stochastic),
- type of approach (optimization, simulation),
- time horizon (short, medium, or long term),
- geographic dimension according to the problem definition.

It must be emphasized that the application of hydrogen to vehicle use serves as an incentive to deploy HSC. Several energy sources are generally considered, whether based on fossil fuels or renewable origins.

The model developed in Almansoori and Shah (2006) can be considered as a precursor model to the optimal design of a network (production, transportation, and storage) for vehicle use in which the network is demand driven. The model was applied to a case study in Great Britain. The model was then extended in 2009 by the same authors (Almansoori and Shah, 2009) to consider the availability of energy sources and their logistics, as well as the variation of hydrogen demand over a long-term planning horizon, leading to phased infrastructure development as well as the possibility of selecting different scales of production and storage technologies. Other works (Almansoori and Shah, 2012) take into account demand uncertainty arising from long-term variation in hydrogen demand using a scenario-based approach. The model adds another echelon, including refueling stations and local distribution of hydrogen, minimizing the total daily cost.

Guillén Gosálbez et al. (2010) design an HSC for vehicle use. The design task is formulated as a bi-criterion MILP problem. A case study in Great Britain is introduced to illustrate the capabilities of the proposed approach. The model optimizes an economic and environmental objective. The economic objective is given by the total discounted cost, and the environmental impact is measured by its contribution to climate change. The problem is decomposed into two levels. The first level is represented by the original MILP model, while the second level refers to the original problem without the variables of production and storage facilities, adding some binary variables to represent the selection of the different technologies. The advantage of this methodology is the reduction of combinatorial complexity of the problem, and thus, its computational effort.

Sabio et al. (2010) also design an HSC for vehicle use. The objective is to determine the optimal design of the production-distribution network. The model is formulated as an MILP problem, controlling the variation of the economic performance of the hydrogen network. A case study in Spain is applied. The uncertainty in the fuel price is introduced into the operating costs of the network.

Recent models have been focusing on the integration of carbon capture and storage technologies, as well as on the utilization of pipelines, resulting in several scenarios of centralized HSCs using fossil fuels instead of renewable energies, if CCS technology is available. One example is the SHIPMod developed by Agnolucci et al. (2013), that is an optimization-based framework involving a multiperiod spatially-explicit MILP formulation, for the design of HSC and CCS pipeline networks over a long planning horizon. These authors have highlighted that varying the level and the spatial pattern of demand has significant impacts on both the optimal supply system and on the overall costs of delivered hydrogen. In the work of Moreno-Benito et al. (2016), the SHIPMod model has included additional options, such as hydrogen imports in the United Kingdom for a multiperiod problem until 2070 in order to minimize the present value of the total infrastructure cost using a discounted cash flow analysis. These works have also developed a hierarchical procedure to reduce the computational time by initializing the solutions in a two-stage approach.

Almansoori and Betancourt-Torcat (2016) propose an approach for the design of the HSC under emission constraints, taking into account the use of carbon capture storage. The problem is formulated as an MILP model. The objective to be optimized is the total network cost, and it was applied to the future supply chain in Germany in the year 2030, showing that the carbon emission target and CO_2 tax are effective strategies for reducing emissions.

Samsatli et al. (2016) present a model that integrates wind-hydrogen electricity networks using an MILP formulation, comprising wind turbines, electrolyzers, fuel cells, and compressors. Some constraints linked to the installation of wind turbines were added. The objective is the total cost of the network, and it was applied in Great Britain, showing the optimal path to install the pipeline network throughout the country.

Some reported works for the optimal design of an HSC involve a mixed integer nonlinear programming (MINLP) formulation. Most of the problems of optimization of the HSC are MINLP because of the nonlinear nature of the objectives, more particularly, the cost objective. For example, Zhou et al. (2013) study the environmental impact of hydrogen consumption, especially the greenhouse gas (GHG) emissions, and propose a hydrogen network integration (HNI) for refinery hydrogen management. They present a systematic mathematical modelling methodology for the optimal synthesis of sustainable refinery hydrogen networks. The proposed mixed integer nonlinear programming (MINLP) model accounts for both the economic and the environmental aspect of the hydrogen network. The total annual cost is employed to evaluate the economic efficiency of the network, while the environmental performance is assessed by the total CO_2 emission of the network. A multiobjective optimization is carried out via the Pareto front generation, which is obtained by an adaptive weighted-sum method. Then a superstructure-based mixed integer optimization methodology is proposed for the integration of the hydrogen network integration aimed at balancing the economic and environmental objectives for sustainable development.

10.3.1.2 Multiobjective Optimization and MCDM

Initially, HSC modelling was mainly tackled by monoobjective optimization (see Table 10.3). In these studies, the objective to be considered has been based on cost considerations. Considering a multiobjective formulation, the problem decision support for HSC design will have to encompass cost consideration, environmental impact, and risk, for which several models exist, such as life cycle costing (LCC), life cycle assessment (LCA), and risk assessment (RA). The literature analysis shows that only significant criteria belonging to each category are considered in the methodological frameworks developed.

Environmental impact in terms of life cycle assessment (LCA) has been considered by Guillén Gosálbez et al. (2010). The risk of combustion or explosion of leaked H_2 in hydrogen infrastructure has been investigated in several studies (Kim and Moon, 2008; Landucci et al., 2010; Rosyid et al., 2007). Kim et al. (2008) have integrated the safety hazard risks with the economic cost of hydrogen supply chains.

The literature survey also reveals that the multiobjective optimization problem is often solved with an ε-constraint method, or less frequently with the weighted-sum method (Zhou et al., 2013), and produces Pareto-optimal curves that reveal the tradeoffs among the three objectives.

Guillén Gosálbez et al. (2010) proposed a bi-criterion formulation that considers simultaneously the total cost and life cycle impact of the hydrogen infrastructure and developed an efficient solution method that overcomes the numerical difficulties associated with the resulting large-scale MILP. More precisely, the cost criterion is represented by the total discounted cost, calculated as the summation of the discounted costs associated with each time period, whereas the environmental impact is measured through its contribution to climate change.

Hugo et al. (2005) represents the financial objectives as the Net Present Value (NPV) and the environmental objective as the greenhouse gas (GHG) emissions. These authors have developed an optimization-based formulation that investigates different hydrogen pathways in Germany. The model identifies the optimal infrastructure in terms of both investment and environmental criteria for many alternatives of H_2 configurations. This model has been extended and considered as a basis for other works (Li et al., 2008) for a case study in China. At the same time, in Iran, Qadrdan et al. (2008) examined a model for the investigation of an optimal hydrogen pathway and the evaluation of the environmental impacts of the hydrogen supply system. Another study also considered hydrogen from water, using electricity from hydro and geothermal power in Iceland for exportation (Ingason et al., 2008).

Sabio et al. (2011) take into account eight environmental indicators in a two-step method based on a combination of MILP multiobjective optimization with a postoptimal analysis by principal component analysis (PCA) to detect and omit redundant environmental indicators.

The work of Dagdougui (2011) describes the risk hazards (delimitation and explanation of potential risks in some parts of the hydrogen infrastructure: pipeline and storage tank) to demonstrate the consequences of a hydrogen accident in the case of future infrastructure operation. The risk is integrated into the HSC to minimize the global risk to population and the environment. The model is applied to regional case studies in the region of Liguria (North of Italy) and Morocco. A GIS based methodology was coupled based on the clean feedstock for hydrogen production. Then, the minimization of the cost of installation of new onsite hydrogen refueling stations, the cost of conversion of existing gasoline to hydrogen stations, and the cost of transporting hydrogen fuel to offsite stations is taken into account. The objective of this work was to develop a decision support system for the localization of hydrogen refueling stations, considering the potential of production within a specific boundary region.

The work of De-León Almaraz et al. (2013) involves a formulation-based on mixed integer linear programming with a multicriteria approach in which three objectives have to be optimized simultaneously, i.e., cost, global warming potential, and safety risk, either at the national or regional scale. This problem is solved by implementing lexicographic and Ɛ-constraint methods. The solution consists of a Pareto front, corresponding to different design strategies in the associated variable space. Multiple choice decision making based on M-TOPSIS (Modified Technique for Order Preference by Similarity to Ideal Solution) analysis is then selected to find the best compromise. The mathematical model is applied to a case study in Great Britain reported in Almansoori and Shah (2006) for validation purposes, comparing the results between the mono and multiobjective approaches. In the regional case, the modelling and optimization of the HSC in the Midi-Pyrénées region was carried out in the framework of the project "*H_2 as a green fuel*" (see Fig. 10.3). The sensitivity of geographical scale was analyzed in De-León Almaraz et al. (2014) to solve a real problem of the HSC in the former "Midi-Pyrénées" region in France. In order to analyze the economies of scale and the real geographical implications, a comparison between a regional and a national case for France is discussed in De-León Almaraz et al. (2015).

The decision making trial and evaluation laboratory (DEMATEL) method, used to study and solve the complicated and intertwined problem group, has been used by Ren et al. (2013) to analyze the cause-effect relationships among the factors that influence the sustainability of hydrogen supply. The interest of

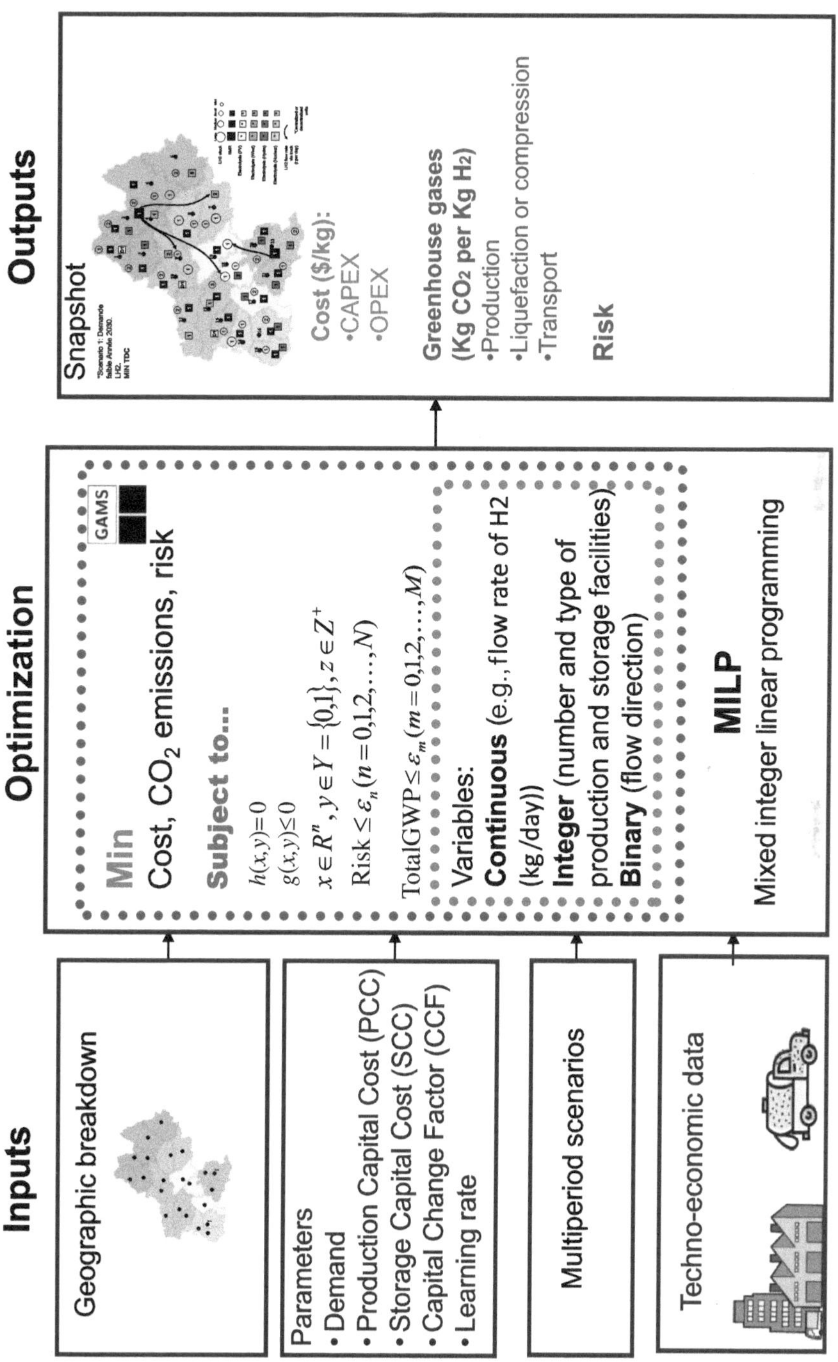

FIG. 10.3 Methodology framework for the "Green H_2 fuel" project (Midi-Pyrénées).

this method, based on graph theory, is to divide multiple criteria into a cause-and-effect group, and the causal relationships in a network relationship map. Four aspects were evaluated: economic, technological, environmental, and societal. A study case was developed in China, and the results are consistent with the current conditions.

Although the weighted-sum method and the ε-constraint method are the most used when solving multiobjective problems, it must be emphasized that assigning a set of compatible objectives, as cost efficiency and safety level, is difficult without knowledge of their possible values (Han et al., 2013), leading to a vague final objective and thus an invalid solution. To overcome this difficulty, fuzzy linear programming with multiple objectives constitutes an interesting alternative. In general, in a fuzzy set methodology, it is assumed that there may be a fuzzy goal for each of the objective functions (Sakawa, 2012). The fuzzy set concept can be adopted to provide a clearer theoretical analysis than the others methods (Han et al., 2013). The fuzzy set method consists of minimizing the distance between the ideal and the desired solutions. Following this approach, Han et al. (2013) designed the HI (H_2 infrastructure) considering economic cost efficiency, safety, and low CO_2 emissions simultaneously. An optimization modelling approach is thus proposed to address such multiple objectives in the HI design. The proposed model employs fuzzy multiple objective programming to compute a compromising solution among multiple objectives. Three objective functions are considered: (1) minimization of the total supply cost of the H_2 of the HSC, (2) minimization of the total relative risk of the HSC, (3) minimization the total mitigation cost of CO_2 for the HSC.

The potential of genetic algorithms (GA) via a variant of NSGA-II has also been explored to cope with the multiobjective formulation, in order to produce compromise solutions automatically (Ochoa Robles et al., 2016). In this work, cost and global warming potential have been simultaneously optimized so that the Pareto Front has been directly obtained. The interest of such an approach is that nonlinearities that may be involved in the formulation of the HSC problem can also be taken into account with a generic framework.

10.3.1.3 Multiperiod Nature

Initially, static models for HSC were developed (Agnolucci et al., 2013; Almansoori and Betancourt-Torcat, 2016; Almansoori and Shah, 2006, 2009, 2012; Gondal and Sahir, 2013; Kamarudin et al., 2009; Kim et al., 2008; Samsatli et al., 2016). Beyond these static models, planning models over multiple periods have been developed. HSC design has to be considered under dynamic conditions with demand exhibiting different realization at each period from deployment to maturity. This is generally modelled through a multiperiod approach in which the demand profiles vary from one period to another, capturing market dynamics.

In energy supply planning and design, coupling between investment/design decisions and operating decisions constitutes an interesting challenge due to the multiscale nature of the problem. Capacity investment and design decisions are typically made on a much longer time scale than operation decisions, while the operation of existing capacities requires decisions on a much faster time scale. The coupling between these decisions makes the overall decision problem a multiscale one (Lee, 2014).

10.3.1.4 HSC Supply Chain Uncertainty

Stochastic modelling tools are classical approaches to the incorporation of different sources of uncertainty into the decision making process. The demand has been the most studied source of uncertainty (Almansoori and Shah, 2012; Chen and Lee, 2004; Dayhim et al., 2014; Gupta and Maranas, 2003; Jung et al., 2004; Kim et al., 2008; Nunes et al., 2015; Peidro et al., 2009; Tsiakis et al., 2001; You and Grossmann, 2008), whereas other uncertainties, especially those appearing in the coefficients of the objective function (product prices, operating cost, etc.) have received much less attention (Sabio et al., 2010).

Traditionally, stochastic models that consider the variability of the uncertain parameters typically optimize the expected economic performance of the system. These approaches can lead to solutions that perform well on average but have a high probability of unfavorable solutions.

The modelling of the uncertainty represents a major issue in the HSC because it deals with the lack of information or forecasts due to the novel nature of the network.

For example, to solve the stochastic mathematical model, the scenario-based approach is employed in Kim et al. (2008). The scenarios emerge from the assumption that the hydrogen demands are "above average," "average," or "below average." Numerically, "above average" and "below average" scenarios are assumed as +20% and −20% of the average values, respectively. First-stage decisions are generally hydrogen production quantities. All the other decision variables are considered as second-stage decisions, which are defined for each scenario.

Nunes et al. (2015) propose the sample average approximation (SAA) technique to manage the large number of scenarios, which enables the calculation of estimates for the objective function value using Monte Carlo simulations while providing statistically certified quality. This technique consists of repeatedly optimizing a set of random sample scenarios, generating different possible solutions for the problem. Then, these solutions are evaluated using new scenarios to allow the calculation of the statistical properties, evaluating their quality regarding the optimality of the problem.

The previous studies of Almansoori and Shah (2012) and Dayhim et al. (2014) proposed to represent the logistic infrastructure, considering the uncertainty in the demand forecast and sought to evaluate different investment alternatives. Comparing the new model with that presented by Almansoori and Shah (2012), there is a reduction in the number of variables, which illustrates the effect of the proposed reformulation of the mathematical model and the data analysis performed.

10.3.1.5 Sensitivity Analysis

The parameter values and assumptions of any model are subject to change and error. Sensitivity analysis is the investigation of these potential changes and errors and their impacts on conclusions to be drawn from the model (Pannell, 1997).

A detailed sensitivity analysis using design of experiments methodology is presented in Ochoa Robles et al. (2015). The importance of hydrogen demand is significantly highlighted, because this factor strongly conditions the optimization criterion of the HSC model. Because the demand for the future HSC is not yet known, its uncertainty is an important issue to be taken into account. The Production Capital Cost is, at a lower level, another significant factor on hydrogen production cost.

Woo et al. (2016) present a new optimization-based approach for design and operation of a renewable hydrogen system from diverse types of biomass, mostly because some works only evaluate technologies that use renewable energy sources in their models, for example, renewable electricity Kim and Kim (2016). The model is tested on an upcoming biomass-to-hydrogen (B_2H_2) supply chain for HFCVs at Jeju Island, South Korea by estimating the expected hydrogen demand in 2040. A sensitivity analysis is conducted to provide insights into the efficient management of the B_2H_2 supply chain.

10.3.1.6 Geographical Information System (GIS)

Literature review reveals that few researchers have used the spatial dimension to construct the infrastructure for hydrogen. In that context, Geographic information systems (GIS), massive software packages providing a range of functions for creating, acquiring, integrating, transforming, visualizing, analyzing, modelling, and archiving information about the surface and near-surface of the earth (Goodchild, 2009), constitute a powerful tool to develop energy supply chain models. Some examples of geographic approaches include the study of Ball et al. (2006) who developed the MOREHyS (Model for Optimization of Regional Hydrogen Supply) approach to the energy system with the integration of geographic aspects in the analysis by the GIS-based method for Germany. This model identifies the cost-optimal way for constructing and implementing an (initial) hydrogen supply infrastructure, as well as possible tradeoffs between hydrogen production and electricity generation within a country-specific context (high degree of regionalization) (Ball et al., 2006).

Johnson et al. (2008) also used GIS for modelling regional hydrogen infrastructure deployment using detailed spatial data and applied the methodology to the case study of a potential coal-based hydrogen transportation system in Ohio with CCS. The objective of this work was to optimize hydrogen infrastructure design for the entire state. The MARKAL model has been applied to the UK and used to develop a GIS-based spatial model to represent the layout of hydrogen infrastructure (Yang and Ogden, 2013).

In the model proposed by Gondal and Sahir (2013), the pipeline network of the natural gas distribution companies has been interfaced with a GIS system. The objective function used is based on profit maximization. An integrated renewable hydrogen model based on a MINLP formulation has been developed based on biomass feedstocks as the input material for hydrogen production in Pakistan because of the strong agrarian economy there. The model involves a statistical database and an up-to-date geographical information system to present accurate and logical results for effective energy planning.

There are very few contributions that have reported to date on hydrogen infrastructure modelling across spatial scales, even if the resulting hydrogen network would depend heavily on the country/region-specific conditions. The framework proposed in De-León Almaraz et al. (2015) has addressed the national and regional scales by linking geographic constraints found by the GIS model to the MILP model.

10.4 CONCLUSIONS

A key point in the development of the hydrogen supply chain is the demonstration of the feasibility of its infrastructure, while many technical, economic, and social obstacles must be overcome. Some strategic roadmaps are currently published about the energy potentialities of hydrogen at the European, national, and regional levels. Their main objective is to evaluate some industrial, technological, environmental, and social issues and to identify the main obstacles associated with the hydrogen economy. A literature review of recent dedicated scientific publications revealed that authors agree on the need to develop systemic studies in order to demonstrate the feasibility of the sector and to validate the technical and economic interest in the production and recovery of hydrogen produced from renewable sources. Such works involve the development of models based on economic scenarios for hydrogen deployment.

Following these guidelines, this chapter has presented the various existing approaches to modelling and optimization of the hydrogen supply chain. Designing the hydrogen supply chain in not a trivial task because different alternatives to produce, store, and distribute H_2 exist.

Most works devoted to hydrogen supply chain modelling are based on mathematical programming approaches and are generally limited to monoobjective (cost minimization) or bi-criteria assessment, generally based on either cost-environment or cost-safety. This is not enough when sustainable development must be taken into account in the strategic stage of any new project, when social, economic, and environmental impacts are interconnected. The spatial- or

GIS-based approach cannot be considered as a general methodology for finding the optimal HSC configuration but can be coupled to mathematical programming to design the HSC. Very few contributions have reported to date on hydrogen infrastructure modelling across spatial scales.

Stochastic methods and genetic algorithms, in particular, have been used more recently for HSC optimization and are well suited to handle multiobjective optimization problems because they are able to search for Pareto solutions simultaneously. More efforts toward robust and sophisticated methods are necessary to deal with demand uncertainty, which is a significant parameter in HSC design.

REFERENCES

Agarwal, A., Shankar, R., Tiwari, M.K., 2006. Modelling the metrics of lean, agile and leagile supply chain: an ANP-based approach. Eur. J. Oper. Res. 173, 211–225. https://doi.org/10.1016/j.ejor.2004.12.005.

Agnolucci, P., Akgul, O., McDowall, W., Papageorgiou, L.G., 2013. The importance of economies of scale, transport costs and demand patterns in optimising hydrogen fuelling infrastructure: an exploration with SHIPMod (spatial hydrogen infrastructure planning model). Int. J. Hydrog. Energy 38, 11189–11201. https://doi.org/10.1016/j.ijhydene.2013.06.071.

Akgul, O., Mac Dowell, N., Papageorgiou, L.G., Shah, N., 2014. A mixed integer nonlinear programming (MINLP) supply chain optimisation framework for carbon negative electricity generation using biomass to energy with CCS (BECCS) in the UK. Int. J. Greenh. Gas Control 28, 189–202. https://doi.org/10.1016/j.ijggc.2014.06.017.

Almansoori, A., Betancourt-Torcat, A., 2016. Design of optimization model for a hydrogen supply chain under emission constraints—a case study of Germany. Energy 111, 414–429. https://doi.org/10.1016/j.energy.2016.05.123.

Almansoori, A., Shah, N., 2006. Design and operation of a future hydrogen supply chain. Chem. Eng. Res. Des. 84, 423–438. https://doi.org/10.1205/cherd.05193.

Almansoori, A., Shah, N., 2009. Design and operation of a future hydrogen supply chain: multi-period model. Int. J. Hydrog. Energy 34, 7883–7897. https://doi.org/10.1016/j.ijhydene.2009.07.109.

Almansoori, A., Shah, N., 2012. Design and operation of a stochastic hydrogen supply chain network under demand uncertainty. Int. J. Hydrog. Energy 37, 3965–3977. Portable Fuel Cells – Fundamental and Applications (ISPFC2010). https://doi.org/10.1016/j.ijhydene.2011.11.091.

Bai, Y., Ouyang, Y., Pang, J.-S., 2012. Biofuel supply chain design under competitive agricultural land use and feedstock market equilibrium. Energy Econ. 34, 1623–1633. https://doi.org/10.1016/j.eneco.2012.01.003.

Ball, M., Wietschel, M., Rentz, O., 2006. Integration of a hydrogen economy into the German energy system: an optimising modelling approach. Int. J. Hydrog. Energy 32, 1355–1368. https://doi.org/10.1016/j.ijhydene.2006.10.016.

Banasik, A., Bloemhof-Ruwaard, J.M., Kanellopoulos, A., Claassen, G.D.H., van der Vorst, J.G.A.J., 2016. Multi-criteria decision making approaches for green supply chains: a review. Flex. Serv. Manuf. J. 1–31. https://doi.org/10.1007/s10696-016-9263-5.

Bellman, R.E., Dreyfus, S.E., 1962. Applied Dynamic Programming. Princeton University Press, Princeton.

Boix, M., 2011. Optimisation multicritère de réseaux d'eau [WWW Document]. http://ethesis.inp-toulouse.fr/archive/00001637/. Accessed 17 September 2013.

Buffett, S., Scott, N., 2004. An Algorithm for Procurement in Supply Chain Management. In: AAMAS 2004 Workshop on Trading Agent Design and Analysis.

Busse, S., Brümmer, B., Ihle, R., 2012. Price formation in the German biodiesel supply chain: a Markov-switching vector error-correction modeling approach. Agric. Econ. 43, 545–560. https://doi.org/10.1111/j.1574-0862.2012.00602.x.

Bussiec, M.R., Pruessner, A., 2003. Mixed-Integer Nonlinear Programming. GAMS Dev. Corp., SIAG/OPT Newsletter: Views & News 14 (1), 19–22.

Chen, C.-L., Lee, W.-C., 2004. Multi-objective optimization of multi-echelon supply chain networks with uncertain product demands and prices. Comput. Chem. Eng. 28, 1131–1144.

Chinneck, J.W., 2006. Practical Optimization: A Gentle Introduction. Syst. Comput. Eng. Carlet. Univ. Ott. Httpwww Sce Carlet, Cafacultychinneckpo Html.

Choi, J., Realff, M.J., Lee, J.H., 2006. Approximate dynamic programming: application to process supply chain management. AIChE J 52, 2473–2485.

Coello, C.A.C., Lamont, G.B., Veldhuisen, D.A.V., 2007. Evolutionary Algorithms for Solving Multi-Objective Problems, Vol. 5. Springer, New York.

Collette, Y., Siarry, P., 2003. Multiobjective Optimization: Principles and Case Studies. Springer Science & Business Media, Berlin.

Dagdougui, H., 2011. Decision Support Systems for Sustainable Renewable Energy Systems and Hydrogen Logistics: Modelling, Control and Risk Analysis. (UNIGE - Italy – CRC – MINES ParisTech- France).

Davis, T., 1993. Effective supply chain management. Sloan Manag. Rev. 34, 35.

Dayhim, M., Jafari, M.A., Mazurek, M., 2014. Planning sustainable hydrogen supply chain infrastructure with uncertain demand. Int. J. Hydrog. Energy 39, 6789–6801. https://doi.org/10.1016/j.ijhydene.2014.02.132.

De-León Almaraz, S., Azzaro-Pantel, C., Montastruc, L., Pibouleau, L., Senties, O.B., 2013. Assessment of mono and multi-objective optimization to design a hydrogen supply chain. Int. J. Hydrog. Energy 38, 14121–14145. https://doi.org/10.1016/j.ijhydene.2013.07.059.

De-León Almaraz, S., Azzaro-Pantel, C., Montastruc, L., Domenech, S., 2014. Hydrogen supply chain optimization for deployment scenarios in the Midi-Pyrénées region, France. Int. J. Hydrog. Energy 39, 11831–11845.

De-León Almaraz, S., Azzaro-Pantel, C., Montastruc, L., Boix, M., 2015. Deployment of a hydrogen supply chain by multi-objective/multi-period optimisation at regional and national scales. Chem. Eng. Res. Des. 104, 11–31. https://doi.org/10.1016/j.cherd.2015.07.005.

García, R.G., Moreno, A.V., 2000. Técnicas de diseño de Algoritmos. Universidad Servicio de Publicaciones.

García-Cascales, M.S., Lamata, M.T., 2012. On rank reversal and TOPSIS method. Math. Comput. Model. 56, 123–132. https://doi.org/10.1016/j.mcm.2011.12.022.

Geletu, A., 2008. GAMS - Modeling and Solving Optimization Problems. Institute of Mathematics, Department of Operations Research & Stochastic, Ilmenau University of Technology.

Gigler, J.K., Hendrix, E.M., Heesen, R.A., Van Den Hazelkamp, V.G.W., Meerdink, G., 2002. On optimisation of agri chains by dynamic programming. Eur. J. Oper. Res. 139, 613–625.

Gondal, I.A., Sahir, M.H., 2013. Model for biomass-based renewable hydrogen supply chain. Int. J. Energy Res. 37, 1151–1159. https://doi.org/10.1002/er.2970.

Goodchild, M.F., 2009. Geographic information system. In: Encyclopedia of Database Systems. Springer, USA, pp. 1231–1236.

Govindan, K., Fattahi, M., Keyvanshokooh, E., 2017. Supply chain network design under uncertainty: a comprehensive review and future research directions. Eur. J. Oper. Res. 263, 108–141.

Grossmann, I.E., Guillén-Gosálbez, G., 2010. Scope for the application of mathematical programming techniques in the synthesis and planning of sustainable processes. Comput. Chem. Eng. 34, 1365–1376.

Grossmann, I.E., Caballero, J.A., Yeomans, H., 2000. Advances in mathematical programming for the synthesis of process systems. Lat. Am. Appl. Res. 30, 263–284.

Guillén Gosálbez, G., Mele, F.D., Grossmann, I.E., 2010. A bi criterion optimization approach for the design and planning of hydrogen supply chains for vehicle use. AIChE J 56, 650–667. https://doi.org/10.1002/aic.12024.

Gupta, A., Maranas, C.D., 2003. Managing demand uncertainty in supply chain planning. Comput. Chem. Eng. 27, 1219–1227.

Haeseldonckx, D., D'haeseleer, W., 2011. Concrete transition issues towards a fully-fledged use of hydrogen as an energy carrier: methodology and modelling. Int. J. Hydrog. Energy 36, 4636–4652. https://doi.org/10.1016/j.ijhydene.2011.01.113.

Haimes, Y., Lasdon, L., Wismer, D., 1971. On a bicriterion formulation of the problems of integrated system identification and system optimization. IEEE Trans. Syst. Man Cybern. 1, 296–297. https://doi.org/10.1109/tsmc.1971.4308298.

Han, J.-H., Ryu, J.-H., Lee, I.-B., 2013. Multi-objective optimization design of hydrogen infrastructures simultaneously considering economic cost, safety and CO2 emission. Chem. Eng. Res. Des. 91, 1427–1439. Special Issue: Computer Aided Process Engineering (CAPE) Tools for a Sustainable World. https://doi.org/10.1016/j.cherd.2013.04.026.

Haq, A.N., Kannan, G., 2006. Fuzzy analytical hierarchy process for evaluating and selecting a vendor in a supply chain model. Int. J. Adv. Manuf. Technol. 29, 826–835. https://doi.org/10.1007/s00170-005-2562-8.

Hernandez-Rodriguez, G., 2011. Multiobjective Optimization of Natural Gas Transportation Networks. (Phd).

Hillier, F., Lieberman, G., 2001. Introduction to Operations Research, seventh ed. vol. 1. McGraw-Hill, Michigan University.

Hugo, A., Rutter, P., Pistikopoulos, S., Amorelli, A., Zoia, G., 2005. Hydrogen infrastructure strategic planning using multi-objective optimization. Int. J. Hydrog. Energy 30, 1523–1534. https://doi.org/10.1016/j.ijhydene.2005.04.017.

Hwang, C.L., Yoon, K., 1981. Multiple Attribute Decision Making: Methods and Applications: A State-of-the-Art Survey, Lecture Notes in Economics and Mathematical Systems. Springer-Verlag, Berlin, Heidelberg.

Ingason, H.T., Pall Ingolfsson, H., Jensson, P., 2008. Optimizing site selection for hydrogen production in Iceland. Int. J. Hydrog. Energy 33, 3632–3643. https://doi.org/10.1016/j.ijhydene.2008.04.046.

Johnson, N., Yang, C., Ogden, J., 2008. A GIS-based assessment of coal-based hydrogen infrastructure deployment in the state of Ohio. Int. J. Hydrog. Energy 33, 5287–5303. https://doi.org/10.1016/j.ijhydene.2008.06.069.

Jung, J.Y., Blau, G., Pekny, J.F., Reklaitis, G.V., Eversdyk, D., 2004. A simulation based optimization approach to supply chain management under demand uncertainty. Comput. Chem. Eng. 28, 2087–2106.

Kamarudin, S.K., Daud, W.R.W., Yaakub, Z., Misron, Z., Anuar, W., Yusuf, N.N.A.N., 2009. Synthesis and optimization of future hydrogen energy infrastructure planning in peninsular Malaysia. Int. J. Hydrog. Energy 34, 2077–2088. https://doi.org/10.1016/j.ijhydene.2008.12.086.

Khorram, E., Zarepisheh, M., Ghaznavi-ghosoni, B.A., 2010. Sensitivity analysis on the priority of the objective functions in lexicographic multiple objective linear programs. Eur. J. Oper. Res. 207, 1162–1168. https://doi.org/10.1016/j.ejor.2010.05.016.

Kim, M., Kim, J., 2016. Optimization model for the design and analysis of an integrated renewable hydrogen supply (IRHS) system: application to Korea's hydrogen economy. Int. J. Hydrog. Energy 41, 16613–16626. https://doi.org/10.1016/j.ijhydene.2016.07.079.

Kim, J., Moon, I., 2008. Strategic design of hydrogen infrastructure considering cost and safety using multiobjective optimization. Int. J. Hydrog. Energy 33, 5887–5896. https://doi.org/10.1016/j.ijhydene.2008.07.028.

Kim, J., Lee, Y., Moon, I., 2008. Optimization of a hydrogen supply chain under demand uncertainty. Int. J. Hydrog. Energy 33, 4715–4729. https://doi.org/10.1016/j.ijhydene.2008.06.007.

Kim, J., Realff, M.J., Lee, J.H., 2011. Optimal design and global sensitivity analysis of biomass supply chain networks for biofuels under uncertainty. Comput. Chem. Eng., Energy Syst. Eng. 35, 1738–1751. https://doi.org/10.1016/j.compchemeng.2011.02.008.

Kurata, H., Liu, J.J., 2007. Optimal promotion planning—depth and frequency—for a two-stage supply chain under Markov switching demand. Eur. J. Oper. Res. 177, 1026–1043. https://doi.org/10.1016/j.ejor.2006.01.009.

Landucci, G., Tugnoli, A., Cozzani, V., 2010. Safety assessment of envisaged systems for automotive hydrogen supply and utilization. Int. J. Hydrog. Energy 35, 1493–1505. https://doi.org/10.1016/j.ijhydene.2009.11.097.

Lee, J.H., 2014. Energy supply planning and supply chain optimization under uncertainty. J. Process Control 24, 323–331.

Li, Z., Gao, D., Chang, L., Liu, P., Pistikopoulos, E.N., 2008. Hydrogen infrastructure design and optimization: a case study of China. Int. J. Hydrog. Energy 33, 5275–5286. https://doi.org/10.1016/j.ijhydene.2008.06.076.

Liu, S., Papageorgiou, L.G., 2013. Multiobjective optimisation of production, distribution and capacity planning of global supply chains in the process industry. Omega 41, 369–382. Management Science and Environmental Issues. https://doi.org/10.1016/j.omega.2012.03.007.

Majumder, P., Groenevelt, H., 2001. Competition in remanufacturing. Prod. Oper. Manag. 10, 125–141.

Mansini, R., Ogryczak, W., Speranza, M.G., 2015. Linear and Mixed Integer Programming for Portfolio Optimization. Springer International Publishing.

Marx-Gomez, J., Rautenstrauch, C., Nürnberger, A., Kruse, R., 2002. Neuro-fuzzy approach to forecast returns of scrapped products to recycling and remanufacturing. Knowl.-Based Syst. 15, 119–128.

Mavrotas, G., 2007. Generation of efficient solutions in Multiobjective Mathematical Programming problems using GAMS. Effective implementation of the ε-constraint method. Lect. Lab. Ind. Energy Econ. Sch. Chem. Eng. Natl. Tech. Univ. Athens.

Mavrotas, G., 2009. Effective implementation of the ε-constraint method in multi-objective mathematical programming problems. Appl. Math. Comput. 213, 455–465. https://doi.org/10.1016/j.amc.2009.03.037.

Momoh, J.A., 2008. Electric Power System Applications of Optimization. second ed.CRC Press, Boca Raton.

Moreno-Benito, M., Agnolucci, P., Papageorgiou, L.G., 2016. Towards a sustainable hydrogen economy: optimisation-based framework for hydrogen infrastructure development. Comput. Chem. Eng. 102, 110–127. https://doi.org/10.1016/j.compchemeng.2016.08.005.

Myerson, R.B., 2013. Game Theory. Harvard University Press, Cambridge, MA.

Nagurney, A., Toyasaki, F., 2005. Reverse supply chain management and electronic waste recycling: a multitiered network equilibrium framework for e-cycling. Transp. Res. Part E Logist. Transp. Rev. 41, 1–28.

Nagurney, A., Dong, J., Zhang, D., 2002. A supply chain network equilibrium model. Transp. Res. Part E Logist. Transp. Rev. 38, 281–303. https://doi.org/10.1016/S1366-5545(01)00020-5.

Nepal, B., Monplaisir, L., Famuyiwa, O., 2011. A multi-objective supply chain configuration model for new products. Int. J. Prod. Res. https://doi.org/10.1080/00207543.2010.511294.

Nunes, P., Oliveira, F., Hamacher, S., Almansoori, A., 2015. Design of a hydrogen supply chain with uncertainty. Int. J. Hydrog. Energy 40, 16408–16418.

Ochoa Robles, J., De-León Almaraz, S., Azzaro-Pantel, C., 2015. In: Design of experiments for sensitivity analysis in multiobjective optimization of hydrogen supply chain.Presented at the ECOS 2015, Pau, France.

Ochoa Robles, J., De-León Almaraz, S., Azzaro-Pantel, C., 2016. Optimization of a hydrogen supply chain network design by multi-objective genetic algorithms. In: Bogataj, Z.K. (Ed.), Computer Aided Chemical Engineering, 26th European Symposium on Computer Aided Process Engineering. Elsevier, pp. 805–810. https://doi.org/10.1016/B978-0-444-63428-3.50139-9.

Pannell, D.J., 1997. Sensitivity analysis: strategies, methods, concepts, examples. Agric. Econ. 16, 139–152.

Patay, E., 2008. Déploiement de la chaîne logistique de l'hydrogène pour le marché des carburants en 2050: Conception et développement d'un outil d'aide à la décision pour l'analyse de scénarios. Ecole Centrale Paris-Laboratoire Génie Industriel, Paris.

Peidro, D., Mula, J., Poler, R., Verdegay, J.-L., 2009. Fuzzy optimization for supply chain planning under supply, demand and process uncertainties. Fuzzy Sets Syst. 160, 2640–2657.

Perea-López, E., Ydstie, B.E., Grossmann, I.E., 2003. A model predictive control strategy for supply chain optimization. Comput. Chem. Eng. 27, 1201–1218. 2nd Pan American Workshop in Process Systems Engineering. https://doi.org/10.1016/S0098-1354(03)00047-4.

Pishvaee, M.S., Rabbani, M., Torabi, S.A., 2011. A robust optimization approach to closed-loop supply chain network design under uncertainty. Appl. Math. Model. 35, 637–649. https://doi.org/10.1016/j.apm.2010.07.013.

Prawda, J., 2004. Métodos y modelos de investigacion de operaciones I: Modelos deterministicos. Limusa, Mexico.

Qadrdan, M., Saboohi, Y., Shayegan, J., 2008. A model for investigation of optimal hydrogen pathway, and evaluation of environmental impacts of hydrogen supply system. Int. J. Hydrog. Energy 33, 7314–7325. https://doi.org/10.1016/j.ijhydene.2008.09.031.

Ravi, V., Shankar, R., Tiwari, M.K., 2005. Analyzing alternatives in reverse logistics for end-of-life computers: ANP and balanced scorecard approach. Comput. Ind. Eng. 48, 327–356.

Ren, L., Zhang, Y., Wang, Y., Sun, Z., 2007. Comparative Analysis of a NovelM-TOPSISMethod and TOPSIS. Appl. Math. Res. Express. 2007,10 pages. https://doi.org/10.1093/amrx/abm005.

Ren, J., Manzardo, A., Toniolo, S., Scipioni, A., 2013. Sustainability of hydrogen supply chain. Part I: Identification of critical criteria and cause-effect analysis for enhancing the sustainability using DEMATEL. Int. J. Hydrog. Energy 38, 14159–14171. https://doi.org/10.1016/j.ijhydene.2013.08.126.

Rosyid, O.A., Jablonski, D., Hauptmanns, U., 2007. Risk analysis for the infrastructure of a hydrogen economy. Int. J. Hydrog. Energy 32, 3194–3200. International Symposium on Solar-Hydrogen-Fuel Cells 2005. https://doi.org/10.1016/j.ijhydene.2007.02.012.

Roy, B., 1968. Classement et choix en présence de points de vue multiples. RAIRO - Oper. Res. - Rech. Opérationnelle 2, 57–75.

Saaty, T.L., 1980. The Analytic Hierarchy Process: Planning, Priority Setting, Resource Allocation. McGraw-Hill, New York; London.

Saaty, T.L., 1996. Decision Making with Dependence and Feedback: The Analytic Network Process. RWS Publications, Pittsburgh.

Sabio, N., Gadalla, M., Guillén-Gosálbez, G., Jiménez, L., 2010. Strategic planning with risk control of hydrogen supply chains for vehicle use under uncertainty in operating costs: a case study of Spain. Int. J. Hydrog. Energy 35, 6836–6852.

Sabio, N., Kostin, A., Guillén-Gosálbez, G., Jiménez, L., 2011. Holistic minimization of the life cycle environmental impact of hydrogen infrastructures using multi-objective optimization and principal component analysis. Int. J. Hydrog. Energy 37, 5385–5405. https://doi.org/10.1016/j.ijhydene.2011.09.039.

Sakawa, M., 2012. Genetic Algorithms and Fuzzy Multiobjective Optimization. Springer Science & Business Media, New York.

Samsatli, S., Staffell, I., Samsatli, N.J., 2016. Optimal design and operation of integrated wind-hydrogen-electricity networks for decarbonising the domestic transport sector in great Britain. Int. J. Hydrog. Energy 41, 447–475. https://doi.org/10.1016/j.ijhydene.2015.10.032.

Sarkis, J., 1999. A methodological framework for evaluating environmentally conscious manufacturing programs. Comput. Ind. Eng. 36, 793–810.

Shabani, N., Sowlati, T., 2013. A mixed integer non-linear programming model for tactical value chain optimization of a wood biomass power plant. Appl. Energy 104, 353–361. https://doi.org/10.1016/j.apenergy.2012.11.013.

Shaw, K., Shankar, R., Yadav, S.S., Thakur, L.S., 2012. Supplier selection using fuzzy AHP and fuzzy multi-objective linear programming for developing low carbon supply chain. Expert Syst. Appl. 39, 8182–8192. https://doi.org/10.1016/j.eswa.2012.01.149.

Simangunsong, E., Hendry, L.C., Stevenson, M., 2012. Supply-chain uncertainty: a review and theoretical foundation for future research. Int. J. Prod. Res. 50, 4493–4523.

Soylu, A., Oruç, C., Turkay, M., Fujita, K., Asakura, T., 2006. Synergy analysis of collaborative supply chain management in energy systems using multi-period MILP. Eur. J. Oper. Res. 174, 387–403. https://doi.org/10.1016/j.ejor.2005.02.042.

Srivastava, S.K., 2007. Green supply-chain management: a state-of-the-art literature review. Int. J. Manag. Rev. 9, 53–80.

Tabkhi, F., 2007. Optimisation de réseaux de transport de gaz, Ph.D. thesis. Institut National Polytechnique de Toulouse, Toulouse, France.

Toktaş-Palut, P., Ülengin, F., 2010. Modeling a supply chain as a queuing system.IFAC Proc. Vol., 5th IFAC Conference on Management and Control of Production Logistics. 43, pp. 242–247. https://doi.org/10.3182/20100908-3-PT-3007.00047.

Tseng, M.-L., Chiang, J.H., Lan, L.W., 2009. Selection of optimal supplier in supply chain management strategy with analytic network process and choquet integral. Comput. Ind. Eng. 57, 330–340. Collaborative e-Work Networks in Industrial Engineering. https://doi.org/10.1016/j.cie.2008.12.001.

Tsiakis, P., Papageorgiou, L.G., 2008. Optimal production allocation and distribution supply chain networks. Int. J. Prod. Econ. 111, 468–483. Special Section on Sustainable Supply Chain. https://doi.org/10.1016/j.ijpe.2007.02.035.

Tsiakis, P., Shah, N., Pantelides, C.C., 2001. Design of multi-echelon supply chain networks under demand uncertainty. Ind. Eng. Chem. Res. 40, 3585–3604.

van der Vorst, J.G.A.J., Beulens, A.J.M., 2002. Identifying sources of uncertainty to generate supply chain redesign strategies. Int. J. Phys. Distrib. Logist. Manag. 32, 409–430. https://doi.org/10.1108/09600030210437951.

van Dyken, S., Bakken, B.H., Skjelbred, H.I., 2010. Linear mixed-integer models for biomass supply chains with transport, storage and processing. Energy 35, 1338–1350. https://doi.org/10.1016/j.energy.2009.11.017.

Williams, J.F., 1983. A hybrid algorithm for simultaneous scheduling of production and distribution in multi-echelon structures. Manag. Sci. 29, 77–92.

Woo, Y., Cho, S., Kim, J., Kim, B.S., 2016. Optimization-based approach for strategic design and operation of a biomass-to-hydrogen supply chain. Int. J. Hydrog. Energy 41, 5405–5418. https://doi.org/10.1016/j.ijhydene.2016.01.153.

Yang, C., Ogden, J.M., 2013. Renewable and low carbon hydrogen for California—modeling the long term evolution of fuel infrastructure using a quasi-spatial TIMES model. Int. J. Hydrog. Energy 38, 4250–4265. https://doi.org/10.1016/j.ijhydene.2013.01.195.

You, F., Grossmann, I.E., 2008. Design of responsive supply chains under demand uncertainty. Comput. Chem. Eng. 32, 3090–3111.

Zhou, L., Liao, Z., Wang, J., Jiang, B., Yang, Y., Hui, D., 2013. Optimal design of sustainable hydrogen networks. Int. J. Hydrog. Energy 38, 2937–2950. https://doi.org/10.1016/j.ijhydene.2012.12.084.

Chapter 11

Multiobjective Life Cycle Optimization of Hydrogen Supply Chains

Michael Ehrenstein and Gonzalo Guillén-Gosálbez
Centre for Process Systems Engineering (CPSE), Imperial College London, London, United Kingdom

NOTATION

Indices

b	environmental burdens
i	hydrogen form
g	grid zones
l	transportation mode
p	manufacturing technologies
s	storage technologies
t	time period

Sets

$IL(l)$	set of hydrogen forms that can be transported via transportation mode l
$IS(s)$	set of hydrogen forms that can be stored via technology s
$IP(p)$	set of hydrogen forms that can be produced by manufacturing technology p
$PI(i)$	set of manufacturing technologies that can produce hydrogen form i
$LI(i)$	set of transportation modes that can transport hydrogen form i
$SI(i)$	set of storage technologies that can store hydrogen form i

Parameters

av_l	availability of transportation mode l
ccl_t	capital cost of transport mode l in period t

Hydrogen Supply Chain. https://doi.org/10.1016/B978-0-12-811197-0.00011-7

$\mathbf{cudl}_t$ maintenance cost of transportation mode l in period t per unit of distance travelled

$\boldsymbol{Dg_t}$ total demand of hydrogen in grid g in period t

$\mathbf{distance}_{gg'}$ average distance travelled between grids g and g'

dsat demand satisfaction level to be fulfilled

$\mathbf{fuelc}_l$ fuel consumption of transportation mode l

$\mathbf{fuelp}_{lt}$ price of the fuel consumed by transportation mode l in period t

$\mathbf{ge}_{lt}$ general expenses of transportation mode l in period t

ir interest rate

$\mathbf{lutime}_l$ loading/unloading time of transportation mode l

$\underline{\mathbf{PC}_p^{PL}}$ upper bound on the capacity expansion of manufacturing technology p

$\overline{\mathbf{PC}_p^{PL}}$ lower bound on the capacity expansion of manufacturing technology p

$\overline{\mathbf{QC}_{gg'l}}$ upper bound on the flow of materials between grids g and $g0$ via transportation model l

$\overline{\mathbf{QC}_{gg'l}}$ lower bound on the flow of materials between grids g and $g0$ via transportation model l

$\overline{\mathbf{SC}_s^{ST}}$ upper bound on the capacity expansion of storage technology s

$\underline{\mathbf{SC}_s^{ST}}$ lower bound on the capacity expansion of storage technology s

$\mathbf{speed}_l$ average speed of transportation mode l

$\mathbf{tcap}_l$ capacity of transport mode l

$\mathbf{UB}_{gpt}^{PL}$ upper bound on the number of plants of type p installed in grid g in period t

$\mathbf{UB}_{gst}^{ST}$ upper bound on the number of storage facilities of types installed in grid g in period t

$\mathbf{UB}_{lt}^{TR}$ upper bound on the number of transportation units of type l purchased in period t (integer variable)

$\mathbf{upc}_{igpt}$ unit production cost of hydrogen form i produced via technology p in grid g in period t

$\mathbf{usc}_{igst}$ unit storage cost of hydrogen form i stored via technology s in grid g in period t

$\mathbf{wage}_{lt}$ driver wage of transportation mode l in period t

α_{gpt}^{PL} fixed investment term associated with manufacturing technology p installed in grid g in period t

α_{gst}^{ST} fixed investment term associated with storage technology s installed in grid g in period t

β_{gpt}^{PL} variable investment term associated with manufacturing technology p installed in grid g in period t

β_{gst}^{ST} variable investment term associated with storage technology s installed in grid g in period t

ω_{bp}^{PR} emissions of chemical b associated with the production of one unit of hydrogen via technology p

$\omega_{bi}^{\mathrm{ST}}$ emissions of chemical b associated with the compression of one unit of hydrogen into physical form i

$\omega_{bl}^{\mathrm{TR}}$ emissions of chemical b per unit of mass transported one unit of distance with technology l

θ average storage period

τ minimum desired percentage of the capacity that must be utilized

m_b damage factor associated with product b

Variables

C_{gpt}^{PL} capacity of manufacturing technology p in grid g in period t

C_{gst}^{ST} capacity of storage technology s in grid g in period t

$\mathrm{CE}_{gpt}^{\mathrm{PL}}$ capacity expansion of manufacturing technology p in grid g in period t

$\mathrm{CE}_{gst}^{\mathrm{ST}}$ capacity expansion of storage technology s in grid g in period t

D_{igt} amount of hydrogen form i distributed in grid g in period t

DAM damage in human health due to climate change

FC_t fuel cost in period t

FCC_t facility capital cost in period t

FOC_t facility operating cost in period t

GC_t general cost in period t

LC_t labor cost in period t

LCI_b life cycle inventory of emissions of chemical b

MC_t maintenance cost in period t

N_{gpt}^{PL} number of plants of type p installed in grid g in period t (integer variable)

N_{gst}^{ST} number of storage facilities of types installed in grid g in period t (integer variable)

N_{lt}^{TR} number of transportation units of type l purchased in period t (integer variable)

PR_{igpt} production of hydrogen mode i via technology p in period t in grid g

$\mathrm{Q}_{igg'lt}$ flow of hydrogen mode i via transportation mode l between grids g and $g0$ in period t

S_{igst} amount of hydrogen in physical form i stored via technology s in grid g in period t

TC_t total amount of money spent in period t

TCC_t total transportation capital cost in period t

TDC total discounted cost

TMC_{lt} transportation capital cost of mode l in period t

TOC_t transportation operating cost in period t

$\mathrm{X}_{gg'lt}$ binary variable (1 if a link between grids g and $g0$ using transportation technology l is established, 0 otherwise)

11.1 INTRODUCTION

The lack of infrastructures to produce, store, and deliver hydrogen can become a major obstacle when transitioning toward a cleaner energy system. Hydrogen supply chains encompass a set of production, storage, and distribution echelons that all together produce hydrogen and deliver it to the final customers in the right location and at the right time. At present, a wide variety of hydrogen technologies are being investigated, including steam methane reforming, coal gasification, water electrolysis, and biomass gasification, among others. Likewise, hydrogen can be stored in different ways and the same applies to its transportation. All these production, storage, and distribution technologies differ in CAPEX and OPEX expenditures as well as in economic and environmental performance (Guillén-Gosálbez et al., 2010; Balat and Kirtay, 2010; Smitkova et al., 2011). Hence, to design a cost effective and environmentally friendly hydrogen network, one needs to assess all such technologies to ultimately select the best in terms of some specific criteria. Mathematical techniques can be applied in this context in order to automate the generation and screening of thousands of alternatives considering simultaneously technical, physical, and legal constraints, as well as multiple sustainability criteria.

The development of mathematical models for optimizing hydrogen supply chains has been the focus of substantial research in the last years. The preferred approach has relied on mixed-integer linear programming formulations (MILP), in which binary variables denote the selection of technologies and establishment of transportation links, while continuous ones denote mass and energy flows, capacities of the SC nodes, and cost and environmental performance metrics. These decision variables are optimized subject to mass balance constraints, capacity limitations, and objective function calculations. The resulting models are often NP-hard and therefore tend to lead to very large CPU times, making the solution of large-scale instances impractical. Hence, the mathematical formulation must be accompanied, in many instances, by a customized decomposition algorithm that expedites the search for optimal solutions by exploiting the underlying mathematical formulation (Grossmann and Guillén-Gosálbez, 2010).

One critical aspect of these MIP models concerns the selection of an appropriate objective function to drive the optimization task. The economic performance has been traditionally the objective function of choice. As an example, Almansoori and Shah (2009) developed a model that seeks to minimize the total cost of the future UK hydrogen SC. Ingason et al. (2008) optimized the design of a hydrogen network in Iceland in terms of its annual cost. The model of Lin et al. (2008) minimizes also the total cost of a hydrogen supply chain in South California. Kim et al. (2008) optimized the total daily cost in a South Korean hydrogen SC considering demand uncertainty. Sabio et al. (2010) developed a stochastic model to optimize the cost of a hydrogen SC in Spain. The supply chain design models mentioned above can be either deterministic, if all the parameters are assumed to be perfectly known in advance, or stochastic, when some of them can take values within a given interval. The preferred approach to deal with the second type of problem has been stochastic programming (as in Kim et al., 2008; Sabio et al., 2010), where decisions are made in several stages so as to be adjusted to the realization of the uncertain events.

The recent trend toward the development of more sustainable processes has recently resulted in the need to enlarge the scope of the analysis carried out in hydrogen supply chains beyond the economic performance. Indeed, optimizing hydrogen networks based solely on economic metrics may lead to solutions that do not fully exploit the environmental benefits of a hydrogen-based economy (Sabio et al., 2012). To identify more sustainable designs, it is necessary to include environmental concerns (in addition to economic aspects) into the design problem. By framing the design task in this way, one inevitably faces a multicriteria decision-making problem in which conflicting goals must be harmonized so as to identify the solution that best meets the stakeholders' preferences. In this context, multicriteria decision-support tools, and, more particularly, multiobjective optimization provides a conceptual and computational framework to tackle these problems effectively.

Along these lines, Hugo et al. (2005) proposed an MILP model to address the long-term strategic planning of a multiechelon hydrogen network considering economic as well as environmental criteria (greenhouse gas (GHG) emissions). Guillén-Gosálbez et al. (2010) minimized the damage to human health caused by climate change in the design of a hydrogen SC. Li et al. (2008) developed an MILP model to optimize the GHG emissions and profit in the design of a hydrogen network. More recently, De-León Alamaraz and coworkers addressed the design of a hydrogen SC in the Midi-Pyrénées regions considering several economic and environmental objectives simultaneously (De-León Almaraz et al., 2014; De-León Almaraz et al., 2015).

Note that most of these works have focused on optimizing only two objectives, one economic and one environmental. What motivates this simplification is the fact that the complexity of multiobjective models grows rapidly with the number of objectives from the viewpoints of generation and analysis of the Pareto solutions. One possible manner to overcome this limitation consists of applying dimensionality reduction methods, which allow identifying and eliminating redundant objectives without losing information (Brockhoff and Zitzler, 2006). This is indeed the approach followed in (Sabio et al., 2012), which accounted for several environmental metrics in the design of a hydrogen SC in Spain. To deal with several environmental objectives, the authors applied a dimensionality reduction technique based on Principal Component Analysis to identify and eliminate redundant metrics from the analysis.

In this chapter we address the multiobjective optimization of hydrogen networks. We first formally state the problem of interest and then introduce a MILP formulation to tackle it that accounts for economic and environmental criteria. Some numerical results are discussed next, and the conclusions of the work are finally drawn in the last section of the chapter.

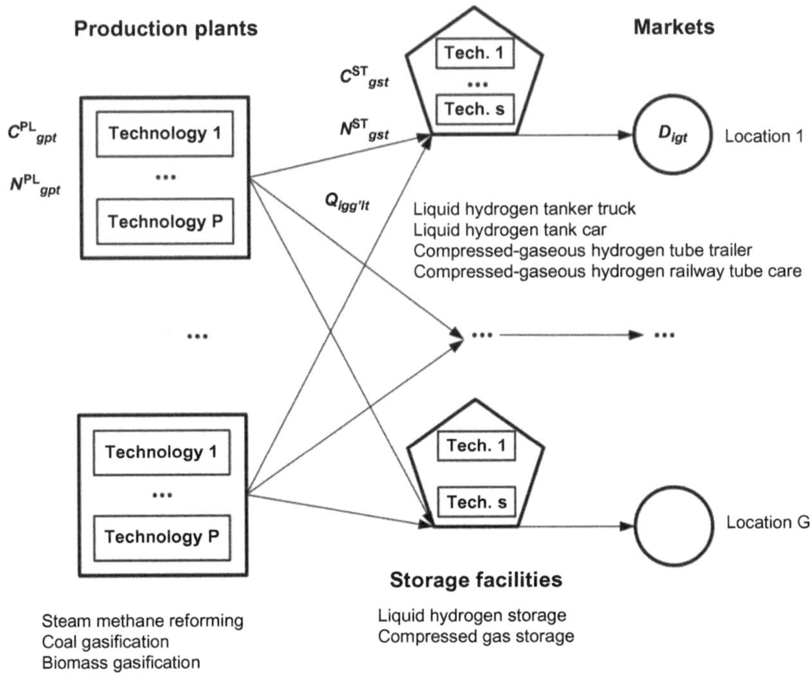

FIG. 11.1 Supply chain configuration considered when deriving the mathematical model.

11.2 MATHEMATICAL FORMULATION

11.2.1 Problem Statement

We address the design of a standard supply chain that produces hydrogen for vehicle use as the one sketched in Fig. 11.1. We are given a region of interest that is divided into a set of subregions. Each such subregion features a hydrogen demand that needs to be covered by a hydrogen supply chain encompassing a set of production and storage facilities as well as a set of transportation units that connect the different subregions. We are also given cost and environmental data of each such technology as well as availability of raw materials, capacity bounds, and mass balance coefficients. The goal of the analysis is to determine the optimal network configuration, including the selection of technologies, their expansion in capacity over time and the transportation flows between subregions, such that hydrogen demand is met while optimizing both the economic and environmental performance of the network. The section that follows provides a mathematical formulation that effectively solves this problem.

11.2.2 Model Equations

To solve the problem stated above, we derive an MILP model that essentially contains mass balances, capacity limitations, and objective function equations.

To construct this MILP, we consider a standard network "production-storage-markets," in which every echelon can be established in any subregion of a wider region of interest (i.e., grid). This spatially explicit model is defined over a number of time periods in order to cover a hydrogen demand pattern that changes over time. An outline of the model including its main equations is provided next, while the reader can find further details on the MILP and the problem data in previous publications (Guillén-Gosálbez et al., 2010).

11.2.2.1 Mass Balance

The mass balance ensures that for every grid g, time period t, and product form i, the inventory S_{igst}, sales D_{igt}, and outgoing transport $Q_{igg'lt}$ match the inventory left from the previous time period S_{igst-1}, plus the production PR_{igpt} and the incoming transport $Q_{ig'glt}$. Indices s, p, and l denote storage technologies, plant technologies, and methods of transportation, respectively. The set $SI(i)$ contains the storage technologies that can store hydrogen of form i, while set $PI(i)$ denotes plant types that can produce hydrogen of form i.

$$\sum_{s\in SI(i)} S_{igst} + D_{igt} + \sum_{g'\neq g}\sum_{l} Q_{igg'lt} = \sum_{s\in SI(i)} S_{igst-1} + \sum_{p\in PI(i)} \mathrm{PR}_{igpt} + \sum_{g'\neq g}\sum_{l} Q_{ig'glt} \quad \forall\ i,g,t \tag{11.1}$$

The sales D_{igt} in each location and time period must be lower than the demand $\overline{D_{gt}}$, and higher than a certain demand satisfaction percentage *dsat*.

$$\overline{D_{gt}}\mathrm{dsat} \leq \sum_{i} D_{igt} \leq \overline{D_{gt}} \quad \forall\ g,t \tag{11.2}$$

11.2.2.2 Capacity Constraints

The total production rate cannot exceed the installed plant capacity C^{PL}_{gpt}, and should be higher than a given ratio τ of the installed capacity. Here, set $IP(p)$ represents hydrogen forms that can be manufactured with technology p.

$$\tau C^{\mathrm{PL}}_{gpt} \leq \sum_{i\in IP(p)} \mathrm{PR}_{igpt} \leq C^{\mathrm{PL}}_{gpt} \quad \forall\ g,p,t \tag{11.3}$$

In every time period, there is the option of expanding the plant capacity by an amount $\mathrm{CE}^{\mathrm{PL}}_{gpt}$.

$$C^{\mathrm{PL}}_{gpt} = C^{\mathrm{PL}}_{gpt-1} + \mathrm{CE}^{\mathrm{PL}}_{gpt} \quad \forall\ g,p,t \tag{11.4}$$

$\underline{\mathrm{PC}^{\mathrm{PL}}_{p}}$ and $\overline{\mathrm{PC}^{\mathrm{PL}}_{p}}$ denote lower and upper bounds, respectively, on the capacity expansion of each technology. These bounds are multiplied with the number of plants of every type that are established in each grid and time period, N^{PL}_{gpt}.

$$\underline{\mathrm{PC}^{\mathrm{PL}}_{p}} N^{\mathrm{PL}}_{gpt} \leq \mathrm{CE}^{\mathrm{PL}}_{gpt} \leq \overline{\mathrm{PC}^{\mathrm{PL}}_{p}} N^{\mathrm{PL}}_{gpt} \quad \forall\ g,p,t \tag{11.5}$$

As with production, inventory levels at each location and time cannot exceed the installed storage capacities C^{ST}_{gst}:

$$\sum_{i \in IS(s)} S_{igst} \leq C^{\mathrm{ST}}_{gst} \quad \forall\, g,s,t \tag{11.6}$$

In this equation, set $IS(s)$ denotes product forms i that can be stored using storage technology s. We further assume that the storage capacity needed for storing and handling of the products is at least twice the average inventory level, which is calculated from the storage period θ and the amount delivered to customers.

$$2\left(\theta D_{igt}\right) \leq \sum_{s \in SI(i)} C^{\mathrm{ST}}_{gst} \quad \forall\, i,g,t \tag{11.7}$$

Both the expansion of the storage capacity as well as the bounds on these expansions are modeled similarly to the equivalent production constraints.

$$C^{\mathrm{ST}}_{gst} = C^{\mathrm{ST}}_{gst-1} + \mathrm{CE}^{\mathrm{ST}}_{gst} \quad \forall\, g,s,t \tag{11.8}$$

The storage capacity is determined from the number of storage facilities as follows:

$$\underline{\mathrm{SC}^{\mathrm{ST}}_{s}} N^{\mathrm{ST}}_{gst} \leq \mathrm{CE}^{\mathrm{ST}}_{gst} \leq \overline{\mathrm{SC}^{\mathrm{ST}}_{s}} N^{\mathrm{ST}}_{gst} \quad \forall\, g,s,t \tag{11.9}$$

Variables $\underline{\mathrm{SC}^{\mathrm{ST}}_{s}}$ and $\overline{\mathrm{SC}^{\mathrm{ST}}_{s}}$ represent the lower and upper capacity bounds, while the integer variable N^{ST}_{gst} stands for the number of storage units.

11.2.2.3 Transport Flows

Eq. (11.10) employs binary variable $X_{gg'lt}$ to define whether there exists a transport connection between two locations. If the variable takes on a value of one, the amount of material moved ($Q_{igg'lt}$) is constrained between lower and upper limits, denoted by $\underline{\mathrm{QC}_{lgg'}}$ and $\overline{\mathrm{QC}_{lgg'}}$, respectively.

$$\begin{aligned} \mathrm{FCC}_t = & \sum_{g}\sum_{p}\left(\alpha^{\mathrm{PL}}_{gpt} N^{\mathrm{PL}}_{gpt} + \beta^{\mathrm{PL}}_{gpt} \mathrm{CE}^{\mathrm{PL}}_{gpt}\right) \\ & + \sum_{g}\sum_{p}\left(\alpha^{\mathrm{ST}}_{gst} N^{\mathrm{ST}}_{gst} + \beta^{\mathrm{ST}}_{gst} \mathrm{CE}^{\mathrm{ST}}_{gst}\right) \quad \forall\, t \end{aligned} \tag{11.10}$$

This model only allows for one-way traffic between two locations, because grids that require imports to satisfy their demand will not have spare product to export. This reasoning is expressed in Eq. (11.11).

$$X_{gg'lt} + X_{g'glt} \leq 1 \quad \forall\, g,g'(g \neq g'),l,t \tag{11.11}$$

11.2.2.4 Objective Function Calculations

The model must minimize the total cost and environmental impact. These objectives are described in detail next.

Total Cost

The final costs are calculated by summing up the discounted costs for each time period, where TC_t is the cost in period t, and ir is the interest rate. In turn, the costs for every period can be calculated as the summation of the facility capital costs FCC_t, transport capital costs TCC_t, facility operating costs FOC_t and transport operating costs TOC_t.

$$\mathrm{TCOST} = \sum_t \frac{\mathrm{TC}_t}{(1+\mathrm{ir})^{t-1}} \tag{11.12}$$

$$\mathrm{TC}_t = \mathrm{FCC}_t + \mathrm{TCC}_t + \mathrm{FOC}_t + \mathrm{TOC}_t \quad \forall\, t \tag{11.13}$$

Facility capital costs are determined in Eq. (11.14), with the first and second half representing plant and storage costs, respectively. Parameters α and β denote fixed and variable investment terms.

$$\begin{aligned}\mathrm{FCC}_t = &\sum_g \sum_p \left(\alpha^{\mathrm{PL}}_{gpt} N^{\mathrm{PL}}_{gpt} + \beta^{\mathrm{PL}}_{gpt} \mathrm{CE}^{\mathrm{PL}}_{gpt}\right) \\ &+ \sum_g \sum_p \left(\alpha^{\mathrm{ST}}_{gst} N^{\mathrm{ST}}_{gst} + \beta^{\mathrm{ST}}_{gst} \mathrm{CE}^{\mathrm{ST}}_{gst}\right) \quad \forall\, t\end{aligned} \tag{11.14}$$

Capital costs for transport, which account for trucks and railcars, are determined from the number of purchased transport units of each type and the associated unitary costs cc_{lt}.

$$\mathrm{TCC}_t = \sum_l N^{\mathrm{TR}}_{lt} \mathrm{cc}_{lt} \quad \forall\, t \tag{11.15}$$

Eq. (11.16) ensures that the total transportation flow can be managed by the number of transport units purchased before the current time period t. Furthermore, for every mode of transport, parameters for transport availability (av_t), container capacity (tcap_l), (un)loading time (lutime_l), and speed are employed, while the distance between two locations g and g' is multiplied by two in order to account for the return journey.

$$\sum_{t' \le t} N^{\mathrm{TR}}_{lt'} \ge \sum_{i \in IL(l)} \sum_g \sum_{g' \ne g} \frac{Q_{igg'lt}}{\mathrm{av}_t \mathrm{tcap}_l} \left(\frac{2\mathrm{distance}_{gg'}}{\mathrm{speed}_l} + \mathrm{lutime}_l\right) \quad \forall\, l, t \tag{11.16}$$

Facility operating costs are separated into costs for production and storage, with upc_{igpt} and usc_{igst} respectively denoting unit production and storage costs. The average inventory levels are given as the product of the storage period θ and the demand D_{igt}.

$$\mathrm{FOC}_t = \sum_i \sum_g \sum_{p \in PI(i)} \mathrm{upc}_{igpt} \mathrm{PR}_{igpt} + \sum_i \sum_g \sum_{s \in SI(i)} \mathrm{usc}_{igst} \theta D_{igt} \quad \forall\, t \tag{11.17}$$

On the other hand, the transport operating costs are calculated as follows:

$$\mathrm{TOC}_t = \mathrm{FC}_t + \mathrm{LC}_t + \mathrm{MC}_t + \mathrm{GC}_t \quad \forall\, t \tag{11.18}$$

Where each of the terms accounting for fuel, labor, maintenance, and general costs, is calculated as follows:

$$\mathrm{FC}_t = \sum_i \sum_g \sum_{g' \neq g} \sum_{l \in LI(i)} \mathrm{fuelp}_{lt} \frac{2\mathrm{distance}_{gg'} Q_{igg'lt}}{\mathrm{fuelc}_l \mathrm{tcap}_l} \quad \forall\, t \tag{11.19}$$

$$\mathrm{LC}_t = \sum_i \sum_g \sum_{g' \neq g} \sum_{l \in LI(i)} \mathrm{wage}_{lt} \left[\frac{Q_{igg'lt}}{\mathrm{tcap}_l} \left(\frac{2\mathrm{distance}_{gg'}}{\mathrm{speed}_l} + \mathrm{lutime}_l \right) \right] \quad \forall\, t \tag{11.20}$$

$$\mathrm{MC}_t = \sum_i \sum_g \sum_{g' \neq g} \sum_{l \in LI(i)} \mathrm{cud}_{lt} \left(\frac{2\mathrm{distance}_{gg'} Q_{igg'lt}}{\mathrm{tcap}_l} \right) \quad \forall\, t \tag{11.21}$$

$$\mathrm{GC}_t = \sum_l \sum_{t' \leq t} \mathrm{ge}_{lt} N_{lt'}^{\mathrm{TR}} \quad \forall\, t \tag{11.22}$$

Environmental Impact

The environmental impact is quantified following life cycle assessment principles. More precisely, the eco-indicator 99 is the method of choice. This metric is calculated from the amount of hydrogen produced and stored and the transportation flows between SC echelons. The calculations can be performed in two different ways. One option is to follow two steps, as illustrated in Fig. 11.2. Here, the life cycle inventory entries of feedstocks requirements, emissions, and waste are first obtained from the production, storage, and transport tasks, while the impact is calculated afterward from this information.

In this work, however, we calculate the impact straight away from the production, storage, and transport flows using Eqs. (11.23)–(11.27), where parameters ω denote the impact per functional unit selected.

$$\mathrm{EI99}^{\mathrm{tot}} = \sum_d \mathrm{WN}_d \mathrm{EI99}_d \tag{11.23}$$

$$\mathrm{EI99}_d = \mathrm{EI99}_d^{\mathrm{PL}} + \mathrm{EI99}_d^{\mathrm{ST}} + \mathrm{EI99}_d^{\mathrm{TR}} \quad \forall\, d \tag{11.24}$$

$$\mathrm{EI99}_d^{\mathrm{PL}} = \sum_i \sum_g \sum_p \sum_t \mathrm{PR}_{igpt} \omega_{dp}^{\mathrm{PL}} \quad \forall\, d \tag{11.25}$$

$$\mathrm{EI99}_d^{\mathrm{ST}} = \sum_i \sum_g \sum_p \sum_t \mathrm{PR}_{igpt} \omega_{di}^{\mathrm{ST}} \quad \forall\, d \tag{11.26}$$

$$\mathrm{EI99}_d^{\mathrm{TR}} = \sum_i \sum_g \sum_{g \neq g'} \sum_{l \in \mathrm{LI}(i)} \sum_t Q_{igg'lt} \mathrm{distance}_{gg'} \omega_{dl}^{\mathrm{TR}} \quad \forall\, d \tag{11.27}$$

In essence, these equations link the continuous variables of the problem, denoting the amount of hydrogen produced, stored, and transported to the associated environmental impact. To this end, they include parameters ω, whose

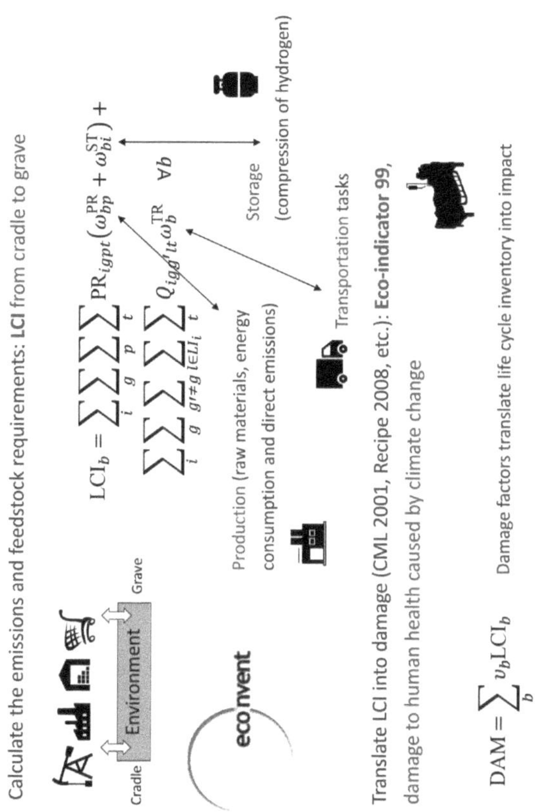

FIG. 11.2 Details on the LCA calculations.

values are retrieved from environmental databases. Such parameters provide the amount of emissions/impact per unit of functional unit (e.g., impact per kg of hydrogen produced). Further details on how the impact is calculated are provided elsewhere (Guillén-Gosálbez et al., 2010).

Finally, the MILP model can be expressed in compact form as follows:

$$\min_{x,y,z}\left(TCOST, EI99^{TOT}\right)$$
$$s.t. constraints\ 1-27$$
$$x \in \mathbb{R}, y \in \{0,1\}, z \in \mathbb{N}$$

where x, y, and z denote the continuous, binary, and integer variables of the MILP model introduced above, which are described in detail in the notation section. Because the model is multiobjective, its solution is given by a set of Pareto points rather than by a single optimal network design. The next section provides details on how these Pareto optimal designs are generated.

11.2.3 Solution Procedure

The solution of a multiobjective problem is given by a set of Pareto points, each achieving a unique combination of objective function values (Ehrgott, 2008). These Pareto solutions feature the property that they cannot be improved simultaneously in all the criteria without necessarily worsening at least one of them. Here we apply the epsilon constraint method to solve the MILP, which consists of calculating a series of single-objective problems in which one objective is kept in the objective function while the others are transferred to auxiliary constraints that impose epsilon values on them. The model is then solved iteratively for different epsilon bounds in order to generate a well-spread set of solutions. These solutions are then passed to decision makers who should identify the one that best meets their preferences.

Note that solving the single-objective models can already be challenging, as they can contain a large number of binary and integer variables that increase their combinatorial complexity. Hence, alternative decomposition methods, such as bi-level or Lagrangian algorithms, might be required to expedite the calculations. The reader is referred to (Corsano et al., 2014) for further details on this topic.

11.3 NUMERICAL RESULTS

The capabilities of the MILP approach are illustrated through its application to the design of the future hydrogen network for vehicle use in the United Kingdom. The problem data was taken from (Guillén-Gosálbez et al., 2010). To simplify the calculations, no lower bounds were defined on the transportation flows. This simplification allows removing the binary variables (not the integer ones) from the formulation, thereby expediting the model resolution.

Furthermore, the model seeks to minimize the impact on human health caused by climate change as a unique criterion (rather than the whole set of impact categories provided by the Eco-indicator 99).

Hence, the goal of the analysis is to design a SC capable of covering the future UK demand while minimizing the total cost and the impact on human health caused by climate change. The model was implemented in GAMS 24.4 interfacing with the solver CPLEX 12.6 on an Intel Core i5-4570 3.20 GHz computer. The CPU time required to close a gap of 0.5% was on the order of minutes for all the instances solved.

Fig. 11.3 shows a set of Pareto solutions that trade off the total cost versus the environmental impact. As seen, there is a clear tradeoff between the two objectives, as when the cost is minimized the impact increases and the other way around. The slope of the curve is quite smooth as we start moving from the minimum cost to the minimum impact solution, but at a certain point becomes very steep. This can be explained by the technological choices made along the curve, which are needed in order to get adapted to more stringent environmental regulations. Note that the minimum impact that can be achieved is negative, as we assume that biomass growth captures more CO_2 than that released during its transformation into hydrogen.

With regard to the technologies selected in each case, in the minimum cost solution, the model decides to implement steam methane reforming,

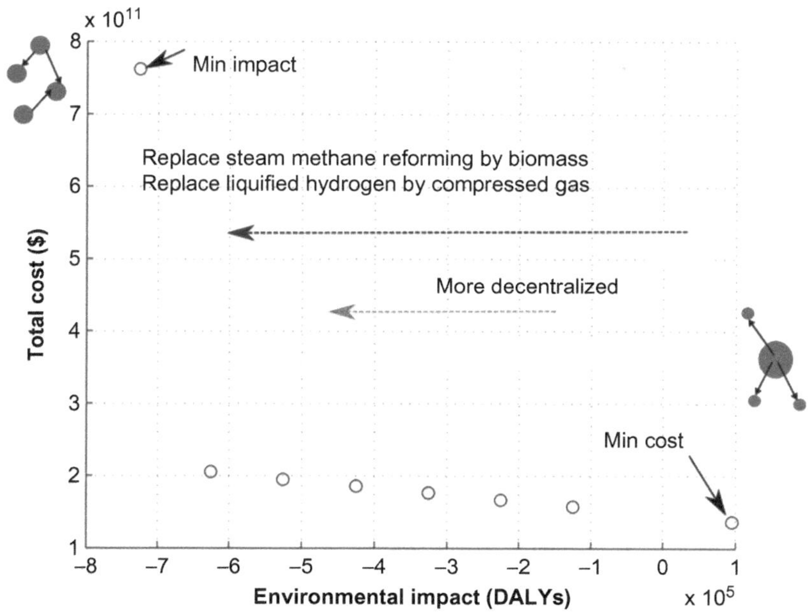

FIG. 11.3 Set of Pareto points minimizing the total cost and the environmental impact simultaneously.

liquefied storage, and tanker trucks (which transport hydrogen in liquid form). In contrast, in the minimum impact, hydrogen is produced via biomass gasification, and stored and transported in compressed form using tube trailers and railway tube cars. While the latter technologies reduce the environmental impact significantly, they also lead to a much higher cost, which results in a very poor economic performance compared to the minimum cost solution.

Fig. 11.4 shows a breakdown of the cost, in which it can be seen how the main contributors in the minimum cost solution are the capital cost of production and storage facilities, followed by the operating cost of both facilities, and finally the transportation cost. In the minimum environmental impact solution, the capital cost increases significantly, mainly due to the move from steam methane reforming to biomass gasification as well as from liquefied hydrogen to compressed gas hydrogen. In both cases, the transportation costs are rather small, mainly due to the fact that we consider full availability of raw materials in every grid (note that if this was not the case, the results would change drastically).

In terms of environmental impact, in the minimum cost solution the impact is mainly caused by the production of hydrogen via steam methane reforming, followed by the storage and finally the transportation. In the minimum impact solution, however, the biomass growth offsets the impact associated with the storage and the transportation, thereby leading to negative values of impact resulting from the amount of net carbon captured by the biomass.

With regard to the network structure, the minimum cost solution represents a more centralized network that exploits further the economies of scale in order to bring the capital costs down, but this is achieved at the expense of increasing the

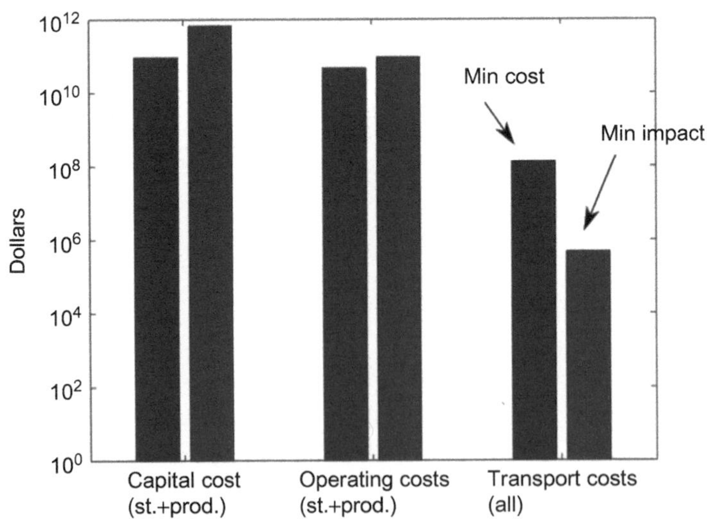

FIG. 11.4 Breakdown of the total cost in the extreme solutions.

transportation tasks. On the other hand, the minimum environmental impact solution leads to a more decentralized network that minimizes the transportation tasks and therefore the associated impact. This, however, leads to larger capital cost as the network cannot exploit economies of scale to the same extent as in the minimum cost design. Note that the solutions obtained and discussed here depend to a large extent on the data used in the analysis, particularly on the values of the economic and environmental parameters.

11.4 CONCLUSIONS

In this chapter we have addressed the design of supply chains that produce and distribute hydrogen for vehicle use. A MILP model was developed to tackle this problem, which contains continuous, binary, and integer variables subject to a set of technical, economic, and environmental constraints. This MILP seeks to optimize the economic and environmental performance of the supply chain simultaneously. The solution of such a formulation is given by a set of Pareto points, each achieving a unique combination of objective function values. From these solutions, decision makers should select the one that best matches their preferences.

The capabilities of this approach were illustrated through its application to the design of the future hydrogen supply chain in the United Kingdom. We show how there is a clear tradeoff between the economic and environmental performance of the network, as when the cost is minimized the impact grows and vice versa. The slope of the curve is smooth as one starts to move from the minimum cost to the minimum impact solution and, at some point, becomes very steep. Hence, there is a region in which significant environmental benefits can be attained at a marginal increase in cost. The minimum cost solution employs steam methane reforming, liquefied hydrogen, and a more centralized network, while the minimum impact one relies on biomass gasification, compressed gas hydrogen and a more decentralized network.

Overall, we argue that mathematical programming tools provide a general framework to tackle the design of complex energy systems in which multiple constraints and conflicting objectives must be accounted for. The ultimate goal is to ensure that the best technological solutions are implemented in the right location and at the right time in the transition toward a cleaner energy system.

REFERENCES

Almansoori, A., Shah, N., 2009. Design and operation of a future hydrogen supply chain: multi-period model. Int. J. Hydrog. Energy 34 (19), 7883–7897.

Balat, M., Kirtay, E., 2010. Major technical barriers to a "hydrogen economy" Energ. Source Part A 32 (9), 863–876.

Brockhoff, D., Zitzler, E., 2006. Are all objectives necessary? On dimensionality reduction in evolutionary multiobjective optimization. Parallel Problem Solving from Nature – PPSN IX. pp. 533–542.

Corsano, G., Guillén-Gosálbez, G., Montagna, J.M., 2014. Computational methods for the simultaneous strategic planning of supply chains and batch chemical manufacturing sites. Comput. Chem. Eng. 60, 154–171.

De-León Almaraz, S., et al., 2014. Hydrogen supply chain optimization for deployment scenarios in the Midi-Pyrénées region, France. Int. J. Hydrog. Energy 39 (23), 11831–11845.

De-León Almaraz, S., et al., 2015. Deployment of a hydrogen supply chain by multi-objective/multi-period optimisation at regional and national scales. Chem. Eng. Res. Des. 104, 11–31.

Ehrgott, M., 2008. Multiobjective optimization. AI Mag. 29, 47–57.

Grossmann, I.E., Guillén-Gosálbez, G., 2010. Scope for the application of mathematical programming techniques in the synthesis and planning of sustainable processes. Comput. Chem. Eng. 34 (9), 1365–1376.

Guillén-Gosálbez, G., Mele, F.D., Grossmann, I.E., 2010. A bi-criterion optimization approach for the design and planning of hydrogen supply chains for vehicle use. AICHE J. 56 (3), 650–667.

Hugo, A., et al., 2005. Hydrogen infrastructure strategic planning using multi-objective optimization. Int. J. Hydrog. Energy 30 (15), 1523–1534.

Ingason, H.T., Pall Ingolfsson, H., Jensson, P., 2008. Optimizing site selection for hydrogen production in Iceland. Int. J. Hydrog. Energy 33 (14), 3632–3643.

Kim, J., Lee, Y., Moon, I., 2008. Optimization of a hydrogen supply chain under demand uncertainty. Int. J. Hydrog. Energy 33 (18), 4715–4729.

Li, Z., et al., 2008. Hydrogen infrastructure design and optimization: a case study of China. Int. J. Hydrog. Energy 33 (20), 5275–5286.

Lin, Z., et al., 2008. The least-cost hydrogen for Southern California. Int. J. Hydrog. Energy 33 (12), 3009–3014.

Sabio, N., et al., 2010. Strategic planning with risk control of hydrogen supply chains for vehicle use under uncertainty in operating costs: a case study of Spain. Int. J. Hydrog. Energy 35 (13), 6836–6852.

Sabio, N., et al., 2012. Holistic minimization of the life cycle environmental impact of hydrogen infrastructures using multi-objective optimization and principal component analysis. Int. J. Hydrog. Energy 37, 5385–5405.

Smitkova, M., Janíček, F., Riccardi, J., 2011. Life cycle analysis of processes for hydrogen production. Int. J. Hydrog. Energy 36 (13), 7844–7851.

Chapter 12

Robust Engineering Strategy for Solving Optimization Problems of Refinery Hydrogen System

Liao Zuwei, Wang Jingdai and Yang Yongrong
State Key Laboratory of Chemical Engineering, Department of Chemical and Biological Engineering, Zhejiang University, Hangzhou, PR China

12.1 INTRODUCTION

In order to maintain business efficiency, refineries are increasingly concerned with improving the operational management of their complex process. In fact, there has been much effort that addresses problems in the management of production systems and steam power systems of refineries. Moro et al. (1998) developed a nonlinear planning model for refinery production, which is able to represent a general refinery topology and allowed the implementation of nonlinear process models as well as blending relations. This framework was later extended to support sequence decisions at the scheduling level by Pinto et al. (2000) and Joly et al. (2002). In the work of Neiro and Pinto (2005), uncertainties related to petroleum and product prices as well as demand are included in the model as a set of discrete probabilities. Based on a fast and robust online data reconciliation method, operational optimization of a utility system in a petrochemical plant was developed by Lee et al. (1998) and the fuel cost was reduced by 5.4%–9.2% compared with the former operation. In order to deal with the prediction errors of energy demands in multiperiod operational planning, Yi et al. (2003) proposed an integration methodology of periodic replanning and hierarchical repair. Micheletto et al. (2008) presented a conceptual modeling framework for operational optimization of utility systems. The MILP (mixed-integer linear programming) model was integrated with the refinery database for the planning of a refinery utility system and effectively optimized the financial performance of the thermoelectric plant without any capital investment.

In recent years, the management of refinery gas systems, such as fuel gas systems (Hasan et al., 2011; Iyer and Grossmann, 1997; Jagannath et al.,

Hydrogen Supply Chain. https://doi.org/10.1016/B978-0-12-811197-0.00012-9

2012; Zhang and Rong, 2008; Zhang et al., 2010) and hydrogen networks (Ahmad et al., 2010; Alves and Towler, 2002; Elkamel et al., 2011; Hallale and Liu, 2001; Jiao et al., 2012; Liao et al., 2011; Liao et al., 2010; Liu and Zhang, 2004; Van den Heever and Grossmann, 2003; Xuan et al., 2010; Zhou et al., 2014; Zhou et al., 2012; Zhou et al., 2013), has drawn more and more attention. This is because they are the primary energy source in the refinery, while hydrogen is critical for improving product quality. However, relatively little research has focused on optimizing these systems, especially the hydrogen system. Hydrogen is continuously generated from the reformer and PSA units and is supplied to most of the hydrogen consumers around the refinery via compressors. Therefore, effective operational optimization of the hydrogen system can be very effective for energy cost reduction and emission reduction in the refining process. Liu and Zhang (2004) established simplified mathematical models for various purification devices and proposed a superstructure model of hydrogen network design with an integrated hydrogen purification scheme. In order to capture a richer network structure space, Liao et al. (2010) introduced a state space superstructure for integrating purifiers with compressors. Jiao et al. (2012) decomposed the optimization problem into two parts: hydrogen purification network and hydrogen supply network. They also linearized the original MINLP model into a MILP model.

The scheduling of hydrogen networks has also attracted wide attention. Van den Heever and Grossmann (2003) established two MINLP models to optimize the production planning and scheduling of the hydrogen systems. Xuan et al. (2010) and Ahmad et al. (2010) introduced a multiperiod hydrogen network optimization model based on their previous single period models. Elkamel et al. (2011) incorporated a hydrogen network with rigorous process models. Zhou et al. (2014) proposed detailed hydrogen pipeline models for the scheduling problem.

The rest of the paper is organized as follows: Section 12.2 describes the problem statement as well as the general features of the hydrogen system in a refinery. Section 12.3 presents the mathematical formulation of the operational optimization problem. Section 12.4 illustrates a systematic procedure to implement the method of Section 12.3 to a large-scale refinery. Conclusions of this work are given in Section 12.5.

12.2 PROBLEM STATEMENT AND DESCRIPTION OF THE HYDROGEN SYSTEM

Typical hydrogen-related streams in a refinery are shown in Fig. 12.1. The lines in red or blue in the figure represent the gas flow. For a hydrogen pipeline system, there are generally two categories: low-pressure hydrogen pipeline networks (LP network) and high-pressure hydrogen pipeline networks (HP network). The hydrogen inside an LP network can be compressed to the HP network, while the hydrogen from a HP network can be purged to the LP network

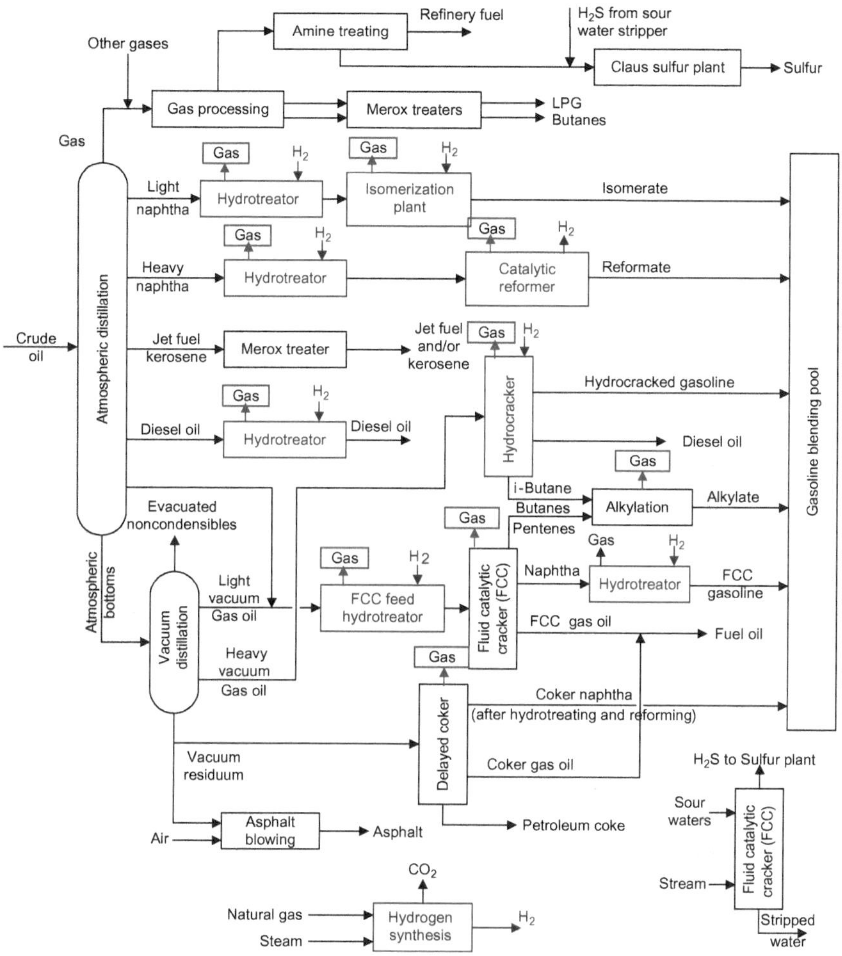

FIG. 12.1 Schematic representation of a refinery process.

via valves. Consequently, the LP network and HP network could be balanced by this two-way connection. Both an LP network and a HP network are critical to the hydrogen supply in a refinery. Therefore, the operational optimization of a hydrogen pipeline system should include both LP and HP networks. The optimization problem can be stated as follows: given a set of hydrogen generating/consuming processes and a network of interconnections among the processes, it is desired to determine a scheduling scheme so that the overall annualized cost is minimized while the processes receive adequate hydrogen resource.

Fig. 12.2 presents a simplified flowchart of an HP network system in which the pipeline network is represented by the red lines and each unit connected to the network is represented by a rounded rectangle that contains its identification.

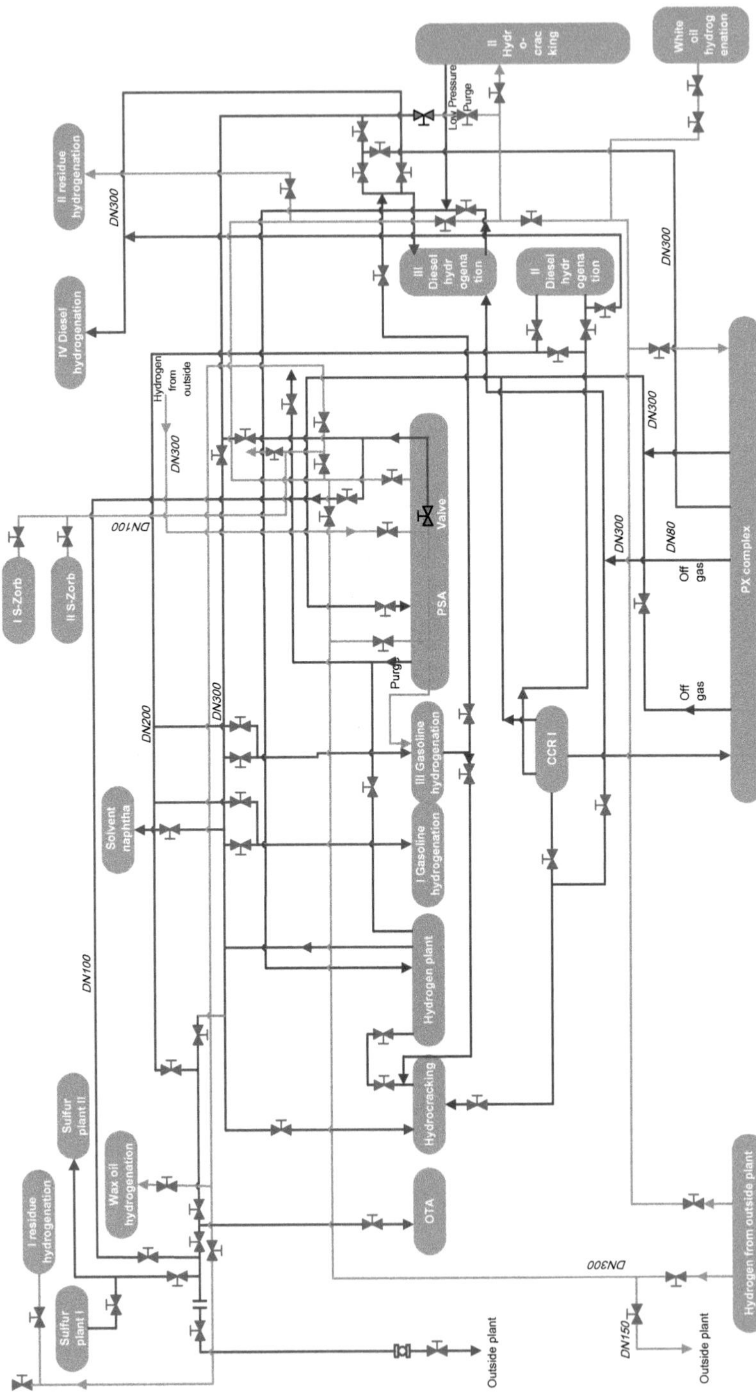

FIG. 12.2 Simplified flowchart of refinery's HP network system.

As illustrated in Fig. 12.2, the units in the HP network include a gasoline hydrotreating unit (GHU), diesel hydrotreating unit (DHU), hydrocracker (HC), s-zorb unit (SZU), wax oil hydrotreating (WHU), residue hydrotreating (RHU), and a hydrogen plant (HP). Despite the hydrogen plant, the units connected to the HP network only consume hydrogen, and they are called "hydrogen consuming units." The hydrogen consuming units are depicted in dark color, while the hydrogen producer—hydrogen plant is depicted in light color.

The units in the LP network include continuous reforming unit (CRU), pressure swing adsorption (PSA), hydrocracker (HC), diesel hydrotreating (DHU), and gasoline hydrotreating unit (GHU). The consumers of LP network are mainly supplied by hydrogen produced from CRU and PSA units. It should be noted that the HC, DHU, and GHU units can receive hydrogen from both the HP and LP network. Thus, we have a degree of freedom for practical hydrogen allocation and balance. The balancing tools are the compression of the surplus hydrogen of the LP network to the HP network and the relief of the surplus hydrogen of the HP network to the LP network. Because compressors consume a large amount of energy, they are the most important equipment for keeping the balance of the hydrogen pipeline network. If the above two tools cannot balance the hydrogen production and consumption instantly, an additional schedule to relieve hydrogen to the fuel gas system or to cut down the production rate of certain units will be employed. Of course, the additional schedules would cause economic loss, as they are not expected.

Hydrogen produced by the CRU cannot satisfy the demand of the whole refinery. The deficit of hydrogen demand is compensated by the hydrogen plant, which converts hydrocarbon resources, such as natural gas, liquefied petroleum gas (LPG), and refinery off gas, into hydrogen and carbon dioxide. As a result, it is desirable to minimize the hydrogen plant production for economic profit.

12.3 MATHEMATICAL MODEL AND SOLVING METHOD

In a previous work (Zhou et al., 2014), a hydrogen pipeline schedule model is proposed to optimize equipment operation. A modeling method called mathematical programming with equilibrium constraint (MPEC) is introduced to model the operation of compressors and the flow of a branching structure pipeline network, which replaces the MINLP formulation with a NLP problem and improves the solution efficiency.

The formulation of the operational optimization model of a hydrogen pipeline network in this research is based on the hydrogen pipeline network of Fig. 12.2, the scheduling model is formulated as follows.

12.3.1 Objective Function

The goal of the multiperiod scheduling model of hydrogen systems is to minimize the total cost. Because the hydrogen from the reforming unit is a byproduct, we only consider the hydrogen cost from the hydrogen plant. The objective function is as follows:

$$\text{Min} \quad obj = \sum_h \sum_t e_{h,t} F_{h,t} \tag{12.1}$$

where $F_{h,\ t}$ represents the production amount of the hydrogen plant in period t and e_{ht} stands for the production cost of the hydrogen plant. The efficiency and unit cost of the hydrogen plant depend on the production rate. In order to minimize the total cost and to keep the hydrogen plant running as smoothly and efficiently as possible, the operating schedule of the hydrogen plant should be optimized according to the efficiency curve. Let us introduce a penalty term, $e^{penalty}$, to the cases that are outside the optimal production range. The production cost of the hydrogen plant can be described by a segmentation function as follows:

$$e_{h,t} = \begin{cases} e + e^{penalty1} & \text{if} F_{h,t} < F_h{}^{Low} \\ e & \text{if} F_h{}^{Low} \le F_{h,t} \le F_h{}^{Up} \\ e + e^{penalty2} & \text{if} F_{h,t} > F_h{}^{Up} \end{cases} \tag{12.2}$$

12.3.2 Hydrogen Pipeline Model

A hydrogen pipeline network not only transfers hydrogen, but also stores hydrogen. Hydrogen transport in the pipeline should satisfy the following balance equations: mass balance equation, momentum balance equation, and energy balance equation. Because the hydrogen pipeline network is operated at room temperature, we assume the whole procedure is isothermal. Consequently, the energy balance of the pipeline can be ignored.

The mass balance equation reads:

$$\frac{MwA}{ZRT}\frac{\partial P}{\partial t} + \frac{\partial F}{\partial z} = 0$$

where Mw is the molecular weight of the gas flow inside the pipe, A is the cross-sectional area of the pipe, Z is the gas compression factor, R is a constant, T is the temperature inside the pipe, P is pressure, and F is the gas flowrate inside the pipeline. After manipulating the above formula, we obtain:

$$\frac{MwA_a \ell_a}{ZRT}\left(\overline{P}_{a,t+1} - \overline{P}_{a,t}\right) = \int_t^{t+1} (F_{a,in} - F_{a,out})dt \quad \forall a \in A, t \in T$$

where ℓ is the length of the pipeline and $P_{a,t+1}$ is the average pressure of the pipeline in period t. In order to obtain a numerical mass balance equation, we integrate the above equation by the trapezoidal integral method. The result is as follows:

$$\frac{MwA_a\ell_a}{ZRT}\left(\overline{P}_{a,t+1}-\overline{P}_{a,t}\right)=\frac{1}{2}[(F_{a,in,t}-F_{a,out,t})+(F_{a,in,t+1}-F_{a,out,t+1})](t_{t+1}-t_t)$$
$$\forall a\in A, t\in T$$

Note that in the above equation, Mw does not vary with location a and period t. This equation is only suitable for gas transport systems with stable components. We have more than one hydrogen source in the hydrogen pipeline network and the flowrate and purity of the hydrogen injection varies with production period. Therefore, the Mw inside the hydrogen pipeline network is not constant. Accordingly, the above equation can be modified as follows:

$$\frac{A_a\ell_a}{RT}\left(\frac{Mw_{a,t+1}}{Z_{a,t+1}}\overline{P}_{a,t+1}-\frac{Mw_{a,t}}{Z_{a,t}}\overline{P}_{a,t}\right)$$
$$=\frac{1}{2}[(F_{a,in,t}-F_{a,out,t})+(F_{a,in,t+1}-F_{a,out,t+1})](t_{t+1}-t_t)\quad \forall a\in A, t\in T \tag{12.3}$$

where $Mw_{a,t}$ can be calculated by the composition of the source hydrogen streams:

$$Mw_{a,t}=\frac{1}{\sum\limits_m {}^{y_{a,m,t}}/_{Mw_m}}\qquad \forall a\in A, t\in t \tag{12.4}$$

The compression factor of the mixed gas can be calculated by several methods. Because its hydrogen content is >80%, we calculate the compression factor based on the hydrogen content:

$$Z_{a,t}=\frac{\overline{P}_{a,t}Vm_{a,t}}{RT}\qquad a\in A, t\in T \tag{12.5}$$

where $Vm_{a,t}$ is the gas volume. The relationship between the average pressure and the gas volume is presented by the Peng-Robinson equation:

$$\overline{P}_{a,t}=\frac{RT}{Vm_{a,t}-b^{P-R}}-\frac{a^{P-R}\alpha^{P-R}}{Vm_{a,t}{}^2+2b^{P-R}Vm_{a,t}-(b^{P-R})^2}\qquad a\in A, t\in T \tag{12.6}$$

where a^{P-R} and b^{P-R} are coefficients.

$$a^{P-R}=0.45724\frac{R^2Tc^2}{Pc} \tag{12.7}$$

$$b^{P-R}=0.0778\frac{RTc}{Pc} \tag{12.8}$$

$$\alpha^{P-R}=\left(1+\left(1-\sqrt{T/Tc}\right)\left(0.37464+1.54226f^{Acentricity}-0.26992\left(f^{Acentricity}\right)^2\right)\right)^2 \tag{12.9}$$

The above simplification has been validated in our previous work.

12.3.3 Momentum Balance Equation

As the hydrogen pipe diameter is relatively large, the gas flow in the pipeline can be recognized as a one-dimensional flow, and the momentum conservation equation may be expressed as:

$$\frac{dP}{dz} = \frac{-fZRTF|F|}{2DA^2MwP}$$

If we only examine the pressure and flow changes at the inlet and outlet of the pipe section, then the above equation can be written as:

$$\frac{dP}{dz}_{a,k,t} = \frac{-f_{a,k,t}Z_{a,t}RTF_{a,k,t}|F_{a,k,t}|}{2D_aA_a{}^2Mw_{a,t}P_{a,k,t}} \quad \forall a \in A, t \in T, k \in K \tag{12.10}$$

where $K = \{\text{in, out}\}$ stands for the inlet and outlet of the pipeline section, respectively. Note that the absolute value is employed for reverse flow phenomena.

$$F_{a,in,t}F_{a,out,t} \geq 0 \quad \forall a \in A, t \in T \tag{12.11}$$

The average pressure of the pipeline is calculated as:

$$\overline{P}_{a,t} = \frac{\int_0^{L_a} P(z')dz'}{\int_0^{L_a} dz} \quad \forall a \in A, t \in T$$

We use the collocation method to solve this momentum balance equation:

$$P_{a,in,t} = h_{a,t}{}^1 \quad \forall a \in A, t \in T \tag{12.12}$$

$$\frac{dP}{dz}_{a,in,t} = h_{a,t}{}^2 \quad \forall a \in A, t \in T \tag{12.13}$$

$$P_{a,out,t} = h_{a,t}{}^3 \quad \forall a \in A, t \in T \tag{12.14}$$

$$\frac{dP}{dz}_{a,out,t} = h_{a,t}{}^4 \quad \forall a \in A, t \in T \tag{12.15}$$

$$h_{a,t}{}^3 = h_{a,t}{}^1 + \frac{L_a}{2}\left(h_{a,t}{}^2 + h_{a,t}{}^4\right) \quad \forall a \in A, t \in T \tag{12.16}$$

where $h^1_{a,t}$, $h^2_{a,t}$, $h^3_{a,t}$ and $h^4_{a,t}$ are auxiliary variables. Now the average pressure calculation equation can be written as:

$$\overline{P}_{a,t} = \frac{\int_0^{L_a} P(z')dz'}{\int_0^{L_a} dz} = \frac{(1/2)h_{a,t}{}^1 + (1/12)h_{a,t}{}^2L_a + (1/2)h_{a,t}{}^3 - (1/12)h_{a,t}{}^4L_a}{L_a}$$

$$\forall a \in A, t \in T \tag{12.17}$$

So far, we have modeled the mass balance and momentum balance of the gas flow in the pipeline section. It should be pointed that the hydrogen streams inside the refinery are multicomponent streams. In addition, there is a composition distribution inside the pipeline network. Therefore, we need to calculate the composition of the hydrogen pipeline network in real time.

Assuming that the pipeline is filled with hydrogen at the beginning of the first period:

$$y_{a,m,0} = y_{a,m,0}{}^{in} \qquad \forall a \in A, m \in M \tag{12.18}$$

In actual production, the pipelines are connected by a flange, which forms physical nodes. For modeling convenience, we denote the flange by node $n \in N$, assuming that the pipelines and the hydrogen sources and demands are all connected by the nodes, as shown in Fig. 12.3. Because the flange (2-way or 3-way) is used only as a connection tool for the pipeline, no material accumulation occurs.

$$y_{i,m,t} = y_{n,m,t} \qquad \forall (i,n) \in Sup(I,N), m \in M, t \in T \tag{12.19}$$

$$y_{j,m,t} = y_{n,m,t} \qquad \forall (j,n) \in Dem(J,N), m \in M, t \in T \tag{12.20}$$

The concentration of the node is equal to the concentration of the stream that is flowing into the node.

$$y_{n,m,t} = \begin{cases} y_{a,m,t}{}^{in} & \text{if } F_{a,in,t} > 0 \quad \forall (a,n) \in A \text{ from } N(A,N), t \in T \\ y_{a,m,t}{}^{in} & \text{if } F_{a,out,t} < 0 \quad \forall (a,n) \in A \text{ to } N(A,N), t \in T \end{cases} \tag{12.21}$$

The mass balance equation at the node and the component mass balance equation are as follows:

$$\sum_{a:(a,n)\in AtoN(A,N)} F_{a,out,t} + \sum_{i:(i,n)\in Sup(I,N)} F_{i,t} = \sum_{a:(a,n)\in AfromN(A,N)} F_{a,in,t} + \sum_{j:(j,n)\in Dem(J,N)} F_{j,t} \quad \forall n \in N \tag{12.22}$$

$$\sum_{a:(a,n)\in AtoN(A,N)} F_{a,out,t} y_{a,m,t} + \sum_{i:(i,n)\in Sup(I,N)} F_{i,t} y_{i,m,t} = \sum_{a:(a,n)\in AfromN(A,N)} F_{a,in,t} y_{a,m,t} + \sum_{j:(j,n)\in Dem(J,N)} F_{j,t} y_{j,m,t} \quad \forall n \in N, m \in M \tag{12.23}$$

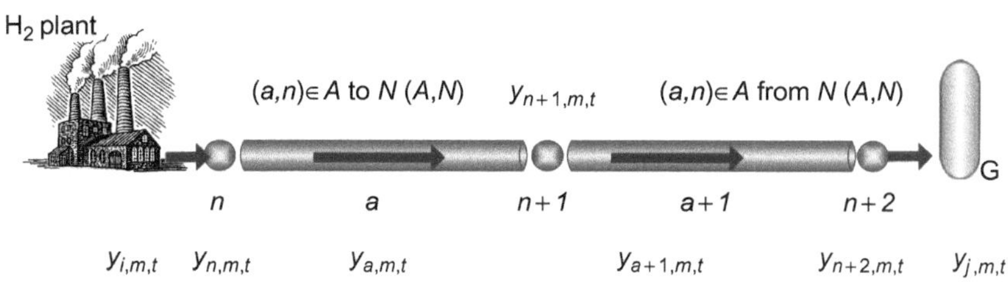

FIG. 12.3 Sketch map for the pipe segments, nodes, and the corresponding variables.

where $y_{a,m,t}$ denotes the concentration of component m at pipeline a in period t.

Once $t>0$, there will be hydrogen streams with different compositions that flow into the pipeline, assuming that the pipeline hydrogen concentration is always uniform. As there may be reverse flow in the pipeline section, the pipeline composition calculation is divided into two cases:

(1) $F_{a,in,t+1}>0$

$$y_{a,m,t+1}\left(F_{a,in,t+1}\Delta t+Inventory_{a,t}\right)=y_{a,m,t+1}{}^{in}F_{a,in,t+1}\Delta t+y_{a,m,t}Inventory_{a,t} \quad \forall a\in A,m\in M,t\in T \tag{12.24}$$

(2) $F_{a,out,t+1}<0$,

$$y_{a,m,t+1}\left(\left|F_{a,out,t+1}\right|\Delta t+Inventory_{a,t}\right)=y_{a,m,t+1}{}^{in}\left|F_{a,out,t+1}\right|\Delta t+y_{a,m,t}Inventory_{a,t}\forall a\in A,m\in M,t\in T \tag{12.25}$$

Component concentration normalization equation:

$$\sum_m y_{a,m,t}{}^{in}=1 \qquad \forall a\in A,t\in T \tag{12.26}$$

$$\sum_m y_{a,m,t}=1 \qquad \forall a\in A,t\in T \tag{12.27}$$

$$\sum_m y_{n,m,t}=1 \qquad \forall n\in N,t\in T \tag{12.28}$$

$$\sum_m y_{d,m,t}=1 \qquad \forall d\in D,t\in T \tag{12.29}$$

where $Inventory_{a,t}$ indicates the amount of hydrogen stored in the pipeline section. The hydrogen pipe network has two major roles: (1) to move the hydrogen from the hydrogen source to the hydrogen trap transport; and (2) as the system of hydrogen buffer equipment. When the hydrogen production is greater than the consumption of a certain amount of hydrogen storage, the system hydrogen production is less than the consumption of hydrogen released. The initial hydrogen storage capacity in the pipe section is calculated according to the average pressure in the pipe section of the initial production cycle. After the start of the dispatching cycle, the hydrogen accumulation in the pipe section is the difference between the hydrogen injection and the consumption at the inlet and outlet.

$$Inventory_{a,0}=\frac{A_a\ell_a Mw_{a,0}\overline{P}_{a,0}}{RT} \qquad \forall a\in A$$

$$Inventory_{a,t+1}=\begin{cases}Inventory_{a,t}+\left(F_{a,in,t+1}-F_{a,out,t+1}\right)\Delta t & \text{if } F_{a,in,t}>0\\ Inventory_{a,t}-\left(F_{a,in,t+1}-F_{a,out,t+1}\right)\Delta t & \text{if } F_{a,in,t}<0\end{cases}$$

$$\forall a\in A,t\in T \tag{12.30}$$

The total hydrogen storage in the hydrogen pipeline network is:

$$Total\_inventory_t = \sum_{a \in A} Inventory_{a,t} \quad \forall t \in T \tag{12.31}$$

In order to ensure the reliability of the scheduling results, to prevent the excessive storage of hydrogen stored in the pipeline, we specify that the ending hydrogen storage should be equal to or larger than the starting storage amount.

$$Total\_inventory_0 \leq Total\_inventory_t \quad \forall t = t^{End} \tag{12.32}$$

When the flowrate in the pipeline changes, the flow pattern may also change. The friction factor varies with flow pattern and the pressure drop varies either. Therefore, it is necessary to describe and calculate the flow pattern inside the pipe network system. The fluid flow pattern and its transformation can be judged by a dimensionless number—the Reynolds number.

$$\mathrm{Re}_{a,k,t} = \frac{|F_{a,k,t}| Dia_a}{\mu Area_a} \quad \forall a \in A, k \in K, t \in T \tag{12.33}$$

When $Re_{a,k,t} < 2300$, the lamina friction factor should be calculated as:

$$f_{a,k,t}{}^{lam} = \frac{64}{\mathrm{Re}_{a,k,t}} \quad \forall a \in A, k \in K, t \in T \tag{12.34}$$

When $Re_{a,k,t} > 2300$, the turbulent friction factor should be calculated by the Colebrook-White equation:

$$f_{i,k,t}{}^{turb} = 1.326 \left[\ln \left(\frac{1}{\varepsilon_a/(3.7Dia_a) + 2.51/\left(\mathrm{Re}_{i,k,t} \sqrt{f_{i,k,t}{}^{turb}} \right)} \right) \right]^{-2}$$
$$\forall a \in A, k \in K, t \in T \tag{12.35}$$

Similar to the concentration definition, the pressure of the pipeline network is defined as follows:

$$P_{a,out,t} = P_{n,t} \quad \forall (a,n) \in A \text{ to } N(A,N), t \in T \tag{12.36}$$

$$P_{a,in,t} = P_{n,t} \quad \forall (a,n) \in A \text{ from } N(A,N), t \in T \tag{12.37}$$

$$P_{i,t} = P_{n,t} \quad \forall (i,n) \in Sup(I,N), t \in t \tag{12.38}$$

$$P_{j,t} = P_{n,t} \quad \forall (j,n) \in Dem(J,N), t \in T \tag{12.39}$$

12.3.4 Constraints for Hydrogen Sources and Demands

The flowrate upper bounds for hydrogen plants and other hydrogen sources are:

$$F_{h,t} \leq F_{h,t}{}^{Capacity} \quad \forall h \in H, t \in T \tag{12.40}$$

$$F_{i,t} \leq F_{i,t}{}^{Max} \qquad \forall i \in I, t \in T \tag{12.41}$$

The flow rate, purity, and pressure constraints for the hydrogen consumption unit are:

$$F_{j,t} = F_{j,t}{}^{De} \qquad \forall j \in J, t \in T \tag{12.42}$$

$$y_{j,H_2,t} \geq y_{j,H_2}{}^{De} \qquad \forall j \in J, t \in T \tag{12.43}$$

$$P_{j,t} \geq P_j{}^{De} \qquad \forall j \in J, t \in T \tag{12.44}$$

12.3.5 The MPEC Method

The above mathematical model contains a large number of noncontinuous expressions, such as the sectional calculation of the hydrogen cost, the calculation of the friction factor, and the absolute value function. To solve this problem, the discontinuous variables in the complementary constrained model are introduced.

The MPEC method has been a research focus in recent years in the field of mathematical programming. Mathematical Programming with Equilibrium Constrained problems can be treated as two-level programming problems with variational inequalities or complementary constraints. The problems dealt with in this paper are complementary constraints.

The absolute functions have the following equality forms:

$$|F_{a,k,t}| = F_{a,k,t}{}^{positive} + F_{a,k,t}{}^{negative} \qquad \forall a \in A, k \in K, t \in T \tag{12.45a}$$

$$F_{a,k,t} = F_{a,k,t}{}^{positive} - F_{a,k,t}{}^{negative} \qquad \forall a \in A, k \in K, t \in T \tag{12.45b}$$

$$0 \leq F_{a,k,t}{}^{positive} \perp F_{a,k,t}{}^{negative} \geq 0 \qquad \forall a \in A, k \in K, t \in T \tag{12.CR-1}$$

where $F_{a,k,t}^{positive}$ and $F_{a,k,t}^{negative}$ are supplemental variables, Eq. (12.CR-1) denotes the relationship between $F_{a,k,t}^{positive}$ and $F_{a,k,t}^{negative}$. Namely, at least one of them should equal the boundary zero value.

The friction factor $f_{a,k,t}$ is calculated in two flow patterns: advection and turbulence.

$$f_{a,k,t} =^{a,k,ta,k,t} \begin{cases} f_{a,k,t}{}^{lam} & \text{Re} < 2300 \\ f_{a,k,t}{}^{turb} & \text{Re} > 2300 \end{cases} \qquad \forall a \in A, k \in K, t \in T$$

Converting the above equation to the complementary constraint form:

$$f_{a,k,t} = switch\_f_{a,k,t} f_{a,k,t}{}^{lam} + (1 - switch\_f_{a,k,t}) f_{a,k,t}{}^{turb} \qquad \forall a \in A, k \in K, t \in T \tag{12.46a}$$

$$(\text{Re}_{a,k,t} - 2300) - \lambda\_f1_{a,k,t} + \lambda\_f2_{a,k,t} = 0 \qquad \forall a \in A, k \in K, t \in T \tag{12.46b}$$

$$0 \leq switch\_f_{a,k,t} \perp \lambda\_f1_{a,k,t} \geq 0 \quad \forall a \in A, k \in K, t \in T \tag{12.CR-2}$$

$$0 \leq (1 - switch\_f_{a,k,t}) \perp \lambda\_f2_{a,k,t} \geq 0 \quad \forall a \in A, k \in K, t \in T \tag{12.CR-3}$$

where switch $-f_{a,k,t}$ is the shift variable, which is determined by the Reynolds number. When $Re < 2300$, it is laminar flow. If Eq. (12.46b) holds, the supplemental variable $\lambda\_f2_{a,k,t}$ should be positive. Therefore, we have $1 - \text{switch}\_f_{a,k,t} = 0$, $\text{switch}\_f_{a,k,t} = 0$, $f_{a,k,t} = f^{lam}_{a,\ k,\ t}$. Similarly, when $Re > 2300$ we can obtain $\text{switch}\_f_{a,k,t} = 0$ and $f_{a,\ k,\ t} = f^{turb}_{a,\ k,\ t}$.

The equivalent constraint form of the hydrogen cost calculation for the hydrogen plant is:

$$e_{H_2} = switch\_p1_{i,t}(e + e^{penalty1}) + switch\_p2_{i,t}e + switch\_p3_{i,t}(e + e^{penalty2}) \\ \forall i \in I, t \in T \tag{12.47a}$$

$$\left(F_{i,t} - F_i^{Low}\right) + \gamma\_p_{i,t} - \lambda\_p1_{i,t} = 0 \quad \forall i \in I, t \in t \tag{12.47b}$$

$$\left(F_{i,t} - F_i^{Low}\right)\left(F_{i,t} - F_i^{Up}\right) + \gamma\_p_{i,t} - \lambda\_p2_{i,t} = 0 \quad \forall i \in I, t \in T \tag{12.47c}$$

$$\left(F_i^{Up} - F_{i,t}\right) + \gamma\_p_{i,t} - \lambda\_p3_{i,t} = 0 \quad \forall i \in I, t \in T \tag{12.47d}$$

$$\gamma\_p_{i,t} \geq 0 \quad \forall i \in I, t \in T \tag{12.47e}$$

$$switch\_p1_{i,t} + switch\_p2_{i,t} + switch\_p3_{i,t} = 1 \quad \forall i \in I, t \in T \tag{12.47f}$$

$$0 \leq switch\_p1_{i,t} \perp \lambda\_p1_{i,t} \geq 0 \quad \forall i \in I, t \in T \tag{12.CR-4}$$

$$0 \leq switch\_p2_{i,t} \perp \lambda\_p2_{i,t} \geq 0 \quad \forall i \in I, t \in T \tag{12.CR-5}$$

$$0 \leq switch\_p3_{i,t} \perp \lambda\_p3_{i,t} \geq 0 \quad \forall i \in I, t \in T \tag{12.CR-6}$$

12.4 ROBUST IMPLEMENTATION STRATEGY

The above modeling procedure is effective in small-scale refinery hydrogen systems, as illustrated in our previous work. However, applying it directly to large-scale refineries may cause instability problems. For instance, the solution procedure might have difficulty finding a feasible solution, and the obtained result may be far from optimal. The instability problems are induced by the parameter settings during the solution procedure. Therefore, in order to guarantee the performance of the proposed methods in large-scale refineries, we need a robust engineering strategy to make the solution procedure stable. In this paper, a systematic procedure is proposed to ensure the feasibility, necessity, and effectiveness of the implemented hydrogen system optimization. As shown in Fig. 12.4, the procedure involves a number of steps, which are informed by the real status and real decision parameters.

The first step is precision validation of the model in the proposed validation structure. It is the foundation of any industrial application of operational optimization. Reliability of the optimal decisions will be guaranteed only when the

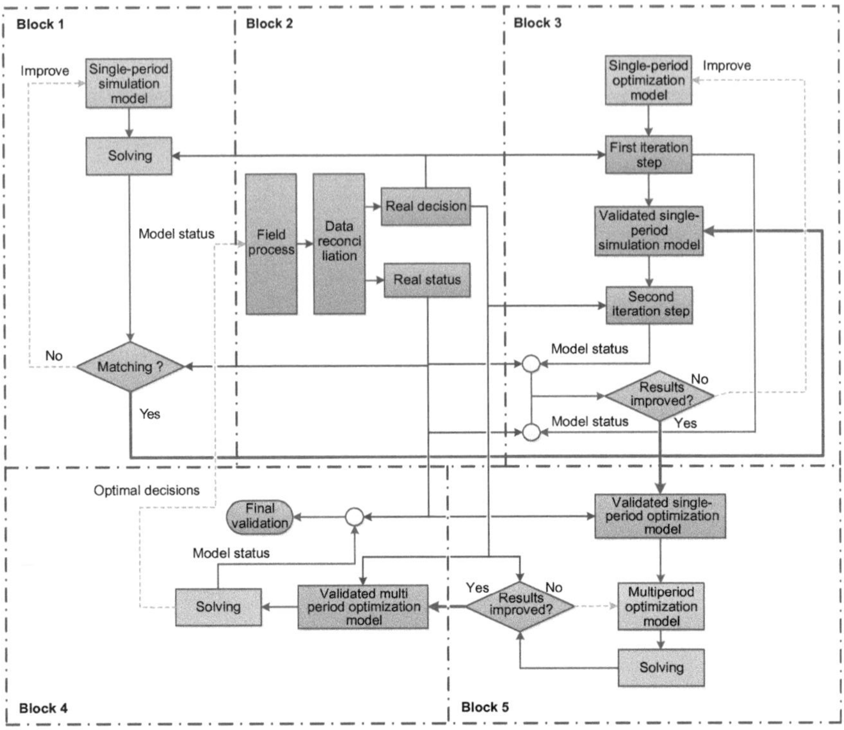

FIG. 12.4 Schematic representation of proposed implementation procedure.

precision of the model meets the requirement, and we need to check the precision of the collocation method. The second step is efficiency validation of the solution method. The proposed approach should be examined by real-time data in this section to explain its value for industrial application. The third step is effect validation of the operational optimization. It is used to indicate the potential economic benefit to refineries if operational optimization of the hydrogen system is executed. After the above three validation processes, the necessity of implementing operational optimization of the hydrogen system can be interpreted. However, these three validation processes are offline ones. In fact, online validation is needed to ensure its effect in industrial application. Based on offline validation, online validation called as "execution validation on field" is finally carried out to interpret the effectiveness of implementing operational optimization of the hydrogen system. The detailed procedures of these four steps are given in the following.

12.4.1 Precision Validation of the Model

The aim of precision validation of the model is to verify whether the optimization model can express the real hydrogen system. The procedure of the validation is illustrated in block 1 of Fig. 12.4. As shown in the figure, the precision validation involves four substeps.

1. Execute data reconciliation to eliminate measurement error in the initial data so that the processed data, which is called validation data, will satisfy mass and energy balances;
2. Input the validated real decisions into the single-period model;
3. Solve the model;
4. Compare the model status derived from step 1 with the validated real status. The precision of the model will be satisfied with the application requirement if and only if the error between these two statuses is small enough, and then this validation can be ended. If this is not the case, the model should be improved according to the comparison result, and then the validation procedure should be executed again.

12.4.2 Efficiency Validation of the Solution Method

Efficiency validation of the solution method is explained here to verify its value for industrial application by the real-time data. The procedure of the validation is illustrated in block 3 of Fig. 12.4 and is described in detail as follows.

1. Develop the multiperiod operational optimization model for the hydrogen system and extract the single-period optimization model from the multiperiod model;
2. Input the validated real decision data into the single-period optimization model and solve the problem.
3. Employ the result of the previous step as the initial value of the model to generate the new model parameters;
4. Refresh the new parameters in the optimization model and run the problem a second time;
5. Improve the single-period optimization model and repeat steps 2 to 4 until the average error tolerance $\overline{\varepsilon}_F$ and $\overline{\varepsilon}_H$ meet the requirements;
6. Examine the comparison results in each iteration step to verify the convergence of the solution method.

12.4.3 Effect Validation of the Operational Optimization

The destination of performing an operational optimization of the hydrogen system is to bring profit to the refinery by reducing its operating cost. The effects of the operational optimization are examined to illustrate whether the proposed optimal operation in the refinery can achieve this destination. The procedure of the validation is illustrated in block 5 of Fig. 12.4 and is described in detail as following.

7. Extend the validated single-period optimization model to the multiperiod model;

8. Input the real status, which consists of initial values of unit status, the production amount of hydrogen, and the energy demand of the units, to the multiperiod model of step 1;
9. Solve the validation model by NLP solvers to obtain the optimal decisions;
10. Compare the optimal decisions with the real decisions to investigate the variation of operating cost.

12.4.4 Execution Validation on Field

The above three offline validations verify the feasibility and necessity of the operational optimization of the hydrogen system in the refinery. However, online validation is required to guarantee its effect in industrial application. After these offline validations, execution validation on field is finally carried out to illustrate the effectiveness of implementing the operational optimization in the refinery. The procedure of the validation is illustrated in block 4 of Fig. 12.4, which involves the following 5 steps:

11. Install the operational optimization of the hydrogen system on field;
12. Obtain the hydrogen production flow rate and energy demand of each unit needed by the optimization model from the real-time data base.
13. Solve the problem by the proposed approach, and save the model status derived during the solution procedure;
14. Execute the optimal decisions on field;
15. Compare the model status derived from step 3 with the real status under the optimal decisions to investigate the effectiveness of implementing the operational optimization in the refinery.

12.4.5 Demonstration of the Execution Validation on Field

The above model and implementation strategy has been applied to a large-scale refinery case, as shown in Fig. 12.2. Due to the limited space, we briefly present the compressor scheduling result in Fig. 12.5. From the figure, we can see a more stable and relatively smaller load on the compressor. The compressor load is 7% smaller than the real-time case. The corresponding hydrogen production from the hydrogen plant is also reduced 5%.

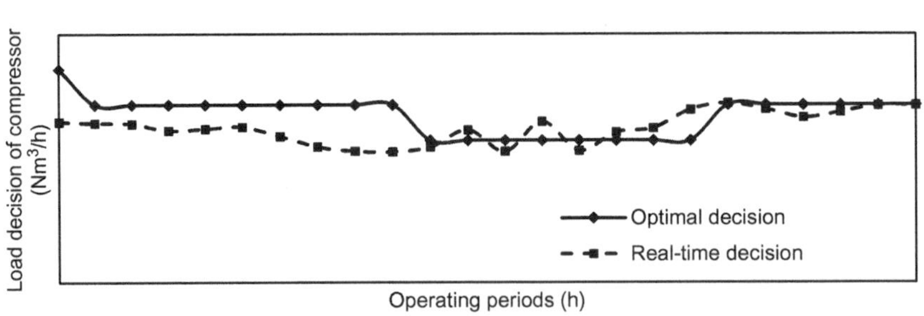

FIG. 12.5 Comparison of load decision of compressors.

12.5 CONCLUSION

This paper introduced a robust engineering strategy for implementing the previously developed optimization method to a hydrogen system in a refinery. Using four kinds of validation procedure, the feasibility, necessity, and effectiveness of implementing the operational optimization was presented. The proposed method has now been developed as a software system and integrated with the refinery database to effectively support the operational optimization of the hydrogen system of an oil refinery. A future effort will be to incorporate a more precise model of the compressor into the proposed formulation so that the work status of the hydrogen system can be expressed much more reasonably.

ACKNOWLEDGMENTS

The financial support provided by the Project of the National Science Fund for Distinguished Young Scholars (21525627), the National Natural Science Foundation of China (Key program 91434205; General program 61590925), and the International S&T Cooperation Projects of China (2015DFA40660) are gratefully acknowledged.

REFERENCES

Ahmad, M.I., Zhang, N., Jobson, M., 2010. Modelling and optimisation for design of hydrogen networks for multi-period operation. J. Clean. Prod. 18 (9), 889–899.

Alves, J.J., Towler, G.P., 2002. Analysis of refinery hydrogen distribution systems. Ind. Eng. Chem. Res. 41 (23), 5759–5769.

Elkamel, A., Alhajri, I., Almansoori, A., Saif, Y., 2011. Integration of hydrogen management in refinery planning with rigorous process models and product quality specifications. Int. J. Proc. Syst. Energ. 1 (3), 302–330.

Hallale, N., Liu, F., 2001. Refinery hydrogen management for clean fuels production. Adv. Environ. Res. 6 (1), 81–89.

Hasan, M.M.F., Karimi, I.A., Avison, C.M., 2011. Reliminary synthesis of fuel gas networks to conserve energy and preserve the environment. Ind. Eng. Chem. Res. 50 (12), 7414–7427.

Iyer, R.R., Grossmann, I.E., 1997. Optimal multiperiod operational planning for utility systems. Comput. Chem. Eng. 21 (8), 787–800.

Jagannath, A., Hasan, M.M.F., Al-Fadhli, F.M., Karimi, I.A., Allen, D.T., 2012. Minimize flaring through integration with fuel gas networks. Ind. Eng. Chem. Res. 51 (39), 12630–12641.

Jiao, Y.Q., Su, H.Y., Hou, W.F., Liao, Z.W., 2012. A multiperiod optimization model for hydrogen system scheduling in refinery. Ind. Eng. Chem. Res. 51 (17), 6085–6098.

Joly, M., Moro, L., Pinto, J., 2002. Planning and scheduling for petroleum refineries using mathematical programming. Braz. J. Chem. Eng. 19 (2), 207–208.

Lee, M.H., Lee, S.J., Han, C., Chang, K.S., Kim, S.H., You, S.H., 1998. Hierarchical on-line data reconciliation and optimization for an industrial utility plant. Comput. Chem. Eng. 22, S247–S254.

Liao, Z.W., Rong, G., Wang, J.D., Yang, Y.R., 2011. Rigorous algorithmic targeting methods for hydrogen networks. Part II: Systems with one hydrogen purification unit. Chem. Eng. Sci. 66 (5), 821–833.

Liao, Z.W., Wang, J.D., Yang, Y.R., Rong, G., 2010. Integrating purifiers in refinery hydrogen networks: a retrofit case study. J. Clean. Prod. 13 (3), 233–241.

Liu, F., Zhang, N., 2004. Strategy of purifier selection and integration in hydrogen networks. Chem. Eng. Res. Des. 82 (A10), 1315–1330.

Micheletto, S.R., Carvalho, M.C., Pinto, J.M., 2008. Operational optimization of the utility system of an oil refinery. Comput. Chem. Eng. 32 (1), 170–185.

Moro, L., Zanin, A., Pinto, J., 1998. A planning model for refinery diesel production. Comput. Chem. Eng. 22, S1039–S1042.

Neiro, S.M., Pinto, J.M., 2005. Multiperiod optimization for production planning of petroleum refineries. Chem. Eng. Commun. 192 (1), 62–88.

Pinto, J., Joly, M., Moro, L., 2000. Planning and scheduling models for refinery operations. Comput. Chem. Eng. 24 (9), 2259–2276.

Van den Heever, S.A., Grossmann, I.E., 2003. A strategy for the integration of production planning and reactive scheduling in the optimization of a hydrogen supply network. Comput. Chem. Eng. 27 (12), 1813–1839.

Xuan, J., Liao, Z.W., Rong, G., Yang, Y.R., 2010. Hydrogen network retrofit design in refinery based on stochastic programming. CIESC J. 61 (2), 398–404.

Yi, H.S., Kim, J.H., Han, C., Jung, J.H., Lee, M.Y., Lee, J.T., 2003. Periodical replanning with hierarchical repairing for the optimal operation of a utility plant. Control. Eng. Pract. 11 (8), 881–894.

Zhang, J., Rong, G., 2008. An MILP model for multi-period optimization of fuel gas system scheduling in refinery and its marginal value analysis. Chem. Eng. Res. Des. 86 (2), 1141–1151.

Zhang, J., Rong, G., Hou, W., Huang, C., 2010. Simulation based approach for optimal scheduling of fuel gas system in refinery. Chem. Eng. Res. Des. 88 (1), 87–99.

Zhou, L., Liao, Z., Wang, J., Jiang, B., Yang, Y., 2012. Hydrogen sulfide removal process embedded optimization of hydrogen network. Int. J. Hydrog. Energy 37 (23), 18163–18174.

Zhou, L., Liao, Z., Wang, J., Jiang, B., Yang, Y., Hui, D., 2013. Optimal design of sustainable hydrogen networks. Int. J. Hydrog. Energy 38 (7), 2937–2950.

Zhou, L., Liao, Z.W., Wang, J.D., 2014. MPEC strategies for efficient and stable scheduling of hydrogen pipeline network operation. Appl. Energy 119, 296–305.

Chapter 13

Optimal Design of Refinery Hydrogen System With Purification Unit

Chun Deng* **and Xiao Feng**†
State Key Laboratory of Heavy Oil Processing, College of Chemical Engineering, China University of Petroleum, Beijing, China, †School of Chemical Engineering & Technology, Xi'an Jiaotong University, Xi'an, China

13.1 INTRODUCTION

The amount of inferior crude oil being processed has been increasing yearly all over the world. For example, Sinopec, one of the main Chinese refinery companies, imported up to 70 million tons of high-sulfur crude oil in 2010 and the annual growth rate is around 17%. The environmental regulations and policies on sulfide contained in product oil have become tighter recently. Both the European Standard (EN 228+A1-2017) and Chinese standard (GB 17930-2016) specify unleaded petrol with a maximum sulfur content of 0.001%. Refineries have been increasing the processing ratio of hydrotreating and hydrocracking processes, which consume a large amount of hydrogen. The hydrogen deficit aggravates the fresh hydrogen shortage in refineries, making fresh hydrogen a more and more expensive resource for modern refineries. Hydrogen production technologies, such as steam reforming of natural gas or methane, are commonly utilized to produce hydrogen to supplement the deficit. However, hydrogen production is a typical high energy consumption process. Synthesis of refinery hydrogen network has been an effective tool to recover hydrogen and reduce the capacity of the hydrogen plant. The methodologies on synthesis of a refinery hydrogen network can be categorized into two types: insight-based pinch technique and the superstructure-based mathematical programming approach.

This chapter is organized as follows. In the following section, we provide a brief introduction of the hydrogen system of a refinery plant. After that, the pinch technique is presented for the targeting of a hydrogen network

Hydrogen Supply Chain. https://doi.org/10.1016/B978-0-12-811197-0.00013-0

with purification reuse/recycle. Then, the superstructure-based mathematical programming approach for the optimal design of a hydrogen network is introduced before the conclusion of the chapter.

13.2 HYDROGEN SYSTEM OF REFINERY PLANT

In this section, an overview of a refinery hydrogen system and its elements is briefly introduced. This is followed by the introduction of a process flow schematic diagram for typical consumers (i.e., hydrocracking) and hydrogen producers (i.e., hydrogen plant). The purification processes, including cryogenic separation, pressure swing adsorption, membrane, light hydrocarbon recovery, and desulfurization, are summarized at the end of this section. An overall view of a refinery hydrogen system is illustrated in Fig. 13.1 and it includes the hydrogen consumers (i.e., hydrocracking, hydrotreating units) and hydrogen sources (i.e., catalytic reforming, hydrogen plant), as well as the purification units (i.e., H_2S removal and PSA).

13.2.1 Typical Hydrogen Consumers

In a modern refinery the demand for petroleum products has shifted from diesel fuel and heating oil to high ratios of gasoline and jet fuel. Environmental regulations limiting sulfur and aromatic contents in motor fuels are stricter.

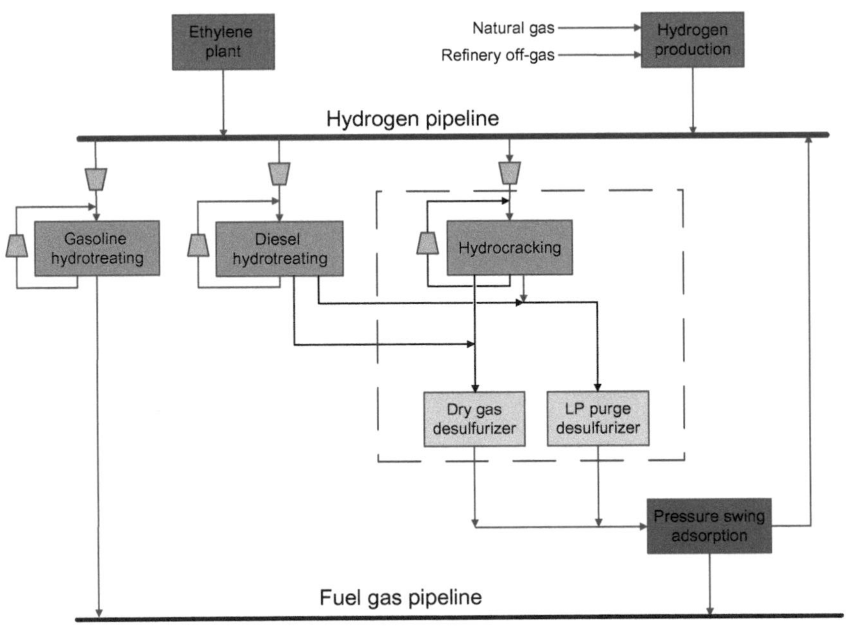

FIG. 13.1 Overall schematic view of a refinery hydrogen network.

These factors lead to the widespread use of hydrocracking in a modern refinery. Catalytic cracking superimposed with hydrogenation is the mechanism of hydrocracking. Catalytic cracking is the breaking of a carbon–carbon single bond, and hydrogenation is the addition of hydrogen to a carbon–carbon double bond. Typical hydrocracker feedstocks include kerosene, straight-run diesel, atmospheric gas oil, vacuum gas oil, fluid catalytic cracking (FCC), light cycle oil (FCC LCO) and heavy cycle oil (HCO), Light Coker Gas Oil (LCGO), and Heavy Coker Gas Oil (HCGO). The aromatic cycle oils and coker distillates are very refractory and resist catalytic cracking. The higher pressures and hydrogen atmosphere make them relatively easy to hydrocrack. The products of hydrocracker mainly include naphtha, and/or jet fuel, diesel, and lube oil (Gary and Handwerk, 2001).

Hydrotreating processes are to catalytically stabilize petroleum products and/or remove objectionable elements from products or feedstocks by the reaction in the presence of hydrogen. Stabilization usually indicates that the unsaturated hydrocarbons, such as olefins and unstable diolefins, are converted to paraffin. Objectionable elements, including sulfur, nitrogen, oxygen, halides, and trace metals, are removed by hydrotreating. The process can be applied to a wide range of feedstocks, from naphtha to reduced crude. Once the process is employed specifically for sulfur removal, it is usually called hydrodesulfurization, or HDS. In order to fulfill environmental regulations, hydrogenation may be conducted on aromatic rings to decrease aromatic content by converting aromatics to paraffin via the hydrotreating process (Gary and Handwerk, 2001).

A simplified process diagram for the hydrocracking process is illustrated in Fig. 13.2. Note that it is similar to the process diagram for hydrotreating. The fresh liquid feed is mixed with makeup hydrogen and recycle gas stream, which is rich in hydrogen content and passed through a heat exchanger to recover the heat of the effluent of the reactors. A furnace is applied to heat the mixed feed to the required temperature. If the liquid feed has not been hydrotreated, a guard reactor should be placed before the first hydrocracker. The guard reactor is

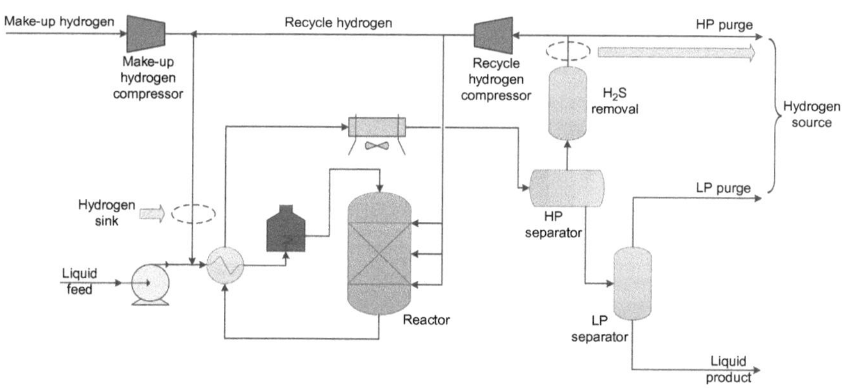

FIG. 13.2 Typical schematic process diagram for hydrocracking process.

usually used to convert organic sulfur and nitrogen compounds to hydrogen sulfide, ammonia, and hydrocarbons to protect the precious metal catalyst found in the following reactors. The hydrocracking reactor(s) is operated at a sufficiently high temperature to convert 40–50vol% of the reactor effluent to material boiling below 205°C. The reactor effluent goes through heat exchangers (may include air cooler) to a high-pressure separator where the hydrogen-rich gases are separated. The gas stream separated from the high-pressure separator may pass through an absorber to remove H_2S prior to being recycled to the first stage for mixing both makeup hydrogen and fresh feed, or it may be fed to intervals of catalyst bed as coolant for temperature control. Portion of the sweetened gas may be purged to the fuel system to avoid the buildup of unfavorable components. The liquid product from the high-pressure separator is sent to a low-pressure separator where the hydrogen gas (so-called low-pressure purge) is separated and then discharged to the fuel system. The liquid effluent from the low-pressure separator is allocated to a distillation column where the C4 and lighter gases are separated overhead, and the light and heavy naphtha, jet fuel, and diesel fuel boiling range streams are removed as liquid side-products. The fractionator bottoms can be partially or totally recycled as the feed of the hydrocracking (Gary and Handwerk, 2001).

Note that, the sweetening high-pressure/low-pressure separator gas streams can be considered as process hydrogen sources. The inlets of the hydrogenation reactors act as process hydrogen sinks, which require hydrogen streams with specified flowrate and pressure as well as hydrogen content.

For the kth hydrogen sink, the inlet flowrate and hydrogen purity can be calculated via Eqs. (13.1), (13.2).

$$F_k^{\text{in}} = F_k^{\text{M}} + F_k^{\text{R}} \tag{13.1}$$

$$y_{k,\text{H}_2}^{\text{in}} = \frac{F_k^{\text{M}} y_{\text{H}_2}^{\text{M}} + F_k^{\text{R}} y_{\text{H}_2}^{\text{R}}}{F_k^{\text{in}}} \tag{13.2}$$

where F_k^{in}, F_k^{M} and F_k^{R} denote the flowrate of total inlet, make-up and recycle hydrogen streams for the kth hydrogen sink, $\text{mol}\,\text{s}^{-1}$. ${y_{k,\,\text{H}_2}}^{\text{in}}$, ${y_{\text{H}_2}}^{\text{M}}$ and ${y_{\text{H}_2}}^{\text{R}}$ represent the hydrogen mole fractions of inlet for the kth hydrogen sink, make-up and recycle hydrogen streams. In addition, the make-up hydrogen and recycle hydrogen compressors are placed to lift the pressure of hydrogen sources to fulfill the pressure requirement of hydrogen sinks.

13.2.2 Typical Hydrogen Producers

The increasing demand of high-octane gasoline has stimulated the use of catalytic reforming. With the implementation of restrictions on the aromatic contents of gasoline, catalytic reforming is expected to decrease. However, the increasing demand for the derivative product of p-Xylene (PX), as its core source, the processing ratio of catalytic reforming keeps increasing. In the catalytic reforming process, the hydrocarbon molecular structures are rearranged to form higher-octane aromatics and a small amount of cracking would occur. The process aims to increase the octane number of gasoline. The hydrogen-rich gas stream from the catalytic reforming process is split into a hydrogen recycle stream and a net hydrogen byproduct that acts as the main hydrogen source for hydrocracking or hydrotreating processes (Gary and Handwerk, 2001).

Additional hydrogen for extensive hydrocracking and hydrotreating processes can be provided via partial oxidation of heavy hydrocarbons (i.e., fuel oil), or stream reforming of methane (natural gas), ethane, or propane. Compared with partial oxidation of fuel oil, the cost for hydrogen production via steam reforming of methane is usually lower and it is more widely used in industry (Gary and Handwerk, 2001). In addition, the syngas from coal gasification serves as another hydrogen source.

Steam-methane reforming (SMR) for hydrogen production includes four steps: reforming, shift conversion, gas purification, and methanation. A simplified flow diagram is presented in Fig. 13.3. Firstly, the catalytic reaction of methane with steam at temperatures in the range of 760–816°C is endothermic and is carried out by passing the gas through catalyst-filled tubes in a furnace. In the second step, more steam is added to convert the CO generated in the first step to an equivalent amount of hydrogen via the shift reaction. It is an exothermic reaction and is carried out in a fixed-bed catalytic reactor at around 343°C. There are multiple catalyst beds in one reactor and external cooling exists between beds to prevent the temperature from getting too high, as that would adversely affect the equilibrium conversion. The generated CO_2 is removed in the third step by adsorption in a circulation amine or hot potassium carbonate solution and the rich solution is regenerated in a stripper. In the fourth step, the remaining small amount of CO and CO_2 are converted to methane via methanation reactions, which are exothermic. This is conducted in a fixed-bed reactor at a temperature of 427°C (Gary and Handwerk, 2001). Note that the gas purification and methanation can be replaced via pressure swing adsorption (PSA). The carbon dioxide and other impurities can be removed via PSA and the purity of the product stream of PSA can reach 99% or even higher.

13.2.3 Industrial Hydrogen Purification Process

A significant amount of the hydrogen-rich gas stream is vented from hydrocracking or hydrotreating processes. Recovery of the hydrogen should be taken

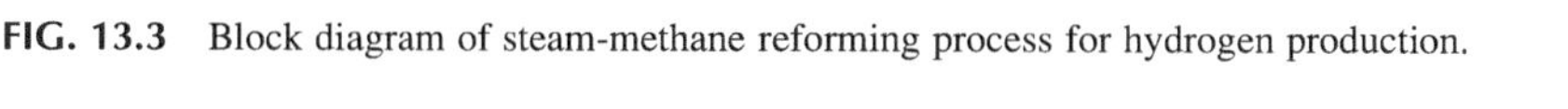

FIG. 13.3 Block diagram of steam-methane reforming process for hydrogen production.

into consideration whenever it is needed to supplement the hydrogen demand. Typical processes for hydrogen recovery include cryogenic phase separation, pressure swing adsorption, and membrane.

In the cryogenic phase separation method, the feed gas stream is cooled to around -129 to $-157°C$ at pressures ranging from 1380 to 3450 kPa. The resulting vapor phase contains 90 mol% hydrogen and the liquid phase includes most of methane and other hydrocarbons. The carbon dioxide, hydrogen sulfide, and water vapor must be removed from the feed gas prior to chilling (Gary and Handwerk, 2001).

In the pressure swing adsorption process, hydrocarbon is absorbed from the gas on a solid absorbent (i.e., molecular sieve) and hydrogen leaves the absorber at the desired purity. Several adsorbent columns are used, and the feed gas flow is periodically switched from one column to another so that the adsorbent can be regenerated. The adsorbed methane and other impurities are released from the adsorbent by simple pressure decline and purging (Gary and Handwerk, 2001).

In the membrane process, a membrane composed of synthetic hollow fibers, which allow hydrogen to permeate, separates hydrogen from methane and other components. The driving force is the difference between the hydrogen partial pressures on each side of the membrane. Thus a significant pressure drop must be imposed in order to achieve high recovery (Gary and Handwerk, 2001).

The most economical technique for hydrogen recovery depends on the volume of the gas stream to be processed, the desired hydrogen recovery and purity, and the types of components to be separated (Gary and Handwerk, 2001).

The valuable C_3, C_4, C_5, and C_6 components contained in the gas streams can be recovered via the gas processing units, so-called light hydrocarbon recovery. Typical light hydrocarbon recovery processes include absorber-deethanizer, sponge absorber, depropanizer, debutanizer, and naphtha splitter. A brief introduction to the process can be found in the literature (Gary and Handwerk, 2001).

In addition, gas streams from various processes in the refinery processing of sour crudes contain hydrogen sulfide. Some dissolved carbon dioxide may be contained in the gas streams. Those components (hydrogen sulfide and carbon dioxide) are typically called acid gases. They can be removed by many processes, which can be categorized into three kinds: chemical solvent (i.e., monoethanolamine (MEA), diethanolamine (DEA), methyl-diethanolamine (MDEA), diglycolamine (DGA), hot potassium carbonate), physical solvent (i.e., selexol, propylene carbonate, sulfinol, rectisol), and dry adsorbents processes (i.e., molecular sieve, activated charcoal, iron sponge, zinc oxide). Generally, the diethanolamine process has been the most widely used for treating refinery gas. This process uses an aqueous solution of diethanolamine with concentrations of the DEA in the range of 15–30 wt% (Gary and Handwerk, 2001).

13.3 TARGETING HYDROGEN NETWORK VIA PINCH TECHNIQUE

The Pinch technique has been accepted as an effective tool for the targeting and design of a hydrogen network. Typically, the Pinch technique includes two steps: targeting and design. Alves and Towler (2002) first introduced the Hydrogen Surplus Diagram to identify the pinch and locate the minimum flowrate of hydrogen utility prior to detailed network design. Many other insight-based pinch techniques, such as Material Recovery Pinch Diagram (El-Halwagi et al., 2003), Source Composite Curve (Bandyopadhyay, 2006), Gas Cascade Analysis (Foo and Manan, 2006), Material Surplus Composite Curve (Saw et al., 2011), Composite Algorithm Table (Agrawal and Shenoy, 2006), and extended Limiting Composite Curve (Agrawal and Shenoy, 2006), have been developed to determine the flowrate targets for hydrogen networks. On the basis of the extension of the Composite Algorithm Table, we presented the Improved Problem Table (IPT) (Deng et al., 2015) to target the hydrogen network with purification units.

13.3.1 Model for Hydrogen Network With One Purifier

Prior to the flowrate targeting procedure, the mass balance for a hydrogen network with purification reuse/recycle is addressed (Deng et al., 2015). Fig. 13.4 illustrates the mass flows for a hydrogen network with purification reuse/recycle. The mass balance around the purifier is given by Eqs. (13.3), (13.4). In addition, the hydrogen recovery ratio (*RR*) is defined by Eq. (13.5).

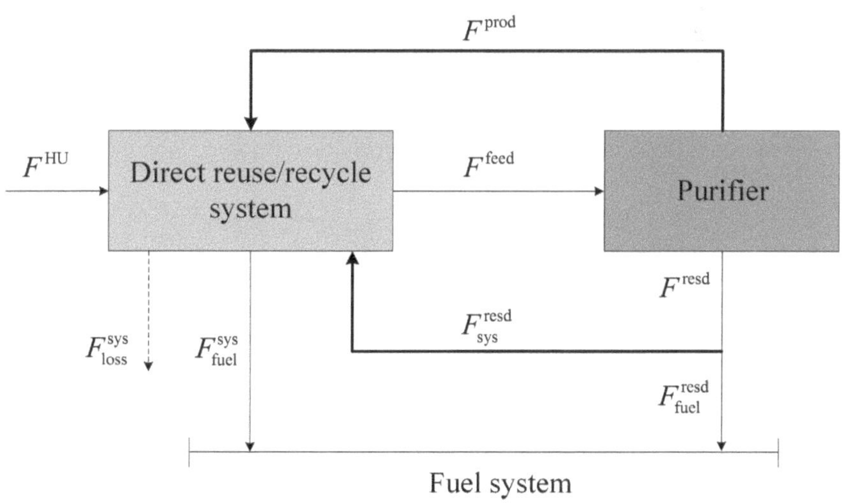

FIG. 13.4 Schematic diagram for mass flows of the hydrogen network with purification reuse/recycle (Deng et al., 2015).

$$F^{feed} = F^{prod} + F^{resd} \tag{13.3}$$

$$F^{feed} y_{H_2}^{feed} = F^{prod} y_{H_2}^{prod} + F^{resd} y_{H_2}^{resd} \tag{13.4}$$

$$RR = \frac{F^{prod} y_{H_2}^{prod}}{F^{feed} y_{H_2}^{feed}} \tag{13.5}$$

where F^{feed}, F^{prod}, and F^{resd} denote the feed, product, and residual flowrate of the purifier and their hydrogen purities are specified as ${y_{H_2}}^{feed}$, ${y_{H_2}}^{prod}$ and ${y_{H_2}}^{resd}$. The optimal F^{prod} can be determined via IPT (introduced in Step 6) with the specified ${y_{H_2}}^{prod}$. There are two options for the residual flowrate of the purifier: reuse/recycled by hydrogen sinks in the direct reuse/recycle system (F_{sys}^{resd}) or discharged to the fuel system (F_{fuel}^{resd}) and the flowrate balance is given by Eq. (13.6).

$$F^{resd} = F_{sys}^{resd} + F_{fuel}^{resd} \tag{13.6}$$

The overall mass balance is given by Eq. (13.7).

$$F^{HU} = F_{loss}^{sys} + F_{fuel}^{sys} + F_{fuel}^{resd} \tag{13.7}$$

where F_{fuel}^{sys} and F_{fuel}^{resd} denote the flowrates that discharged from the direct reuse/recycle system and residual of the purifier to the fuel system. F_{loss}^{sys} denotes the total flowrate loss for the direct reuse/recycle system and it can be determined via Eq. (13.8). It is identical with the net flowrate in the last impurity interval determined in Step 2 of IPT. Note that, F_{loss}^{sys} keeps unchanged if the system is selected. To minimize the flowrate of the hydrogen utility, F_{fuel}^{sys} and F_{fuel}^{resd} should be minimized as well.

$$F_{loss}^{sys} = \sum_{k}^{NSK} F_{SKk} - \sum_{i}^{NSR} F_{SRi} \tag{13.8}$$

The overall mass balance around the direct reuse/recycle system is given by Eq. (13.9).

$$F^{HU} + F^{prod} + F_{sys}^{resd} = F_{loss}^{sys} + F_{fuel}^{sys} + F^{feed} \tag{13.9}$$

13.3.2 Improved Problem Table

Next, IPT (Deng et al., 2015) is used to locate the targets for a refinery hydrogen network with purification reuse/recycle scheme. The targets include the minimum flowrates of hydrogen utility, feed and product flowrate of the purifier, and the optimal feed purity of the purifier. The optimal placement of the purifier corresponds to the optimization of its feed purity/impurity. We use an example to show the IPT. The limiting data shown in Table 13.1 are extracted from the literature (Elkamel et al., 2011). The hydrogen purity for the product of

TABLE 13.1 Limiting Data of Case 1

Hydrogen Sources	Purity (Fraction)	Flowrate (Nm^3/h)	Hydrogen Sinks	Purity (Fraction)	Flowrate (Nm^3/h)
HC	0.8	60,678	HC	0.8671	93,306
CR	0.8	17,303	GOHT	0.8358	82,656
GOHT	0.75	55,281	RHT	0.8257	39,164
RHT	0.75	25,870	DHT	0.7230	12,472
DHT	0.65	8004	NHT	0.7148	5726
NHT	0.6	3840			
Hydrogen utility	0.95	89,304 (current)			

the purifier is given as 0.9 and hydrogen recovery is defined as 0.9. The maximum inlet flowrate for the purifier is assumed to be 35,712 Nm^3/h. Because the purity of the purification product is known, the product of the purifier can be treated as an external hydrogen source.

Step 1: Purities and impurities arrangement: All of the purities of process hydrogen sources and sinks, as well as hydrogen utility and product stream of purifier, are arranged in decreasing order in the first column (Table 13.2). One value for duplicate purities (if any) is kept listed in the column. The arbitrary purity is added as the last entry of the column such that it is the smallest value, i.e., 0.55 in the first column. The arbitrary purity serves to provide an endpoint and facilitates the plotting of the last segment of the LCC. The second column shows the impurities (y^{ν}), which are determined by $1-{y_{H_2}}^{\nu}$. The impurities in the second column fulfil the relationship shown as Eq. (13.10).

$$y^1 < y^2 < \cdots < y^{\nu} < \cdots < y^{arbitrary} \tag{13.10}$$

Step 2: Net flowrate deficits targeting: The net flowrates F_{net}^{ν} in the third column (Table 13.2) are calculated by subtracting the summation of the flowrates of the hydrogen sources from those of the hydrogen sinks in each impurity interval. Once the impurities of the hydrogen sources and sinks are less than y^{ν}, these hydrogen sources and sinks will appear in the impurity interval (y^{ν}, $y^{\nu-1}$), and the net flowrate of this impurity interval can be calculated by Eq. (13.11). For instance, within the impurity interval (0.2, 0.25), there exist two process hydrogen sources (HC (0.2) and CR (0.2)) and three process hydrogen sinks (HC (0.1329), GOHT (0.1642) and RHT (0.1743)) in the impurity interval. The net flowrate can be calculated as 137,146 Nm^3/h by solving Eq. (13.11).

$$F_{net}^{\nu} = \sum_k F^{SKk} - \sum_i F^{SRi} \quad \forall y^{SRi},\ y^{SKk} < y^{\nu} \tag{13.11}$$

Note that the last entry of the third column of Table 13.2 is obtained by subtracting the summation of all flowrates of process hydrogen sources from that of process hydrogen sinks and it can be defined as the total flowrate deficit for a network. For a given hydrogen network, the total flowrate deficit is a constant and it indicates that at least such a flowrate of external hydrogen sources shall be supplemented for the network. For this case, the minimum flowrate for external hydrogen sources is 62,349 Nm^3/h, as indicated by the last entry in the third column.

Step 3: Net mass loads targeting: The net mass loads in the fourth column (Table 13.2) are calculated using Eq. (13.12). The net mass load for each impurity interval is the product of the net flowrate and the impurity difference of the corresponding interval.

$$\Delta M_{net}^{\nu} = F_{net}^{\nu}\left(y^{\nu} - y^{\nu-1}\right) \tag{13.12}$$

TABLE 13.2 Implementation of IPT for Case 1 With Purification Reuse/Recycle (Preliminary Solution) (Deng et al., 2015)

Purity (Fraction)	Impurity (Fraction)	Net Flowrate (Nm^3/h)	Net Load (Nm^3/h)	Cumulative Load (Nm^3/h)	Hydrogen Utility (Nm^3/h)	Flowrate for Purification Product (Nm^3/h)	Flowrate Above Pinch (Nm^3/h)	Flowrate Above Pinch (Nm^3/h)	Flowrate Above Pinch (Nm^3/h)	Flowrate for Waste Hydrogen Stream (Nm^3/h)
0.95	0.05			0	0					
		0	0							
0.9	0.1			0	0					
		0	0							
0.8671	0.1329			0	0	0				
		93,306	2917							
0.8358	0.1642			2917	25,549	45,454				
		175,962	1783							
0.8257	0.1743			4700	37,808	63,247				
		215,127	5526							
0.8	0.2			10,226	68,175	102,262				
		137,146	6857							
0.75	0.25			17,084	85,418	113,890				45,082
		55,995	1512							

Continued

TABLE 13.2 Implementation of IPT for Case 1 With Purification Reuse/Recycle (Preliminary Solution) (Deng et al., 2015)—cont'd

Purity (Fraction)	Impurity (Fraction)	Net Flowrate (Nm^3/h)	Net Load (Nm^3/h)	Cumulative Load (Nm^3/h)	Hydrogen Utility (Nm^3/h)	Flowrate for Purification Product (Nm^3/h)	Flowrate Above Pinch (Nm^3/h)	Flowrate Above Pinch (Nm^3/h)	Flowrate Above Pinch (Nm^3/h)	Flowrate for Waste Hydrogen Stream (Nm^3/h)
0.7230	0.277			18,596	81,917	105,057	55,994			
		68,467	562							
0.7148	0.2852			19,158	81,448	103,436	58,901			
		74,193	4807							
0.65	0.35			23,964	79,881	95,858	68,808			2619
		66,189	3309							
0.6	0.4			27,274	77,925	90,913	67,935	66,189		3840
		62,349	3117							
0.55	0.45			30,391	75,978	86,832	66,538	64,269	62,349	
Eqs. (13.12), (13.11)		$F^{feed}y_{H_2}{}^{feed}$		$F^{prod}y_{H_2}{}^{prod}$	$F^{resd}y_{H_2}{}^{resd}$	$y_{H_2}{}^{feed}$	F^{feed}	$\sum F^{WH}$		$F^{feed}y_{H_2}{}^{feed}$
		113,890		*102,501*	*11,389*	**0.734**	*155,218*	51,542		*103,677*
Eq. (13.9)		F_{HU}		F^{prod}	F^{resd}_{sys}	F^{sys}_{loss}	F^{feed}	F^{sys}_{fuel}		$\sum F^{WH} - F^{feed}$
		0		**113,890**	**0**	**62,349**	**51,542**	**0**		0

Step 4: Cumulative mass loads targeting: The cumulative mass loads are calculated in the fifth column (Table 13.2) using Eq. (13.13). The first entry for the column has no cumulative mass load so that it equals zero. The cumulative mass load of the νth row is determined by the summation of the mass loads for all earlier rows.

$$\Delta M^{\nu}_{cum} = 0 \qquad \nu = 1$$
$$\Delta M^{\nu}_{cum} = \sum_{t=1}^{t=\nu} \Delta M^{t}_{net} \quad \nu > 1 \tag{13.13}$$

The LCC can be constructed via plotting the cumulative load column as abscissa against the impurity column as ordinate, as shown in the bold line in Fig. 13.5. The reciprocal of the slope of a segment on the LCC corresponds to the net flowrate of the same impurity interval determined in Step 2. Note that the reciprocal of the slope of the last segment of the LCC is identical with the total flowrate deficit of the network.

Step 5: Flowrate targeting for hydrogen utility: The possible supply flowrates for hydrogen utility at each impurity level (y^{ν}) in the sixth column are calculated using Eq. (13.14).

$$F^{HU} = \frac{\Delta M^{\nu}_{cum}}{y^{\nu} - y^{HU}} \tag{13.14}$$

where ΔM^{ν}_{cum} and y^{ν} denotes the cumulative mass load and impurity for the vth impurity level and y^{HU} denotes the impurity of hydrogen utility.

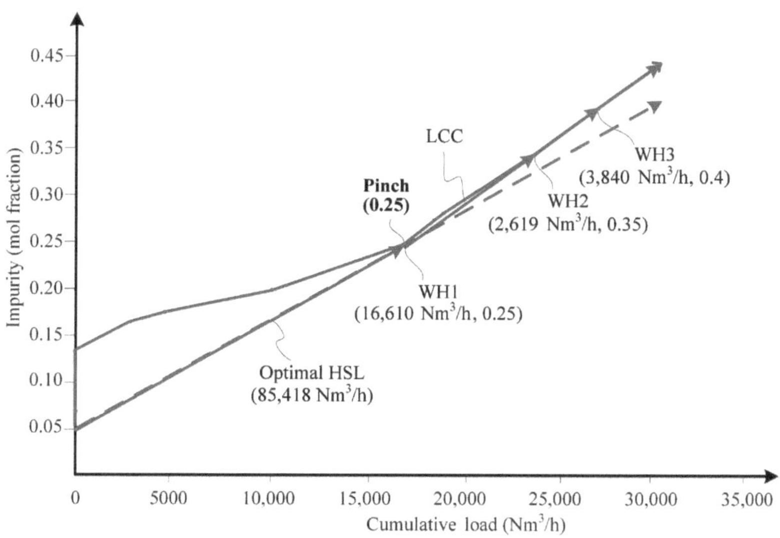

FIG. 13.5 LCC and optimal HSL with direct reuse/recycle (Deng et al., 2015).

The maximum value in the sixth column of Table 13.2 is 85,418 Nm^3/h and it is marked in bold. It is greater than the total flowrate deficit (62,349 Nm^3/h) determined in Step 2 and the maximum value (85,418 Nm^3/h) is the minimum flowrate of hydrogen utility. The corresponding impurity (0.25) is identified as the pinch impurity.

If the maximum value in the sixth column (calculated by solving Eq. 13.14) is smaller than the total flowrate deficit determined in Step 2, the total flowrate deficit is taken as the minimum flowrate of hydrogen utility for the network. In this case, the HSL that touches the LCC would be infeasible. The optimal and feasible HSL should be in parallel with the last segment of the LCC as shown in Fig. 13.6.

Step 6: Flowrate targeting for other external hydrogen sources: All possible flowrates of other external hydrogen sources are calculated in the following columns. Note that if several external hydrogen sources with different impurities are available, an external hydrogen source with higher impurity (typically with lower cost) should be introduced prior to one with lower impurity (typically with higher cost) to fulfill the requirement of the sinks so that refineries can cut down on cost.

The impurity corresponding to the maximum value of all possible flowrates for the first external hydrogen source (HU_1) determined in Step 5 defines its pinch impurity ($y_{HU_1}{}^{pinch}$). The introduction of a second external hydrogen source (HU_2) with impurity of y^{HU_2} ($y^{HU_1} < y^{HU_2}$), if and only if the condition ($y^{HU_2} < y_{HU_1}{}^{pinch}$) is fulfilled, can reduce the flowrate for the first external

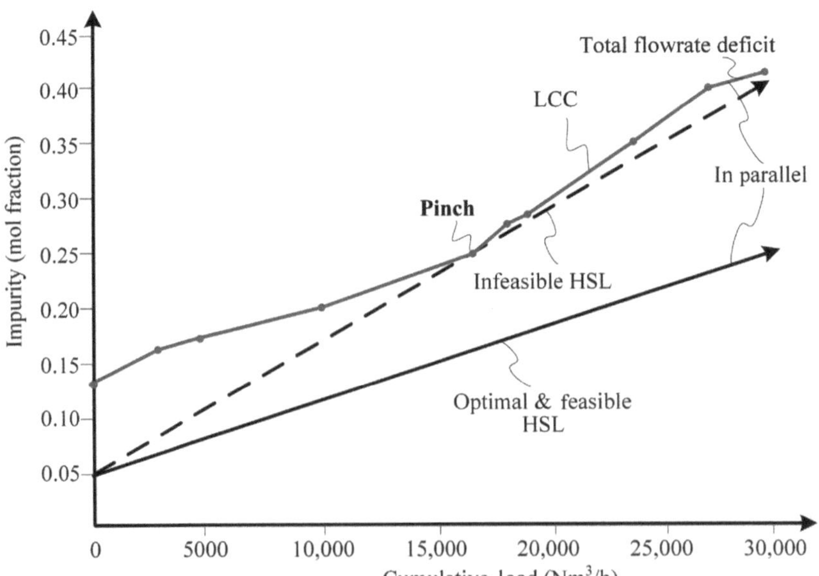

FIG. 13.6 Optimal and feasible HSL (Deng et al., 2015).

hydrogen source (F^{HU_1}) further. Typically, the value for the possible F^{HU_1} at the impurity level of y^{HU_2} is found as the optimal flowrate of the first external hydrogen source. Next, all the possible flowrates of the second external hydrogen source (HU_2) may be calculated using Eq. (13.15). This equation can be explained as follows. The first term of the numerator on the right-hand side (RHS) of Eq. (13.15) denotes the load between impurity interval $[y^{HU_1}, y^{\nu}]$. The load removed by HU_1 is determined by the second term of the numerator on the RHS of Eq. (13.15). The first term on the RHS of Eq. (13.15) is solved to get the summation flowrate of HU_1 and HU_2. Next, the optimal flowrate of HU_1 is subtracted from the summation flowrate of HU_1 and HU_2 and the residual flowrate can be determined as the target of HU_2.

$$F^{HU_2} = \frac{\Delta M^{\nu}_{cum} - F^{HU_1}\left(y^{HU_2} - y^{HU_1}\right)}{y^{\nu} - y^{HU_2}} - F^{HU_1} \tag{13.15}$$

In order to calculate all the possible flowrates for the rth external hydrogen sources, Eq. (13.15) can be generalized to Eq. (13.16).

$$F^{HU_r} = \frac{\Delta M^{\nu}_{cum} - \sum_{n=1}^{r-1} F^{HU_n}\left(y^{HU_r} - y^{HU_n}\right)}{y^{\nu} - y^{HU_r}} - \sum_{n=1}^{r-1} F^{HU_n} \tag{13.16}$$

where n denotes the index for the nth external hydrogen source ($n < r$).

The product of the purifier is considered as an external hydrogen source. The product impurity ($y^{\text{prod}} = 0.1$) fulfills the condition ($y^{\text{prod}} < y^{\text{pinch}}_{\text{HU}}$) and thus the flowrate of hydrogen utility can be further reduced with the introduction of the product of the purifier. All the possible flowrates of the product of the purifier at an impurity level greater than and equal to the product impurity (i.e., $y^{\nu} \geq 0.1$) can be calculated via solving Eq. (13.15). The maximum value (113,890 Nm^3/h) in the seventh column of Table 13.2 may be determined as the optimal flowrate of the product of the purifier and its feasibility should be checked in Step 8.

Step 7: Waste hydrogen streams identification: Waste hydrogen streams discharged from the network are identified. On the pinch (the impurity is 0.25), the accumulated hydrogen flowrate is 85,418 Nm^3/h. It can be considered as an internal hydrogen source with an impurity of 0.25. Then, for each impurity interval above 0.25 (i.e., $y^{\nu} > y^{\text{pinch}}$), the required flowrates can be calculated via Eq. (13.17) and all the possible flowrates for $F^{\text{pinch}}_{\text{above}}$ are listed in the eighth column of Table 13.2 and the maximum value (68,808 Nm^3/h) is the targeted minimum flowrate that would be supplied for the region above an impurity of 0.25. Therefore, only 68,808 Nm^3/h of hydrogen source at an impurity of 0.25 needs to be distributed to the system, and the surplus flowrate 16,610 Nm^3/h (=85,418 − 68,808 Nm^3/h) is identified as the waste hydrogen stream WH1 (0.25). Similarly, the waste hydrogen streams (WH2 and WH3) are identified and the flowrates are shown in the final column of Table 13.2. Their corresponding impurities (the values in the 11th column) are impurities of identified waste hydrogen streams. The identified waste hydrogen streams are discharged to the fuel system in the direct reuse/recycle scheme. However, for the purification reuse/recycle scheme, the identified waste hydrogen streams are either allocated to the purifier as its feed or discharged into the fuel system.

$$F_{above}^{pinch} = \frac{\Delta M_{cum}^{\nu} - \Delta M_{cum}^{pinch}}{y^{\nu} - y^{pinch}} \quad \forall y^{\nu} > y^{pinch} \tag{13.17}$$

Step 8: Mass balance check: The mass balance for a system with purification reuse/recycle shall be performed to check the feasibility. The procedure can be divided into three substeps.

Substep 1: Calculate the values for F^{HU}, F^{prod}, F_{sys}^{resd}, F_{loss}^{sys}, F_{fuel}^{sys}, F^{feed}, and $y_{H_2}{}^{feed}$.

The maximum value in the sixth column of Table 13.2 within the impurity region ($0.05 < y \leq 0.1$) is 0 t/h, which is identified as the minimum flowrate of hydrogen utility ($F^{HU} = 0\,Nm^3/h$) with purification reuse/recycle. The maximum value in the seventh column within the concentration region ($0.1 < y \leq 0.45$) is targeted as the minimum flowrate of the product of the purifier that is allocated to the direct reuse/recycle system ($F^{prod} = 113{,}890\,Nm^3/h$). The total net flowrate for the system (F_{loss}^{sys}) is calculated in Step 2 as 62,349 Nm^3/h. The residual of the purifier with high impurity is not allowed to be reused by any hydrogen sinks and F_{sys}^{resd} is assumed to be zero. In order to maximize the recovery, all the waste hydrogen streams are assumed to be fed to the purifier and thus F_{fuel}^{sys} is assumed to be zero. Next, F^{feed} can be calculated as 51,542 Nm^3/h via solving Eq. (13.9), and this is greater than the capacity of the purifier. It is infeasible and the infeasibility will be removed in substep 3. The total flowrate of all the identified waste hydrogen streams ($\sum F^{WH}$) is calculated as 51,542 Nm^3/h. The difference between $\sum F^{WH}$ and F^{feed} is calculated to validate the assumption that all the waste hydrogen streams are assigned as the feed of the purifier. Next, the purity of the mixed waste hydrogen stream specifies the feed purity for the purifier and $y_{H_2}{}^{feed}$ is calculated as 0.734, which is less than the pinch purity. The values for F^{HU}, F^{prod}, F_{sys}^{resd}, F_{loss}^{sys}, F_{fuel}^{sys}, F^{feed}, and $y_{H_2}{}^{feed}$ are marked in bold in Table 13.2.

Substep 2: Calculate the values for $F^{prod}y_{H_2}{}^{prod}$, $F^{feed}y_{H_2}{}^{feed}$, F^{feed}, $F^{resd}y_{H_2}{}^{resd}$ and ΔF^{feed}.

$F^{prod}y_{H_2}{}^{prod}$ can be calculated as 102,501 Nm^3/h with the given $y_{H_2}{}^{prod} = 0.90$ and determined value $F^{prod} = 113{,}890\,Nm^3/h$. $F^{feed}y_{H_2}{}^{feed}$ is next calculated as 113,890 Nm^3/h via solving Eq. (13.5). Next F^{feed} is determined as 155,218 Nm^3/h (=113,890 Nm^3/h/0.734). Next $F^{resd}y_{H_2}{}^{resd}$ is calculated as 11,389 Nm^3/h via solving Eq. (13.4). The difference between two F^{feed}s (ΔF^{feed}) is determined as 103,677 Nm^3/h. The values for $F^{prod}y_{H_2}{}^{prod}$, $F^{feed}y_{H_2}{}^{feed}$, F^{feed}, $F^{resd}y_{H_2}{}^{resd}$ and ΔF^{feed} are marked in *italic* in Table 13.2.

Substep 3: Check the feasibility and determine the optimal targets.

Note that ΔF^{feed} is 103,677 Nm^3/h and it indicates the infeasibility of the results. The optimal and feasible target for ΔF^{feed} should be zero. The other variables in *substeps 1 and 2* can be related with the variable F_{HU} via the equations mentioned in *substeps 1 and 2*. The Excel Goal Seek feature is utilized to determine the optimal target ΔF^{feed} by changing the value of F^{HU}. In the Excel Goal Seek control box, ΔF^{feed} is set to be the target value of zero and the value of F^{HU} is set to be altered. Once we click the "OK" button on the Excel Goal Seek control box, ΔF^{feed} achieves zero and the values of all other variables are changed to the new results shown in Table 13.3. The optimal flowrate of hydrogen utility with a purification reuse/recycle scheme is determined to be 70,220 Nm^3/h, which is 15,197 Nm^3/h less than that (85,418 Nm^3/h) with a direct reuse/recycle scheme. The optimal flowrate of the product of the purifier is determined to be 20,263 Nm^3/h with the minimum flowrate of hydrogen utility. The optimal flowrate of the feed of the purifier is 28,135 Nm^3/h with an optimal feed purity of 0.72 (i.e., the impurity is 0.28). This indicates that the optimal feed impurity of the purifier is not necessarily to be the pinch impurity as assumed in the literature (2006).

As shown in Fig. 13.7, the first segment of HSL can be constructed within the impurity intervals of 0.05 and 0.1 with its inverse slope corresponding to 70,220 Nm^3/h (F^{HU}). The section of LCC above it is considered as Region 1, where only hydrogen utility is allocated to fulfill the requirement. The second segment of HSL can be constructed with the impurity intervals of 0.1 and 0.45 with its inverse slope corresponding to 90,484 Nm^3/h (=70,220 + 20,263 Nm^3/h). The section of LCC above it is considered as Region 2, where F^{HU} and F^{prod} are referenced to remove the load of this region.

Step 9: Hydrogen network design: Design the hydrogen network via the nearest neighbors algorithm (NNA) (Prakash and Shenoy, 2005). Fig. 13.8 illustrates one optimal hydrogen network with purification reuse/recycle.

In this section, IPT is introduced in detail for the determination of the optimal targets for a hydrogen network with purification reuse/recycle. With the introduction of the product stream of purifier, the flowrate of hydrogen utility can be reduced from 85,418 to 70,220 Nm^3/h. It is worth mentioning that IPT can locate the exact optimal feed impurity of the purifier, which is greater than or equal to the pinch impurity.

13.4 DESIGN OF HYDROGEN NETWORK VIA MATHEMATICAL PROGRAMMING APPROACH

The main feature of the insight-based pinch technique is that the pinch technique can be used to determine the flowrate of hydrogen utility intuitively. However, it has certain difficulties in handling the automated optimal synthesis of a hydrogen network with pressure constraints, multiple impurities, and the annualized cost as the objectives. It leads to the development of superstructure-based mathematical programming approaches. Hallale and Liu (2001) firstly proposed a superstructure embedded with hydrogen sources, sinks, and compressors, and then optimized it mathematically to maximize

TABLE 13.3 Implementation of IPT for Case 1 With Purification Reuse/Recycle (Optimal Solution) (Deng et al., 2015).

Purity (Fraction)	Impurity (Fraction)	Net Flowrate (Nm^3/h)	Net Load (Nm^3/h)	Cumulative Load (Nm^3/h)	Hydrogen Utility (Nm^3/h)	Flowrate for Purification Product (Nm^3/h)	Flowrate Above Pinch (Nm^3/h)	Flowrate Above Pinch (Nm^3/h)	Flowrate Above Pinch (Nm^3/h)	Flowrate for Waste Hydrogen Stream (Nm^3/h)
0.95	0.05			0	0.00					
		0	0							
0.9	0.1			0	70,220					
		0	0							
0.8671	0.1329			0	0	−1,768,904				
		93,306	2917							
0.8358	0.1642			2917	25,549	−79,472				
		175,962	1783							
0.8257	0.1743			4700	37,808	−54,221				
		215,127	5526							
0.8	0.2			10,226	68,175	−3068				
		137,146	6857							
0.75	0.25			17,084	85,418	20,263				21,675
		55,995	1512							

0.7230	0.277			18,596	81,917	15,001	55,995			
		68,467	562							
0.7148	0.2852			19,158	81,448	14,259	58,901			
		74,193	4807							
0.65	0.35			23,964	79,881	11,593	68,808			2619
		66,189	3309							
0.6	0.4			27,274	77,925	8989	67,935	66,189		3840
		62,349	3117							
0.55	0.45			30,391	75,978	6580	66,538	64,269	62,349	

Eqs. (13.12), (13.11)	$F^{feed}{y_{H_2}}^{feed}$	$F^{prod}{y_{H_2}}^{prod}$	$F^{resd}{y_{H_2}}^{resd}$	${y_{H_2}}^{feed}$	F^{feed}	$\sum F^{WH}$	ΔF^{feed}
	20,263	***18,237***	***2026***	***0.72***	***28,135***	***28,135***	***0***
Eq. (13.9)	F_{HU}	F^{prod}	F^{resd}_{sys}	F^{sys}_{loss}	F^{feed}	F^{sys}_{fuel}	$\sum F^{WH} - F^{feed}$
	70,220	20,263	0	62,349	28,135	0	0

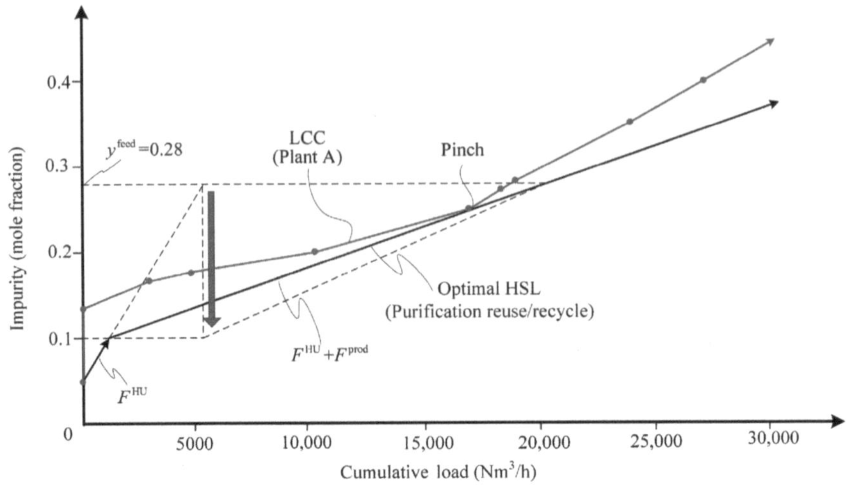

FIG. 13.7 LCC and optimal HSL with purification reuse/recycle (Deng et al., 2015).

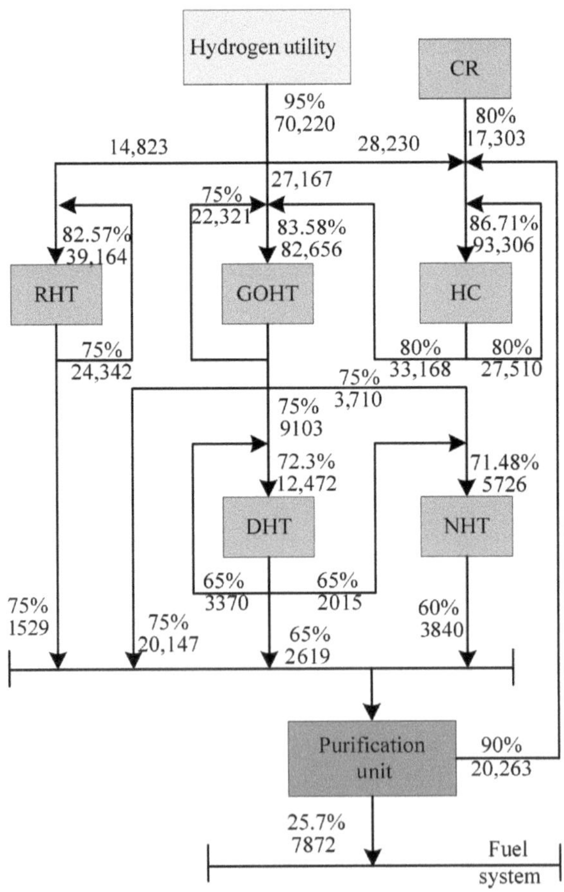

FIG. 13.8 Optimal network for case 1 with purification reuse/recycle (Deng et al., 2015).

hydrogen recovery in the clean fuels production process. Many other mathematical programming approaches were developed and they can be found in the review (Marques et al., 2017). In this section, a mathematical model (Deng et al., 2014) for the synthesis of refinery hydrogen networks is introduced and one case is analyzed to illustrate the application of the introduced model.

13.4.1 Problem Statement

Given a set of internal hydrogen sources (or so-called process hydrogen sources) with the total number of NSR, for each internal hydrogen source ($s \in NSR$) with specified flow rate (FSR_s), find the concentration of the cth component ($y_{s,c}^{\mathrm{out}}$) ($c \in NC$), and pressure (PSR_s). With the appropriate placement of a number of gas compressors ($i \in NI$), those hydrogen sources can be allocated with a set of hydrogen sinks. With a number of hydrogen sinks (NSK), each sink ($k \in NSK$) has its own flow rate requirement ($\mathrm{FK}_k^{\mathrm{in}}$), minimum hydrogen purity, maximum allowable inlet concentration of impurity, and pressure specification ($\mathrm{PK}_k^{\mathrm{in}}$). In addition, a number of external hydrogen sources ($u \in NHU$), so-called hydrogen utilities, from hydrogen plants would be utilized to compensate the internal hydrogen source to fulfill the requirements of hydrogen sinks. Besides, in order to reduce the flow rate of hydrogen utility, a number of purifiers ($p \in NP$) can be placed to upgrade the quality of certain internal hydrogen sources for further utilization by hydrogen sinks. The superstructure of the problem, embedded with potential configurations of interest is shown in Fig. 13.9.

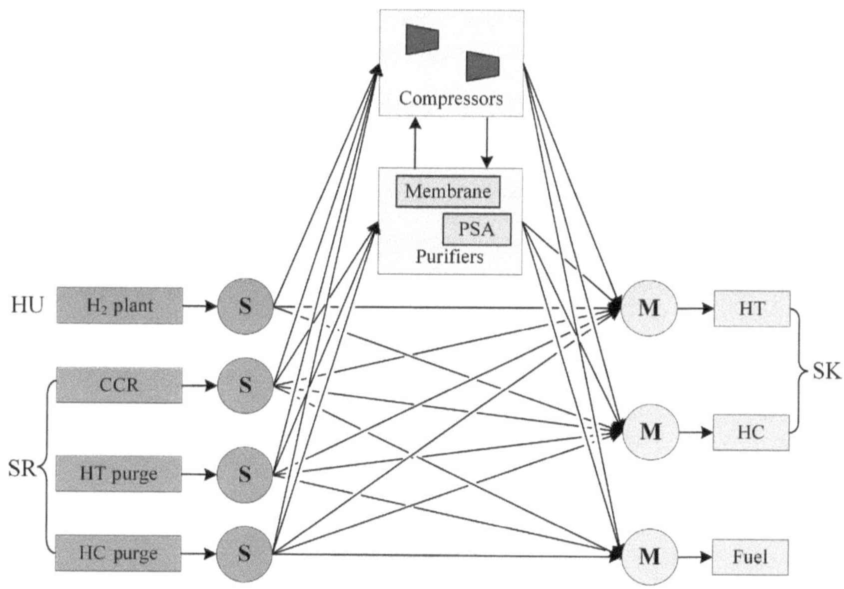

FIG. 13.9 Superstructure for the synthesis of a hydrogen network (Deng et al., 2014).

13.4.2 Mathematical Model

The mathematical formulations (Deng et al., 2014) for the superstructure shown in Fig. 13.9 can be presented as follows.

13.4.2.1 Formulations Related to the uth Hydrogen Utility

A hydrogen production process, such as steam reforming of natural gas, serves as hydrogen utility in the refinery plant. In addition, several available external hydrogen sources from an ethylene plant, fertilizer plant, and coal gasification section would be possible hydrogen utilities. The pressure for the hydrogen utility would not fulfill the pressure requirement of hydrogen sinks and hydrogen compressors would be installed to raise the pressure. Typically, due to the high quality of the hydrogen utility, it would not be allocated to purifiers. In addition, the hydrogen utility would not be allowed to discharge to the fuel system.

The gas stream from hydrogen utility can be allocated to the hydrogen sink if the inlet conditions are satisfied or be allocated to the *i*th hydrogen compressor to lift its pressure to fulfill the requirement of hydrogen sinks or purifiers. The flow rate balance is made on the splitting node after the *u*th hydrogen utility,

$$FHU_u = \sum_{k \in NSK} FUK_{u,k} + \sum_{i \in NI} FUI_{u,i} \quad \forall u \in NHU \tag{13.18}$$

Note that the gas flow rate for the *u*th hydrogen utility cannot exceed its maximum capacity,

$$FHU_u \leq \mathrm{FHU}_\mathrm{u}^\mathrm{UB} \tag{13.19}$$

13.4.2.2 Formulations Related to the sth Hydrogen Source

The continuous catalytic reforming (CCR) process is the typical hydrogen source in a refinery. In addition, the purges of hydrogen treating and hydrogen cracking processes contain a certain amount of hydrogen. They are categorized as internal hydrogen sources and can be reused by other hydrogen sinks (hydrogen treating and hydrogen cracking processes). Alternatively, they may be allocated to purifiers (PSA or Membrane) to upgrade the quality and then assigned to hydrogen sinks. The compressors can be installed to lift the pressure of the hydrogen sources. The surplus hydrogen sources with low purity of hydrogen can be discharged into the fuel system.

The flow rate balance can be made for the splitting node after the *s*th hydrogen source,

$$FSR_s = \sum_{k \in NSK} FSK_{s,k} + \sum_{i \in NI} FSI_{s,i} + \sum_{p \in NP} FSP_{s,p} + \sum_{f \in NF} FSF_{s,f} \quad \forall s \in NSR \tag{13.20}$$

The gas flow rate for the *s*th hydrogen source cannot exceed its upper bounds,

$$FSR_s \leq \mathrm{FSR}_{\mathrm{s}}^{\mathrm{UB}} \tag{13.21}$$

13.4.2.3 Formulations Related to the ith Compressor

The inlet gas stream of the *i*th compressor could come from hydrogen utility, hydrogen source, product of purifiers or other compressors.

The flow rate balance can be made for the mixing node before the *i*th compressor,

$$FI_i^{in} = \sum_{u \in NHU} FUI_{u,i} + \sum_{s \in NSR} FSI_{s,i} + \sum_{p \in NP} FPI_{p,i}^{prod} + \sum_{\substack{i' \neq i \\ i' \in NI}} FII_{i',i} \quad \forall i \in NI \tag{13.22}$$

In addition, the mass balance for the *c*th component can be made for the mixing node before the *i*th compressor,

$$FI_i^{in} \cdot y_{i,c}^{in} = \sum_{u \in NHU} FUI_{u,i} \cdot y_{u,c}^{out} + \sum_{s \in NSR} FSI_{s,i} \cdot y_{s,c}^{out} + \sum_{\substack{i' \neq i \\ i' \in NI}} FII_{i',i} \cdot y_{i',c}^{out} \quad \forall i \in NI \ \forall c \in NC \tag{13.23}$$

The flow rate balance and mass balance for the *c*th component are made around the outlet and inlet of the *i*th compressor,

$$FI_i^{out} = FI_i^{in} \quad \forall i \in NI \tag{13.24}$$

$$y_{i,c}^{out} = y_{i,c}^{in} \quad \forall i \in NI \ \forall c \in NC \tag{13.25}$$

The outlet gas stream of the *i*th compressor could be allocated to hydrogen sinks, the purifier, or other compressors. The flow rate balance can be made for the splitting node after the *i*th compressor,

$$FI_i^{out} = \sum_{k \in NSK} FIK_{i,k} + \sum_{p \in NP} FIP_{i,p} + \sum_{\substack{i' \neq i \\ i' \in NI}} FII_{i,i'} \quad \forall i \in NI \tag{13.26}$$

In addition, the inlet gas flow rate must not exceed its maximum flow rate capacity for the existing compressor,

$$FI_i^{in} \leq \mathrm{FI}_i^{\mathrm{in,UB}} \quad \forall i \in NI_{\mathrm{exist}} \tag{13.27}$$

13.4.2.4 Formulations Related to the pth Purifier

The inlet gas stream for the purifier would include gas streams from hydrogen sources and compressors. Typically, the purifier (PSA and Membrane) has two outlets: product and residual. The product of higher hydrogen purity can be allocated to hydrogen sinks or compressors, if necessary. However, the residual with low hydrogen purity is typically assigned to the fuel system.

The flow rate balance and component mass balance are made for the mixing node before the pth purifier,

$$FP_p^{in} = \sum_{s \in NSR} FSP_{s,p} + \sum_{i \in NI} FIP_{i,p} \quad \forall p \in NP \tag{13.28}$$

$$FP_p^{in} \cdot y_{p,in}^{in} = \sum_{s \in NSR} FSP_{s,p} \cdot y_{u,c}^{out} + \sum_{i \in NI} FIP_{i,p} \cdot y_{i,c}^{out} \quad \forall p \in NP \ c \in NC \tag{13.29}$$

The flow rate balance and component mass balance are made around the pth purifier,

$$FP_p^{in} = FP_p^{prod} + FP_p^{resd} \quad \forall p \in NP \tag{13.30}$$

$$FP_p^{in} \cdot y_{p,c}^{in} = FP_p^{prod} \cdot y_{p,c}^{prod} + FP_p^{resd} \cdot y_{p,c}^{resd} \quad \forall p \in NP \ c \in NC \tag{13.31}$$

The cth component contained in the product divided by that contained in the feed of the pth purifier is defined as the recovery ratio ($RR_{p,c}$). Typically, the hydrogen recovery ratio is specified for each purifier.

$$FP_p^{prod} \cdot y_{p,c}^{prod} = RR_{p,c} \cdot FP_p^{in} \cdot y_{p,c}^{in} \quad \forall p \in NP \ c \in NC \tag{13.32}$$

The flow rate balance is made on the splitting node after the pth purifier,

$$FP_p^{prod} = \sum_{k \in NSK} FPK_{p,k}^{prod} + \sum_{i \in NI} FPI_{p,i}^{prod} \quad \forall p \in NP \tag{13.33}$$

$$FP_p^{resd} = \sum_{f \in NF} FPF_{p,f}^{resd} \quad \forall p \in NP \tag{13.34}$$

13.4.2.5 Formulations Related to the kth Hydrogen Sink

The inlet gas stream for the kth hydrogen sink would be supplied by hydrogen utility, hydrogen sources, purifiers, and compressors.

The flow rate and component mass balance are made on the mixing node before the kth hydrogen sink,

$$\mathrm{FK}_k^{\mathrm{in}} = \sum_{u \in NHU} FUK_{u,k} + \sum_{s \in NSR} FSK_{s,k} + \sum_{p \in NP} FPK_{p,k}^{prod} + \sum_{i \in NI} FIK_{c,k} \quad \forall k \in NSK \tag{13.35}$$

$$\mathrm{FK}_k^{\mathrm{in}} \cdot y_{k,c}^{in} = \sum_{u \in NHU} FUK_{u,k} \cdot y_{u,c}^{out} + \sum_{s \in NSR} FSK_{s,k} \cdot y_{s,c}^{out} + \sum_{p \in NP} FPK_{p,k}^{prod} \cdot y_{p,c}^{prod} + \sum_{i \in NI} FIK_{i,k} \cdot y_{i,c}^{out} \quad \forall k \in NSK \ c \in NC \tag{13.36}$$

The inlet hydrogen purity must be higher than its lower bound specified by the hydrogen sink,

$$y_{k,c}^{in} \geq y_{k,c}^{\mathrm{in,LB}} \quad \forall c = \{\mathrm{H_2}\} \tag{13.37}$$

13.4.2.6 *Formulations Related to the Fuel System*

The surplus internal hydrogen source and the residual gas stream of the purifier would be sent to the fuel system. The flow rate and component mass balance are made on the mixing node before the fuel system,

$$FF^{in} = \sum_{s \in NSR} FSF_s + \sum_{p \in NP} FPF_p^{resd} \tag{13.38}$$

$$FF^{in} \cdot y_{f,c}^{in} = \sum_{s \in NSR} FSF_s \cdot y_{s,c}^{out} + \sum_{p \in NP} FPF_p^{resd} \cdot y_{p,c}^{resd} \quad \forall c \in NC \tag{13.39}$$

Note that, the inlet flow rate of the fuel system (FF^{in}) and inlet concentration for the cth component ($y_{f,c}^{in}$) are two variables and they make the term ($FF^{in} \cdot y_{f,c}^{in}$) a bilinear term. To reduce the number of bilinear terms in the mathematical model, Eq. (13.39) is not included in the constraints, but it will be solved after the optimization.

13.4.2.7 *Connection and Pressure Constraints*

The binary variables $z_{a,b}$ are introduced to indicate the connection status between supplier (hydrogen utility, hydrogen source, outlet of compressor, product, and residual of purifier) and receiver (hydrogen sink, inlet of compressor, inlet of purifier, fuel system). The necessary and sufficient condition for a connection extension is that the flow rate is nonzero. The flow rates serving as continuous variables are related with binary variables by Eq. (13.40),

$$\begin{aligned}
&F_{a,b} - z_{a,b} \cdot \mathrm{F_{a,b}^{UB}} \leq 0 \\
&F_{a,b} + (1 - z_{a,b}) \cdot \mathrm{F_{a,b}^{UB}} \geq \mathrm{F_{a,b}^{LB}} \\
&z_{a,b} \in \left\{ \begin{array}{l} zUK_{u,k}, zUP_{u,p}, zUI_{u,i} \\ zSK_{s,k}, zSP_{s,p}, zSI_{s,i}, zSF_{s,f} \\ zPK_{p,k}^{prod}, zPI_{p,i}^{prod}, zPF_{p,f}^{resd} \\ zIK_{i,k}, zIP_{i,p}, zII_{i,i'} \end{array} \right\} \\
&F_{a,b} \in \left\{ \begin{array}{l} FUK_{u,k}, FUP_{u,p}, FUI_{u,i} \\ FSK_{s,k}, FSP_{s,p}, FSI_{s,i}, FSF_{s,f} \\ FPK_{p,k}^{prod}, FPI_{p,i}^{prod}, FPF_{p,f}^{resd} \\ FIK_{i,k}, FIP_{i,p}, FII_{i,i'} \end{array} \right\}
\end{aligned} \tag{13.40}$$

where $F_{a,b}^{UB}$ and $F_{a,b}^{LB}$ are the upper and lower bounds for the flow rate variable $F_{a,b}$.

Once the pressure level of a hydrogen sink or purifier is higher than that of a hydrogen source, the hydrogen source cannot be fed to the hydrogen sink or purifier directly and the hydrogen compressor is necessary to upgrade the pressure level of the hydrogen source. It indicates that the flow rate from the hydrogen source to the sink or purifier with higher pressure level equals zero. The condition can be fulfilled by Eqs. (13.40), (13.41).

$$
\begin{aligned}
&(P_b - P_a) \cdot F_{a,b} \leq 0 \\
&P_a \in \left\{ \mathrm{PHU}_u^{out}, \mathrm{PSR}_s^{out}, \mathrm{P}_p^{prod}, \mathrm{P}_p^{resd}, \mathrm{PI}_i^{out} \right\} \\
&P_b \in \left\{ \mathrm{PK}_k^{in}, \mathrm{PP}_p^{in}, \mathrm{PI}_i^{in} \right\} \\
&F_{a,b} \in \left\{ \begin{array}{l} FUK_{u,k}, FUP_{u,p}, FUI_{u,i} \\ FSK_{s,k}, FSP_{s,p}, FSI_{s,i}, FSF_{s,f} \\ FPK_{p,k}^{prod}, FPI_{p,i}^{prod}, FPF_{p,f}^{resd} \\ FIK_{i,k}, FIP_{i,p}, FII_{i,i'} \end{array} \right\}
\end{aligned}
\tag{13.41}
$$

The pressure for the mixed stream of hydrogen sources is determined by the minimum pressure of the hydrogen sources associated with the mixed stream.

$$
P^{mix} = \min\{\mathrm{P_a}\} \quad \forall \mathrm{P_a} \in \left\{ \mathrm{PHU}_u^{out}, \mathrm{PSR}_s^{out}, \mathrm{P}_p^{prod}, \mathrm{P}_p^{resd}, \mathrm{PI}_i^{out} \right\} \tag{13.42}
$$

The total number of connections (N_{total}) is an important parameter for the demonstration of network complexity. It is the summation of all the connection variables, and given as Eq. (13.43).

$$
\begin{aligned}
N_{total} = & \sum_u \sum_k zUK_{u,k} + \sum_u \sum_p zUP_{u,p} + \sum_u \sum_i zUI_{u,i} \\
& + \sum_s \sum_k zSK_{s,k} + \sum_s \sum_p zSP_{s,p} + \sum_s \sum_i zSI_{s,i} + \sum_s \sum_i zSF_{s,f} \\
& + \sum_p \sum_k zPK_{p,k}^{prod} + \sum_p \sum_i zPI_{p,i}^{prod} + \sum_p \sum_f zPF_{p,f}^{resd} \\
& + \sum_i \sum_k zIK_{i,k} + \sum_i \sum_p zIP_{i,p} + \sum_i \sum_{i'} zII_{i,i'}
\end{aligned}
\tag{13.43}
$$

13.4.2.8 Objective Functions

The flow rate of hydrogen utility is strongly related to the capacity of the hydrogen plant and its operating cost. Therefore, the objective function can be formulated to minimize the consumption of hydrogen utility.

$$\min FHU = \sum_{u \in NHU} FHU_u \tag{13.44}$$

In addition, the minimum total number of connections can be determined by setting an upper bound. The upper bound can be reduced one by one until the optimal flow rate of hydrogen utility tends to increase. The corresponding total number of connections is considered as the minimum target.

Besides, the total annualized cost (TAC) can also be employed as an optimization objective. It can be found in the literature (Deng et al., 2014).

13.4.3 Case Study

Fig. 13.10 shows the current hydrogen network for case 2, which is adopted from the base case of Elkamel et al. (2011)). As shown, the existing fresh hydrogen consumption from the hydrogen plant is reported as 89,280 Nm^3/h (80 MMscfd) (Elkamel et al., 2011). The current process data for this example are listed in Table 13.4. As shown, there are five hydrogen consuming processes, namely, hydrocracker unit (HCU), gas oil hydrotreater (GOHT), residue hydrotreater (RHT), diesel hydrotreater (DHT), and naphtha hydrotreater (NHT). The inlets for these hydrogen consumers serve as internal hydrogen sinks and their outlets would be considered as internal hydrogen sources.

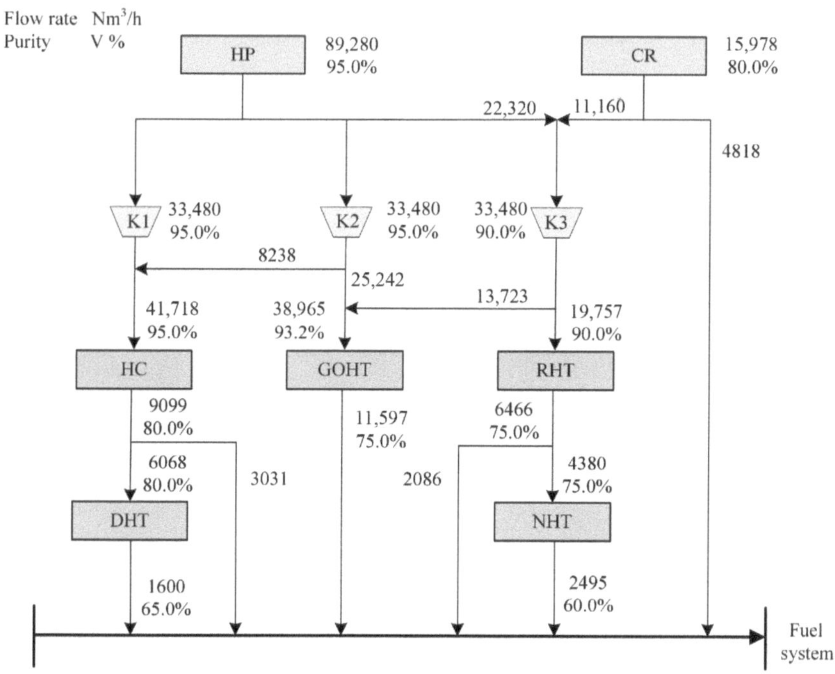

FIG. 13.10 Current hydrogen network for Case 2 (Elkamel et al., 2011).

TABLE 13.4 Limiting Data for Case 2

	Sinks									
Sources	HCU	GOHT	RHT	DHT	NHT	Fuel	Current Flow Rate (Nm^3/h)	Maximum Flow Rate (Nm^3/h)	Purity (V% H_2)	Pressure (MPa)
HP	41,718	38,965	8597				89,280	89,280	95	2.069
CRU			11,160			4818	15,978	16,182	80	2.069
HCU				6068		3031	9099	9099	80	8.276
GOHT						11,597	11,597	11,597	75	2.414
RHT					4380	6466	10,846	10,846	75	2.759
DHT						1600	1600	1600	65	2.414
NHT						2495	2495	2495	60	1.379
Flow rate (Nm^3/h)	41,718	38,965	19,757	6068	4380	30,008				
Purity (V % H_2)	95.00	93.20	90.00	80.00	75.00					
Pressure (MPa)	13.79	3.448	4.138	3.448	2.069					

The purity and pressure requirements for the sinks and the purity and pressure conditions for the sources are listed in Table 13.4. Note that all the hydrogen consumers have internal recycle compressors. The recycle hydrogen streams would not serve as hydrogen sources. In addition, a catalytic reforming unit (CRU) is an internal hydrogen source with specified hydrogen purity of 80%, outlet pressure of 2.069 MPa, and maximum capacity of 16,182 Nm^3/h. Its maximum capacity is assumed in this paper according to the optimized results in the literature (Elkamel et al., 2011). In order to reduce the operating cost for the hydrogen plant, the internal hydrogen sources should be fully utilized before the hydrogen utility from the hydrogen plant is considered. In this case, the hydrogen utility is specified with a hydrogen purity content of 95%, outlet pressure of 2.069 MPa, and maximum capacity of 89,280 Nm^3/h. Besides, the fuel system operates at low pressure (1.379 MPa), which would receive unused internal hydrogen sources. The case will be used to investigate the influences of the placement of the purifier on the consumption of hydrogen utility.

In order to reduce the consumption of hydrogen utility further, the purification unit (PSA) is incorporated into the hydrogen network to upgrade the quality of internal hydrogen sources and the product of the PSA will be assigned to hydrogen sinks if the pressure requirement is fulfilled.

To model the placement of compressor and purifier, the objective function is selected as Eq. (13.44) and it is subjected to the constraints in Eqs. (13.18)–(13.38) and connection constraint (13.40) and pressure constraint (13.41). Note that the connection constraint (13.40) and the bilinear terms ($FI_i^{in} \cdot y_{i,c}^{in}$) and ($FP_p^{in} \cdot y_{p,c}^{in}$) contained in Eqs. (13.23), (13.29) make the model a MINLP problem. It is solved in GAMS software (Rosenthal, 2010) using DICOPT as solver (based on the PC specification: Intel Core i3-2100 CPU 3.10 GHz, 4 GB RAM). Note that the solver CPLEX is defined to solve MIP problems as well as KNITRO for NLP problems.

The process data for the purification unit PSA are assumed in this paper. The inlet pressure for PSA is set to 2.069 MPa. The pressure for the product of PSA is assumed to be 2.069 MPa and the slight pressure drop is neglected. The hydrogen purity for its product is given as 95% and the hydrogen recovery is defined as 90%. Besides, the pressure for its residue is set to 1.379 MPa. The maximum inlet flow rate for PSA is assumed to be 35,712 Nm^3/h. Once the above data, as well as the data listed in Tables 13.4 and 13.5, were input to the model, the model was solved in 0.02 CPUs and the optimal flow rate of hydrogen utility was found to be 72,444 Nm^3/h with a minimum number of connections of 20. The optimized results are listed in Table 13.6 and Fig. 13.11 shows the optimized hydrogen network with the placement of compressors and purifier. As shown, the product of PSA is assigned to compressor K1 and then allocated to hydrogen sinks. The total inlet flow rate and purity for PSA is determined as 23,235 Nm^3/h and 76.022%.

In order to investigate the influence of the feed hydrogen purity for PSA on the hydrogen utility, feed hydrogen purity for PSA is given in the region [70, 80]

TABLE 13.5 Process Data for the Make-Up Hydrogen Compressor

Compressor	Inlet Pressure (MPa)	Discharge Pressure (MPa)	Maximum Capacity (Nm^3/h)
K1	2.069	13.79	35,154
K2	2.069	13.79	35,154
K3	2.069	4.138	35,154

with an increment of one. Fig. 13.12 shows the variation tendencies of feed flow rate of PSA, optimum flow rate of hydrogen utility, and minimum total number of connections with the increase of feed hydrogen purity for PSA. As shown in Fig. 13.12, the minimum flow rate of hydrogen utility is 72,444 Nm^3/h with 76.022% of feed hydrogen purity for the product of PSA. The feed flow rate of PSA increases with the increment of feed hydrogen purity of PSA. The minimum total number of connections is kept at 20 until the feed hydrogen purity of PSA reaches 76.022%. Once the feed hydrogen purity of PSA is less than 76.022%, the minimum total number of connections will be greater than 20 and fluctuate with the increase of the feed hydrogen purity of PSA. And the minimum total number of connections is kept at 22 when the feed hydrogen purity of PSA is greater than 76.022%. For this case, the minimum flow rate of hydrogen utility is targeted to be 72,444 Nm^3/h with 20 connections and 76.022% of feed hydrogen purity of PSA and the feed flow rate of PSA is determined as 23,235 Nm^3/h.

13.5 CONCLUSION

In this chapter, both the Pinch technique and mathematical programming approaches are introduced for the synthesis of a hydrogen network with a purification unit. A generalized Improved Problem Table (IPT) is presented and mass balance equations are incorporated to determine the optimal targets (i.e., the optimal flowrates of hydrogen utility and product of the purifier, optimal feed impurity of purifier). The optimal feed impurity of the purifier is greater than or equal to the pinch impurity with the direct reuse/recycle scheme. The limiting composite curves (LCCs) and optimal hydrogen supply lines (HSL) are plotted vividly to show the insights of the proposed approach. Next, a superstructure-based optimization model is introduced and a comprehensive superstructure is embedded with hydrogen utility, process hydrogen sources, hydrogen sinks, fuel system, compressor, purifier, and all the feasible interconnections between them. The optimization results show that the optimal flow rate of hydrogen utility will be increased with the reduction of the total number of

TABLE 13.6 Optimum Results for Case 2 With the Placement of Compressors (Purification Reuse/Recycle) (Deng et al., 2014)

	Sinks													
Sources	HCU	GOHT	RHT	DHT	NHT	K1	K2	K3	PSA	Fuel System	Optimized Flow Rate (Nm^3/h)	Maximum Flow Rate (Nm^3/h)	Purity (V% H_2)	Pressure (MPa)
HP						16,914	28,606	26,924			72,444	89,280	95	2.069
CRU								8230	7952		16,182	16,182	80	2.069
HCU			3031	6068							9099	9099	80	8.276
GOHT					4380				7217		11,597	11,597	75	2.414
RHT									6466		6466	6466	75	2.759
DHT									1600		1600	1600	65	2.414
NHT										2495	2495	2495	60	1.379
K1	13,112	18,993	1544								33,649	35,154	95	13.79
K2	28,606										28,606	35,154	95	13.79
K3		19,972	15,182								35,154	35,154	91.488	4.138
PSA Product						16,735							95	2.069
PSA Residue										6501			27.171	1.379
Flow rate (Nm^3/h)	41,718	38,965	19,757	6068	4380	33,649	28,606	35,154	23,235	8996			36.277	
Purity (V % H2)	95.00	93.20	90.00	80.00	75.00	95.00	95.00	91.488	76.022	36.277				
Pressure (MPa)	13.79	3.448	4.138	3.448	2.069	2.069	2.069	2.069	2.069	1.379				

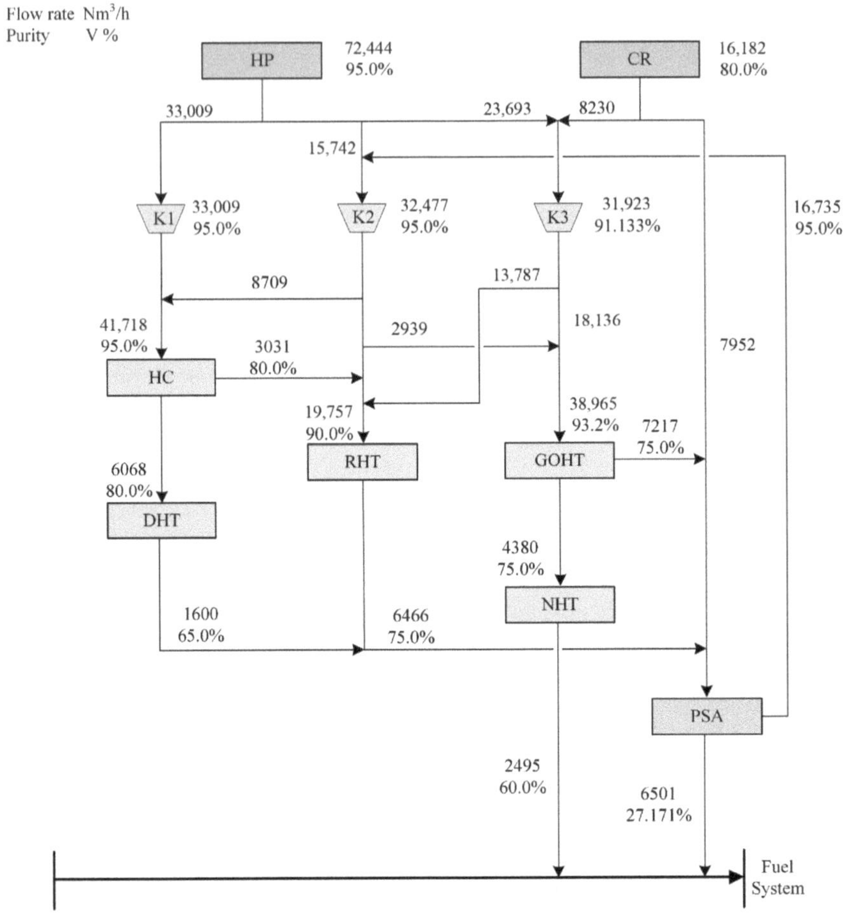

FIG. 13.11 Optimal hydrogen network for Case 2 with the placement of PSA and compressors (Deng et al., 2014).

connections. The decision can be determined according to the plotted Pareto front. With the limitation for the capacity of the purifier, the optimal feed purity for the purifier and number of connections can be determined via minimizing the flow rate of hydrogen utility.

ACKNOWLEDGMENTS

Financial support provided by the National Natural Science Foundation of China (No. 21576287) is gratefully acknowledged. The research is also supported by Science Foundation of China University of Petroleum, Beijing (No. 2462015BJB02 and 2462018BJC003).

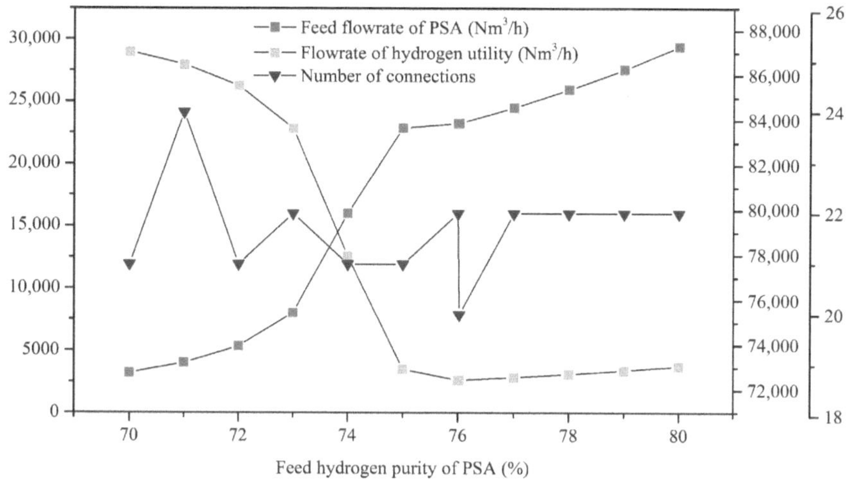

FIG. 13.12 Variation tendency of optimal flow rate of hydrogen utility, feed flow rate of PSA, and total number of connections with the increase of feed hydrogen purity of PSA (Deng et al., 2014).

NOMENCLATURE

Sets and indices

NC	set of components
NF	set of fuel systems
NHU	set of external hydrogen sources or hydrogen utilities
NI	set of compressors
NP	set of hydrogen purifiers
NSK	set of process hydrogen sinks
NSR	set of process hydrogen sources
c	index for component
f	index for fuel system
i	index for compressor
k	index for hydrogen sink
p	index for purifier
s	index for process/internal hydrogen source
u	index for hydrogen utility
ν	index for impurity level

Parameters

$\mathbf{F}_{a,b}^{\mathbf{UB}}$	upper bound for flow rate allocated from *a* to *b*
$\mathbf{F}_{a,b}^{\mathbf{LB}}$	lower bound for flow rate allocated from *a* to *b*
$\mathbf{FHU}_{u}^{\mathbf{UB}}$	upper bound for the flow rate of *u*th hydrogen utility
$\mathbf{FI}_{i}^{\mathbf{in,\ UB}}$	upper bound for the inlet flow rate of *i*th hydrogen compressor

$\mathbf{FK}_k^{\mathbf{in}}$ inlet flow rate of *k*th hydrogen sink
$\mathbf{FSR}_s^{\mathbf{UB}}$ upper bound for the flow rate of *s*th hydrogen source
F_{loss}^{sys} total flowrate loss for direct reuse/recycle system, Nm^3/h
$RR_{p,c}$ recovery ratio for *c*th component in *p*th purifier
RR hydrogen recovery of purifier
y^{HU} impurity of hydrogen utility, volume fraction
$y_{\mathbf{H}_2}^{arbitrary}$ arbitrary hydrogen purity, volume fraction
$y_{SK_k}^{\mathbf{lim}}$ lower limit of inlet hydrogen purity for hydrogen sinks, volume fraction
$y_{\mathbf{H}_2}^{prod}$ product purity of purifier, volume fraction
$y_{k,c}^{\mathbf{in,\ LB}}$ lower bound for inlet concentration for *c*th component in *k*th hydrogen sink
$y_{u,c}^{\mathbf{out}}$ outlet concentration for *c*th component in *u*th hydrogen utility
$y_{s,c}^{\mathbf{out}}$ outlet concentration for *c*th component in *s*th hydrogen source
$y_{p,c}^{\mathbf{prod}}$ concentration for *c*th component in the product of *p*th purifier
λ correction coefficient

Variables

$F_{\mathbf{above}}^{\mathbf{pinch}}$ flowrate needed above pinch point, Nm^3/h
F_{net}^{ν} net flowrate in *ν*th impurity level, Nm^3/h
ΔM_{cum}^{ν} cumulative mass load of the *ν*th impurity level, Nm^3/h
F^{feed} feed flowrate of purifier, Nm^3/h
F^{prod} product flowrate of purifier, Nm^3/h
F^{resd} residual flowrate of purifier, Nm^3/h
$y_{\mathbf{H}_2}^{feed}$ feed purity of purifier, volume fraction
$y_{\mathbf{H}_2}^{resd}$ residual purity of purifier, volume fraction
F_{sys}^{resd} residual flowrate of purifier allocated to the direct reuse/recycle system, Nm^3/h
F_{fuel}^{resd} residual flowrate of purifier discharged to the fuel system, Nm^3/h
$Q_{\mathbf{eq}}^{\mathbf{HU}}$ the amount of equivalent energy consumption conservation for the reduction of hydrogen utility, kW

Continuous variables

$F_{a,b}$ flow rate allocated from *a* to *b*
FF^{in} inlet flow rate for *f*th fuel system
$FII_{i',i}$ flow rate allocated from *i*′th hydrogen compressor to *i*th hydrogen compressor
FI_i^{out} outlet flow rate for *i*th hydrogen compressor
$FIK_{i,k}$ flow rate allocated from *i*th hydrogen compressor to *k*th hydrogen sink
$FIP_{i,p}$ flow rate allocated from *i*th hydrogen compressor to *p*th purifier
FI_i^{in} inlet flow rate for *i*th hydrogen compressor
$FPI_{p,i}^{prod}$ flow rate allocated from the product of *p*th purifier to *i*th hydrogen compressor

FP_p^{in} inlet flow rate for pth purifier
FP_p^{prod} flow rate for the product of pth purifier
FP_p^{resd} flow rate for the residual of pth purifier
$FPK_{p,k}^{prod}$ flow rate allocated from the product of pth purifier to kth hydrogen sink
$FPI_{p,i}^{prod}$ flow rate allocated from the product of pth purifier to ith hydrogen compressor
$FPF_{p,f}^{resd}$ flow rate allocated from the residual of pth purifier to fth fuel system
FSR_s flow rate allocated from sth hydrogen source
$FSI_{s,i}$ flow rate allocated from sth hydrogen source to ith hydrogen compressor
$FSK_{s,k}$ flow rate allocated from sth hydrogen source to kth hydrogen sink
$FSP_{s,p}$ flow rate allocated from sth hydrogen source to pth purifier
$FSF_{s,f}$ flow rate allocated from sth hydrogen source to fth fuel system
$FUK_{u,k}$ flow rate allocated from uth hydrogen utility to kth hydrogen sink
$FUI_{u,i}$ flow rate allocated from uth hydrogen utility to ith hydrogen compressor
$y_{i,c}^{in}$ inlet concentration for cth component in ith hydrogen compressor
$y_{i,c}^{out}$ outlet concentration for cth component in ith hydrogen compressor
$y_{p,c}^{in}$ inlet concentration for cth component in pth purifier
$y_{p,c}^{resd}$ concentration for cth component in the residual of pth purifier
$y_{k,c}^{in}$ inlet concentration for cth component in kth hydrogen sink
$y_{f,c}^{in}$ inlet concentration for cth component fth fuel system

Binary variables

$z_{a,b}$ connection variable from a to b
$zIK_{i,k}$ connection variable from ith hydrogen compressor to kth hydrogen sink
$zIP_{i,p}$ connection variable from ith hydrogen compressor to pth purifier
$zII_{i,i'}$ connection variable from i'th hydrogen compressor to ith hydrogen compressor
$zPK_{p,k}^{prod}$ connection variable from the product of pth purifier to kth hydrogen sink
$zPI_{p,i}^{prod}$ connection variable from the product of pth purifier to ith hydrogen compressor
$zPF_{p,f}^{resd}$ connection variable from the residual of pth purifier to fth fuel system
$zSK_{s,k}$ connection variable from sth hydrogen source to kth hydrogen sink
$zSP_{s,p}$ connection variable from sth hydrogen source to pth purifier
$zSI_{s,i}$ connection variable from sth hydrogen source to ith hydrogen compressor
$zSF_{s,f}$ connection variable from sth hydrogen source to fth fuel system

$zUK_{u,k}$ connection variable from uth hydrogen utility to kth hydrogen sink
$zUP_{u,p}$ connection variable from uth hydrogen utility to pth purifier
$zUI_{u,i}$ connection variable from uth hydrogen utility to ith hydrogen compressor

Subscripts/Superscripts

cum cumulative
feed feed of purifier
LB Lower bound
UB Upper bound
prod product of purifier
resd residual of purifier
in inlet
out outlet
lim limiting value
max maximum
min minimum
net net flowrate or load
pinch pinch point
prod product of purifier
resd residual of purifier
SRi ith process hydrogen source
SKk kth process hydrogen sink

Abbreviations

CNHT cracked naphtha hydrotreater
CCR continuous catalytic reforming
GCA gas cascade analysis
GOHT gas oil hydrotreater
IPT Improved Problem Table
HCU hydrocracker unit
LCC Limiting Composite Curve
LP linear programming
MILP mixed integer linear programming
MINLP mixed integer nonlinear programming
MSCC material surplus composite curve
MRPD material surplus composite curve
NLP nonlinear programming
NHT naphtha hydrotreater
NNA nearest neighboring algorithm
PSA pressure swing adsorption
RHS right hand side
RHT residue hydrotreater
TAC total annualized cost
WH waste hydrogen stream

REFERENCES

Agrawal, V., Shenoy, U.V., 2006. Unified conceptual approach to targeting and design of water and hydrogen networks. AICHE J. 52 (3), 1071–1082.

Alves, J.J., Towler, G.P., 2002. Analysis of refinery hydrogen distribution systems. Ind. Eng. Chem. Res. 41 (23), 5759–5769.

Bandyopadhyay, S., 2006. Source composite curve for waste reduction. Chem. Eng. J. 125 (2), 99–110.

Deng, C., Pan, H., et al., 2014. Comparative analysis of different scenarios for the synthesis of refinery hydrogen network. Appl. Therm. Eng. 70 (2), 1162–1179.

Deng, C., Zhou, Y., et al., 2015. Systematic approach for targeting interplant hydrogen networks. Energy 90 (Part 1), 68–88.

El-Halwagi, M.M., Gabriel, F., et al., 2003. Rigorous graphical targeting for resource conservation via material recycle/reuse networks. Ind. Eng. Chem. Res. 42 (19), 4319–4328.

Elkamel, A., Alhajri, I., et al., 2011. Integration of hydrogen management in refinery planning with rigorous process models and product quality specifications. Int. J. Process. Syst. Eng. 1 (3), 302–330.

Foo, D.C.Y., Manan, Z.A., 2006. Setting the minimum utility gas flowrate targets using cascade analysis technique. Ind. Eng. Chem. Res. 45 (17), 5986–5995.

Gary, J.H., Handwerk, G.E., 2001. Petroleum Refining: Technology and Economics. Marcel Dekker, New York.

Hallale, N., Liu, F., 2001. Refinery hydrogen management for clean fuels production. Adv. Environ. Res. 6 (1), 81–98.

Marques, J.P., Matos, H.A., et al., 2017. State-of-the-art review of targeting and design methodologies for hydrogen network synthesis. Int. J. Hydrog. Energy 42 (1), 376–404. https://www.sciencedirect.com/science/article/pii/S0360319916329020.

Prakash, R., Shenoy, U.V., 2005. Targeting and design of water networks for fixed flowrate and fixed contaminant load operations. Chem. Eng. Sci. 60 (1), 255–268.

Rosenthal, R.E., 2010. GAMS-A User's Guide. GAMS Development Corporation, Washington, DC.

Saw, S., Lee, L., et al., 2011. An extended graphical targeting technique for direct reuse/recycle in concentration and property-based resource conservation networks. Clean Techn. Environ. Policy 13 (2), 347–357.

Chapter 14

Metamodeling of Hydrogen Supply Chains: A Programmable Structure Based Representation

Monika Varga and Bela Csukas
Research Group on Process Network Engineering, Institute of Methodology, Kaposvar University, Kaposvar, Hungary

14.1 INTRODUCTION

14.1.1 Basic Approaches for Modeling of Hydrogen Supply Chains

The need for the energy transition is an accepted fact worldwide. According to the concise summary of the Hydrogen Council (2017), "hydrogen can be used as fuel for power or in industry as feedstock, it can be produced from (renewable) electricity and from carbon-abated fossil fuels, it produces zero emissions at point of use, it can be stored and transported at high energy density in liquid or gaseous form, and it can be combusted or used in fuel cells to generate heat and electricity." In transforming energy systems, hydrogen can serve as a clean and safe energy carrier to pave the way toward a low-carbon economy.

Considering the slow but definite development toward a hydrogen economy, the optimal design, planning, and operation of a hydrogen supply chain (HSC) has come to the forefront of research interest in the last decade. Several design and planning case studies have been carried out in the past years. This Chapter does not intend to give a detailed review, because many really comprehensive review papers are already available on this topic. This introduction refers to some of them in the context of the challenges for model-based analysis and problem solving.

Dagdougui (2012) reviews the state of the art of the generally applied modeling approaches in the planning and design of HSC with almost 100 references. The review distinguishes three main methodological approaches, namely the mathematical optimization methods, the spatial (GIS based) models and frameworks, as well as the assessment plans/transition scenarios. On the basis of the comprehensive review, the author concludes that mathematical

Hydrogen Supply Chain. https://doi.org/10.1016/B978-0-12-811197-0.00014-2

programming–based optimization methods are widely used for the design and planning of future HSCs.

Five years later, in a recent paper Maryam (2017), reviews again the applied modeling approaches. From the almost 100 references overviewed, around 50% published in the period of 2013–17. Focusing on the context of the United Kingdom, similar methodologies (various mathematical optimization methods, such as MILP, MINLP, MPP, MOP; GIS based approaches; transition models) support the tracking of progress during the last 5 years. The review confirms that the mathematical programming–based optimization methods are still the most widely used tools in planning and design, besides the GIS-based tools. The author also refers to some alternatively applied System's Dynamic and Agent Based approaches in the overview.

As a general review (not only HSC focused) of supply chain network design, a recent paper of Govindan et al. (2017) can be mentioned. It gives a comprehensive overview of more than 260 actual research papers and case studies for supply chain network design in various fields. As uncertainty is a critical factor of viable design decisions, the authors put the emphasis on the consideration of uncertainty in the model and in the optimization procedure. Uncertainty, both on the supply and demand sides, seems to be the most significant challenge also in the design of HSCs.

Turning to other challenges in HSC optimization, managing the multiple spatial and temporal scales can be mentioned. As Samsatli et al. (2016) highlight, the need for the combined consideration of various temporal scales also presents a notable challenge. While the representation of, for example, energy storage requires a short time scale (e.g., hourly), the consideration of a long planning horizon of many years may result in computationally intractable or very computationally intensive models.

Consideration of multiple criteria (e.g., costs, environmental sustainability, safety, etc.) is an obvious expectation during "green" supply chain design, but the incorporation of these issues generates another interesting challenge to be solved (Nurjanni et al., 2017).

14.1.2 Typical Elements of Hydrogen Supply Chains

Having studied some comprehensive case studies (De-Leon Almaraz, 2014; De-Leon Almaraz and Azzaro-Pantel, 2017), the typical elements of Hydrogen Supply Chains are allocated at geographically determined sites or districts that can be identified by an appropriate grid or by specific districts in a GIS. Many elements are time-specifically controlled by the meteorological or seasonal characteristics. The typical elements can be classified, as follows.

- Primary renewable resources,
- Primary accumulated resources,
- Processing of primary resources,

- Intermediate energy containing products and byproducts,
- Intermediate transportations and transformations of the energy containing products and byproducts,
- Hydrogen production sources,
- Hydrogen production processes,
- Produced raw hydrogen,
- Conditioning processes of the produced hydrogen,
- Production site storage of hydrogen,
- Transportation of hydrogen,
- Consumption site storage of hydrogen,
- Utilization of hydrogen,
- Hydrogen-based produced energy and materials,
- Car fuel demand,
- Other pools to be taken into consideration.

Primary renewable resources of hydrogen production are solar radiation, the wind, and hydropower generated by the environmental system. The utilization of these resources (especially of the solar energy) needs *natural or cultivated land* that is a special finite resource shared with other uses (agriculture for food, etc.). Renewable resources and land are distributed in the districts of the investigated territory.

The *primary nonrenewable, accumulated resources* are the nuclear raw materials, the natural gas, and the coal. Nuclear raw materials need vehicles to transport them to the specifically located power stations, while natural gas is available from pipelines. In contrast, coal-based processing should be installed near the mining sites.

Processing of the primary resources can be solved by many well-developed and promising new methods, but all of them depend on hourly and seasonally changing environmental conditions (e.g., sunshine). Direct *utilization of the solar energy* is solved by various photovoltaic processes with a limited efficiency and capacity. There is intensive research and development for many new solutions, with special organisms (e.g., microalgae) or with artificial biosystems (e.g., artificial chloroplast). Some of them might cause revolutionary change; accordingly, the long-term economic evaluation that is currently applied may lead to incorrect decisions. The indirect utilization of solar energy is realized by cultivation of energy plants, which competes with land use for food production. The products appear seasonally.

Wind power stations and some smaller hydropower stations produce electrical energy in certain (often randomly changing) periods, so their effective use needs onsite conversion to electrical energy. Large hydropower and nuclear stations produce easily transportable electrical energy that can contribute to buffered hydrogen production by remote electrolysis.

The most frequently applied Steam Methane Reforming (SMR) competes with other uses of natural gas but can produce hydrogen-containing gas anywhere in the vicinity of pipelines. In contrast, coal gasification has to be conducted near the resource, while it produces more expensive hydrogen-containing gas from a resource that has less convenient traditional applications.

There are *intermediate energy containing products and byproducts* in HSC networks, which originate from many distributed districts. Energy producing biomass (energy plants), as well as some agricultural byproducts and waste appear seasonally, however municipal biological wastes are formed permanently.

These intermediate energy-containing materials usually need *transportation, followed by transformation* into hydrogen-containing gas. Biomasses are processed by gasification, while biological wastes produce biogas by anaerobic fermentation. There are some experimental results on how to enrich the hydrogen content in the fermentation process (e.g., Wang and Yin, 2017).

The energy containing, *direct hydrogen producing sources* are the electrical energy, resulting from various chains, and the hydrogen-containing gas, coming from SMR and gasification processes. The two corresponding kinds of *hydrogen producing processes* are electrolysis and the different methods of gas separation for hydrogen enrichment (e.g., membrane separation, pressure swing adsorption, etc.).

The *produced raw hydrogen has to be conditioned* at the production site by liquefaction or by compression. The resulting liquid or compressed hydrogen product needs *production site storage* of liquid and gaseous hydrogen, respectively.

The *transportation of hydrogen* from the production to the consumption sites can be solved by tanker trucks, tube trailers, railway tankers, railway tube cars, or pipelines. Optionally the gases of high hydrogen content can be added to the available natural gas distribution pipelines.

Pure liquid or gaseous hydrogen requires *consumption site storage*. The *utilization of hydrogen* is connected to these storage places. In addition to the fueling of cars and other vehicles (e.g., aircraft), the hydrogen can be used for the generation of thermal energy, as well as for various industrial processes (e.g., ammonia production, synthesis of organic products, metal processing, etc.).

The logistics process is controlled by the *site and time dependent demands* for the fuels, heating, and other uses.

The evaluation of the various HSC process systems requires the calculation of the effects on balances of some *globally important pools*, such as carbon dioxide, oxygen, water, and thermal energy.

14.1.3 Challenges of Process Modeling for Hydrogen Supply Chain Design and Operation

In the literature, the basic approach to HSC design, planning, and operation is mathematical programming–based optimization. In these solutions, a collaboration with uncertainty model is embedded in the usually superstructure-based multiobjective optimization program. In thinking about an alternative, unconventional approach, we try to summarize the major challenges to be addressed by the potential methodologies. Based on the literature and the above discussed characteristics of the typical elements in the Hydrogen Supply Chain, the challenges can be summarized, as follows:

- Considering the intensive scientific and technical development (especially the emerging new bio-based hydrogen production methods), modeling has to prepared for modifications involving optional new structures and process elements;
- Extensibility of the model is another important issue. Accordingly, the stepwise structural and functional changes of supplies and demands have to be taken into consideration without the reformulation of the model;
- Easy consideration of multiple sites and districts (compartments) is necessary (e.g., via optional connection of spatial coordinates of the elements with an external GIS tool);
- The methodology has to tolerate also case-specific multiscale extensions (i.e., part of the elements needs a more detailed model and/or finer spatial or temporal resolution);
- Part of the elements and structural connections need a time-driven execution at given points of time and/or during prescribed intervals;
- Material and energy balances have to be combined with signal and rule-based event-driven processes;
- The model and its elements must have built-in capabilities for robust communication with external (e.g., meteorological) databases and with collaborating (e.g., metaheuristic optimizing) programs;
- Considering the extraordinary uncertainty of the cost coefficients and some other parameters in a longer time-horizon, the model has to support the handling of the uncertain coefficients and data, and the respective collaboration with the external optimizer;
- The embedded elementary evaluations and evaluating elements have to be prepared for the multiobjective case, with a comprehensive set of time-invariant natural objective functions.

14.2 PROGRAMMABLE STRUCTURE BASED REPRESENTATION OF PROCESS SYSTEMS

The hybrid, multiscale models may be more complex than any specific mathematical apparatus. The presented *Programmable Structures* are motivated by the need that engineering analysis, design and control of the complex, multidisciplinary process systems have easily modifiable, extensible, and connectable methodologies, with common representation of functional and structural features.

14.2.1 Methodology of Direct Computer Mapping Based Programmable Structures

In *Direct Computer Mapping* (DCM) of process models (Csukas, 1998; Csukas et al., 2011) the natural building blocks of the elementary states, actions, and connections are mapped onto a unified set of building elements, which determine an executable program code, directly, without any specified mathematical apparatus. This principle was successfully applied for various processes from technological process systems (e.g., Csukas et al., 1999; Temesvári et al., 2004) up to the multisectoral agri-food processes (e.g., Varga et al., 2012).

Based on these previous applications and on the *recently developed multidisciplinary applications,* it is obvious that the structure of processes is more sophisticated than the networks in the sense of network science, on the one hand. However, the case-specific local functionalities of the elementary building blocks of these structures are less complicated than the systems of integral and partial differential equations, combined with discrete events, on the other hand. As an intermediate solution, we developed *Programmable Structures* (Varga et al., 2016a,b, 2017; Varga and Csukas, 2017) that can be generated from a network description and from two functional metaprototypes. The metaprototypes and the derived process structures are prepared for the semantically distinguished, but syntactically uniform representation of the "*model specific conservation law based, additive*" measures, and of the "*over-writable*" signals.

In this approach, the usual network or net-based formalization of process structures is extended to a special structure of unified state and transition elements, containing dedicated input and output slots for additive measures and signals, as well as parameter slots for parameters. The actual structure is defined by standardized connecting elements, corresponding to increases and decreases of (extensive) measures, as well as to reading and overwriting of the intensive properties and other qualitative or quantitative signals.

In various applications many state and transition elements can usually be modeled with the same local programs. Accordingly, we distinguish the actual elements and the program defining prototypes. The prototype elements contain symbolic input, parameter, and output variables, as well as local programs, using these variables and a limited number of global data. During the simulation, the actual elements start with the initial conditions and parameters, and the output values are recalculated stepwise with the knowledge of input and parameter data, according to the associated local program prototype. The distinguished input and output slots for extensive/intensive data and for signals support the combined execution of the balance-based and rule-based functionalities.

The data flow between the elements is determined by the connections. Both the actual elements, as well as the receiving and sending sides of the connections may be specified by a spatial identifier, determining the compartment and/or the level and/or the scale of the given entity. Also, both the elements and the connections may have timing that defines temporal constraints for the execution, supporting the combined event-driven and time-driven execution of models.

Summarizing the methodology, the *programmable structure* of process models is generated from the description of the (optionally multilevel or multiscale) process network and from two functional metaprototypes. The case-specific, actual functional prototypes, containing the local models may be copied and edited from the general metaprototypes. The actual state and transition elements are parameterized and initialized concerning their case-specific prototypes. The actual model elements are executed by their associated prototypes. This execution and the connection-based communication between the state and transition elements of the *programmed structure* are solved by a general-purpose kernel program.

In the proposed model representation, the detailed description of possibility (design) space, as well as the measurements and the evaluating functions of the optionally multiple objectives, may be embedded in the state and transition elements. Accordingly, in identification and optimization tasks, the robust simulator may easily be combined with external metaheuristic optimizers.

The *major advantages* of the developed methodology come from the inherent and plausible coupling between the structural and functional knowledge, as well as from the extensibility and connectivity of the models.

In line with the well-known trade-off principle between being well-structured and effective, the *major disadvantage* of the methodology is less effectiveness compared with the case-specifically developed and implemented procedural methods and tools. These disadvantages might be compensated by the forthcoming micro-granularly parallel hardware/software implementations because the well-structured representation of local program prototypes supports these implementations.

14.2.2 Recent Implementation of Programmable Structures

The code of the recent implementation is written in declarative, logical Prolog language (GNU Prolog 1.4.4), which is a limited subset of first order predicate calculus. Prolog makes the easy description, and the unification-based execution, of the local program prototypes possible. This feature, combined with the applied, restricted, and standardized method of model description, supports the application of a general kernel for the quite different actual models. In addition, this declarative language offers a formal description thereby, which also fosters possible future implementations.

The automatic generation of the programmable structures starts from the GraphML-based graphical description of state and transition metaprototypes, as well as from the textual or graphical description of the process network. During generation, metaprototype elements are multiplied according to the description of the process network, supplemented with the connections between them.

It is to be noted, that the expert user can also start the model building with a GraphML editor from scratch. The graphical model can be edited further by various graph editors (actually yEd Graph Editor). In this step, the expert user makes copies from the metaprototypes and next declares the case-specific programs. This is followed by initialization and parameterization of the actual elements. Next, a Prolog program interprets the GraphML description into the facts and clauses, declaring the dynamic model. Finally, the general-purpose Prolog kernel program executes the various modes of dynamic simulation. This means a cyclically repeated processing of state elements, state → transition connections, transition elements and transition → state connections.

The *declarative model implementation* makes possible the application of list structures and recursive unification at any level of the description. Accordingly, the general description of the model is prepared for an optional number of list members, without the previous declaration of their length.

14.2.3 Previous and Ongoing Applications of Programmable Structures

Programmable Structures were successfully applied for quite different complex, multidisciplinary processes from low-scale cellular biosystems (Varga et al., 2017) through ecological (Varga and Csukas, 2017) and technological process systems (Csukas et al., 2013) up to Recirculating Aquaculture Systems (Varga et al., 2016a), and environmental process networks (Varga et al., 2016b). The advantage of this approach is that it gives flexibility for both structural and functional modification and extension of the process model.

The methodology supports the unified generation of reusable process models. The generated models are reconfigurable and extensible, and the models can be incremented from any "final" state. The robustness of dynamic simulation is based on the built-in control mechanisms, initiated by the automatic checking of feasibility bounds for the underlying "model specific conservation law based" measures.

The robust simulation fosters the dynamic model-based (sub)optimization with various, externally implemented metaheuristic methods (e.g., Genetic Algorithm). The basic principle, in accordance with the engineering way of thinking, is that instead of searching for an exact global optimum with a simplified model, it might be better to generate multiobjective suboptimal solutions from a more detailed model.

14.3 PROCESS NETWORK AND EXAMPLE FOR A SIMPLIFIED HYDROGEN SUPPLY CHAIN

14.3.1 Network and Net Representations of Process Systems

Starting from a simplified structural analysis of a process system, first we can recognize some relationships between the various state entities within a given spatial compartment or between the various locations. These simplest structures can be represented by *networks*, defined as a relation in the set of state elements, as follows:

$$R \subset S \times S, \quad S_i \in S \tag{14.1}$$

where R = a relation, S = the set of state characteristics, and = S_i is the ith state property.

In graphical representation the elements of S correspond to the nodes, while R designates the edges between them. It is worth mentioning that the nowadays very popular "network science" and "network analysis" basically work with nondirected edges (as can be seen in Fig. 14.1A) that correspond to symmetric (compatibility) relations, while the Cartesian product according to Eq. (14.1) assumes ordered pairs (i.e., directed edges). Considering that the transformations, transportations, and rules in process systems are usually directed, in the process systems directed relations and edges are usually applied (see Fig. 14.1B), while the bidirectional changes have to be symbolized by two edges of opposite direction.

Moreover, in the process systems the transformations and rules can be represented by causally and/or stoichiometrically coupled edges (Varga et al., 2017). Accordingly, we must define these coordinated changes by introducing a second set (i.e., a second kind) of nodes, designating the transitions. The respective net structures were defined by (Petri, 1977), as follows:

$$N \subset (S \times T) \cup (T \times S), \quad S_i \in S, \; T_j \in T \tag{14.2}$$

where N = a net relation, S = the set of state elements, T = the set of transition elements, S_i = the ith state property, and T_j = the jth transition property. The well-known original Petri Nets (Petri, 1962) represent a special case of these nets.

In the graph representation of the net structure (Eq. 14.2), the state and transition nodes are distinguished, for example, by dots and bars, as illustrated in Fig. 14.1C. In these *process flux networks*, the transition nodes (signed with bars) correspond to the coordinated transportations, transformations, and rules, while the state nodes (signed with dots) represent the underlying state entities.

The net structure discussed above correctly represents the *flux structure of reaction or transportation* networks, in which the state → transition and transition → state edges of conservational processes follow the respective decreasing and increasing fluxes, but the causalities are not visible. However, in the *state/transition nets of signals and rules* the state → transition and transition → state edges follow the causalities, determined by antecedent and consequent signals.

Considering that in process modeling we use both additive quantities and signals, *in adequate process structures we must distinguish the additive measure decreasing and increasing, as well as the intensive property and other signal transferring connections*. In this representation of conservation law-based (conservational) processes (see Fig. 14.1D), the transitions read (dotted lines)

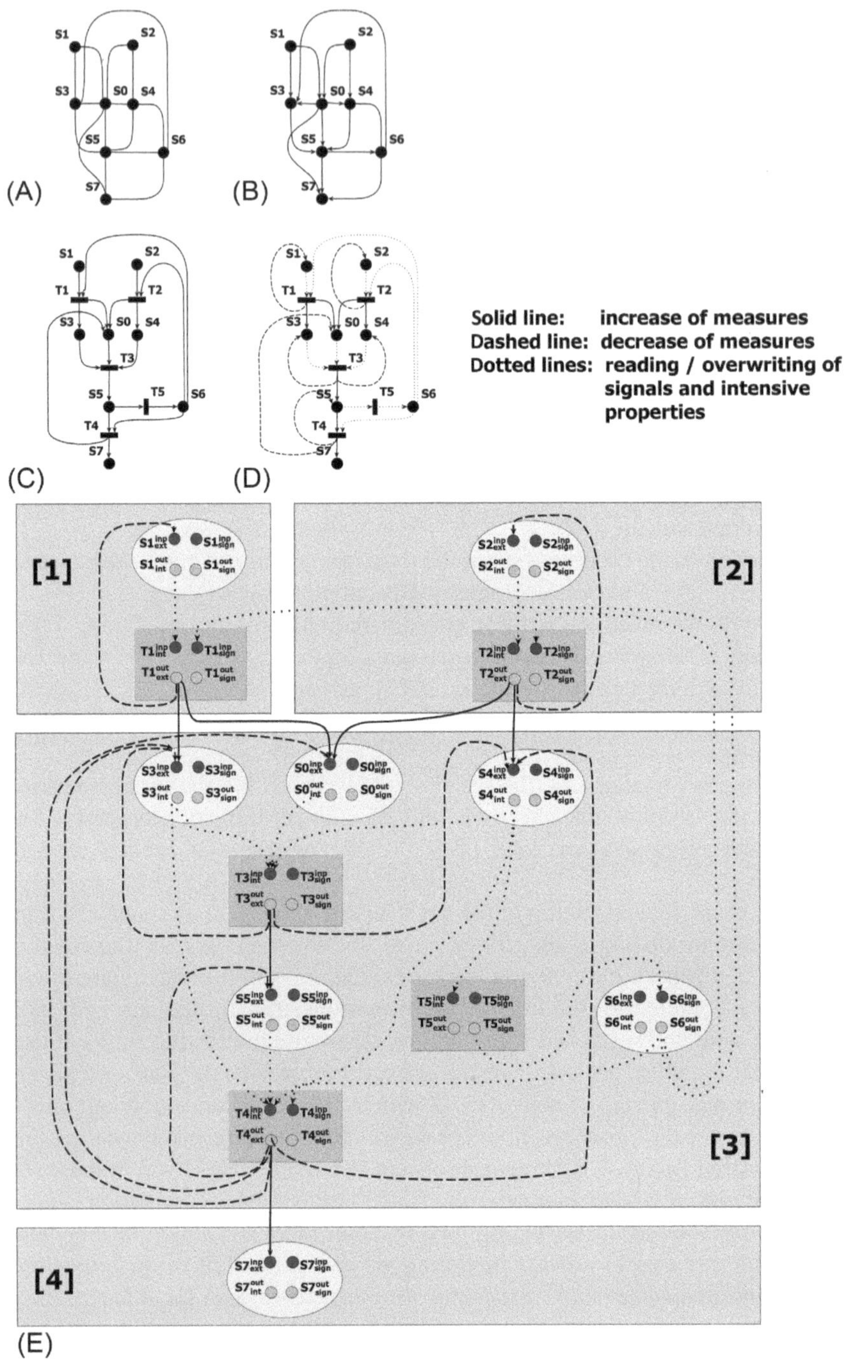

FIG. 14.1 Various structural representations of process systems. (A) Nondirected network, (B) directed network, (C) state-transition network/net, (D) process network/net, and (E) programmable structure representation of process network/net, involving compartments [1], [2], [3], and [4].

the rate determining intensive properties of the state elements, as well as increase (solid lines) and decrease (dashed lines) the extensive measures of the modified states. Similarly, in the rule-based signaling (informational) processes, the transitions read the antecedent signals and overwrite the consequent ones. It results in a uniform and causally right structural description for both conservational and informational processes (Varga et al., 2017; Varga and Csukas, 2017).

Accordingly, both state and transition elements must be *prepared for receiving input from, and sending output to, additive measures and signs.* These input and output connectors of the state and transition elements are called *slots* and are depicted by small circles, as shown in Fig. 14.1E. In this more sophisticated structure, the unified representation of the conservation law–based (quantitative) and rule-based (qualitative) edges, as well as the input/output characteristics of the nodes can be described uniformly and causally right. Accordingly, the above outlined *process nets* can be defined by the relation.

$$N \subset \left(S_{int}^{out} \times T_{int}^{inp}\right) \cup \left(S_{sign}^{out} \times T_{sign}^{inp}\right) \cup \left(T_{decr}^{out} \times S_{ext}^{inp}\right) \cup \left(T_{incr}^{out} \times S_{ext}^{inp}\right) \cup \left(T_{sign}^{out} \times S_{sign}^{inp}\right) \tag{14.3}$$

where in the indexes.

inp = input;
out = output;
int = intensive (derived from measures);
ext. = extensive (additive measure);
incr = increase (of an additive measure);
decr = decrease (of an additive measure),
sign = signal.

14.3.2 Illustration of a Simple Hydrogen Supply Chain Process Network

The process structure of a simple Hydrogen Supply Chain, in the sense of Eq. (14.3) and Fig. 14.1, is shown in Fig. 14.2. This is a small, enlarged part of the example structure, discussed in Section 14.4. The names, IDs and locations of the illustrated state and transition elements are summarized in Table 14.1. The districts, designated by 1, 2, 3, and 4 correspond to four fictitious, geographically outlined territories (e.g., according to a GIS shape file, as we combined it with a programmable structure of an environmental model (Varga et al., 2016b).

In the process structure, the state and transition elements are symbolized by ellipses and rectangles (Fig. 14.2), respectively. In both kind of elements, there are two input slots (signified with small red upper dots within the elements in

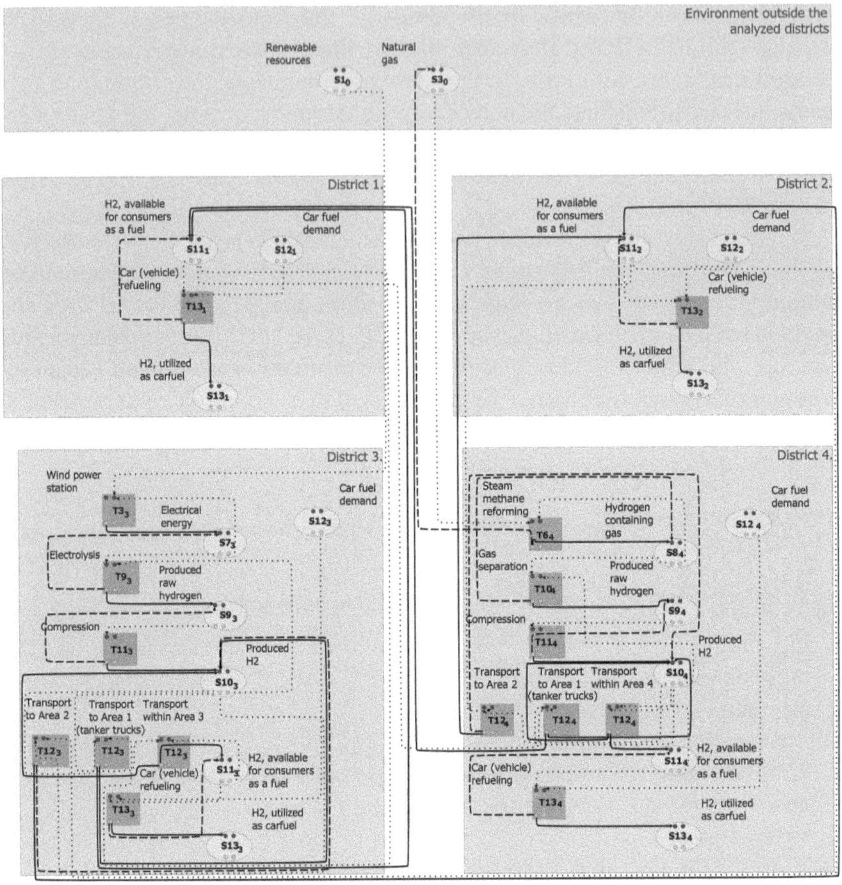

FIG. 14.2 Illustration of the process structure of a simplified part of a Hydrogen Supply Chain.

Fig. 14.2) for the input additive measures and over-writable signals, as well as two output slots (signified with small yellow lower dots within the elements in Fig. 14.2) for the output measure related properties and signals.

There are five kinds of connections, as follows:

- Increasing of (model-specific conservation law–based) measures: solid edges from measure change output slots of transitions to measure input slots of states;
- Decreasing of (model-specific conservation law–based) measures: dashed edges from measure change output slots of transitions to measure input slots of states;
- Reading/writing of measure-related (e.g., intensive) properties: dotted edges from measure-related output slots of states to the measure related input slots of transitions;

TABLE 14.1 Elements of the Simple Example Structure

	Name of Element	ID	District, Where the Given Material or Process is Involved
State elements	Renewable resources	S1	0
	Natural gas	S3	0
	Electrical energy	S7	3
	Hydrogen-containing gas	S8	4
	Produced raw hydrogen	S9	3, 4
	Produced compressed H2	S10	3, 4
	Compressed H2, available for consumers as a fuel	S11	1, 2, 3, 4
	Car fuel demand	S12	1, 2, 3, 4
	H2, utilized as car fuel	S13	1, 2, 3, 4
Transition elements	Wind power station	T3	3
	Steam methane reforming	T6	4
	Electrolysis	T9	3
	Gas separation	T10	4
	Compression	T11	3, 4
	Transport (tanker truck)	T12	3, 4
	Car (vehicle) refueling	T13	1, 2, 3, 4

- Reading/writing of transition output signals: dotted edges from signaling output slots of transitions to signaling input slots of states;
- Reading/writing of state output signals: dotted edges from signaling output slots of states to signaling input slots of transitions.

14.4 GENERATION OF THE PROGRAMMABLE STRUCTURE FOR A SIMPLE EXAMPLE HYDROGEN SUPPLY CHAIN

Hydrogen Supply Chains can usually be modeled by a large number of holistically connected state and transition elements, while their detailed functionalities can be described by a significantly lower number of prototyped programs. Considering this, we start the model generation from the detailed graphical implementation of one state and one transition metaprototype, as well as from the description of the process network. Generation means multiplying the prototyped state and transition elements, as well as adding the connections between them, according to the actual structure, automatically. Actually, the graphically editable models can be generated from the GraphML definition of metaprototypes and from the textual description of the network structure, similarly to other process systems (Varga et al., 2017; Varga and Csukas, 2017).

In the following sections we shall use a limited Prolog syntax, in which:

upper case alphanumerical symbol = variable;
lower case alphanumerical symbol = constant;
Anything* = list of Anything;
[H|T] = list of any elements or functors with a head H and a list of tail T;
[] = empty list;
full stop = end of logical sentence;
if (or :-) implication in the sense of Horn clauses;
comma in the program body of clauses = and.

14.4.1 Declaration of the Metaprototypes

The structures of Hydrogen Supply Chains discussed above are built from state and transition elements. Also, there are two major classes of characteristics, namely the (often model-specific conservation law–based) extensive/intensive properties and the signals (signs). Accordingly, the primary metaprototypes, used for the generation of the actual structures should be prepared for the input and output of both extensive/intensive properties and signals.

The graphical representation of the general state and transition metaprototypes is shown in Fig. 14.3. In line with the representation of process models in Figs. 14.1E and 14.2, the state and transition prototypes are depicted by an ellipse and a rectangle, respectively. The outside communication and inside functionality related data structures, as well as the place for the local program codes are contained in the slots, as follows:

- balance measure and signal related input and data storage slots (small red dots),
- predefined general classes of local parameters (small red rectangles),
- place for the locally executable (actually declarative) program code of the various prototypes (small yellow diamonds), and
- balance measure and signal related data storage and output slots (small yellow dots),
- the reporting slot of the transitions that makes it possible to forward case specific data to the output stream of the simulation.

	State metaprototype	Transition metaprototype
	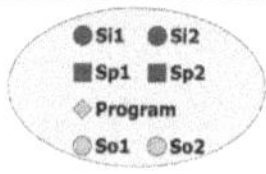	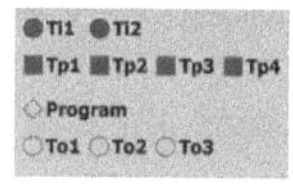
Heading	{e(p,y,state,'Spatial',state,'Temporal','Possibilities','Evaluations')}	{e(a,y,trans,'Spatial',trans,'Temporal','Possibilities','Evaluations')}
Inputs	Si1 {i(comp,dl,'InpComps')} Si2 {i(inp,dl,'InpSigns')}	Ti1 {i(conc,dl,'InpConcs')} Ti2 {i(inp,dl,'InpSigns')}
Parameters	Sp1 {c(param,dl,'Parameters')} Sp2 {c(cond,dl,'Conditions')}	Tp1 {c(param,dl,'Parameters')} Tp2 {c(antecedent,dl,'Antecedent')} Tp3 {c(consequence,dl,'Consequence')} Tp4 {c(cond,dl,'Conditions')}
Outputs	So1 {o(conc,dl,'OutConcs')} So2 {o(out,dl,'OutSigns')}	To1 {o(comp,dl,'OutComps')} To2 {o(out,dl,'OutSigns')} To3 {o(report,dl,'Reports')}
Program	{program(' ')}	{program(' ')}

FIG. 14.3 Metaprototypes of state and transition elements.

The e(.) heading of the elements specifies:

- the type (p = passive state or a = active transition),
- the switchable existence (y = yes or n = no),
- the name of the metaprototype (for this application 'state' or 'trans'),
- the actual spatial scale (variable Spatial, unifying with a list of integers),
- the name of the prototype (in the metaprototype it replies the name of the metaprototype that has to be changed for the name of the derived prototypes in the course of editing the generated structure),
- the timing (variable Temporal),
- the design space (Possibilities, an empty list for the optional description of possibilities, associated with the given elements) and
- a list of data for the calculation of objectives (Evaluations).

The input, parameter, and output slots contain i(), c(), and o() functors. The i(), c (), and o() functors are described by a name, a data type (actually dl), and a list of the respective data. In case of data type dl, the data sets are described by lists of triplets.

```
d(Identifier, Variable_list, Instructions)
```

where.

Identifier = a symbol or a list, determining an identifying name;
Variable_list = a list of variables; and.
Instructions = instruction for the use (e.g., dimension, specification, etc.).

The program slot contains an empty place, starting with the *program ('* and ending with the *')* character sequences. In the course of the editing process, this place has to be filled with the declaration of case-specific prototypes.

The slots of the actually used *state metaprototype* (state, see Fig. 14.3 left) are the following:

- The balance measure input slot "comp" declares an empty or predefined list InpComps, for an optional number of d(.) triplets for the balance measure variables that summarize the increases and decreases, for the same d() triplet, as well as optionally new triplets that can be added or removed. The respective data come from the various transitions, via the connections, as can be seen in Fig. 14.2;
- The signaling input slot "inp" declares an empty or predefined list InpSigns, for an optional number of d(.) triplets for the sign variables that may be overwritten by new signs, as well as optionally new triplets that can be added or removed. The respective data come from the various transitions, via the connections, as can be seen in Fig. 14.2;
- The parameter slot "param" declares an empty list for the optional state parameters (described by d(.) triplets in the list Parameters);
- The condition slot "cond" declares an empty list for the optional conditions (described by d(.) triplets in the list Conditions);
- The measure-related output slot "conc" declares an empty list OutConcs, for an optional number of d(.) triplets for the calculated properties (e.g., available quantities, concentrations, temperature, etc.). The respective data are forwarded to the various transitions via the connections, as illustrated in Fig. 14.2. In addition, all of the output data can be optionally forwarded to the output recording;
- The signaling output slot "out" declares an empty list OutSigns for an optional number of d(.) triplets for the determined output signs. The respective data are forwarded to the various transitions via the connections, as illustrated in Fig. 14.2. In addition, all of the output data can optionally be forwarded to the output recording;
- The program code is to bind (to determine) all of the output variables (OutComps, OutSigns, Report) with the knowledge of the bound input (Spatial, Temporal, InpConcs, InpSigns) and parameter (Parameters, Antecedent, Consequence, Conditions) values. The expert-defined local prototype programs have to be prepared for the actual cases in edition of the generated programmable structure (see Section 14.5).

The slots of the actually used *transition metaprototype* (transition, see Fig. 14.3 right) are the following:

- The measure-related input slot "conc" declares an empty list InpConcs for an optional number of d(.) triplets for the variables, used for the calculation of the transitions. The respective data come from the various states, via the connections, as can be seen in Fig. 14.2;
- The signaling input slot "inp" declares an empty or predefined list InpSigns for an optional number of d(.) triplets for the sign variables, used for the

calculation of the transitions. The respective data come from the various transitions, via the connections, as can be seen in Fig. 14.2;

- The parameter slot "param" declares an empty list for the optional state parameters (described by d(.) triplets in the list Parameters);
- The antecedent slot "antecedent" declares an empty list for the optional antecedent signals (described by d(.) triplets in the list Antecedent);
- The consequence slot "consequence" declares an empty list for the consequence signals (described by d(.) triplets in the list Consequence);
- The condition slot "cond" declares an empty list for the optional conditions (described by d(.) triplets in the list Conditions);
- The measure change related output slot "comp" declares an empty list OutComps for an optional number of d(.) triplets for the calculated increases and decreases of measures. The respective data are forwarded to the various states via the connections, as illustrated in Fig. 14.2;
- The signaling output slot "out" declares an empty list OutSigns for an optional number of d(.) triplets for the determined output signs. The respective data are forwarded to the various states via the connections, as illustrated in Fig. 14.2;
- The reporting slot "report" declares an empty list Report for an optional number of d(.) triplets to forward case-specific data to the output stream of the simulation;
- The program code is to bind (to determine) all of the output variables (OutComps, OutSigns, Report) with the knowledge of the bound input (Spatial, Temporal, InpConcs, InpSigns) and parameter (Parameters, Antecedent, Consequence, Conditions) values. The expert-defined local prototype programs have to be prepared for the actual cases in edition of the generated programmable structure (see later on in Section 14.5).

The definition of the metaprototypes can be saved in GraphML files for further processing. We can also define other metaprototypes, however, the solution described above proved to be quite general in various applications (e.g., Varga et al., 2017).

14.4.2 Definition of an Example Network

The names, IDs, short symbolic names, and locations of the building elements of an illustrative, fictitious HSC example network are summarized in Table 14.2. The short symbolic names, used in the databases and programs, are written according to the Prolog syntax with lower case capitals. The districts, designated by 1, 2, 3, and 4 correspond to four fictitious, GIS represented territories.

The declaration of the process structure follows the definition Eq. (14.3), as well as the architecture of the metaprototypes discussed above. Accordingly, to

TABLE 14.2 State and Transition Elements of the Fictitious Example HSC

	Name of Element	ID	Short Symbol in the Model	District, Where the Given Material or Process is Involved
State elements	Renewable resources	S1	raw_renew	0
	Cultivated land	S2	raw_land	2, 4
	Natural gas	S3	raw_gas	0
	Nuclear energy raw materials	S4	raw_nucl	0
	Energy producing biomass	S5	biomass	2, 4
	Biological wastes	S6	biowaste	2, 4
	Electrical energy	S7	eleng	1, 2, 3, 4
	Hydrogen-containing gas	S8	h2contgas	2, 4
	Produced raw hydrogen	S9	hydrogen	1, 2, 3, 4
	Produced compressed H2	S10	h2prod	1, 2, 3, 4
	Compressed H2, available for consumers as a fuel	S11	h2cons	1, 2, 3, 4
	Car fuel demand	S12	demand	1, 2, 3, 4
	H2, utilized as car fuel	S13	h2carfuel	1, 2, 3, 4
	Heating energy	S14	h2heating	0
	Carbon-dioxide	S15	co2	0
	Oxygen	S16	o2	0

TABLE 14.2 State and Transition Elements of the Fictitious Example HSC —cont'd

	Name of Element	ID	Short Symbol in the Model	District, Where the Given Material or Process is Involved
Transition elements	Land cultivation	T1	t_cultivation	2, 4
	Photovoltaic process	T2	t_photovolt	1, 2, 3, 4
	Wind power station	T3	t_wind	2, 3
	Hydropower station	T4	t_hydro	3, 4
	Nuclear energy station	T5	t_nucl	1
	Steam methane reforming	T6	t_smr	2, 4
	Biomass gasification	T7	t_gasify	2, 4
	Anaerobic fermentation	T8	t_ferment	2, 4
	Electrolysis	T9	t_electrolysis	1, 2, 3, 4
	Gas separation	T10	t_gassep	2, 4
	Compression	T11	t_compression	1, 2, 3, 4
	Transport (tanker truck)	T12	t_tankertruck	1, 2, 3, 4
	Car (vehicle) refueling	T13	t_h2refuel	1, 2, 3, 4
	Heat production	T14	t_h2heat	1, 2, 3, 4

the applied *transition-based model description* (e.g., Varga and Csukas, 2017) the actual structure can be declared by the listing of state and transition elements, as well as by the declaration of the individual transitions by the state elements being connected to the different input and output slots of them. In the simplest models, all of the state and transition elements are in the same spatial location (with the same Spatial identifier, say Spatial = []), accordingly the structure can be defined in Prolog syntax as follows:

```
states([],Statelist).
          Statelist = State*
transitions([],Transitionlist).
          Transitionlist = Transition*
trans(Transition,InpConcs,OutComps,InpSigns,OutSigns).
          InpConcs; InpSigns; OutComps; OutSigns = Part_of_Statelist
          Part_of_Statelist = State*
```

InpConcs, InpSigns, OutComps and OutSigns contain the names of states connected with measure input, signal input, measure output and signal output slots of transition.

In generation of multicompartment and/or multilevel and/or multiscale models, the states and transitions are declared with their respective Spatial coordinates, as follows:

```
states(Coord1,Statelist1).
states(Coord2,Statelist1).
...etc.
transitions(Coord1,Transitionlist1).
transitions(Coord2,Transitionlist2).
...etc.
```

Accordingly, the transition-based declaration of the structures is as follows:

```
trans(Transition,InpConcs,OutComps,InpSigns,OutSigns).
          InpConcs; InpSigns; OutComps; OutSigns = Specified_State*
          Specified_State = n(Coord,State)
```

This solution makes possible the use of the same state names in the various compartments, levels, or scales, because they are identified by the spatial coordinate and the name together.

The declaration of the fictitious example HSC, using the short symbolic names from Table 14.2 is as follows:

```
   states([0],[raw_renew,raw_gas,raw_nucl,h2heating,co2,o2]).
   states([1],[h2cons,demand,h2carfuel,eleng,hydrogen,h2prod]).
   states([2],[h2cons,demand,h2carfuel,raw_land,biomass,biowaste,
eleng,h2contgas,hydrogen,h2prod]).
   states([3],[eleng,hydrogen,h2prod,h2cons,demand,h2carfuel]).
   states([4],[h2contgas,hydrogen,h2prod,h2cons,demand,h2carfuel,
raw_land,biomass,biowaste,eleng]).

   transitions([0],[]).
   transitions([1],[t_h2refuel1,t_photovolt1,t_nucl1,t_electrolysis1,
             t_compression1,t_tankertruck11,t_h2heat1]).
```

```
transitions([2],[t_h2refuel2,t_cultivation2,t_photovolt2,t_wind2,
                t_smr2,t_gasify2,t_ferment2,t_electrolysis2,
                t_gassep2,t_compression2,t_tankertruck22,
                t_h2heat2]).
transitions([3],[t_wind3,t_electrolysis3,t_compression3,
                t_h2refuel3,t_tankertruck33,
t_tankertruck31,t_tankertruck32,t_photovolt3,t_hydro3,t_h2heat3]).
transitions([4],[t_smr4,t_gassep4,t_compression4,t_h2refuel4,
                t_tankertruck44,t_tankertruck41,t_tankertruck42,
                t_cultivation4,t_photovolt4,t_hydro4,t_gasify4,
                t_ferment4,t_electrolysis4,t_h2heat4]).

trans(t_h2refuel1,[n([1],h2cons)],[n([1],h2cons),n([1],h2carfuel)],
          [n([1],demand)],[]).
trans(t_h2refuel2,[n([2],h2cons)],[n([2],h2cons),n([2],h2carfuel)],
          [n([2],demand)],[]).
trans(t_h2refuel3,[n([3],h2cons)],[n([3],h2cons),n([3],h2carfuel)],
          [n([3],demand)],[]).
trans(t_h2refuel4,[n([4],h2cons)],[n([4],h2cons),n([4],h2carfuel)],
          [n([4],demand)],[]).
trans(t_tankertruck11,[n([1],h2prod)],[n([1],h2prod),n([1],
          h2cons)],[n([1],h2cons)],[]).
trans(t_tankertruck22,[n([2],h2prod)],[n([2],h2prod),n([2],
          h2cons)],[n([2],h2cons)],[]).
trans(t_tankertruck33,[n([3],h2prod)],[n([3],h2prod),n([3],
          h2cons)],[n([3],h2cons)],[]).
trans(t_tankertruck31,[n([3],h2prod)],[n([3],h2prod),n([1],
          h2cons)],[n([1],h2cons)],[]).
trans(t_tankertruck32,[n([3],h2prod)],[n([3],h2prod),n([2],
          h2cons)],[n([2],h2cons)],[]).
trans(t_tankertruck44,[n([4],h2prod)],[n([4],h2prod),n([4],
          h2cons)],[n([4],h2cons)],[]).
trans(t_tankertruck41,[n([4],h2prod)],[n([4],h2prod),n([1],
          h2cons)],[n([1],h2cons)],[]).
trans(t_tankertruck42,[n([4],h2prod)],[n([4],h2prod),n([2],
          h2cons)],[n([2],h2cons)],[]).
trans(t_compression1,[n([1],hydrogen)],[n([1],hydrogen),
          n([1],h2prod)],[],[]).
trans(t_compression2,[n([2],hydrogen)],[n([2],hydrogen),
          n([2],h2prod)],[],[]).
trans(t_compression3,[n([3],hydrogen)],[n([3],hydrogen),
          n([3],h2prod)],[],[]).
trans(t_compression4,[n([4],hydrogen)],[n([4],hydrogen),
          n([4],h2prod)],[],[]).
trans(t_electrolysis1,[n([1],eleng)],[n([1],eleng),n([1],
           hydrogen)],[n([1],h2prod)],[]).
```

```
trans(t_electrolysis2,[n([2],eleng)],[n([2],eleng),n([2],
            hydrogen)],[n([2],h2prod)],[]).
trans(t_electrolysis3,[n([3],eleng)],[n([3],eleng),n([3],
            hydrogen)],[n([3],h2prod)],[]).
trans(t_electrolysis4,[n([4],eleng)],[n([4],eleng),n([4],
            hydrogen)],[n([4],h2prod)],[]).
trans(t_photovolt1,[],[n([1],eleng)],[],[n([0],raw_renew)]).
trans(t_photovolt2,[],[n([2],eleng)],[],[n([0],raw_renew)]).
trans(t_photovolt3,[],[n([3],eleng)],[],[n([0],raw_renew)]).
trans(t_photovolt4,[],[n([4],eleng)],[],[n([0],raw_renew)]).
trans(t_wind2,[],[n([2],eleng)],[],[n([0],raw_renew)]).
trans(t_wind3,[],[n([3],eleng)],[],[n([0],raw_renew)]).
trans(t_nucl1,[n([0],raw_nucl)],[n([0],raw_nucl),n([1],eleng)],
        [n([1],eleng)],[]).
trans(t_hydro3,[],[n([3],eleng)],[],[n([0],raw_renew)]).
trans(t_hydro4,[],[n([4],eleng)],[],[n([0],raw_renew)]).
trans(t_gassep2,[n([2],h2contgas)],[n([2],h2contgas),n([2],
          hydrogen)],[n([2],h2prod)],[]).
trans(t_gassep4,[n([4],h2contgas)],[n([4],h2contgas),n([4],
          hydrogen)],[n([4],h2prod)],[]).
trans(t_smr2,[n([0],raw_gas)],[n([0],raw_gas),n([2],h2contgas)],
          [n([2],h2contgas)],[]).
trans(t_smr4,[n([0],raw_gas)],[n([0],raw_gas),n([4],h2contgas)],
        [n([4],h2contgas)],[]).
trans(t_ferment2,[n([2],biowaste)],[n([2],biowaste),n([2],
           h2contgas)],[n([2],h2contgas)],[]).
trans(t_ferment4,[n([4],biowaste)],[n([4],biowaste),n([4],
           h2contgas)],[n([4],h2contgas)],[]).
trans(t_gasify2,[n([2],biomass)],[n([2],biomass),n([2],h2contgas)],
          [n([2],h2contgas)],[]).
trans(t_gasify4,[n([4],biomass)],[n([4],biomass),n([4],h2contgas)],
          [n([4],h2contgas)],[]).
trans(t_cultivation2,[n([2],raw_land)],[n([2],raw_land),n([0],co2),
            n([0],o2),n([2],biomass)],[n([2],biomass)],[]).
trans(t_cultivation4,[n([4],raw_land)],[n([4],raw_land),n([0],co2),
            n([0],o2),n([4],biomass)],[n([4],biomass)],[]).
trans(t_h2heat1,[n([1],hydrogen)],[n([1],hydrogen),
        n([0],h2heating)],[],[]).
trans(t_h2heat2,[n([2],hydrogen)],[n([2],hydrogen),
        n([0],h2heating)],[],[]).
trans(t_h2heat3,[n([3],hydrogen)],[n([3],hydrogen),
        n([0],h2heating)],[],[]).
trans(t_h2heat4,[n([4],hydrogen)],[n([4],hydrogen),
        n([0],h2heating)],[],[]).
```

14.4.3 Generation of the Programmable Structure Into a Graphml File

The generation of the editable graphical model utilizes the GraphML description of metaprototypes and the textual definition of the structure discussed above. The generating algorithm first takes the list of state elements and appends the GraphML file with the appropriately modified copies of the state metaprototype. Next, the algorithm takes the list of transition elements and appends the GraphML file with the appropriately modified copies of the transition metaprototype. Finally, the program takes the individual transitions, one after the other, and appends the GraphML file with the description of the edges. The resulting GraphML file contains the two prototypes and the "empty" structure of the whole process model.

The example structure, generated from the two metaprototypes and from the example structure (having already extended by some edited prototypes), can be seen in Fig. 14.4. The GraphML-based graphical model can be edited further by various graph editors (actually yEd Graph Editor).

For better understanding and visualization, the edited process structure is shown in Fig. 14.5.

14.5 PROGRAMMING AND INITIALIZATION OF THE EXAMPLE STRUCTURAL MODEL

The generated GraphML file can be further edited, for example, with the yEd graph editor. The GraphML representation of the process structure supports any graphical modifications and extensions of the model, as well as the addition of editable texts, declaring the local programs, actualizing of initial values and of the various model parameters.

If we start from the automatic generation, then first we have to prepare the editable prototypes from the general metaprototypes. This requires making copies from the metaprototypes, followed by the declaration of the local program of the prototypes. The definition of the state and transition prototype elements needs the creative knowledge of the modelers (advantageously the collaborating field and model experts) about the investigated process. Nevertheless, the implementation is supported by the metaknowledge embedded in the metaprototypes, as well as in the generated process structure.

The variables in prototypes are described by alphanumeric symbols, starting with a capital letter (and written within apostrophes in the graphical interface). The i(), c(), and o() functors are described by a name, by a data type (actually dl), and by a list of the respective variables. Recently, the variables have been described by lists of triplets *d(Identifier, Variable_list, Instructions)*, as was shown in Section 14.4.1.

In the case of the case-specific prototypes, the content of the state and transition elements has to be actualized. This can be done by setting the initial

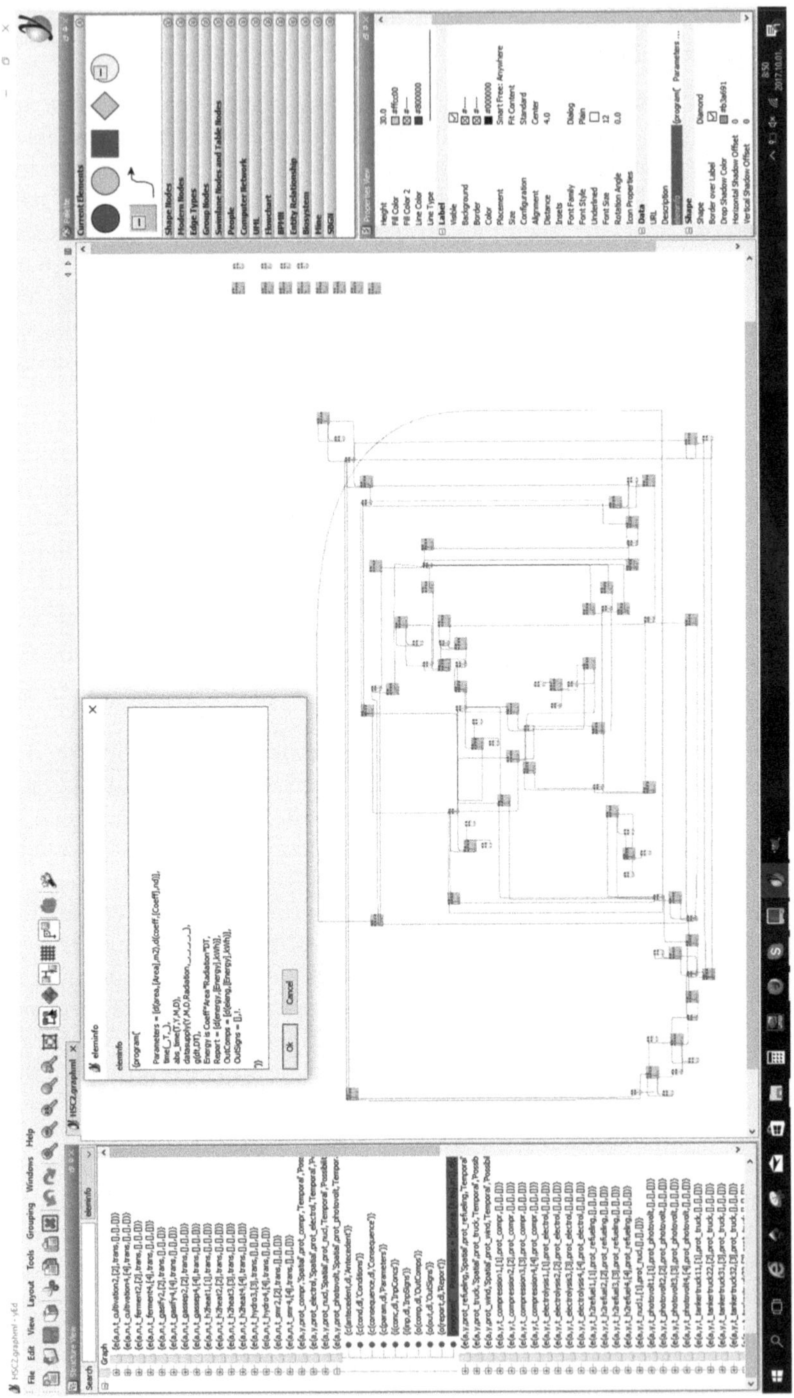

FIG. 14.4 Generated programmable structure of the fictitious example HSC.

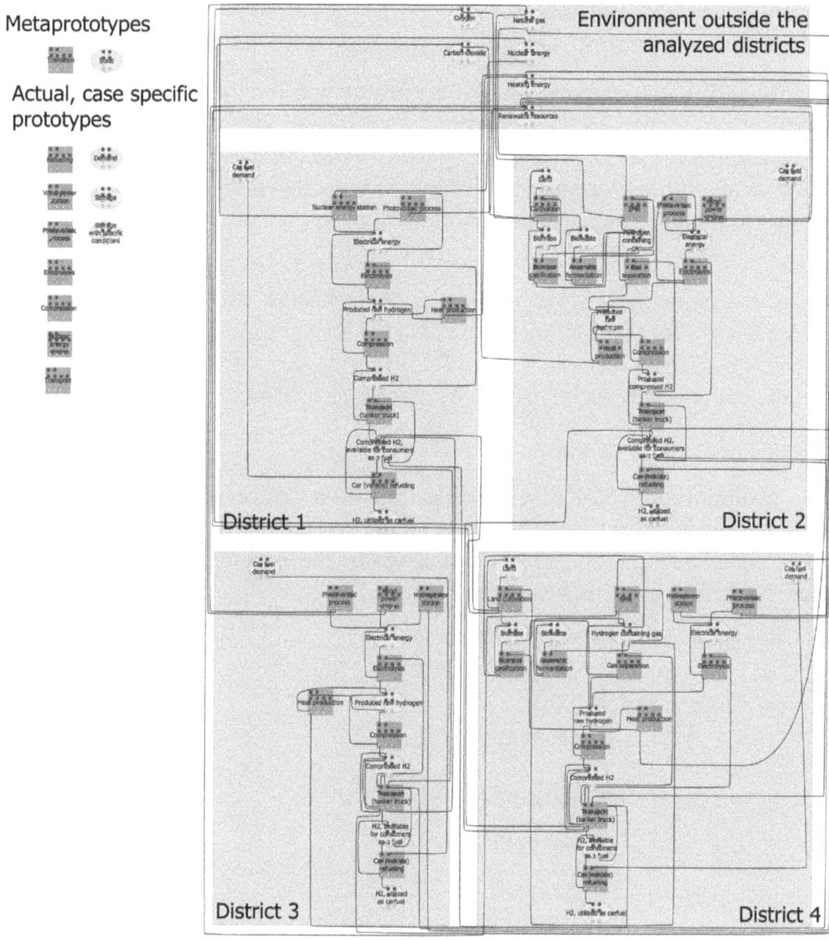

FIG. 14.5 Edited process structure of the fictitious example HSC.

conditions (i.e., the input parameters of the state elements), as well as by defining the parameters (i.e., the actual values of the parameter slots both in state and in transition elements). Also, the name of the prototype that was actually used has to be overwritten in the heading e(.) of each element. This name refers to the local program to be executed for the given element.

The original metaprototypes and the a few actual prototypes of the fictitious example process are also shown in Figs. 14.4 and 14.5. In the illustrative HSC example given, the metaprototypes and prototypes are the following:

- Transition metaprototype: trans;
- Transition prototypes: prot_compr, prot_electrol, prot_nucl, prot_hotovolt, prot_refueling, prot_truck, prot_wind;

- State metaprototype: state;
- State prototypes: prot_demand, prot_storage, prot_condstorage.

In the following, the programming of the prototypes and the initialization of the respective actual elements will be illustrated by simplified examples.

Programming and initialization of transition prototypes and elements.

The transition metaprototype corresponds to the following Prolog clause with an "almost empty" program in the body:

```
v(y,trans,Spatial,trans,Temporal,Possibilities,Evaluations,
[c(antecedent,dl,Antecedent),c(consequence,dl,Consequence),c
(cond,dl,Conditions),c(param,dl,Parameters)],
[i(inp,dl,InpSigns),i(conc,dl,InpConcs)],
[o(comp,dl,OutComps),o(out,dl,OutSigns),o(report,dl,Report)]) :-

 OutComps = [], OutSigns = [], Report = [].
```

When programming of a case-specific transition prototype via the graphical interface, the identifier "trans" has to be replaced by the prototype name (in case of a wind power station "prot_wind"), while the empty program has to be replaced for the respective code, as follows:

```
Parameters = [d(coeff,[Coeff],nd),d(number,[Nu],nd)],
time(_,T,_),
abs_time(T,Y,M,D),
meteo(Y,M,D,_,Windspeed,_,_,_,_),
g(dt,DT),
E is Nu*Coeff*((0.62*Windspeed)**3)*DT,
Report = [d(energy,[E],kWh)],
OutComps = [d(eleng,[E],kWh)],
OutSigns = [].
```

(It is to be noted that in Prolog syntax "is" signifies assignment, "=" refers to unification, and g(dt,DT) calls for the time step.)

Simultaneously in every actual wind power station defining element, the prototype name and the parameters used in the program have to be overwritten (supported by the graph editor, of course), as is signified by bold italics in the following example:

```
a(y,t_wind3,[3],prot_wind,[],[],[],
[c(antecedent,dl,[]),c(consequence,dl,[]),c(cond,dl,[]),c
(param,dl,[d(coeff,[30],nd),d(number,[200],nd)])],
[i(inp,dl,[]),i(conc,dl,[])],
[o(comp,dl,[]),o(out,dl,[]),o(report,dl,[])]).
```

Programming and initialization of state prototypes and elements.

The state metaprototype corresponds to the following Prolog clause with an "almost empty" program in the body:

```
m(y,Spatial,state,Temporal,Possibilities,Evaluations,
[c(param,dl,Parameters),c(cond,dl,Conditions)],
[i(comp,dl,InpComps),i(inp,dl,InpSigns)],
[o(conc,dl,OutConcs),o(out,dl,OutSigns)],) :-
OutConcs = [], OutSigns = [].
```

When programming of a case-specific state prototype via the graphical interface, the identifier "state" has to be replaced by the prototype name (in the case of conditional storage "prot_condstorage"), while the empty program has to be replaced for the respective code, as follows:

```
Conditions = [d(limit,[Limit],Dim),d(destinations,[Place],nd)],
InpComps = [d(Material,[M],Dim)],
needed(Limit,M,Need),
OutConcs = [d(Material,[M],Dim)],
OutSigns = [d(Place,[Need],Dim)].

needed(Limit,M,Need) :-
 Limit > M,
 Need is Limit-M,!.
needed(_,_,0).
```

Simultaneously, in every actual element executed according to this program, the prototype name, the parameters, and the necessary initial values used in the program have to be overwritten, as is signified by bold italics in the following example:

```
p(y,h2prod,[3],prot_condstorage,[],[],[],
[c(param,dl,[]),c(cond,dl,[d(limit,[5000],kg),d(destinations,
[electrolysis3],nd)])],
[i(comp,dl,[d(h2prod,[2000],kg)]),i(inp,dl,[])],
[o(conc,dl,[]),o(out,dl,[])]).
```

14.6 INTERPRETATION OF THE EXAMPLE MODEL AND PREPARATION FOR SIMULATION-BASED PROBLEM SOLVING

The programming and initialization of the process structure is followed by its interpretation into the executable dynamic databases of the given model. In this step, the facts and clauses of the executable Prolog model definitions are generated from the edited GraphML file automatically. First, the general interpreter reads the GraphML text, while it generates and saves:

- all of the facts describing the actual state and transition elements in the "user" file,
- all of the facts determining state → transition and transition → state connections in the "user" file, and
- all of the clauses, declaring the program prototypes in the "expert" file.

The resulting user and expert files contain the detailed description of the process model, including the locally executable program prototypes and the actual elements and connections containing the initial data and the case-specific parameters.

Having finished the model generation, the executable code has to be supplemented by the (optionally scale-specific) global data. They can be declared for example by the following global facts:

```
g(timescale,day).            declares the applied absolute (day, hour,
minute, second) or relative (rel) time (in multiscale application scale-
specific time steps may be applied);
g(starttime,at(2020,1,1,0,0,0)).   declares the absolute starting
time;
g(direction,f ).             declares the direction of the simulation
(f = forward, b = backward);
g(dt,1).                     the initial time step;
g(dur,3652).                 the duration of the simulation;
g(ndto,1).                   the time step of the output data recording
(NDTO >= DT);
```

Also, the simulator is prepared for the communication with the necessary optional external databases and applications. For example, in the studied Hydrogen Supply Chain application the model has to communicate:

- with a Geographical Information System, specifying the various districts and locations, as well as assisting with the optionally embedded routing algorithm;
- with a market related database, containing data about the time varying and location specific demand for the hydrogen, and for the optional other products of the model investigated;
- with a meteorological (or climate scenario) database, supplying location and time depending meteorological data for the consideration of the natural resource related processes; and
- with an optional external metaheuristic optimizer (see more detailed in Section 18.8).

An example architecture for combining Programmable Structures with GIS and meteorological databases was published in (Varga et al., 2016b).

14.7 EXECUTION OF THE DYNAMIC SIMULATION OF PROGRAMMABLE STRUCTURE

Dynamic simulation is organized by a general executing kernel, by default comprising the cyclically repeated processing of state elements, state → transition connections, transition elements and transition → state connections.

The simulation starts at the globally determined relative or absolute initial time with the previously initiated data (i.e., input variables of states, parameters of states and transitions) and with the initial time step. The algorithm checks the Temporal constraints of the given elements and only the allowed procedures are executed. The Temporal constraints may determine prescribed time intervals, discrete points of time, and specific time steps for the given elements, as follows:

```
   Temporal = Timing* (may contain any number of timings)
        Timing = t(Start, End, Timelist,TimeStep)
                 Start = the prescribed starting time
                 End = the prescribed ending time
                 Timelist = Time* (may contain any number of discrete
times)
                          Time = prescribed discrete points of time for
execution
                 TimeStep = the prescribed time interval of the execution
```

The balance checking of the "model specific conservation law–based measures" helps to recognize the infeasible states (e.g., negative amounts) that may cause serious numerical problems. The sequentially executed state and transition elements, connected by the sequentially executed standalone connections, make it possible to develop robust models, as well as to fine-tune the discretized execution in runtime (e.g., with an automatic time step control, based on feasibility bounds). In some logistic problem solving, the negative values may also be meaningful, and accordingly the time step control is switched off.

During the execution of the local state programs, the state facts are matched with their respective state prototypes. The unification of input and parameter variables is followed by the calculation of the local program, resulting in the values for the output variables.

Analogously, during the execution of the local transition programs, the transition facts are matched with their respective transition prototypes. The unification of input and parameter variables is followed by the calculation of the local program, resulting in the values for the output variables.

The applied architecture makes it possible to save the whole model (together with the actual local and global values) in a file, from which the simulation can be continued later. This feature makes it possible to continue the interrupted simulation with optionally modified parameters to test the effect of various changes. The same option also supports the stepwise communication of the simulation model with external applications (e.g., external control algorithms, additional data sources, etc.).

The simulation models, described by additive measures, are suitable for the causally right reversed (backward) simulation. In this case, the simulation can start from a realistic (or simulated) "future" state with decrementing (backward) time steps, while the increases and extensions of additive measures are replaced with decreases and removals, respectively. Similarly, instead of the decreases and removals of additive measures, increases and extensions are calculated, vice versa. It is to be noted that the functionalities are calculated with the same (i.e., causally right) programs.

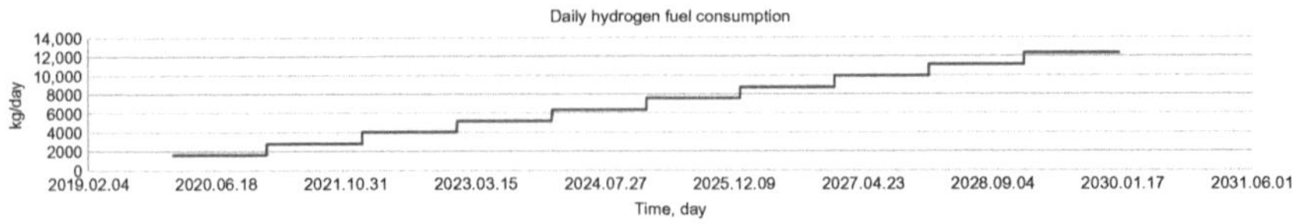

FIG. 14.6 Stepwise increasing demand for hydrogen in the investigated districts.

The detailed output is written in CSV files, to be elaborated further in the case of specifically prepared MS EXCEL workbooks.

In the following we illustrate the results from simulation of a simplified part of the fictitious HSC generated above.

In this case we start from a stepwise, yearly increase of the demand for hydrogen car fueling, as can be seen in Fig. 14.6. Suppose there is an estimated increase in the number of stations and hydrogen-powered cars in the investigated districts in the future time period from 2020 until 2030.

We assume an available capacity of photovoltaic and wind power stations that produce electrical energy, depending on the meteorological situation (solar radiation and wind speed). Unfortunately, the available climate scenarios do not contain data for radiation and wind speed. That is why we used meteorological data from the past 10 years to imitate some realistic fluctuation tendencies. The calculated change of the electrical energy generated in the photovoltaic process is illustrated in Fig. 14.7.

In Fig. 14.8, the contribution of energy generated in the photovoltaic process to the total energy demand of the electrolytic hydrogen production of the necessary car fuel is shown.

Similarly, the calculated change in the electrical energy generated by wind power is illustrated in Fig. 14.9.

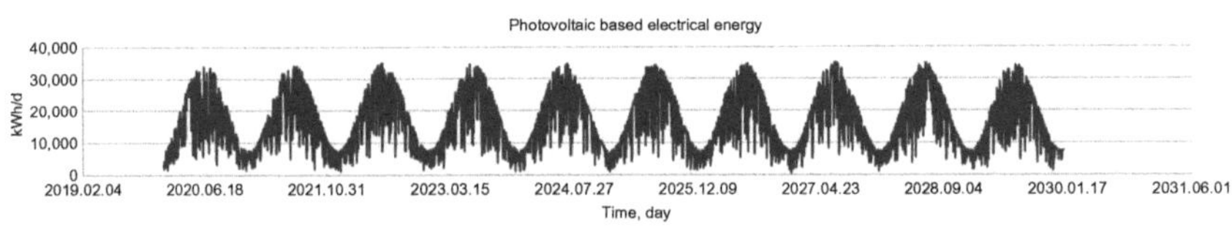

FIG. 14.7 The electrical energy generated in the photovoltaic process in the districts investigated.

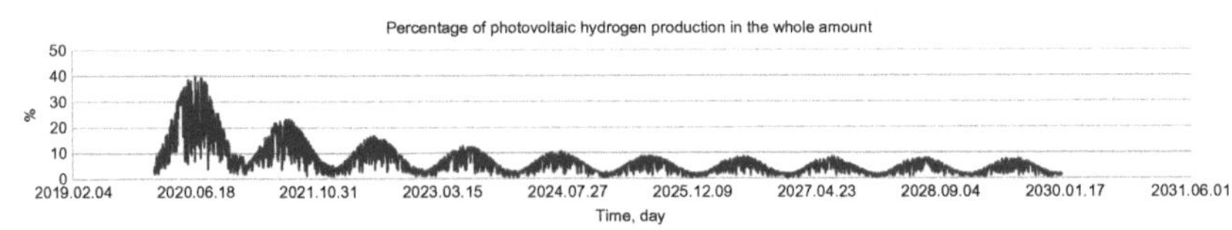

FIG. 14.8 Percentage of photovoltaic-generated energy in the total energy demand of the electrolytic hydrogen production.

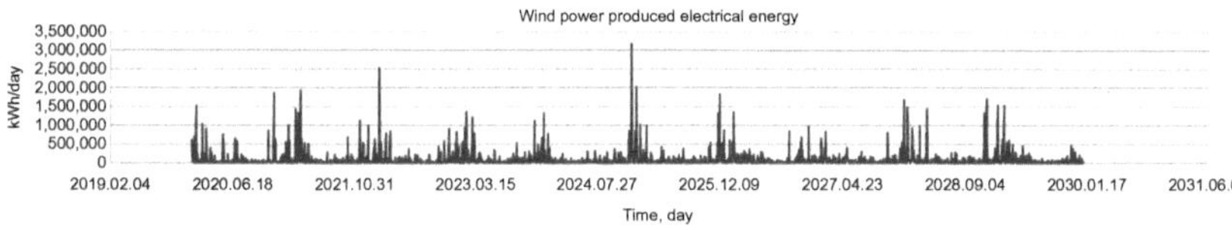

FIG. 14.9 Electrical energy generated by wind power in the investigated districts.

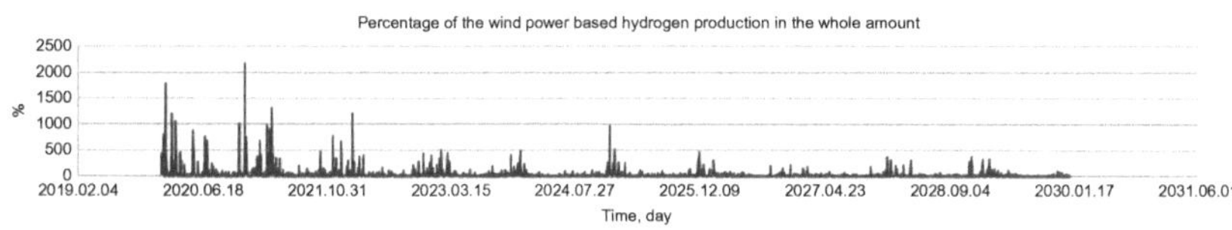

FIG. 14.10 Percentage of electrical energy generated by wind power in the total energy demand of the electrolytic hydrogen production.

In Fig. 14.10 the contribution of energy generated by wind power to the total energy demand of the electrolytic hydrogen production of the necessary car fuel is shown.

It can be seen that in the given districts, wind power can produce more electrical energy than the photovoltaic systems, however, this production fluctuates wildly. In case of lower car fuel demand (in the first years) the solar energy–based renewable resources can produce more energy, and consequently more hydrogen, than the requirement of the car fueling stations. The extraordinary surplus amounts are produced in randomly appearing short periods. The contribution of photovoltaic systems is smoother, with a characteristic seasonal change. Increasing demands for electrolytic hydrogen production requires more renewable energy production or additional electrical energy (e.g., from nuclear power stations, from Steam Methane Reforming, from biomass gasification, etc.)

14.8 REPRESENTATION OF POSSIBILITY (DESIGN) SPACE AND EVALUATIONS IN THE PROGRAMMABLE STRUCTURE

The robust simulation framework makes possible *dynamic model-based (sub) optimization*, for example, with various externally implemented metaheuristic methods (e.g., Genetic Algorithm). The basic principle, in accordance with the engineering way of thinking, is that instead of searching for an exact global optimum with a simplified model, it might be better to generate suboptimal solutions with a more detailed model. This principle was successfully applied for the solution of various design problems by the previous version of the Direct Computer Mapping–based simulator (e.g., Csukas et al., 2013).

Using Programmable Structures, the possibility (design) space and the evaluations can be adequately defined within the model elements. This extended simulation model can communicate with various external optimizing tools by means of special evaluating and configuring connections. The evaluations (i.e., calculated objective functions for the optimizing program) come from the output variables of the evaluating elements, while the new configurations (i.e., the formal suggestions of the optimizer) are processed by the configuring elements of the simulation model. The advantage of this solution is that interfacing between the detailed model and the external optimizer can be organized within the model, with the knowledge of the model-specific details. At the same time, the optimizer may work with a number of evaluating objectives and with an abstract possibility (design) space, formally. The collaboration between the simulator and the optimizer is determined by the respective evaluation feedback.

In *combining Programmable Structure of Hydrogen Supply Chains with metaheuristic optimizers*:

- the possibility (design) space and the components of evaluations are defined within the state and transition elements, while the model is extended with special configuring and evaluating elements;
- the simulator initializes the formal connection with optional optimizers at the beginning of the run;
- the (optionally in-parallel running) model(s) organize the repeated configuration and evaluation of the communication with the metaheuristic optimizer.

The *uncertainty* of model parameters and/or evaluations (e.g., cost coefficients) can be taken into consideration with an additional objective for minimizing the standard deviation of the (e.g., economic) objective, using a set of range-based, randomly generated parameters for each scenario. This method was applied for the optimization of a methane producing anaerobic fermentor of sugar beet slices (Varga, 2009).

14.8.1 Embedding Possibilities in the State and Transition Elements

The *Possibilities*, belonging to a given element, can be declared by the following data set:

```
Possibilities = Possibility*
Possibility  = pos(PosID,VariableIdentifier,Type,Mode,PossibleValues,
                   SuggestedValue)
```

where:

PosID = identifies the given modification in the possibility (design) space,
VariableIdentifier = identifies the variable to be modified in the given model element,

Type = determines the type of the variable (symbol, integer, real or timing).
Mode = defines the method of modification (select or range),
PossibleValues = defines the possibilities (alternatives) for the given variable,
SuggestedValue = the actually proposed upgraded value (initialized by a default value).

The most important configurable characteristics of model elements are the followings:

- existence of state and transition elements (Type = symbol);
- initial input measures of the state elements (Type = real);
- initial input signs of the state elements (Type = symbol);
- state parameters (Type = symbol, integer or real);
- transition parameters (Type = symbol, integer or real);
- timing of transitions: (Type = timing); etc.

In accordance with the above model properties, the possible alternatives can be defined, as follows:

- by a set of symbols to select an actual one (Mode = select);
- by a set of integers to select an actual one (Mode = select);
- by a range of integers, and its initial resolution to determine the prescribed granularity that can be refined stepwise (Mode = range);
- by a range of real numbers, and its initial resolution to determine the prescribed granularity that can be refined stepwise (Mode = range);
- by a set of possible timings in sense of Eq. 6 to select an actual one (Mode = select or range); etc.

In case of (Mode = select) PossibleValues are defined by the set of possible values, while in case of (Mode = range).

```
PossibleValues = [Min,Max,Resolution]
```

where Min = the minimal value, Max = the maximal value and Resolution = the suggested initial resolution of the range (that can be modified by the external optimizer).

At the beginning of the optimizing cycle, the domain of the possibility (design) space is translated for the external optimizer. It should be emphasized that the optimizer does not know about the meaning of the data to be modified and considers only their formal representation. Accordingly, an actual Possibility is transformed into the tuple.

```
(PosID,Number,Min,Max,Resolution)
```

where PosID = the identifier of the change, Number = the number of possible values or 0, Min = the minimal value or 0, Max = the maximal value or 0 and Resolution = the suggested initial resolution within the range or 0.

The decisions from the optimizer come in the form of:

```
(PosID,Ordinal_number,SuggestedValue)
```

where Ordinal_number = the respective element of the set or 0, Suggested Value = the suggested value from the range or 0, while these formal messages are interpreted according to PosID.

14.8.2 Embedding Elementary Evaluations in the State and Transition Elements

In addition to the very uncertain economic objectives, the optimal design and operation of an HSC requires multiple natural objectives (utilized solar energy, carbon-dioxide balance, hydrogen production of biological wastes, the food-related part of land use, etc.). Accordingly, evaluations have to be embedded into the elements of the more detailed dynamic models.

The *Evaluations*, belonging to a given element, can be declared by the following data set:

```
Evaluations=Evaluation*.
Evaluation=ev(EvalID,EvaluationName,VariableIdentifier,Mode,
Measurements).
```

where:

EvalID = an integer identifier of the corresponding objective (used by all of the respective measurements);
EvaluationName = a symbol, describing the objective function;
VariableIdentifier = a simulated variable corresponding to the measured data or nil;
Mode = mode of the use (e.g., ident, max, min, etc.); and.
Measurements = set of ($Time_i$,$Value_i$) pairs for the respective variable, while Value means the measured data at Time.

The individual evaluations are executed by the objective defining transitions that calculate the objective functions individually. Considering the above kinds of objectives, there are three modes of evaluations, as follows:

- for identification according to the supplemented measurements (Mode = ident);
- for maximization of an objective function (Mode = max);
- for minimization of an objective function (Mode = min).

At the beginning of the optimizing cycle, the identifiers of objectives are transformed for the external optimizer. It should be emphasized that the optimizer does not know about the meaning of the objective functions and considers only their formal representation. Accordingly, an Evaluation for the optimizer is described by the pair of.

```
(EvalID,EvaluationName).
```

The most important data of the model elements, used for objective evaluations are the following:

- intensive properties at the output of the state elements;
- special output data of the transition elements: for example, the reported rate of a given transformation or transportation; and
- state and transition parameters for the calculation of objective functions.

The individual evaluating components, identified by an EvalID, are collected during and/or at the end of simulation by the respective evaluating objectives. At the end of the individual simulations these special evaluating elements calculate the values of the objective functions and forward them to the optimizer in the form of:

```
(EvalID,Calculated_Objective)
```

where EvalID = the identifier of the objective and Calculated_evaluation = the value of this objective function after the simulation.

14.9 CONCLUSIONS AND FURTHER WORK

Besides the well-developed mathematical programming–based optimization methods, the design and operation of the large-scale, long-term processes of Hydrogen Supply Chains might also require easily extensible, generic dynamic simulation. We have introduced a nonconventional modeling and simulation methodology, developed in other multidisciplinary fields.

Keeping in mind the challenges of Hydrogen Supply Chain design and operation, we explained Programmable Structure for experts of the given field to promote possible future collaboration. Programmable Structure of process models can be generated from the description of a process network (optionally geographically determined and multiscale) and from two general functional metaprototypes. The case-specific, actual functional prototypes are copied from the metaprototypes, and they are associated with locally executable declarative programs in a graphical editor. Simultaneously, the actual state and transition elements are parameterized and initialized concerning their case-specific prototypes. The execution of the programmed structures is solved by a general-purpose kernel program, while the actual model elements are executed according to their prototypes. In the proposed model representation, the detailed description of possibility (design) space, as well as the components of the evaluating functions may be embedded in the state and transition elements.

The methodology has been illustrated with a fictitious and simple example in the context of Hydrogen Supply chains.

The major points of the work planned for the future, in collaboration with HSC experts, are the following:

- to test the framework for a realistic case study;

- to investigate the application of repeated forward/backward simulation based balancing algorithms;
- to combine Programmable Structures with metaheuristic optimizers for the multiobjective optimization of HSC design and operation.

ACKNOWLEDGMENT

Work was supported by the TÉT_16-1-2016-0116 French-Hungarian bilateral program.

REFERENCES

Csukas, B., 1998. In: Simulation by direct mapping of the structural models onto executable programs. AIChE Annual Meeting 1998, Miami, Paper 239/9.

Csukas, B., Balogh, S., Kovats, S., Aranyi, A., Kocsis, Z., Bartha, L., 1999. Process design by controlled simulation of the executable structural models. In: Comput. Chem. Eng. 23 (Suppl), 569–572.

Csukas, B., Varga, M., Balogh, S., 2011. Direct computer mapping of executable multiscale hybrid process architectures.Proceedings of Summer Simulation Multiconference'2011, Den Haag, The Netherlands, 2011.06.26–2011.06.29, pp. 87–95. ISBN: 1-56555-345-4.

Csukas, B., Varga, M., Miskolczi, N., Balogh, S., Angyal, A., Bartha, L., 2013. Simplified dynamic simulation model of plastic waste pyrolysis in laboratory and pilot scale tubular reactor. Fuel Process. Technol. 106, 186–200. https://doi.org/10.1016/j.fuproc. 2012.07.024.

Dagdougui, H., 2012. Models, methods and approaches for the planning and design of the future hydrogen supply chain. Int. J. Hydrog. Energy 37, 5318–5327. https://doi.org/10.1016/j.ijhydene.2011.08.041.

De-Leon Almaraz, S., 2014. Multi-objective optimisation of a hydrogen supply chain. Ph.D. Thesis, Université de Toulouse.

De-Leon Almaraz, S., Azzaro-Pantel, C., 2017. Chapter 4—Design and optimization of hydrogen supply chains for sustainable future. In: Scipioni, A., Manzardo, A., Ren, J. (Eds.), Hydrogen Economy. Academic Press, Cambridge, MA, pp. 85–120.

Govindan, K., Fattahi, M., Keyvanshokooh, E., 2017. Supply chain network design under uncertainty: a comprehensive review and future research directions. Eur. J. Oper. Res. 263, 108–141. https://doi.org/10.1016/j.ejor.2017.04.009.

Hydrogen Council, 2017. How hydrogen empowers the energy transition. Available from: http://hydrogeneurope.eu/wp-content/uploads/2017/01/20170109-HYDROGEN-COUNCIL-Vision-document-FINAL-HR.pdf. Accessed 21 September 2017.

Maryam, S., 2017. Review of modelling approaches used in the HSC context for the UK. Int. J. Hydrog. Energy 42 (39), 24927–24938. https://doi.org/10.1016/j.ijhydene.2017.04.303.

Nurjanni, K.P., Carvalho, M.S., Costa, L., 2017. Green supply chain design: a mathematical modeling approach based on a multi-objective optimization model. Int. J. Prod. Econ. 183, 421–432. https://doi.org/10.1016/j.ijpe.2016.08.028.

Petri, C.A., 1962. Kommunikation Mit Automaten (Communication with Automata). Ph.D. Thesis, University of Bonn.

Petri, C.A., 1977. General Net Theory.Proc. of 1976 Joint IBM/Univ. of Newcastle upon Tyne Seminar on Computing System Design, Newcastle upon Tyne, Computing Laboratory, 1977. pp. 131–169. http://homepages.cs.ncl.ac.uk/brian.randell/GeneralNetTheory.pdf.

Samsatli, S., Staffell, I., Samsatli, N.J., 2016. Optimal design and operation of integrated wind-hydrogen-electricity networks for decarbonising the domestic transport sector in great Britain. Int. J. Hydrog. Energy 41, 447–475. https://doi.org/10.1016/j.ijhydene.2015.10.032.

Temesvári, K., Aranyi, A., Csukas, B., Balogh, S., 2004. Simulated moving bed separation of a two components steroid mixture. Chromatographia 60, 189–199.

Varga, M., 2009. Economic optimization of sustainable complex processes (model based optimization under uncertain cost parameters for industrial scale anaerobic fermentation of sugar beet slice). PhD Thesis, Kaposvar University, Kaposvar (in Hungarian).

Varga, M., Csukas, B., 2017. Generation of extensible ecosystem models from a network structure and from locally executable programs. Ecol. Model. 364C, 25–41.

Varga, M., Csukás, B., Balogh, S., 2012. Transparent Agrifood Interoperability, Based on a Simplified Dynamic Simulation Model. In: Mildorf, T., Charvat, K. (Eds.), ICT for Agriculture, Rural Development and Environment: Where We Are? Where We will Go?. Czech Centre for Science and Society, Prague. ISBN: 978-80-905151-0-9, pp. 155–174.

Varga, M., Balogh, S., Wei, Y., Li, D., Csukas, B., 2016a. Dynamic simulation based method for the reduction of complexity in design and control of recirculating aquaculture systems. Inf. Process. Agric. 3, 146–156. https://doi.org/10.1016/j.inpa.2016.06.001.

Varga, M., Balogh, S., Csukas, B., 2016b. An extensible, generic environmental process modelling framework with an example for a watershed of a shallow lake. Environ. Model. Softw. 75, 243–262. https://doi.org/10.1016/j.envsoft.2015.10.022.

Varga, M., Prokop, A., Csukas, B., 2017. Biosystem models, generated from a complex rule/reaction/influence network and from two functionality prototypes. Biosystems 152, 24–43. https://doi.org/10.1016/j.biosystems.2016.12.005.

Wang, J., Yin, Y., 2017. Principle and application of different pretreatment methods for enriching hydrogen-producing bacteria from mixed cultures. Int. J. Hydrog. Energy 42 (8), 4804–4823. https://doi.org/10.1016/j.ijhydene.2017.01.135.

Chapter 15

Life Cycle Assessment of Hydrogen Supply Chain: A Case Study for Japanese Automotive Use

Yuki Kudoh and Akito Ozawa
Research Institute of Science for Safety and Sustainability, National Institute of Advanced Industrial Science and Technology, Tsukuba, Japan

15.1 INTRODUCTION

Hydrogen is a flexible energy carrier having the following characteristics in its supply chain. From the hydrogen supply side, it can be produced from various primary and renewable energy resources. This includes hydrogen production from fossil fuels and biomass resources by chemical processes and from renewable electricity by electrolysis of water. From this point of view, hydrogen is expected to play an important role to diversify the energy supply. Another role expected from the supply side is its use as a buffer to stabilize the electricity from intermittent renewable energy, such as wind power and solar photovoltaics, in which the excess renewable power is stored in the form of hydrogen by electrolysis of water. From the hydrogen demand side, hydrogen can be consumed by diverse end-use applications, such as fuel cells, turbines, boilers, and engines to obtain electric, thermal, and kinetic energy across all energy sectors. Because hydrogen itself contains no carbon, another advantage of using hydrogen as energy is that no carbon dioxide (CO_2) is emitted from the energy conversion processes of hydrogen use technologies.

Japan has long focused on the potential of hydrogen energy use. In 1974, right after the 1973 oil crisis, Japan formulated the first national strategic program to promote the long-term research and development of potential energy technologies, including hydrogen energy, to secure and diversify the national energy supply (Kimura and Suzuki, 2006). Since the start of the World Energy Network (WE-NET) Program (Chiba et al., 1998) in 1993, the Japanese government has been continuously implementing national projects related to the

Hydrogen Supply Chain. https://doi.org/10.1016/B978-0-12-811197-0.00015-4

hydrogen value chain and has accumulated plenty of technological know-how in hydrogen infrastructure operations, as well as hydrogen production, mass transport and storage, and utilization technologies. Hence, it is not too much to state that Japan is leading the world in the technologies surrounding hydrogen.

Given this background, hydrogen is regarded as one of the promising clean secondary energies for Japan that can contribute to diversifying the energy basket and reducing greenhouse gas (GHG) emissions. The Japanese government, industry, and academia are now working in tandem according to the Strategic Road Map for Hydrogen and Fuel Cells (Agency for Natural Resources and Energy, Ministry of Economy, Trade and Industry, 2016), which specifies the efforts to be made for the components of the hydrogen supply chain and the timelines necessary to enable realization of a future hydrogen economy. The Japanese government also plans to showcase hydrogen technology during the Tokyo 2020 Olympic and Paralympic Games and pass on a hydrogen society as a legacy of the Tokyo 2020 Games (Tokyo Organising Committee of the Olympic and Paralympic Games, 2017).

Let us place the focus on the environmental benefits of using hydrogen as energy. As described earlier, no CO_2 is generated when obtaining energy from hydrogen. However, it should be emphasized again that hydrogen is an energy carrier, but not an energy source, as it must be produced from other primary and renewable energies. In addition, a supply chain is necessary to transport hydrogen from where it is produced to where it is used. The supply chain comprises hydrogen production, storage, and distribution processes and each process requires energy and material inputs for its operation. This means that hydrogen is responsible for a certain amount of CO_2 and other GHG emissions that are attributed to the energy and material inputs to the supply chain. Hence, if we are to understand the environmental benefits of hydrogen use technology over the conventional counterpart system, the emissions from the whole hydrogen supply chain should be included in the assessment. In this context, life cycle assessment (LCA) must be conducted to evaluate the emissions associated with hydrogen use technology and its counterpart.

LCA is a methodology to evaluate the environmental performance and potential impacts associated with a product system across its entire life cycle. Among various LCA studies that evaluated the environmental benefits of using hydrogen over conventional energy sources, the so-called well-to-wheel (WtW) studies, an LCA approach to evaluate the environmental advantages of alternative fuel vehicles (AFVs) over conventional internal combustion engine vehicles across the entire automotive fuel pathway, are conducted to calculate the life cycle emissions, including the hydrogen supply chains, in the United States (Argonne National Laboratory, 2016), Europe (JEC-Joint Research Centre-EUCAR-CONCAWE Collaboration, 2014), and Japan (Toyota Motor Corporation and Mizuho Information and Research Institute Inc., 2004; Japan Hydrogen and Fuel Cell Demonstration Project Steering Committee, 2011).

However, it should be noted that WtW studies are conducted under specific assumptions, such as the energy supply structure of a certain year, system boundary settings, and the LCA methodology and database used. This means that comparison of the calculated results is valid only within the same WtW study and the observation of the same hydrogen pathway or the WtW results in different studies may not make sense. The distinctive problem surrounding the Japanese WtW studies is that the Japanese energy supply figures have drastically changed due to the Fukushima nuclear accident that occurred after the 2011 Great East Japan Earthquake and Tsunami, but the previous WtW studies (Toyota Motor Corporation and Mizuho Information and Research Institute Inc., 2004; Japan Hydrogen and Fuel Cell Demonstration Project Steering Committee, 2011) were conducted under the energy supply structure of the first decade of the 2000s. As far as the authors can ascertain, the LCA of hydrogen pathways conducted by the Mizuho Information & Research Institute (Mizuho Information and Research Institute Inc., 2016) is the only study that has assessed the life cycle GHG profiles of various hydrogen supply chains under the current energy supply structure, but most of the data used for their calculations are not provided in their report due to confidentiality agreements with their contributors. LCA should be conducted with an equal footing and clear evidence to understand the energy and environmental profiles of the hydrogen supply chains, and to discuss the role of hydrogen in diversifying the energy supply and mitigating global warming.

As a part of establishing a Japanese scenario for a future hydrogen economy, the authors have been conducting technology assessments of hydrogen to understand the role of hydrogen within the energy system and its socioeconomic effects. LCA studies were conducted to assess the environmental benefits of hydrogen use, to identify the environmental hotspots in the hydrogen supply chain, and to examine the opportunities to reduce GHG emissions across the whole supply chain. In addition to our previous results for calculating the life cycle GHG emissions from the hydrogen supply chain (Ozawa et al., 2017), case studies are being conducted for various hydrogen technologies, including the emissions from end-use applications. This chapter introduces the WtW GHG emissions results that were calculated for a Japanese fuel cell vehicle (FCV), and its conventional counterparts. First, it outlines the general procedure for calculating the GHG emissions by life cycle inventory (LCI) analysis. It then introduces the case study results for the life cycle GHG emissions that were calculated for the potential Japanese hydrogen supply chains as automotive fuel and concludes with the potential options for establishing a low-carbon hydrogen supply chain.

15.2 LIFE CYCLE INVENTORY ANALYSIS IN BRIEF

According to ISO 14040 (International Organization of Standardization (ISO), 2006), which standardizes the LCA methodology, the LCA framework comprises goal and scope definition, inventory analysis, impact assessment, and interpretation phases. In brief, the goal and scope of an LCA study are determined first. Secondly, in the inventory analysis, relevant inputs and outputs of a product system are collected and calculated throughout its life cycle. Then, in the impact assessment, the magnitude and significance of the potential environmental impacts of a product system are evaluated. Finally, the findings of either the inventory analysis or the impact assessment, or both, are combined, consistent with the defined goal and scope, in order to reach the conclusions and recommendations in the interpretation phase.

The most important thing when conducting an LCI analysis to calculate the environmental footprint is to collect all the data related to the life cycle of the target product system. As shown in Fig. 15.1, the data necessary for LCI analysis are the foreground and background data. Foreground data are the direct inputs, such as energy and materials, that are specific to the target product system, and are usually collected by those who are going to conduct the LCI analysis. Background data are not directly related to the product system but relate to all the indirect inputs that are induced by the direct inputs to the product system. Because it is difficult for those who are going to conduct LCI analysis to collect all the indirect input data, including the upstream processes data, the industry average data from LCA databases are often used as background data.

The life cycle GHG emissions, E, from a product system can be calculated by Eq. (15.1), where x_i are the foreground data of the input i to the product

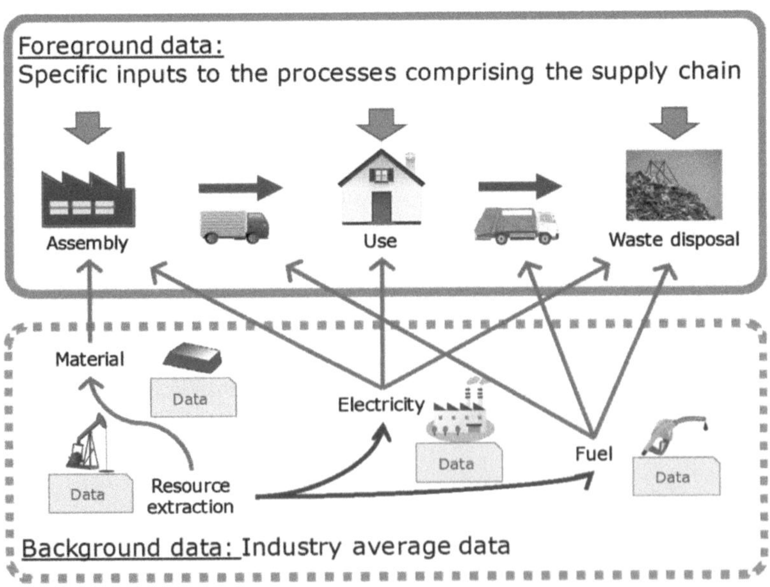

FIG. 15.1 Foreground and background data for LCI analysis.

system and e_i are the background embodied GHG emissions data of the input i from an LCA database.

$$E = \sum_i e_i x_i \tag{15.1}$$

15.3 CASE STUDY FOR JAPANESE WTW EMISSIONS

15.3.1 Overview

This case study calculated the GHG emissions from the entire hydrogen supply chain used for a FCV driven in Japan. A gasoline vehicle (GV) and gasoline hybrid vehicle (HV) were set as the conventional counterparts of the FCV, and the GHG emissions from both the gasoline supply chain and onboard gasoline combustion were calculated for the GV and HV. Fig. 15.2 outlines the target WtW pathways and their system boundary assumed in this study. In the WtW studies, the supply chain of automotive fuel from resource extraction to the vehicle tank is called well-to-tank (WtT), and the energy or environmental performance of vehicles themselves is called tank-to-wheel (TtW).

Among the potential hydrogen supply chains for Japanese FCVs, this study placed the focus upon renewable hydrogen (hydrogen produced by electrolysis using renewable power) and natural gas (NG) reforming hydrogen supply chains. Because the main target of our entire project was to investigate the potential of producing cheap, low-carbon hydrogen for a future Japanese hydrogen economy, it was assumed in this study that the renewable hydrogen was produced abroad and imported into Japan using hydrogen carriers. Thus, the renewable hydrogen supply chain comprised overseas and domestic stages. The overseas chain included renewable power generation, hydrogen production via water electrolysis by renewable power, production and storage of the hydrogen carrier, and ocean transport of the hydrogen carrier to Japan. The domestic chain involved the hydrogen carrier storage and distribution to a hydrogen station by tanker truck, restoration of hydrogen from hydrogen carrier, and hydrogen compression and fueling of a FCV. In the other study conducted as part of our project, the countries and regions that have enough potential to supply renewable electricity (wind and solar photovoltaic (PV) power generation) hydrogen production at a low cost to Japan were identified. Among them, electrolysis hydrogen from Australian wind and solar PV power, and from Norwegian wind power, were chosen for the assessment. The one-way ocean transport distance to Japan was set to 10,000 km from Australia and 20,000 km from Norway. Liquid hydrogen (LH) and methylcyclohexane (MCH) were assumed to be the hydrogen carriers of this case study, due to their fitness for long-term storage, long-range transport, and their relative ease of handling (Gupta et al., 2015). The NG reforming hydrogen supply chain comprised the overseas extraction and liquefaction of NG, ocean transport of liquefied natural gas

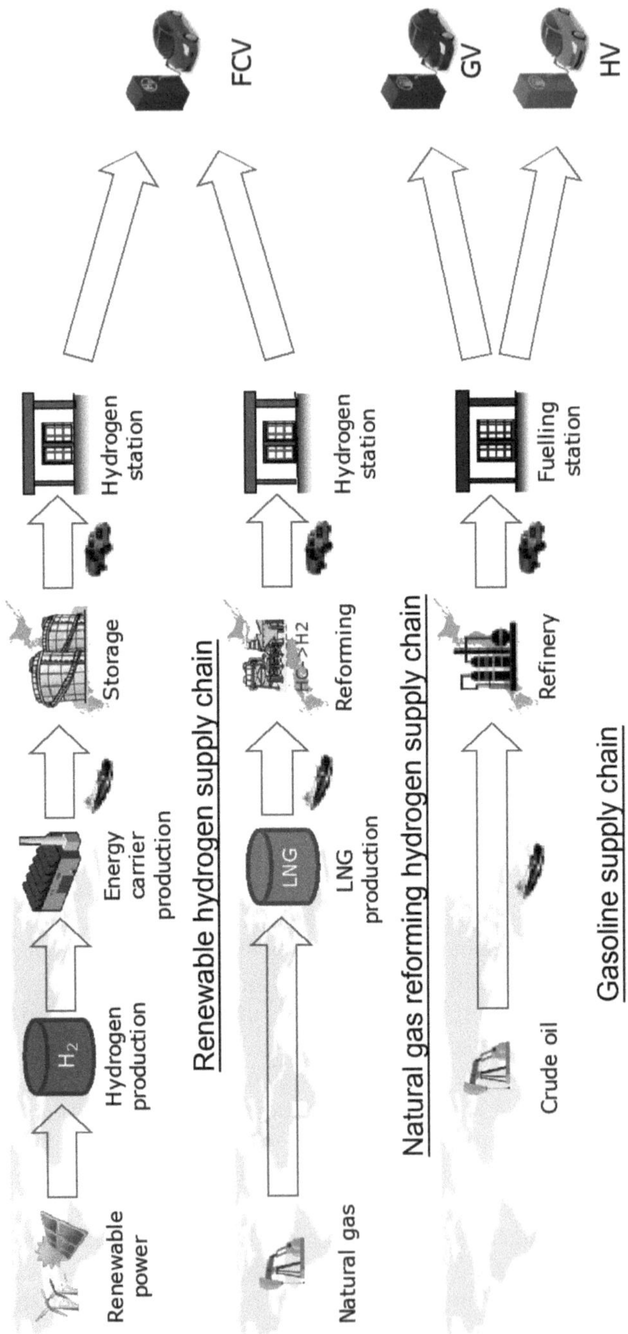

FIG. 15.2 Target WtW pathways and their system boundary outline.

(LNG), domestic hydrogen production by steam reforming of NG, which is obtained by regasification of LNG, compression, and distribution of hydrogen as compressed gaseous hydrogen (CH) to the hydrogen station by tanker truck, and hydrogen compression and fueling of a FCV. The gasoline supply chain comprised the overseas extraction of crude oil, ocean transport of crude oil, domestic refining of crude oil to gasoline, gasoline distribution, and fueling of a GV or HV.

All the foreground data were collected from previous WtW and LCI studies related to the target automotive fuel supply chains in this study. If more than two process options were found for a process comprising the supply chain, due to differences in equipment specifications, the minimum, average, and maximum values were calculated as the foreground input data. The Inventory Database for Environmental Analysis (IDEA) (National Institute of Advanced Industrial Science and Technology, 2015) developed by National Institute of Advanced Industrial Science and Technology was used as the background data.

IDEA is a Japanese LCA database that mainly uses national statistics and process data as its data source and aims to model the environmental footprints associated with all Japanese industries. Since the latest version 2.0 was released in May 2016, which covers all the items in the Japan Standard Commodity Classification, the environmental footprints of any product produced in Japan in the year 2014 are included in the IDEA. This study focused on the embodied GHG emissions of the products that were calculated as the sum of CO_2, CH_4, N_2O, HFCs, PHCs, and SF_6 converted to CO_2 equivalent using the global warming potential (GWP) over 100 years of the IPCC AR4 (Intergovernmental Panel on Climate Change (IPCC), 2007).

The embodied GHG emissions data provided in the IDEA are limited to Japanese industries. In order to approximate the overseas emissions (in this study, the GHG emissions associated with the inputs to Australian and Norwegian processes comprising the renewable hydrogen supply chain) using the IDEA, the following assumptions were made:

- The economic activity level of the target country or region was the same as in Japan. For example, if the same product was produced in Japan or Australia, all the inputs were the same in both countries.
- The inventory data for overseas transport of resources were set to zero. This meant that the approximated emissions of a product did not include those by resource overseas transport.
- All Japanese grid electricity inputs to the processes were substituted for the grid electricity in the target country or region.

It should be noted here, that in the same manner as the other WtT and WtW studies (Argonne National Laboratory, 2016; JEC-Joint Research Centre-EUCAR-CONCAWE Collaboration, 2014; Toyota Motor Corporation and Mizuho Information and Research Institute Inc., 2004; Japan Hydrogen and Fuel Cell Demonstration Project Steering Committee, 2011; Mizuho

Information and Research Institute Inc., 2016), the GHG emissions calculated in this study only focused upon the embodied emissions related to energy and material input of the process inventory data. Hence, the emissions attributed to the life cycle of infrastructure components were out of the system boundary of this study, except for the renewable power in hydrogen producing countries (details explained in Section 15.3.2).

15.3.2 Renewable Power Generation in Hydrogen Producing Countries

It was assumed that the renewable power plants (wind and solar PV in Australia, and wind in Norway) were constructed solely to supply electricity for the hydrogen supply chain. This meant that the hydrogen supply chain was responsible for all the embodied emissions of the renewable power plants. Hence, in this study, all the emissions that could be attributed to the life cycle of renewable powers (materials production, construction, transport, and operation) were included in the system boundary. Table 15.1 shows the specifications assumed and the life cycle GHG emissions from the wind and solar PV power plants used in this case study, which were calculated using the input data by the Central Research Institute of Electric Power Industry (CRIEPI) (Central Research Institute of Electric Power Industry (CRIEPI), 2016) and the IDEA database.

15.3.3 Renewable Hydrogen Production by Water Electrolysis

It was assumed in this process that hydrogen was produced from renewable power by water electrolysis using either polymer electrolyte membranes or alkaline water. The electricity input to this process was calculated as

TABLE 15.1 Specifications and the Life Cycle GHG Emissions From Wind and Solar PV Power Plants

Parameters		Wind	Solar PV
Capacity (MW)		40	10
Load factor (%)		35	18
Auxiliary power ratio (%)		10	3
Amount of electricity generated (MWh/year)		122,640	14,532
Power plant life time (year)		30	30
Life cycle GHG emissions (g-CO_2eq./kWh)	Australia	15.3	66.4
	Norway	10.8	–

4.9 kWh/Nm^3-H_2 on average, with a range of 3.7–6.5 kWh/Nm^3-H_2. The pure water requirement was set to the theoretical value of 0.80 kg/Nm^3-H_2.

15.3.4 Hydrogen Energy Carriers

15.3.4.1 Liquid Hydrogen

Fig. 15.3 shows the renewable hydrogen supply chain using LH that was assumed in this study. Assumptions given for each process comprising the supply chain were described as follows:

- LH production

The LH could be produced by adiabatic expansion of gaseous hydrogen. The amount of electricity required for this process was 0.91 kWh/Nm^3-H_2 on average, with a range of 0.55–1.3 kWh/Nm^3-H_2.

- LH storage at loading port

The LH was stored in stationary insulation tanks at a loading port. It was assumed that the boil-off hydrogen gas during this stage was liquefied back to LH. The electricity input for this stage was set to 0.055 kWh/Nm^3-H_2.

- LH ocean transport by tanker

A LH tanker with 160,000 m^3 tank capacity and 16 knots cruising speed (Mizuno et al., 2017) was assumed for LH transport to Japan. Because LH tankers are still under development, there are no available data for calculating emissions for LH ocean transport. Thus, the GHG emissions were estimated using the data for LNG tankers in the IDEA under the assumption that GHG emissions of a LH tanker per transport volume of LH expressed in ton-kilometer units were equal to those of a LNG tanker per ton-kilometer of LNG. The emissions from both laden and ballast voyages were included in the calculations. It was also assumed that the average boil-off rate of LH released to the atmosphere during the voyage is 0.30% per day, with a range of 0.20%–0.40% per day.

- LH storage at discharging port

It was assumed that the LH was transferred from the tanker to stationary tanks at a domestic discharging port. The boiled-off hydrogen gas from this stage was liquefied back to LH and stored in stationary tanks. The electricity input for this stage was set to 0.055 kWh/Nm^3-H_2.

- Domestic LH distribution by tanker truck

A LH tanker truck with a 23-kL tank capacity and 3.5 km/L-diesel fuel economy was assumed for domestic distribution from the LH storage terminal at the discharging port to a hydrogen station. The one-way distribution distance was set to 50 km. The emissions from both laden and empty trips were included to the calculation.

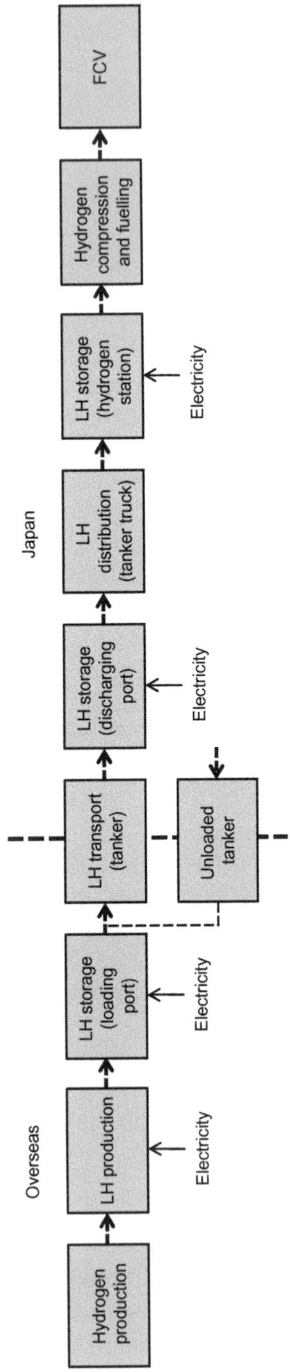

FIG. 15.3 Renewable hydrogen supply chain using LH.

- LH storage at hydrogen station

The LH was transferred from the tanker truck to a stationary tank at a hydrogen station. The electricity input for this stage was set to 0.055 kWh/Nm^3-H_2.

15.3.4.2 Methylcyclohexane

MCH ($CH_3C_6H_{11}$) is produced by the hydrogenation of toluene (TOL: $CH_3C_6H_5$) and releases hydrogen via catalytic dehydrogenation. Fig. 15.4 shows the renewable hydrogen supply chain using MCH that is assumed in this case study. After hydrogenation of TOL at the hydrogen supply site, MCH is transported by tanker and dehydrogenated at the demand site to yield H_2 and the original TOL, which is then returned to the supply site and reused. The dehydrogenation of MCH to TOL requires a large endothermic heat of reaction (205 kJ/mol-MCH or 68.3 kJ/mol-H_2). A variety of catalysts have been investigated for their ability to facilitate an efficient dehydrogenation of MCH. The assumptions given for each process comprising the supply chain are described as follows:

- MCH production

The MCH was produced by the chemical reaction between TOL and hydrogen. The reaction yield of hydrogenation to TOL and the hydrogen consumption rate were set to 99.8% and 97.9%, respectively. The electricity input for this process was 41 kWh/t-MCH on average, with a range of 7.5–93 kWh/t-MCH.

- MCH storage at loading port

The MCH was stored in corn roof tanks at a loading port. The electricity input for this stage was set to 0.92 kWh/t-MCH on average, with a range of 0.83–1.0 kWh/t-MCH.

- MCH ocean transport by tanker

An oil product tanker was assumed to be used for ocean transport of the MCH to Japan. The GHG emissions were calculated using the emissions data for the oil product tankers in the IDEA.

- MCH storage at discharging port

It was assumed that the MCH was transferred from the tanker to corn roof tanks at a domestic discharging port. The electricity input for this stage was assumed to be the same as that required for storage at the loading port.

- Domestic MCH/TOL distribution by tanker truck

A tanker truck with 20 kL tank capacity and 2.3 km/L-diesel fuel economy was assumed for domestic distribution of MCH to a hydrogen station and TOL return from the station. The one-way distribution distance was set to 50 km.

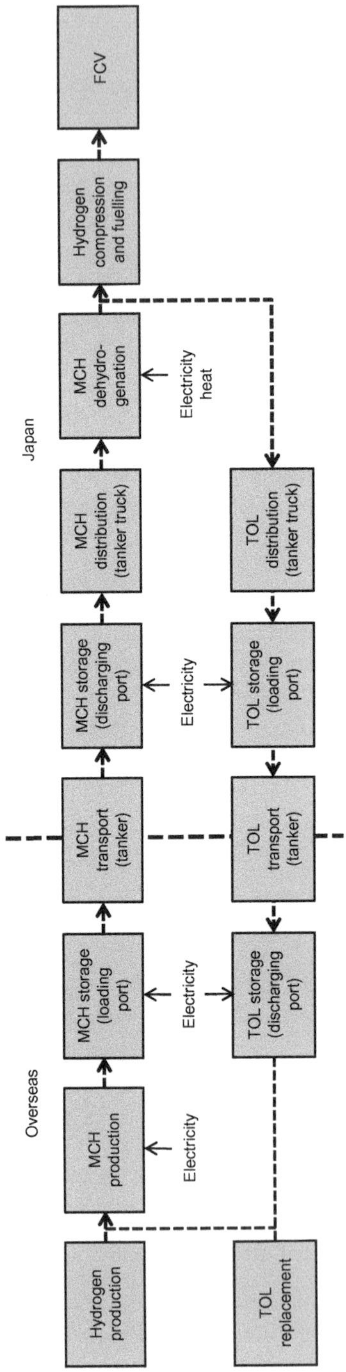

FIG. 15.4 Renewable hydrogen supply chain using MCH.

The emissions from both MCH- and TOL-laden trucks were included in the calculations.

- Dehydrogenation of MCH at hydrogen station

The MCH was converted to hydrogen and TOL by the dehydrogenation reaction at the hydrogen station. The conversion rate, selectivity, and hydrogen yield of the reaction were set to 95.0%, 99.9% and 90.0%, respectively. The electricity input for this stage was set to 0.31 kWh/t-MCH on average, with a range of 0.24–0.35 kWh/t-MCH.

- TOL storage at loading port

After the TOL was transferred from the hydrogen station to a loading port by tanker truck, TOL was stored in corn roof tanks. The electricity input for this stage was set to 0.92 kWh/t-TOL on average, with a range of 0.83–1.0 kWh/t-TOL.

- TOL ocean transport by tanker

An oil product tanker was assumed to be used for ocean transport of the TOL from Japan to the hydrogen producing countries. The GHG emissions were calculated using the emissions data for the oil product tankers in the IDEA.

- TOL storage at discharging port

It was assumed that the TOL was transferred from the tanker to corn roof tanks at an overseas discharging port. The electricity input for this stage was assumed to be the same as that required for storage at the loading port.

- TOL replacement

In the liquid organic hydride cycle of TOL and MCH, it was assumed that TOL from the supply chain could be used in the same manner as virgin TOL, so that the GHG emissions to produce the required amount of TOL were not included in the calculations. However, due to the chemical reactions, such as demethylation, isomerization, cycloreversion, and dimerization, that occur during the iterations of hydrogenation and dehydrogenation reactions, it was assumed that 3% of the initial TOL-loading should be replaced every year. The GHG emissions attributed to produce this amount of TOL were included in the calculations.

15.3.5 Supply Chain for Hydrogen Produced by NG Reforming

Fig. 15.5 shows the supply chain for hydrogen produced domestically by NG reforming assumed in this case study. The embodied GHG emissions of LNG imported to Japan provided in the IDEA were used for the upstream emissions of this supply chain. The assumptions for the rest of the supply chain were set as follows:

- Hydrogen production via steam reforming of NG

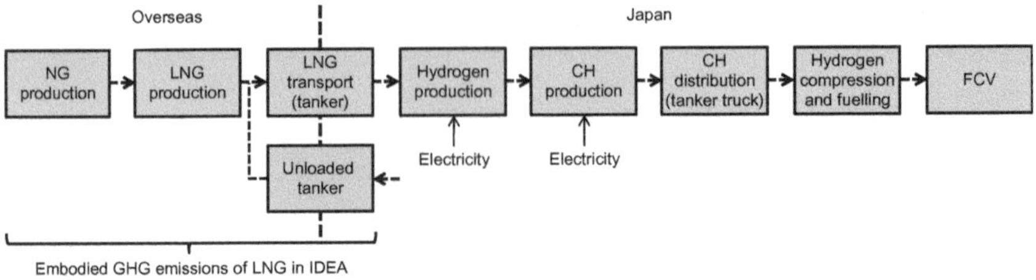

FIG. 15.5 Supply chain for hydrogen produced by NG reforming.

The gaseous hydrogen was assumed to be produced in Japan by reforming NG derived from imported LNG. The average NG input was calculated as 15.1 MJ-LHV/Nm^3-H_2, with a range of 14.8–15.3 MJ-LHV/Nm^3-H_2, while the average electricity input was calculated as 0.440 kWh/Nm^3-H_2, with a range of 0.0998–0.780 kWh/Nm^3-H_2. The process water requirement was set to 0.00196 m^3/Nm^3-H_2.

- Compressed hydrogen (CH) production

The hydrogen produced was assumed to be compressed to 20 MPa and loaded into a gas tanker truck for distribution. The electricity input for CH production was set to 0.272 kWh/Nm^3-H_2 on average, with a range of 0.119–0.440 kWh/Nm^3-H_2.

- Domestic CH distribution by tanker truck

A tanker truck was assumed for domestic distribution of CH to a hydrogen station. The average tank capacity of the truck was calculated as 2330 Nm^3-H_2, with a range of 2200–2460 Nm^3-H_2, while the average fuel economy was calculated as 2.75 km/L-diesel, with a range of 2.5–3.0 km/L-diesel. The one-way distribution distance was set to 50 km. The emissions from both laden and empty trips were included in the calculations.

15.3.6 Hydrogen Fueling of FCVs

At the hydrogen station, hydrogen was compressed to 70 MPa and pumped into the FCV tank. The electricity required for compressing and fueling hydrogen was set to 0.282 kWh/Nm^3-H_2 and 0.0928 kWh/Nm^3-H_2, respectively.

15.3.7 Supply Chain for Gasoline

Fig. 15.6 shows the supply chain for gasoline. The embodied GHG emissions of gasoline that was produced by domestic refining of imported crude oil were used for the upstream emissions of the supply chain. The assumptions for the rest of the supply chain were set as follows:

- Domestic gasoline distribution by tanker truck

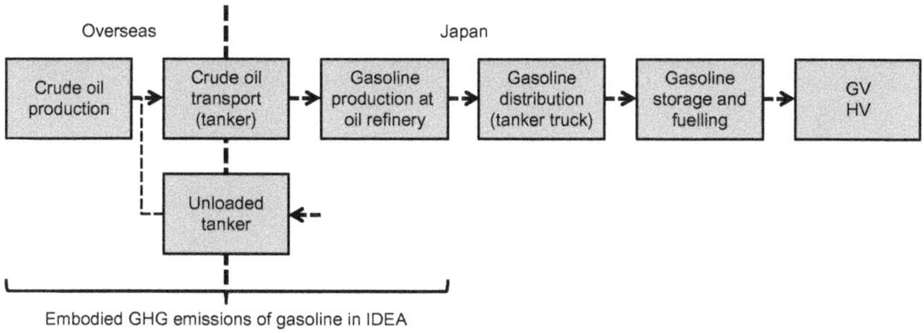

FIG. 15.6 Supply chain for gasoline.

The gasoline from oil refining was distributed and stored at a fueling station using a tanker truck with 20kL tank capacity and 2.34km/L-diesel fuel economy. The one-way distribution distance was set to 50km. The emissions from both laden and empty trips were included in the calculations.

- Gasoline fueling of internal combustion engine vehicles

At the fueling station, the stored gasoline was pumped into the fuel tank of GVs and HVs. The electricity input for gasoline fueling was set to 0.0140kWh/MJ-gasoline.

15.3.8 TtW Performance of the Target Vehicles

Table 15.2 shows the TtW energy performance of the FCV, GV, and HV used in this study that was assumed for the Japanese JC08 mode type-approval test cycle (Fig. 15.7) in the Toyota MIRAI's LCA report (Toyota Motor Corporation, 2015).

TABLE 15.2 Specifications and Life Cycle GHG Emissions From Wind and Solar PV Power Plants

Vehicle Type	TtW Fuel Economy	TtW Energy Consumption
FCV	152.17km/kg-H_2	0.79MJ-LHV/km
GV	11.4km/L-gasoline	2.9MJ-LHV/km
HV	23.2km/L-gasoline	1.4MJ-LHV/km

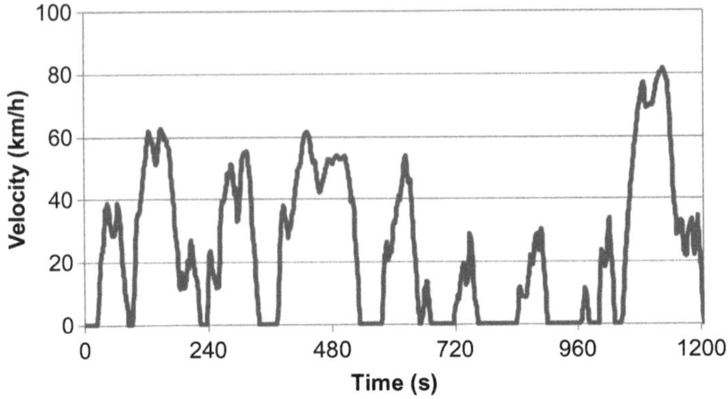

FIG. 15.7 Japanese JC08 mode test cycle. The cycle time, cycle distance, and average velocity of the cycle are 1204 s, 8.17 km, and 24.4 km/h, respectively.

15.4 RESULTS AND DISCUSSION

15.4.1 WtT GHG Emissions for Hydrogen Carriers

15.4.1.1 Liquid Hydrogen

For the renewable hydrogen supply chain using LH as the hydrogen carrier, it was assumed by default that all the electricity inputs to overseas processes were supplied with the grid electricity of the hydrogen producing country, except for the hydrogen production process by water electrolysis using renewable electricity. As shown in Table 15.3, the technological opportunity of implementing a low-carbon electricity case was assumed to reduce the supply chain emissions by utilizing the same renewable electricity as the water electrolysis to all the overseas processes.

Fig. 15.8 shows the average WtT GHG emissions of imported renewable hydrogen using LH. Because a large amount of the electricity required in the LH production process was used to cool the hydrogen below its critical point of 33 K, the emissions from this process in the Australian default cases became

TABLE 15.3 Configuration of Default and Low-Carbon Electricity Cases for Hydrogen Supply Chain Using LH

Case	Description
Default case	Electricity for hydrogen production by water electrolysis was supplied with renewable electricity, while the electricity used in the other overseas processes was supplied with the grid electricity of the hydrogen producing countries
Low-carbon electricity case	Electricity used in all the overseas processes was supplied with the same renewable electricity as the water electrolysis

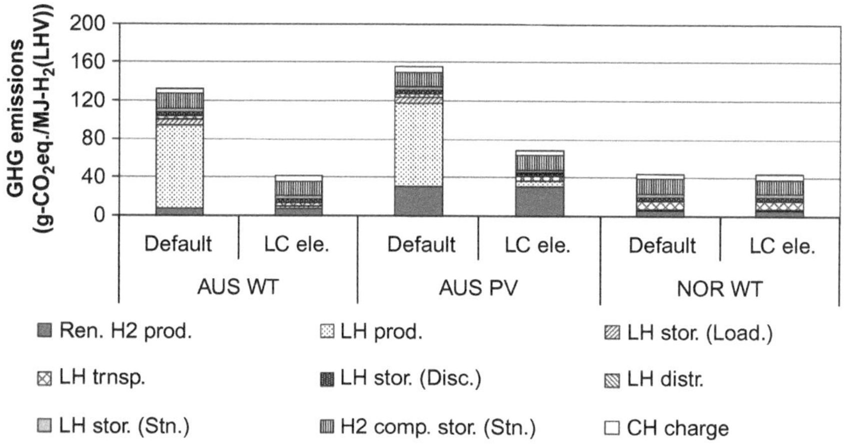

FIG. 15.8 WtT GHG emissions from imported renewable hydrogen using LH.

dominant (66% and 56% of the total for the wind and solar PV supply chains, respectively) due to the high GHG emissions intensity of grid electricity, which depends strongly on coal-fired power generation. By implementing a low-carbon electricity case in the Australian supply chains, the total GHG emissions would be reduced by 69% and 56% for wind power and solar PV cases, respectively. On the other hand, the main source of electricity in Norway is hydropower, and there was only a slight difference between the GHG emissions of wind power and grid electricity. Thus, the reduction effect of the Norwegian low-carbon electricity case only accounted for 1% of the default case emissions.

15.4.1.2 Methylcyclohexane

As described earlier, a large amount of heat is required to release hydrogen from MCH in the hydrogen supply chain, when using MCH as the hydrogen carrier. It was assumed by default that the heat required for dehydrogenation was supplied by combustion of city gas. As one of the options to reduce GHG emissions from this supply chain, a waste heat case that used waste heat instead of city gas combustion was configured as shown in Table 15.4.

TABLE 15.4 Configuration of Default and Waste Heat Cases for Hydrogen Supply Chain Using MCH

Case	Description
Default case	Heat required for dehydrogenation at hydrogen station was supplied by combustion of city gas
Waste heat case	Heat required for dehydrogenation at hydrogen station was supplied by waste heat from nearby plants or facilities

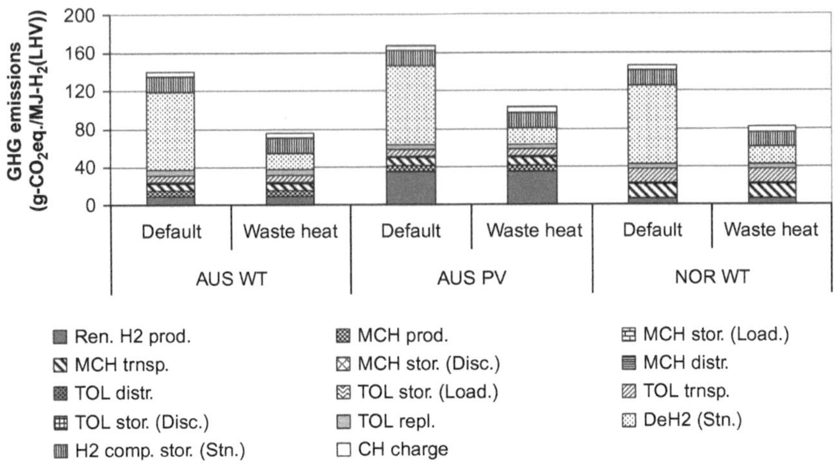

FIG. 15.9 WtT GHG emissions from imported renewable hydrogen using MCH.

Fig. 15.9 shows the average WtW GHG emissions of imported hydrogen using MCH. It can be confirmed that the emissions from dehydrogenation at the hydrogen station were dominant, which accounted for 47%–59% of the total. If the dehydrogenation process at the hydrogen station was designed to take advantage of waste heat utilization, it was expected that the emissions could be reduced by 46%, 39%, and 44% from the default case for Australian wind power, Australian solar PV, and the Norwegian wind power cases, respectively.

15.4.2 Variation of Hydrogen WtT GHG Emissions

Fig. 15.10 shows the variation in hydrogen WtT GHG emissions due to combinations of hydrogen sources and carriers. The bar chart shows the calculated average (same as Figs. 15.8 and 15.9) and the error bars representing the potential emissions range (minimum and maximum values) due to differences in the process inventory data included in the supply chain. It can be confirmed that the WtT GHG emissions from any of the renewable hydrogen pathways assumed in this study could be lower than those from NG reforming hydrogen, if the low-carbon electricity and waste heat cases were implemented for a renewable hydrogen supply chain using LH and MCH, respectively.

15.4.3 WtW GHG Emissions of the Target Vehicles

Fig. 15.11 illustrates the WtW GHG emissions of the FCV using hydrogen supplied with different supply chains, and of the HV and GV. It can be confirmed that the GHG emissions from the FCV were smaller than from the GV. However, it should be noted that the environmental advantage of the FCV over the HV in terms of WtW GHG emissions depends upon the selection of a

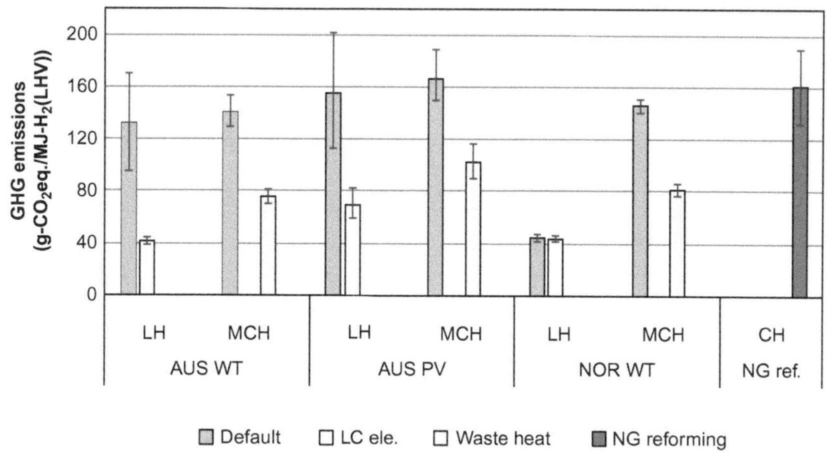

FIG. 15.10 WtT GHG emissions variation of hydrogen.

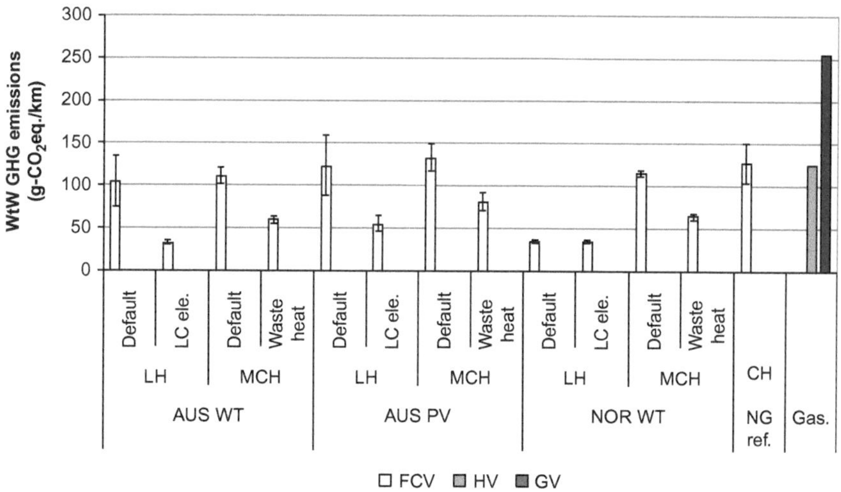

FIG. 15.11 WtW GHG emissions variation of the target vehicles.

low-carbon hydrogen supply chain and the TtW performance of both vehicles. Given the TtW performance of the vehicles shown in Table 15.2 and the calculated WtT GHG emissions from gasoline, it was calculated that the WtT GHG emissions from hydrogen should be lower than 158 g-CO_2eq./MJ, if the FCV was to have a GHG advantage over the HV.

15.5 CONCLUSIONS

In order to understand the role of hydrogen in mitigating GHG emissions in the automobile sector, a Japanese WtW analysis was conducted to evaluate the

GHG emissions profile of a FCV supplied with imported renewable hydrogen and domestically-produced NG reforming hydrogen and compared with the WtW emissions of a HV and GV as conventional counterparts. A LCI analysis was conducted using the energy and material input data to the processes comprising the automotive fuel pathways as the foreground data, and the Japanese LCA database IDEA as the background data.

The results indicate that in terms of GHG emissions the FCV had the advantage over the GV, but whether the FCV could prevail over the HV depends upon the choice of low-carbon hydrogen supply chain and TtW performance. It was identified that the main GHG hotspots of the renewable hydrogen pathways were hydrogen liquefaction for the LH supply chain and dehydrogenation for the MCH supply chain. Thus, innovations in technology development and process design are indispensable for the renewable hydrogen supply chain from overseas, if hydrogen is intended to contribute to GHG mitigation as one of the promising automotive fuels. Another important technological opportunity for the FCV to compete with the GHG profile of the HV is to improve energy consumption performance of the FCV.

In the context of using hydrogen as a secondary energy in the energy system, there are other technological options to produce hydrogen from other resources and transport hydrogen using other hydrogen carriers. In Japan, for example, another potentially feasible option is to produce hydrogen from brown coal with the combination of carbon capture and storage technologies (The Australian, 2016). Using ammonia as a hydrogen carrier or energy is another potential technology to reduce GHG emissions (Ammonia Energy, 2017). Conducting LCI studies of these hydrogen supply chains should prove useful in providing technological opportunities to reduce the environmental footprints from the life cycle perspective.

ACKNOWLEDGMENT

This study was supported by "Advancement of Hydrogen Technologies and Utilization Project" funded by the New Energy and Industrial Technology Development Organization (NEDO).

REFERENCES

Agency for Natural Resources and Energy, Ministry of Economy, Trade and Industry. Compilation of the revised version of the strategic roadmap for hydrogen and fuel cells. http://www.meti.go.jp/english/press/2016/0322_05.html; 2016 (Accessed 17 September 2017).

Ammonia Energy. Japanese Cabinet Office Holds Energy Carriers Symposium. http://www.ammoniaenergy.org/japanese-cabinet-office-holds-energy-carriers-symposium/; 2017 (Accessed 23 September 2017).

Argonne National Laboratory. The greenhouse gases, regulated emissions, and energy use in transportation model. http://greet.es.anl.gov; 2016 (Accessed 17 September 2017).

Central Research Institute of Electric Power Industry (CRIEPI), 2016. Comprehensive Assessment of Life Cycle CO_2 Emissions From Power Generation Technologies in Japan. CRIEPI, Tokyo, Japan [in Japanese].

Chiba, M., Arai, H., Fukuda, K., 1998. WE-NET: Japanese hydrogen program. Int. J. Hydrogen Energy 23 (3), 159–165.

Gupta, R., Basile, A., Angelo, B., Veziroglu, T.N., 2015. Compendium of Hydrogen Energy: Hydrogen Storage, Distribution and Infrastructure. Woodhead Publishing, Cambridge, UK.

Intergovernmental Panel on Climate Change (IPCC), 2007. IPCC fourth assessment report: climate change 2007. Geneva.

International Organization of Standardization (ISO), 2006. ISO 14040: 2006 Environmental Management—Life Cycle Assessment—Principles and Framework. ISO, Geneva.

Japan Hydrogen & Fuel Cell Demonstration Project Steering Committee. Well-to-wheel efficiency study commission, analysis of total efficiency and GHG emissions. http://www.jari.or.jp/portals/0/jhfc/data/seminor/fy2010/pdf/day1_E_20.pdf; 2011 (Accessed 18 September 2017).

JEC-Joint Research Centre-EUCAR-CONCAWE Collaboration, 2014. JEC "WELL-to-WHEELS report" version 4.a.

Kimura O, Suzuki T. 30 years of solar energy development in Japan: co-evaluation process of technology policies, and the market. http://userpage.fu-berlin.de/ffu/akumwelt/bc2006/papers/Kimura_Suzuki.pdf; 2006 (Accessed 17 September 2017).

Mizuho Information & Research Institute Inc., Evaluation report on GHG emission from hydrogen from life cycle perspectives. https://www.mizuho-ir.co.jp/publication/report/2016/pdf/wttghg1612.pdf; 2016 [in Japanese] (Accessed 18 September 2017).

Mizuno, Y., Ishimoto, Y., Sakai, S., Sakata, K., 2017. Economic analysis on international hydrogen energy carrier supply chains. Jpn Soc. Energy Resour. 38 (3), 11–17 [in Japanese].

National Institute of Advanced Industrial Science and Technology. IDEA (Inventory Database for Environmental Analysis). http://idea-lca.jp/index.html>; 2015 (Accessed 18 September 2017).

Ozawa, A., Inoue, M., Kitagawa, N., Muramatsu, R., Anzai, Y., Genchi, Y., et al., 2017. Assessing uncertainties of well-to-tank greenhouse gas emissions from hydrogen supply chains. Sustainability 9 (7), 1101.

The Australian. Japanese to use Latrobe brown coal to produce green hydrogen. http://www.theaustralian.com.au/business/mining-energy/japanese-to-use-latrobe-brown-coal-to-produce-green-hydrogen/news-story/22115a61770d6f25a798e0a629839ca2; 2016 (Accessed 23 September 2017).

Tokyo Organising Committee of the Olympic and Paralympic Games. Tokyo 2020 Olympic and Paralympic Games sustainability plan version 1. https://tokyo2020.jp/en/games/sustainability/data/20170130-sus-plan-1-EN.pdf; 2017 (Accessed 17 September 2017).

Toyota Motor Corporation. The MIRAI life cycle assessment report for communication. http://www.toyota-global.com/sustainability/environment/challenge2/lca-and-eco-actions/pdf/life_cycle_assessment_report.pdf; 2015 (Accessed 19 September 2017).

Toyota Motor Corporation and Mizuho Information & Research Institute Inc., 2004. Well-to-wheel analysis of greenhouse gas emissions of automotive fuels in the Japanese context–Well-to-Tank report. http://www.mizuho-ir.co.jp/english/knowledge/report/pdf/wtwghg041130.pdf. (Accessed 18 September 2017).

Chapter 16

Risk Analysis of Complex Hydrogen Supply Chains

Frank Markert* and Zaza Nadja Lee Hansen†
*Department of Civil Engineering, Section of Building Design, Denmark's Technical University, Kongens Lyngby, Denmark, †Department of Management Engineering, Division of Management Sciences, Denmark's Technical University, Kongens Lyngby, Denmark

16.1 INTRODUCTION

This chapter is about methods to assess safety risks of the future hydrogen-based infrastructure in connection with the design, deployment, and operation of a hydrogen supply chain (HSC). The societal goal is to establish an inherently safer hydrogen-based economy embracing new production and storage facilities as well as transportation. It is part of the broader goal to overcome the inherent problems of an unsustainable crude oil-based energy supply.

Hydrogen production can be achieved by several means and may utilize various energy sources. It is therefore rather flexible as it is not dependent on a certain energy source. As the hydrogen economy is assumed to be a part of the goal to overcome an unsustainable energy supply, the known renewable sources for hydrogen production are of great interest, for example, wind and solar energy. Therefore, hydrogen is regarded as one of the future sustainable fuels for mobile and stationary applications. Hereunder, hydrogen technologies are part of scenarios to store electric power produced by unsteady and fluctuating energy sources, as is the case for wind and solar energy. The benefits of hydrogen are that it is a carbon free energy vector and it is not a greenhouse gas. Another benefit is that hydrogen is an excellent fuel for fuel cells. Therefore, hydrogen can both be produced by electrical power and produce electrical power. This makes hydrogen an appropriate storage medium for electrical power with its ever-increasing demand. This also implies the need to develop new large-scale infrastructure with new connecting supply chains.

Nevertheless, hydrogen technologies are new and under development, which provides some uncertainty with regard to the reliability and robustness of the emerging technology. New and developing technical systems need time to mature and hence may provide some unknown aspects with regard to the

Hydrogen Supply Chain. https://doi.org/10.1016/B978-0-12-811197-0.00016-6

safety risks in the establishment phase. Hydrogen's physical-chemical properties are different from ordinary fuels, such as gasoline, that are common in our daily life. This implies a necessity to rethink the codes for establishing the new infrastructure. It is a learning process for professionals and lay people to handle the new fuel appropriately. Both aspects, the reliability and robustness of the emerging technology and hydrogen's different physical-chemical properties, are important to address in order to develop the best possible future infrastructure and to continue to develop a more inherently safer society relative to the present one. Thus, the overall goal should be to continually reduce the safety risks toward a minimum.

This is not a new thought limited to hydrogen technologies, but is valid for all technologies, and in particular, for energy technologies. It has and always will be a challenge to safely process energy, as it needs appropriate control. Failures in the control may lead to accidents and incidents that may have destructive potential to vulnerable objects, such as people, the environment, and property. This is the essence of the uncontrolled flow of energy (UFOE) model. Energy in this context is to be understood very widely as, for example, kinetic energy, electric energy, heat and radiation, as well as "toxic flow of materials" (Rasmussen and Grønberg, 1997). The UFOE model was originally designed to be used to support accident scenario development in emergency training but is considered to be applicable to also describe HSC-related accidents (Fig. 16.1).

Historically, the application of modern energy using wood, coal, crude oil, natural gas, and LPG has been a long process and throughout the development, many lessons had to be learned as many small, large, and even catastrophic accidents have occurred. For instance, the very beneficial use of furnaces and open fires in our towns has caused substantial and even catastrophic losses. Over the

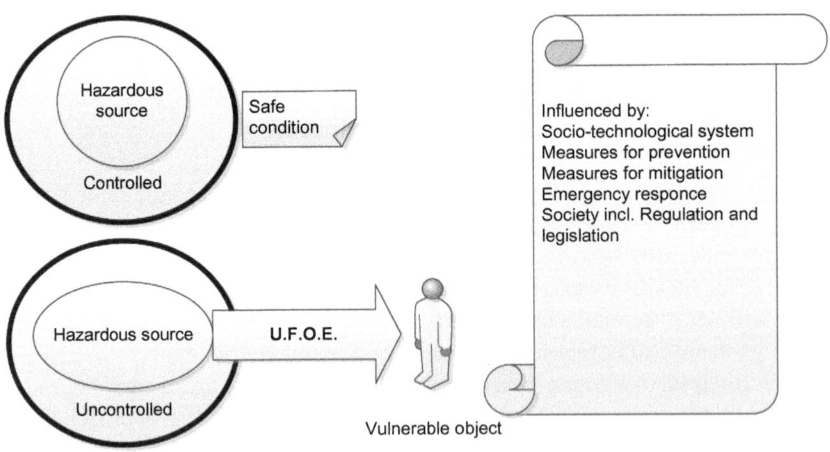

FIG. 16.1 A simple accident model uncontrolled flow of energy.

centuries, disastrous fires occurred in many towns worldwide. They could destroy large parts of towns and cause many fatalities. Well-known examples from historic times are the London fire in 1666 and the Copenhagen fires in 1728 and 1795, to name only a few (Anonymous, 2017b). Triggered by such events, prescriptive fire codes and regulations developed successfully over the centuries to enable the safe use of this type of energy. This was the result of the lessons learned after an accident, as the technical and scientific understanding of central terms, such as "fire" and "fire spread," were only poorly understood. Only recently, a few decades ago, the scientific-based methodologies developed sufficiently to understand and predict the behavior of fires and to have a performance-based approach for the design of new infrastructure. Nowadays, our towns and infrastructure are built with a minimum of fire risks. Unfortunately, some large fire accidents still happen due to design errors, or possibly inappropriate codes and standards, as the 2017 Grenfell-Tower disaster in London (Anonymous, 2017a). The specific choice and installation of the façade's cladding and insulation materials to improve the energy demand of the building, together with other factors, are believed to be the reason behind this accident. Thus, it may be argued that efforts to reduce the building's energy demand were not balanced with an appropriate prescriptive or performance-based fire safety engineering approach. Possibly, an improved and more comprehensive fire risk management, in combination with a sustainability assessment, would have prohibited such a disaster. Any decision support should equally address both the sustainability and the safety risks of technological systems.

The future development of hydrogen technologies and supply chains have the benefit of a much higher understanding of technological systems in general and the availability of science-based methods for consequence modeling (performance-based approach as in fire safety engineering). This higher state-of-the-art knowledge of the systems reliability and the potential consequences as a result of system deviations provides a much better general understanding of causes and effects of accidents involving fire and explosions. It is an excellent basis for risk assessments providing risk-based decision support to establish the HSC. Similar risk assessments have been conducted for large-scale supply chains for LPG and other chemicals (Molag et al., 2004). Parallel to the development of hydrogen applications, prenormative safety risk research for hydrogen technologies is performed within the network activities of NOE HySafe (Thomas et al., 2011) and its successor the International Association HySafe (https://www.hysafe.info). Such research provides important input to the standardization activities. It also provides knowledge support to many different safety-oriented projects concerned with the understanding of the risks of hydrogen and with guidance to establish safer systems (MacIntyre et al., 2007; Luis Aprea, 2008; LaChance et al., 2009; Watanabe et al., 2007; Duijm and Markert, 2009; Marangon et al., 2007), to mention just a few of these activities.

Nevertheless, additional effort is needed, as the HSC grows more and more complex with the developing infrastructure. Therefore, the understanding of the complex system may become insufficient as the possible number of cross-influences increases and the analyst has to face the huge challenge of assessing and evaluating the very large amount of data that a safety risk assessment of complex systems provides. This is discussed by Rasmussen (1997) on the level of risk management comprising a socio-technical system, that is, the technical system together with the legislators, managers, work planners, and system operators. He argues that a system model cannot be built by a bottom-up approach but requires a top-down system-oriented approach based on control theoretic concepts. The reason for this is that *"[...] a system is more than the sum of its elements. Often we found that attempts to improve the safety of a system from models of local features were compensated by people adapting to the change in an unpredicted way."* (Rasmussen, 1997, p. 184). It is similarly argued by Haimes (2018) that common risk assessment of single systems is insufficient at a certain stage of complexity, for instance, for systems of systems. For appropriate risk assessment and management of complex systems, Haimes proposes to extend his 10 guiding principles of risk assessment of single systems to be beneficial for complex systems (Haimes, 2018), as listed in Table 16.1.

This complexity of systems seems to have not been sufficiently addressed in the HSC research field. HSC modeling and analysis provide important decision support for establishing the hydrogen economy. Current research on the HSC only partly focuses on developing a holistic model for supply chain analysis to predict safety risks in the energy sector. There are, though, many different modeling approaches for hydrogen supply chains, as discussed in the reviews by De-León Almaraz et al. (2013, 2014, 2015).

Early research into hydrogen supply chains tended to focus on individual technologies of the supply chain, such as production, storage, and distribution, and also focused on specific areas, for example:

- A Southern California case study to develop a hydrogen vehicle refueling infrastructure (Ogden, 1999; Ogden et al., 1999).
- The feasibility of developing an initial hydrogen infrastructure for refueling hydrogen buses in London, and whether this infrastructure might provide a sufficient and suitable platform for private vehicles (Joffe et al., 2004).

Other authors have used mathematical models to describe and integrate all components of a hydrogen supply chain within a single framework. Examples include:

- The integration of production planning and reactive scheduling for the optimization of a hydrogen supply network (van den Heever and Grossmann, 2003).
- Design of a hydrogen supply chain and creation of a single framework for such a design and analysis of the important tradeoffs in such a supply chain, using OR methods to optimize the supply chain design (Almansoori and Shah, 2006).

TABLE 16.1 Risk Analysis According to Haimes' 10 Guiding Principles and Adoption to Complex Systems of Systems (Haimes, 2018)

Principal		Adaption for Complex Systems
1	Holism is the common denominator that bridges risk analysis and systems engineering	For interdependent and interconnected complex systems of systems the holistic approach must account for the impacts of adverse initiating events on systems with multiple shared states, coping with multiple objectives of the systems and taking account for multiple time horizons associated with each subsystem and the hole system of systems
2	The process of risk modeling, assessment, management, and communication must be methodical, disciplined, systemic, integrated, and commensurate in its comprehensiveness with the criticality of each subsystem and the entire systems of systems	The basic questions central to quantitative risk analysis need updating: (1) What can go wrong? (2) What is the likelihood? (3) What are the consequences? The update need to address the complexity to search for and to understand the nature, configurations, and levels of the interdependent and interconnected complex subsystems
3	Models and state variables are central to quantitative risk analysis	Modeling require the utmost understanding and appreciation of the critical role of shared states by risk practitioners engaged in decision making under risk and uncertainty
4	Multiple models are required to present the essence of the multiple perspectives of complex systems of systems	Central principle: Complex systems of systems cannot adequately be modeled and represented from a single perspective
5	Meta-modeling and subsystems integration must derive from intrinsic states of the system of system	The task is interdisciplinary and the modeler has to learn from other contributors. Thus, he or she has to develop a coherent methodological process that builds on what we know and extend this knowledge forward
6	Multiple conflicting and competing objectives are inherent in risk management	Individuals and organizations often have to deal with multiple, competing, and conflicting objectives, which is a main

Continued

TABLE 16.1 Risk Analysis According to Haimes' 10 Guiding Principles and Adoption to Complex Systems of Systems (Haimes, 2018)—cont'd

Principal		Adaption for Complex Systems
		characteristic of complex systems. Balancing the outcomes of decisions is needed, for example, maximizing the benefits from risky actions, and minimizing the cost resulting from associated risk management
7	Risk analysis must account for epistemic and aleatory uncertainty	This principle is one of the most difficult to identify, to model and address for complex systems of systems
8	Risk analysis must account for risks of low probability with extreme consequences	The common metric of risk: likelihood times consequences. This practice has played a decisive role in dangerously masking the criticality of extreme and catastrophic events. It is important to recognize the averaging of risk (a misuse) when it is used as the sole criterion for risk in decision-making
9	The time frame is central to quantitative risk analysis	The role of the time frame is probably the most important, yet the least recognized, in risk modeling, assessment, management and communication of complex systems of systems, given that each system or subsystem will be commonly driven, or affected by different adverse initiating events
10	Risk analysis must be holistic, adaptive, incremental, and sustainable; and it must be supported with appropriate data collection, metrics with which to measure efficacious progress, and criteria on the basis of which to act	This principle should be the sine qua non for all risk analysts, especially when addressing complex systems of systems

- A case study of a future hydrogen supply chain for Korea to develop a stochastic model to take into account the effect of the uncertainty in the hydrogen activities and examine the total network costs of various configurations of a hydrogen supply chain in an uncertain environment for hydrogen demand (Kim et al., 2008).

Presently, systems safety in the energy sector has been addressed by Caputo et al. (2011). They found a high safety cost for long-range hydrogen transport through densely populated regions. Also Kim and Moon (2008) predicted the safety costs for an optimized Korean infrastructure partly based on renewable energy and Dayhim et al. (2014) implemented risk costs into a multiperiod optimization model with the objective function "minimization of the total daily social cost" of a hydrogen supply chain network. Other discussions focus on topics, such as the potential growth of supply chain networks, optimizing the investment and running costs, and calculating the environmental impacts, by assessing single impacts, for example, the carbon dioxide reduction potentials using energy models (see e.g., Agnolucci, 2007; Andrews and Shabani, 2012).

In the following, a number of methods and emerging ideas for risk assessment of complex systems are presented. They partly relate to Haimes' principals and follow the basic idea of a holistic approach. Hereunder, the methods will show a new approach for establishing meta-models using the concept of "functional modeling" that can support collection and storage of data for systems under development. The concept supports cross-disciplinary assessments needed for a holistic safety and sustainability evaluation. Furthermore, a method is discussed to better handle dynamic and time related events using discrete event simulation. Finally, a case study provides an example of an integrated modeling and simulation of an HSC, including dynamic event simulation.

16.2 LAYOUT OF A MODEL HSC

This section describes a basic hydrogen supply chain taking various modes of production, storage, and transport into account. Hydrogen is produced from various energy sources and is therefore a common platform for many other applications. It is also assumed to be a link between the energy supply sector providing power to industries and households and the transport sector providing fuel and power to fulfill these transport needs. This scenario defines a complex infrastructure as it involves many different processes with many different process plants. It involves complex supply chains to transport and store hydrogen for the retailers and customers. This makes the HSC a complex system of systems.

Part of the infrastructure will be hydrogen production by electrolysis. This may utilize wind power, but hydrogen is not bound to be produced in close proximity to the large wind parks as the electrical power may be cost efficiently transported to other locations and the needed water is available everywhere. Thus, scenarios to distinguish between large-scale remote site production and small-scale on-site (e.g., at hydrogen refueling stations) may be considered (Markert et al., 2007).

Planning such an infrastructure is a challenge and it will need the application of appropriate HSC models to ensure an optimized development. This includes the application of methods to assess the safety risks of such a new and complex supply chain. Such assessment should be part of wider optimization modeling to find a truly optimized sustainable supply chain that comprises the environment, the economy, and the social aspects, which include the potential safety risks. Hereunder, as any complex technology change will not be completed overnight, it is essential to assess the new supply chain with its technologies and applications together with the old one and to ensure a smooth transition from one technology to the other.

In the following, a specific case scenario is drawn to facilitate the discussion of the methods to assess complex safety risks of the new hydrogen economy. This discussion will use a simplified scenario in which the components of a hydrogen supply chain are divided into four categories, including production, storage, transportation, and delivery to the end users. Such an HSC can be seen in Fig. 16.2. As mentioned before, hydrogen may be produced by several means. The most mature technology is steam reforming using hydrocarbons. This technology, though, is not to be considered a sustainable hydrogen production method but enables production of hydrogen in cost-efficient large quantities.

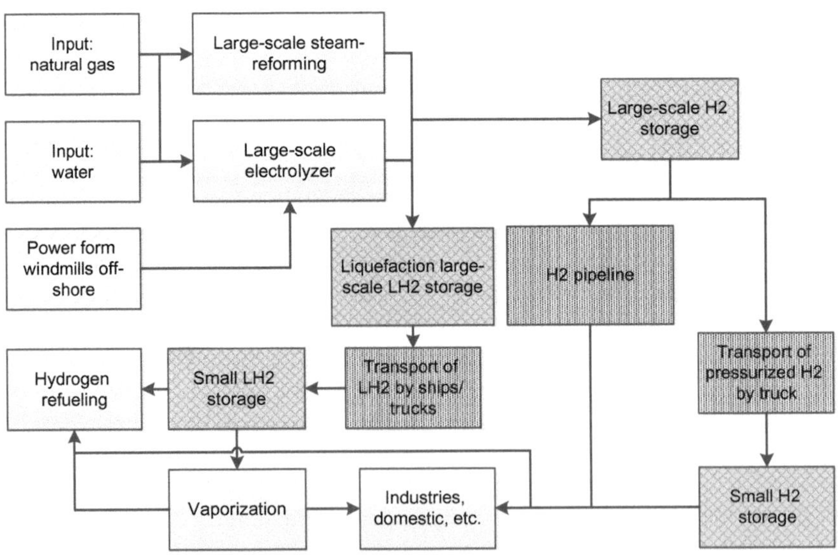

FIG. 16.2 An example of a HSC (Markert et al., 2017).

A sustainable production method is included in the model using wind energy to electrolyze water.

The model assumes that in the introduction phase of the hydrogen economy the demand will be low and therefore hydrogen will preferably be transported in pressurized containers or be produced in decentralized small-scale units placed at the refueling stations. In the long term, the demand for hydrogen is assumed to increase and therefore hydrogen is assumed to be produced in large-scale centralized units and then transported as liquefied hydrogen to the refueling stations by trucks, rail, or pipelines. The last is shown in Fig. 16.2.

Decentralized hydrogen production can take advantage of the existing and widely available grids for electricity and natural gas for hydrogen production. Moreover, delocalized hydrogen production reduces or even eliminates costs linked to hydrogen transportation and delivery. However, small-scale production units require effective process control and high safety standards mainly due to the localization of the production units in inhabited areas (Drennen and Rosthal, 2007). Natural gas reforming and small-scale water electrolysis are the current processes used for small-scale hydrogen production.

Centralized production units allow large-scale production of about 750 ton day^{-1} (U.S. Department of Energy, 2016). Reforming, electrolysis, and recovery of industrial hydrogen byproduct are the most commonly used processes to produce hydrogen in large-scale quantities. The main advantage of centralized large-scale hydrogen production is the lower production cost (economy of scale). However, hydrogen needs to be transported and delivered to the end users resulting in considerable capital investments (Drennen and Rosthal, 2007).

The production of hydrogen leaves many design choices that affect the risks to the supply chain. Some of the main ones include:

- Production in urban areas or not—the closer production is, the less risk of the transportation but the greater risks associated with the production facility;
- Transport by various means (local to regional), including transport by large-scale transporters (ship, H_2 liquid, train, and other means)—transportation means and distance will also affect risks and therefore the security of supply;
- Storage facilities (large to small scale): The larger the storage facility, the less risk associated with the frequency of transportation to the facility, but greater risk will be connected to the facility itself.

Safety and security of supply are central topics in the establishment of a future hydrogen economy, as these address the basic demands of any society.

16.3 FUNCTIONAL MODELING

The future HSC will have to fulfill several objectives, as it will likely supply the fueling system for transport and the system for power production. Such a complex and large infrastructure is a challenge when conducting risk assessment and performing risk management, because of the increasing interactions between the subsystems of the overall socio-technical system (Haimes, 2018; Rasmussen, 1997). The challenge for risk analysts is to treat a number of different threads in a dynamic and emerging system, as described by Markert et al. (2017).

In order to make the necessary and appropriate strategic decisions to establish the new infrastructure, quantitative risk assessment (QRA) may be applied in Europe and worldwide. It is a valuable tool to measure and evaluate the individual and societal risk for new infrastructure, that is, for process plants, storage, and transport routes, including tunnels, for hazardous goods transport (Baesi et al., 2013; Pasman and Reniers, 2014; Vianello and Maschio, 2014). It supports risk informed land-use planning and therefore directly supports the overall goal of moving toward inherently safer systems. This is, of course, only a part of the broader decision support needed to introduce a new technology. For planning, design, and establishment of an HSC, many different aspects have to be considered, which concern energy economy, efficiency, and security of supply. It also concerns technical reliability, safety risks, and security, as well as the general sustainability of an HSC. Usually, different experts at different times conduct these assessments for a wide variety of customers. Implicitly, there is a certain time delay between these assessments. In the meantime, the hydrogen systems being evaluated may have developed in technology and may not be fully comparable. Therefore, it would be beneficial to align these different assessments in one way or another, as a result from one study may influence the assumptions of another.

Conducting a comprehensive QRA in connection with other assessments imply, of course, some serious challenges. An important one is the handling of large amounts of data and the necessary set of assumptions used to conduct the quantitative risk assessment (Markert et al., 2017). Such data also need to reflect the aleatory and epistemic uncertainties of the parameters, the models, and the systems boundaries (Leveson, 2004; Bedford and Cooke, 2001; Aven and Reniers, 2013).

The method of "Functional modeling," discussed below, shows a potential way to analyze complex systems and to create a meta-model to store data for the different assessment methods.

16.3.1 Functional Modeling of Hydrogen Supply Chains

Functional modeling is a high-level hazard identification method (Rasmussen and Whetton, 1997) that is capable of analyzing large socio-technical systems. The methodology enables identifying hazards already during the planning and design of systems in the future hydrogen infrastructure, as exemplified in Fig. 16.2. It uses a functional breakdown followed by a high-level hazard identification method.

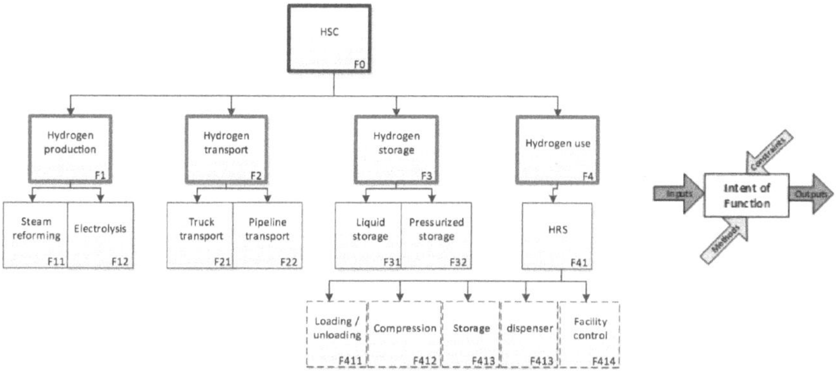

FIG. 16.3 Functional modeling of an HSC. Each function has Inputs, Outputs, Methods, and Constraints.

The idea of a functional breakdown is that a set of functions is needed to establish systems, such as a plant or HSC. The socio-technical system implements hardware, software, operations, work organization, and many other aspects. The principal of the functional modeling assumes that each function "Fx" of a system is an object that has a certain "Intent" or goal to fulfill. The function is associated with "Methods" that are necessary to establish the function and with "Constraints" that need to be regarded for safe operation.

Each method or constraint can itself similarly be regarded as an object and can be further decomposed into a hierarchy of other lower level intents. This is shown in Fig. 16.3 and in Table 16.2.

The starting point for the functional breakdown is the whole HSC, indicated in Fig. 16.3 as F0. It represents the HSC drawn up in Fig. 16.2. The inputs are the natural gas, water, and electricity needed to produce hydrogen. The output would here be the hydrogen that is dispersed at the hydrogen refueling station HRS.

The methods for the HRS are the production plants, storage, and means of transportation. These have constraints, for example, technical limits and potential safety risks. Each function is further detailed, as well as each method and constraint. This procedure should be continued until the system's hazardous areas can be identified with reasonable precision using hazard identification methods. Thus, already during the design stage, a comprehensive assessment of the system can be initiated and the assessment can be continued and extended when new information is available due to, for example, new methods being implemented or new constraints being discovered. Hereunder, the functional modeling comprises detailed technical and organizational solutions of the socio-technical HSC system (Markert et al., 2017).

16.3.2 Hazard Identification

Having established the functional model, the basis for several types of assessments in the framework of decision support can be provided. The first type is the

TABLE 16.2 Outcome Example of the Functional Breakdown

Code	Inputs	Intent	Method	Constraints	Outputs
F12	Electrical power Water Etc.	Hydrogen production	Electrolyzer	Max. pressure Availability of cheap power sources Hydrogen purity Etc.	Hydrogen Oxygen Etc.
F21	Hydrogen gas Engine fuel Etc.	Hydrogen transport	Truck	Max. pressure Route planning ADR regulation Etc.	Hydrogen gas Engine pollutants Etc.
F3	Hydrogen gas Energy Etc.	Hydrogen storage at large amounts	Cryogenic storage Pressurized storage	Max. pressure Temperature control Evaporation control	Hydrogen gas/liquid Engine pollutants Etc.
F4141	Data Power Etc.	(HRS) Remote control signals	Internet/software HRS safety functions Surveillance: Detection and Alarm → Decision → Action Communication Training	On-line uninterrupted power supply Knowledge on specific HRS Intercultural understanding Etc.	Control of HRS

identification of possible safety risks using the methodology of QRA. Hereunder the first step is the hazard identification, which may apply many different methods. Some of these are developed for certain purposes, as HazOp and FMEA, for example. HazOp (hazard and operability study) is developed within chemical industries and is very suitable to analyze process flows, while failure mode and effect analysis (FMEA) is an excellent method for analyzing component failures and their effects on the system (IEC, 2001, 2006). Following these basic procedures, accident scenarios are defined and more detailed analysis performed. The analysis applies methods, such as Fault Tree and Event Tree (ISO, 2010), which may be combined with additional methods, such as the safety barrier diagrams, which focus more on the safety barriers of a system and provide some advantages in terms of readability and communication to non-experts (Duijm, 2009; Duijm and Markert, 2009).

For assessing the high-level functional breakdown, general methods, such as Checklists or What-If questions, are suitable. The method of "Concept Hazard Analysis" (CHA) is applied to identify the main hazards, as suggested by Wells et al. (1993) and applied by Rasmussen and Whetton (1997). The CHA may be used in the early stages of planning and design of supply chains and need only block diagrams or preliminary process flow diagrams as input. The assessment is based on a list with generic keywords agreed upon by the group of analysts of the specific systems (see Table 16.3).

The outcome for the functional breakdown is exemplified in Table 16.2, while the correlation with the Hazard identification method is shown in Table 16.4. These results can be combined with other methods and types of assessments as shown in the following chapters.

16.3.3 Support by Geographic Information Systems

An important issue when analyzing hydrogen supply and distribution networks, is the knowledge about the specific geographical positions of the hazardous areas to evaluate social risk criteria. This is closely related to decisions on additional preventive and mitigating measures to ensure the acceptance criteria of a given installation. It is important to know the population density, the environmental vulnerability, and the location of hospitals and emergency services, among others, along the networks. For this Geographic Information System (GIS) is a very efficient and valuable tool (Verter and Kara, 2001; Rigina, 2002) as it allows for superimposing thematic maps and analyzing the population density for any geographical position, for example. It is straightforward to model the hydrogen supply and distribution networks with a GIS environment using established geographical maps and CAD drawings from the planning state of the networks.

As functional modeling defines objects (the intents), it is possible to attach one or several graphical object(s) from the GIS describing the HSC to a functional model intent to preserve its geographical position. In that way, it enables

TABLE 16.3 CHA- Examples for Generic Keywords

Category	Keyword	Category	Keyword
Flammables	Ignition Fire Explosion/ detonation	Mechanical hazards	Structural hazards Collapse, drop
Chemicals	Toxicity Corrosion Off-specification	Mode of operation	Start-up/shutdown Maintenance Abnormal Emergency
Pollutants	Emissions Effluents Ventilation	Release of material	Release on rupture Release by discharge Fugitive emissions Periodic emissions Handling/Entry
Health hazards	Chemical contact Noise Illumination	Loss of services	Electricity Water Other services
Electrical/radiation hazards	Electrical Radiation Laser	External threats	Accidental impact; drop/fall Extreme weather External interferences Loosening/vibration Sabotage/theft External energetic event External toxic event External contamination Corrosion/erosion
Thermodynamic hazards	Over-/under pressure Over-/under-temperature		

TABLE 16.4 Analysis of the Found Intents (I), Methods (M), and Constraints (C) Using Selected Keywords From CHA

Function			Concept Hazard Analysis				
Ref	**T**	**Description**	**Keyword**	**Main Variance**	**Consequences**	**Mitigation**	**Notes**
F12	M	Water electrolysis	Chemicals: corrosion	Release → Fire	Heat radiation on equipment	ATEX	
F3	I	Hydrogen storage	External: accidental impact due to obstacle collision	Structural damage: → Leakage → Insulation	Release of hydrogen/ overpressure in cryogenic system	Fences authorization to enter	
F4141	C	On-line with data connection	Mode of operation: abnormal	Off-line → Loss of control of HRS	Possible escalation of minor events	High SIL level local operation	HRS shuts automatically down on loss of data connection

the assessment of GIS databases together with the results from the functional modeling, including the results from the hazard identification.

For a quantitative risk assessment, data on the system state (amounts, pressures, temperature, etc.) could be attached as well to the graphical objects supporting consequence assessments, while respective thematic maps could provide necessary weather, population densities, and other data.

16.3.4 Combination With Methods for Sustainability Assessment

In order to establish sustainable hydrogen supply and distribution networks, there are the environmental, economic, and social aspects to be evaluated for a comprehensive decision support. Together with the evaluation of the safety aspects, such an approach provides a more comprehensive decision support for complex systems. It is intended to support decisions and to support further communication in order to discuss and reach social acceptance of emerging technologies, for example.

The decision support within sustainability assessment is commonly provided using the well-established and widely used methods within life cycle assessment (LCA) and life cycle costing (LCC). The LCA method has been standardized by several ISO standards, for example (DS/EN ISO 14001, 2015; DS/EN ISO 14040, 2008; DS/EN ISO 14044, 2008). The steps to perform the assessment according to the ISO standard involve, for example, a goal and scope definition, and the definition of the fuel unit (called functional unit, e.g., the amount of hydrogen to fill a fleet of 1000 vehicles) that is followed through the different stages of the life cycle of the fuel, as shown in Fig. 16.4. Each stage will have an environmental impact due to the respective processes, amount of energy, and materials used at each stage. This is followed by establishing a comprehensive inventory for all the materials going into and out of the stages and the energies used. Having established all that, a high-level environmental impact assessment for potential adverse effects on the ecosystem and humans is performed, resulting in scores for defined categories and an overall score. For each step, an interpretation of the results is done. Similarly, the LCC method as described by Hunkeler et al. (2008) models the costs of the stages. The calculation applies the same model that was used for the LCA (Fig. 16.4).

The functional model (Fig. 16.3) and the LCA/LCC model may be structured in the same way to ensure model compatibility. In that way, the functions F1 to F4 are directly comparable to the stages 1–4 of the LCA/LCC (Table 16.5).

16.4 DYNAMIC RISK ANALYSIS

To ensure proper safety management, qualitative and quantitative risk assessments are powerful and widely accepted decision support methods to predict

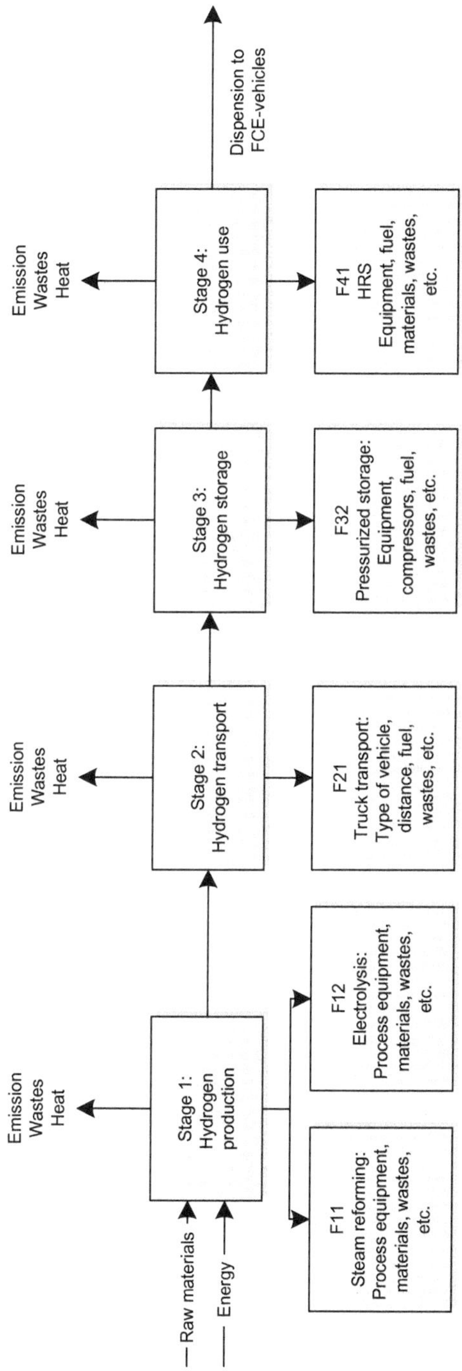

FIG. 16.4 LCA/LCC assessment for a fueling system: Stages and indication of detailed input.

TABLE 16.5 Example of the Combination of Hazard Identification and LCA

Function			Hazard Identification (SA Part)					Inventory (LCA Part)		
Ref	T	Method	k	Keyword	Main Deviation	Consequence	Mitigation	Material	Unit	Quantity
F12	M	Water electrolysis		Chemicals: corrosion	Release → Fire	Heat radiation on equipment	ATEX	Water	Kg	X
								Wind power	kWh	Y

the safety level of processes. Hereunder, application of event and fault trees are very common and widely accepted tools. In this way, the risk analyst records and evaluates the events' probability, their consequences, and final impacts.

The overall objective of a quantified risk assessment (QRA) is to evaluate an installation's safety level based on acceptance criteria, for example, the number of people that potentially may be exposed to intolerable risk levels. The risk level is established by identifying the most important, representative events that contribute to safety risk. The risk evaluation is also used to ensure that the overall safety risk is effectively and efficiently reduced to an accepted risk level by implementing preventive, or at least mitigating, measures.

The widely accepted state-of-the-art risk assessment techniques use static methods to analyze the systems, as (static) fault trees and event trees, for example. Time-dependent input parameters are applied as averages over a period, for example, average failure rates over the installation's lifetime, average ignition probabilities, average numbers of workers, and average escape route distances. This includes also weather data and the process conditions, using initial release data, for example.

In conducting a QRA for complex accident scenarios, some simplifications are made in order to regard events as classes of scenarios that can be treated in a homogeneous way. To calculate the total risk, the combined outputs for these representative scenarios (the consequences and likelihood) may be mapped to a single parameter, the risk indicator. An example is the F-N curve describing the multifaceted aspect of "consequence" to a number of fatalities or financial damage in the form of a cumulative probability distribution. Alternative ways of reporting the QRA output can be by using the concept of individual risk, the location based risk, or the fatal accident rate. Thus, a QRA applies a set of linked models describing possible events and their outcomes. The outcome of the QRA is determined through the models and the way they are linked.

Analyzing complex and dynamic systems, in which the occurrences of concurrent events may be mutually dependent, this static approach is challenged. Therefore, it has been suggested to use a dynamic risk assessment approach (Markert et al., 2017). The mentioned simplifications used to simplify static risk assessment methods are not sufficient to avoid very complex event trees as an outcome for such systems. This complexity in the outcome causes difficulties in comprehensively analyzing and using the assessment for improvements. Moreover, the simplifications do not capture the dynamic nature of the system in a fully convenient manner, leading to conservative assumptions to avoid underestimation of the risks.

The term "dynamic risk assessment," though, provides some ambiguity in the definitions used by different authors. Hakobyan et al. (2008) identified three different interpretations in reviewing the literature. These are:

1. Methods for periodic updates of a probabilistic risk analysis (PRA) to address any changes in a plant configuration (Villa et al., 2015);

2. Updates to account for the aging of equipment;
3. Approaches that include explicit deterministic modeling of dynamic processes combined with stochastic modeling to describe a systems evolution.

The third definition introduces time-dependent variables to describe the plants states when establishing an event tree. It is this third understanding that is applied in the following discussion of methods for dynamic risk assessment. An example is given of modeling a complex system simulating the dynamic interactions (see Figs. 16.5–16.7) of concurrent phenomena and procedures following a loss of containment situation. These are:

- The physical processes (outflow, dispersion, ignition, heat radiation, explosion);
- Detection, alarming, and emergency shutdown;
- Escape and evacuation; and
- Impact on persons, escalation, and impairment of safety functions.

The simulation model, as shown in Fig. 16.6, runs a large number of loss-of-containment scenarios to evaluate the associated stochastic events in time with random delays, durations, instances of occurrences, and others. The output datasets are collected for all the simulated scenarios and are stored in a database. Therefore, these data are accessible for further statistical processing of the results. They may be used to predict the important risk indicators, such as the individual fatality risk (IR), the potential loss of life (PLL), the fatal accident rate (FAR, at workplace level), and the group risk (distribution of number of simultaneous fatalities). The approach also provides a new possibility to find

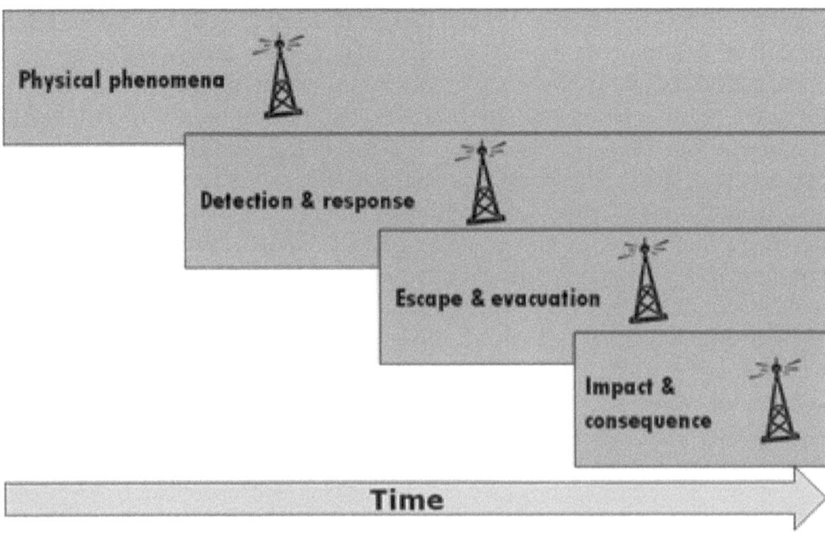

FIG. 16.5 Time dependence of events.

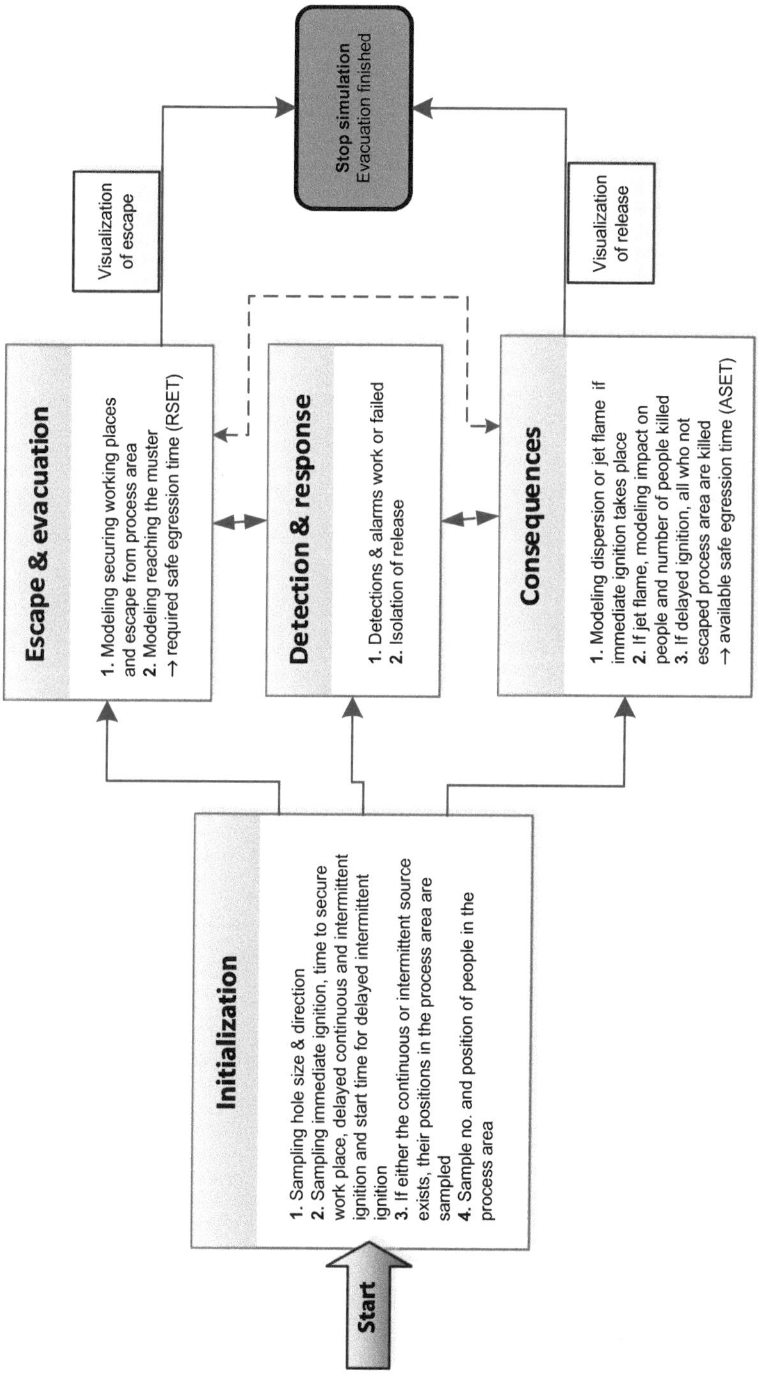

FIG. 16.6 DES model logic.

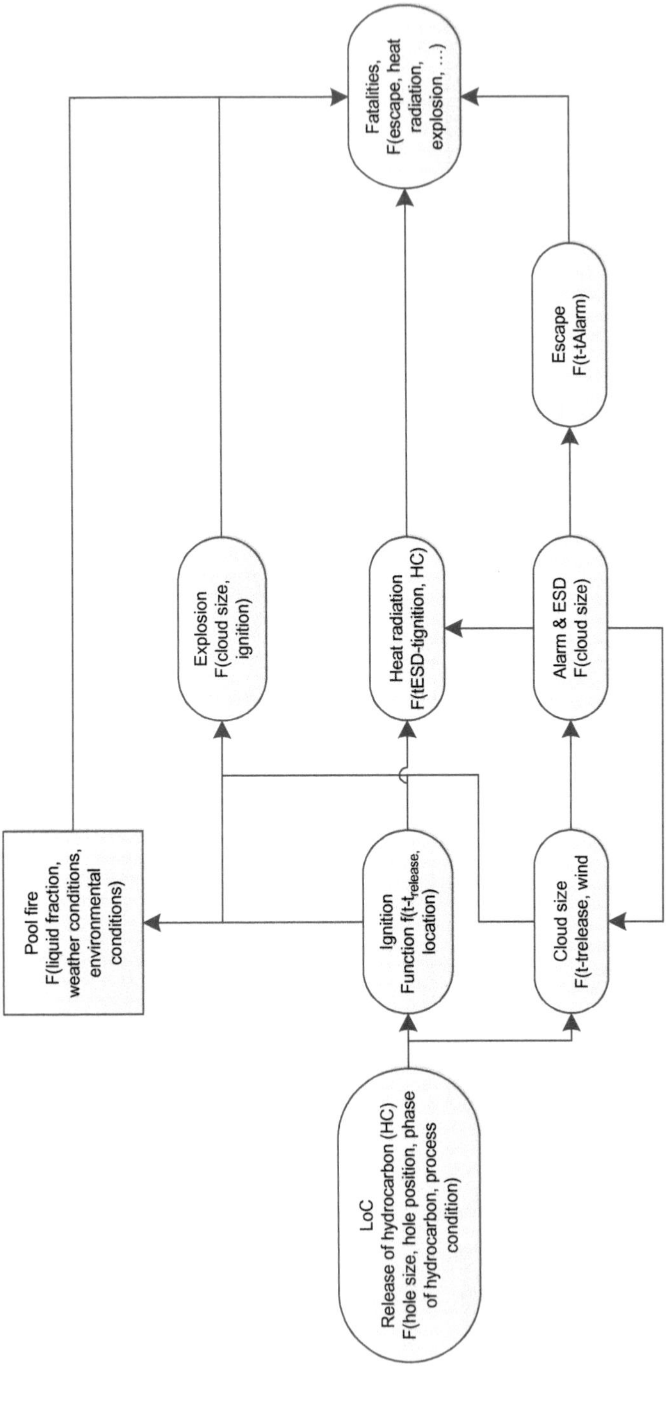

FIG. 16.7 Indication of interdependencies used in the model. The ones in the *squared box* are still optional.

the worst-case scenarios resulting from the very many combinations of parameters and events using the database. The latter is not possible with static risk assessment methods.

This approach allows capturing the specific characteristics of different workplaces, dynamic change of people's responses, and other characteristics. Scenarios with severe consequences can be "played back" to learn from them and can be animated, which, in addition to the learning effect, provides a new method of validation. This also makes the simulation models a good communication tool between system analysts and domain experts.

16.5 HSC MODELING INCLUDING SAFETY RISK

This section describes an example in which HCS and safety risk modeling are combined in order to combine supply chain modeling and dynamic risk assessment simulations. An example is presented with two distinct interconnected models: (1) for the HSC; and (2) for the consequences of a hydrogen release (Belamaric, 2016). The modeling and simulation uses the SimEvents commercial software package for discrete event simulation (DES) provided with the MatLab environment.

This case study assumes a large-scale hydrogen production plant located in a rural area with compressed hydrogen stored onsite in a high-pressure storage tank. On demand, the pressurized hydrogen is transferred to a tube trailer that transports it to a hydrogen refueling station located in a town. At the refueling station, the hydrogen is unloaded and further compressed to be stored in a high-pressure storage tank onsite. On demand, new hydrogen is ordered from the production plant when the amount of stored hydrogen runs low after being dispensed to a number of hydrogen vehicles.

The large-scale production plant produces hydrogen continuously. A part of the production will be delivered to the refueling stations in the HSC. This hydrogen is compressed and stored in high-pressure vessels. Another industry plant buys the major part of the hydrogen production. It is connected through a pipeline system.

A loading gate is located next to a high-pressure vessel that allows hydrogen loading in tube trailers. After arriving at the refueling station's unloading gate, hydrogen is transferred to the high-pressure storage.

The model assumes a varying customer arrival rate at the refueling station dependent on the weekday as well as a varying hydrogen refill amount.

The model structure is made up of four main submodels, namely submodels for: (1) hydrogen storage in the large-scale production plant; (2) hydrogen transportation and storage in the refueling station; (3) hydrogen dispensing to hydrogen cars in the refueling station; and (4) equipment reliability. For submodel 4, it is assumed that all the equipment has a lifetime of 25 years. For securing a realistic outcome, the model parameters for the HSC model are calibrated using data provided by the NREL (2015) that were collected from real observations

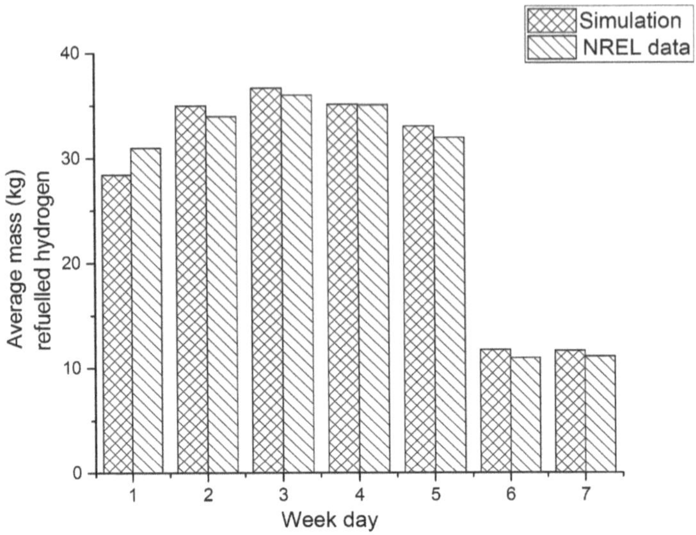

FIG. 16.8 Comparison of simulation results with real-world NREL data.

made on operational hydrogen refueling stations in the United States over the period from 2009 to 2015. Fig. 16.8 shows that the simulation results are in excellent agreement with the real data.

The simulation of a realistic customer demand allows the modeling of the necessary amounts of hydrogen stored and transported within the designed supply chain. Based on the simulation results for the specific scenario with a limited number of hydrogen vehicles, a single dispenser, and storage tanks with a capacity of 350 kg, the hydrogen at the production plant and the refueling stations is enough to meet the customer demand. The mean amount of hydrogen dispensed per day is about 27 kg averaged over a week (see Fig. 16.8).

Jaramillo (2014) describes an algorithm to simulate equipment failure and repair times using SimEvents. The same logic is adapted for this case in submodel 4 and is constituted by seven loops in which each loop simulates the availability of a specific component with their specific mean-time-to failure (MTTF) and mean time to repair (MTTR) parameters, as listed in Table 16.6. One HSC simulation with this model covers a period of 25 years as shown in Fig. 16.9. The simulation starts on a Monday at 00:00 a.m. Looking at the refilling demand of the customers, a uniform refilling rate is observed during the first 5 days. This is due to the uniform intergeneration time of customers between Mondays and Fridays. Then, between day 6 and 7, the rate of vehicles refilled decreases. This corresponds to the weekend, when demand is lower. Then at the start of the second week, the rate of vehicles refilled increases again. This pattern is repeated week by week. It may be seen in Fig. 16.9 that during the days 11 and 23, the rate curve of vehicles refilled is distorted for a

TABLE 16.6 MTBF and MTTR of the Supply Chain's Equipment

Equipment	MTBF (days)	MTTR (h)
Compressor	13	10
Storage	19	2
Loading and unloading gates	25	14
Dispenser	48	6

Adapted from NREL, 2015. Hydrogen fueling infrastructure analysis. Available from: https://www.nrel.gov/hydrogen/hydrogen-infrastructure-analysis.html [Accessed August 8, 2017].

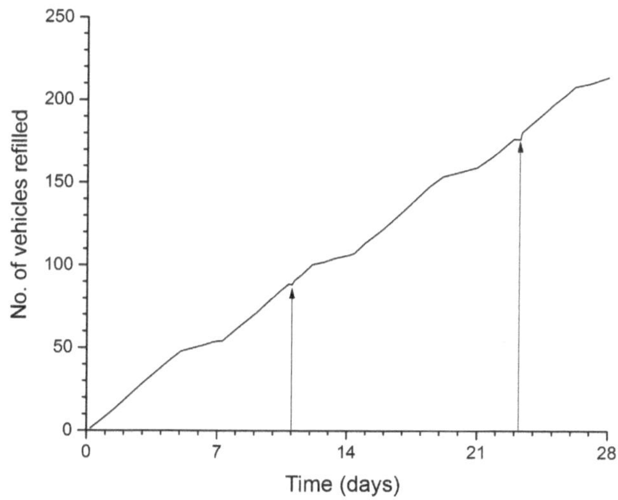

FIG. 16.9 Refuelling of hydrogen cars simulation. The *arrows* indicate reliability issues caused by MTTF and MTTR.

short period. This is due to failures that occurred at the dispenser and at the storage tank of the refueling station, preventing hydrogen dispensing to vehicles.

16.5.1 Risks Analyzed

The initiating event assumed in the case study is a hydrogen release caused by a hole in a high-pressure storage tank, which can be located at different locations within the HSC, for example, at the production plant or at the refueling station. An accident, such as a hydrogen release, may develop into different consequence scenarios as indicated in Fig. 16.10. The release scenarios may result in jet fires, flash fires, or explosion (detonation or deflagration) (Cadwallader

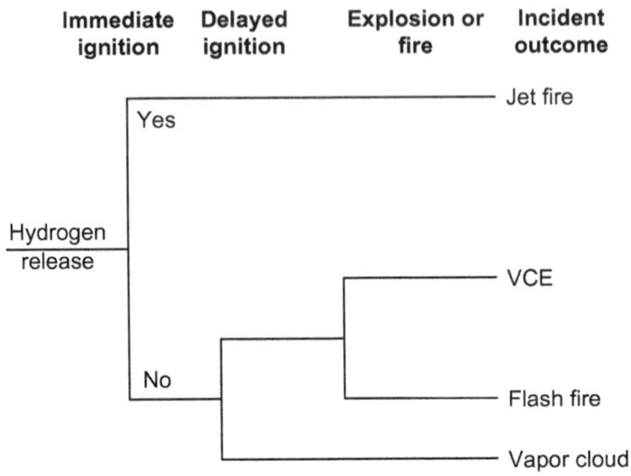

FIG. 16.10 Event tree for gaseous hydrogen release through a hole of a pressurized tank. *(Adapted from Moosemiller, M., 2011. Development of algorithms for predicting ignition probabilities and explosion frequencies. J. Loss Prev. Process Ind. 24(3), 259–265.)*

and Herring, 1999; Moosemiller, 2011). Possible accidents are modeled using the event tree shown in Fig. 16.10. The International Association of Oil and Gas Producers (OGP) (2010) suggests the use of the frequencies presented in Table 16.7 for each range of hole diameters. Estimation of the discharge rate is a crucial step in the evaluation of the consequences analysis.

In order to estimate immediate and delayed ignition in the consequence model according to the event tree as shown in Fig. 16.10, the following probabilities are used: (1) immediate ignition = 0.15; (2) delayed ignition = 0.30; and (3) explosion = 0.40 (Moosemiller, 2011). The general model developed for the case study has four submodels: (1) general parameters; (2) jet fire consequences; (3) VCE consequences; and (4) flash fire (explosion) consequences. One DES entity representing an initiating event is created and receives a number of attributes, as listed below. On demand the attributes can easily be extended if other scenarios are to be calculated.

TABLE 16.7 Pressure Vessel Hole Frequencies (OGP, 2010)

Hole Diameter (mm)	Leak Frequency (Per Vessel Year)
1–3	2.3×10^{-5}
3–10	1.2×10^{-5}
10–50	7.1×10^{-6}
50–150	4.3×10^{-6}
Total	4.6×10^{-5}

- Attribute 1: Hole diameter;
- Attribute 2: Immediate ignition;
- Attribute 3: Delayed ignition;
- Attribute 4: Explosion or fire;
- Attribute 5: Time before ignition;
- Attribute 6: Hole height;
- Attribute 7: People in zone 2;
- Attribute 8: People in zone 3 (rural area); and
- Attribute 9: People in zone 3 (urban area).

In this way, each entity receives a unique set of parameters and will trigger a specific set of consequence models to estimate the outcome of a specific scenario. In order to estimate the fatalities for each accident, the number of people around the facility is generated and assigned to the entity. Three zones around the location of the initiating event are considered, as indicated in Fig. 16.11. The first zone corresponds to the area occupied by the storage vessel with a radius of 2.2 m. This zone is, to a large extent, occupied by the equipment and only a very few persons may conduct maintenance work for a few hours during a year. Zone 2 represents the safety area around the vessel. Main safety distances around a hydrogen storage tank are defined by codes and regulations, such as the National Fire Protection Association (NFPA 50A from 1999) or the California Energy Commission (Venkatesh et al., 2004). These distances vary from 4.6 to 15 m and therefore a mean safety distance of 8 m is assumed in this study. Therefore, Zone 2 corresponds to a circular area with an outer radius of 10.2 m,

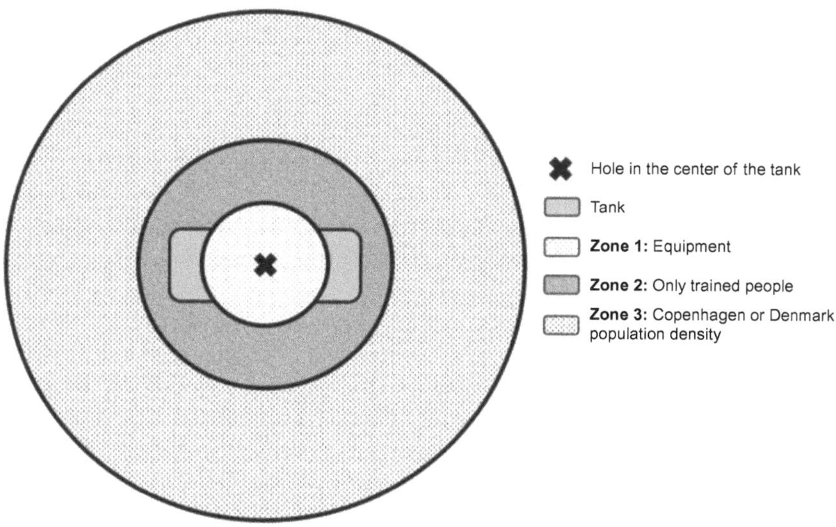

FIG. 16.11 Introduction of zones with different population densities around the LoC (not to scale).

excluding the area of zone 1. The case study assumes between 0 and 4 people may be located in zone 2. Finally, zone 3 assumes a population density typical for the specific wider area for simplicity. The simulation operates here with two scenarios, an urban population and a rural one.

Analyzing the consequences of a continuous hydrogen release through a hole in a high-pressure storage tank, this analysis can now incorporate specific states of the storage or the HSC. Therefore, based on the dynamic inventory data, different risk scenarios may be assessed and the individual and societal risks can be calculated not only for the individual parts of the HCS, but also for the whole chain. In conclusion, the DES models built in this study using SimEvents are powerful tools that allow the analysis of complex systems involving stochastic variables. The main limitation resides in the availability of data. As more data from hydrogen supply chains become available in the future, such models as described in this case study can be improved and provide reasonable prediction of security of supply and safety risk scenarios.

16.6 CONCLUSIONS

This chapter describes the challenges to assess the safety risks of complex hydrogen infrastructure together with possible methodologies to address these challenges. The development of new large-scale infrastructure is a stepwise procedure and it needs different decision support tools, such as cost-benefit assessments, sustainability assessments, optimization of supply chains, the best placements of buildings and process equipment in a growing market, and, last but not least, safety risk assessment and management.

It is argued that complex systems may behave differently, as the overall safety risks are not just the sum of risks from its components. Therefore, a holistic approach is needed, which provides a challenge for any analyst. Some methods are indicated that could help ensure that the different methods for decision support regards the same HSC and that may support comprehensive data collection. For dynamic systems a new modeling approach is suggested that enables the modeling of the time dependence of events. The dynamic approach may also be combined with DES-based supply chain modeling.

The suggestion is to establish basic models based on the functional method, which may store the input data, the set of assumptions, and the system model. The system model may be enlarged in parallel with the growing maturity of the system. Functional modeling can model a supply chain down to small details, depending on the actual state of knowledge. Therefore, it is applicable for developing systems, as applicable already in early design phases. With this as a basis, data important for the system may be stored and updated in a structured way. The data should be stored on a centralized server, and cloud solutions seem to be promising to ensure that each of the assessments uses the correct, updated data, and ensures improved comparability of the assessments, as they will look at the same design stage and the same model.

ACKNOWLEDGMENTS

The research of the different topics described here has been conducted over some years. The authors acknowledge the excellent collaboration with our colleagues world wide contributing to parts of the research as well as the many excellent discussions and feedback from our colleagues at the university. Special thanks to the engagement of our master thesis students who worked in the field and the feedback from our external colleagues. All this was very valuable for us.

REFERENCES

Agnolucci, P., 2007. Hydrogen infrastructure for the transport sector. Int. J. Hydrogen Energy 32 (15), 3526–3544. Available from: http://www.sciencedirect.com/science/article/B6V3F-4NDDSX9-4/2/31cb8a6075121af3e83cd661f3babf43.

Almansoori, A., Shah, N., 2006. Design and operation of a future hydrogen supply chain: snapshot model. Chem. Eng. Res. Des. 84 (6), 423–438.

Andrews, J., Shabani, B., 2012. Re-envisioning the role of hydrogen in a sustainable energy economy. Int. J. Hydrogen Energy 37 (2), 1184–1203. Available from: http://www.sciencedirect.com.globalproxy.cvt.dk/science/article/pii/S0360319911022841.

Anonymous, 2017a. The grenfell tower fire. In: Wikipedia—The Free Encyclopedia. Available from: https://en.wikipedia.org/wiki/Grenfell_Tower_fire. (Accessed 6 August 2017).

Anonymous, 2017b. Town and city fires. In: Wikipedia—The Free Encyclopedia.Available from: https://en.wikipedia.org/wiki/List_of_fires#17th_century. (Accessed 6 August 2017).

Aven, T., Reniers, G., 2013. How to define and interpret a probability in a risk and safety setting. Safety Sci. 51 (1), 223–231.

Baesi, S., et al., 2013. Application of a multi-plant QRA: a case study investigating the risk impact of the construction of a new plant on an existing chemical plant's risk levels. J. Loss Prev. Process Ind. 26 (5), 895–903. Available from: https://doi.org/10.1016/j.jlp.2012.11.005.

Bedford, T., Cooke, R.M., 2001. Probabilistic Risk Analysis-Foundations and Methods. Cambridge University Press, Cambridge.

Belamaric, A.S., 2016. Assessment of a Hydrogen Supply Chain and Evaluation of the Societal Risk Using Discrete Event Simulation. EPFL & DTU.

Cadwallader, L.C., Herring, J.S., 1999. Safety issues with hydrogen as a vehicle fuel. *Contract* (September), pp. 1–13. Available from: http://inl.gov/hydrogenfuels/projects/docs/h2safetyreport.pdf.

Caputo, A.C., Pelagagge, P.M., Salini, P., 2011. Impact of accidents risk on hydrogen road transportation cost. Int. J. Energy Sect. Manage. 5 (2), 215–241. Available from: http://www.scopus.com/inward/record.url?eid=2-s2.0-79960064442&partnerID=40&md5=3e9680b6bcab10e3115a2cc48f107138.

Dayhim, M., Jafari, M.A., Mazurek, M., 2014. Planning sustainable hydrogen supply chain infrastructure with uncertain demand. Int. J. Hydrogen Energy 39 (13), 6789–6801. Available from: https://doi.org/10.1016/j.ijhydene.2014.02.132.

De-León Almaraz, S., et al., 2013. Assessment of mono and multi-objective optimization to design a hydrogen supply chain. Int. J. Hydrogen Energy 38 (33), 14121–14145. Available from: http://linkinghub.elsevier.com/retrieve/pii/S0360319913018065. (Accessed 3 May 2017).

De-León Almaraz, S., et al., 2014. Hydrogen supply chain optimization for deployment scenarios in the Midi-Pyrénées region, France. Int. J. Hydrogen Energy 39 (23), 11831–11845. Available from: http://www.sciencedirect.com/science/article/pii/S0360319914015778. (Accessed 3 May 2017).

De-León Almaraz, S., et al., 2015. Deployment of a hydrogen supply chain by multi-objective/multi-period optimisation at regional and national scales. Chem. Eng. Res. Des. 104, 11–31. Available from: http://www.sciencedirect.com/science/article/pii/S0263876215002506. (Accessed 31 March 2017).

Drennen, T.E., Rosthal, J.E., 2007. Pathways to a Hydrogen Future. Elsevier, Amsterdam.

DS/EN ISO 14001, 2015. Environmental management systems—requirements with guidance for use. Dansk standard. Available from: https://sd-ds-dk.proxy.findit.dtu.dk/.

DS/EN ISO 14040, 2008. Environmental management —life cycle assessment—principles and framework. Dansk standard. Available from: https://sd-ds-dk.proxy.findit.dtu.dk/.

DS/EN ISO 14044, 2008. Environmental management —life cycle assessment—requirements and guidelines. Dansk standard. Available from: https://sd-ds-dk.proxy.findit.dtu.dk/.

Duijm, N.J., 2009. Safety-barrier diagrams as a safety management tool. Reliab. Eng. Syst. Saf. 94 (2), 332–341. Available from: http://www.sciencedirect.com/science/article/B6V4T-4S85DF0-3/2/d0def3350fb579a159cf428f9b7770d4.

Duijm, N.J., Markert, F., 2009. Safety-barrier diagrams as a tool for modelling safety of hydrogen applications. Int. J. Hydrogen Energy 34 (14), 5862–5868. Available from: http://www.sciencedirect.com.globalproxy.cvt.dk/science/article/B6V3F-4VV1B5D-2/2/6f6969abe95b48c795be13a8a12abdc4.

Haimes, Y.Y., 2018. Risk modeling of interdependent complex systems of systems: theory and practice. Risk Anal. 38 (1), 84–98. https://doi.org/10.1111/risa.12804.

Hakobyan, A., et al., 2008. Dynamic generation of accident progression event trees. Nucl. Eng. Des. 238 (12), 3457–3467. Available from: http://www.sciencedirect.com/science/article/pii/S0029549308004470.

Hunkeler, D., Lichtenvort, K., Rebitzer, G., 2008. Environmental Life Cycle Costing. CRC Press SETAC, Pensacola, FL.

IEC, 2001. Hazard and operability studies (HAZOP studies)—application guide. IEC 61882:2001IEC.

IEC, 2006. CENELEC, (Ed.), Analysis Techniques for System Reliability—Procedure for Failure Mode and Effect Analysis (FMEA). CENELEC, Brussels.

International Association of Oil & Gas Producers (OGP), March 2010. Risk Assessment Data Directory—Storage incident frequencies, Report No. 434-3. March 2010.

ISO, 2010. IEC/ISO 31010:2009: Risk management—risk assessment techniques. I. E. C. T. C. 56, Ed., 2010E.

Jaramillo, M.A., 2014. Optimising the rating of offshore and onshore transformers for an offshore wind farm. Wind Energy Master Report.

Joffe, D., Hart, D., Bauen, A., 2004. Modelling of hydrogen infrastructure for vehicle refuelling in London. J. Power Sources 131 (1), 13–22.

Kim, J., Moon, I., 2008. Strategic design of hydrogen infrastructure considering cost and safety using multiobjective optimization. Int. J. Hydrogen Energy 33 (21), 5887–5896. Available from: http://www.sciencedirect.com.globalproxy.cvt.dk/science/article/pii/S0360319908008434.

Kim, J., Lee, Y., Moon, I., 2008. Optimization of a hydrogen supply chain under demand uncertainty. Int. J. Hydrogen Energy 33 (18), 4715–4729.

LaChance, J., Tchouvelev, A., Ohi, J., 2009. Risk-informed process and tools for permitting hydrogen fueling stations. Int. J. Hydrogen Energy 34 (14), 5855–5861. Available from: http://www.sciencedirect.com/science/article/B6V3F-4W232BG-3/2/f36a577a75803139c406d9a5ff3b4e29.

Leveson, N., 2004. A new accident model for engineering safer systems. Safety Sci. 42 (4), 237–270.

Luis Aprea, J., 2008. New standard on safety for hydrogen systems in spanish: keys for understanding and use. Int. J. Hydrogen Energy 33 (13), 3526–3530. Available from: http://www.sciencedirect.com/science/article/B6V3F-4S9FH1G-2/2/33b098f4d05e0804500010561be7389e.

MacIntyre, I., et al., 2007. Canadian hydrogen safety program. Int. J. Hydrogen Energy 32 (13), 2134–2143. Available from: http://www.sciencedirect.com/science/article/B6V3F-4NVC8VP-2/2/f4b9b6f2cef059cbc28a2a1090f47ca0.

Marangon, A., et al., 2007. Safety distances: definition and values. Int. J. Hydrogen Energy 32 (13), 2192–2197. Available from: http://www.sciencedirect.com/science/article/B6V3F-4NX8RB4-1/2/86043f6a1d29f8afc6c381db0c3d5f02.

Markert, F., et al., 2007. Safety aspects of future infrastructure scenarios with hydrogen refuelling stations. Int. J. Hydrogen Energy 32 (13), 2227–2234. Available from: http://www.sciencedirect.com/science/article/B6V3F-4NTHMYB-8/2/f47d0cbe54201d237480aa786171817e.

Markert, F., Marangon, A., Carcassi, M., Duijm, N.J., 2017. Risk and sustainability analysis of complex hydrogen infrastructures. Int. J. Hydrogen Energy 42 (11), 7698–7706. https://doi.org/10.1016/j.ijhydene.2016.06.058.

Molag, M., Ham, K., Elbers, S., 2004. Safety assessment of the ammonia, chlorine and LPG product chains in the Netherlands.In 11th International Symposium Loss Prevention. Praha: Loss Prevention and Safety Promotion in the Process Industries. pp. 1101–1109.

Moosemiller, M., 2011. Development of algorithms for predicting ignition probabilities and explosion frequencies. J. Loss Prev. Process Ind. 24 (3), 259–265.

NREL, 2015. Hydrogen fueling infrastructure analysis. Available from: https://www.nrel.gov/hydrogen/hydrogen-infrastructure-analysis.html. (Accessed 8 August 2017).

Ogden, J.M., 1999. Prospects for building a hydrogen energy infrastructure. Annu. Rev. Energy Environ. 24 (1), 227–279. Available from: http://www.annualreviews.org/doi/10.1146/annurev.energy.24.1.227.

Ogden, J.M., Steinbugler, M.M., Kreutz, T.G., 1999. A comparison of hydrogen, methanol and gasoline as fuels for fuel cell vehicles: implications for vehicle design and infrastructure development. J. Power Sources 79 (2), 143–168.

Pasman, H., Reniers, G., 2014. Past, present and future of Quantitative Risk Assessment (QRA) and the incentive it obtained from Land-Use Planning (LUP). J. Loss Prev. Process Ind. 28, 2–9. Available from: http://linkinghub.elsevier.com/retrieve/pii/S095042301300079X. (Accessed 18 May 2016).

Rasmussen, J., 1997. Risk management in a dynamic society: a modelling problem. Safety Sci. 27 (2–3), 183–213. Available from: http://linkinghub.elsevier.com/retrieve/pii/S0925753597000520. (Accessed 7 August 2017).

Rasmussen, B., Grønberg, C.D., 1997. Accidents and risk control. J. Loss Prev. Process Ind. 10 (5–6), 325–332. Available from: http://linkinghub.elsevier.com/retrieve/pii/S0950423097000223. (Accessed 7 August 2017).

Rasmussen, B., Whetton, C., 1997. Hazard identification based on plant functional modelling. Reliab. Eng. Syst. Saf. 55, 77–84.

Rigina, O., 2002. Regional radiation risk and vulnerability assessment by integration of mathematical modelling and GIS analysis. Environ. Int. 27 (7), 527–540.

Thomas, J., Paul, A., Inaki, A., Daniele, B., Herve, B., Luc, B., Alain, B., Sile, B., Marco, C., Arief, D., Norbert, E., Angunn, E., Espen, F., Eduardo, G., Andrey, G., Erling, H., Marit, H.A., Petra, H.G., Stuart, H., Olaf, J., Armin, K., Alexei, K., Suresh, K., Gesa, L., Stefan, L., Alexander, L., Dmitriy, M., Alessia, M., Frank, M., Prankul, M., Vladimir, M.,

Sandra, N., Efthymia, P., Lionel, P., Ernst-Arendt, R., Ulrich, S., Pierre, S.-C., Michael, S., Aurelie, S., Andrzej, T., Delphine, T., Alexander, V., Karl, V., Nico, V., Ana, V., Manfred, W., Nikolay, Z., 2011. Achievements of the EC network of excellence HySafe. Int. J. Hydrogen Energy. 36(3). https://doi.org/10.1016/j.ijhydene.2010.05.087.

U.S. Department of Energy, 2016. Central versus distributed hydrogen production. Available from: https://energy.gov/eere/fuelcells/central-versus-distributed-hydrogen-production.

van den Heever, S.A., Grossmann, I.E., 2003. A strategy for the integration of production planning and reactive scheduling in the optimization of a hydrogen supply network. Comput. Chem. Eng. 27 (12), 1813–1839. Available from: http://www.sciencedirect.com/science/article/pii/S0098135403001583.

Venkatesh, S., et al., 2004. Failure modes and effects analysis for hydrogen fueling options. California Energy Commission, Sacramento, California.

Verter, V. & Kara, B.Y., 2001. A GIS-based framework for hazardous materials transport risk assessment. Risk Anal., 21(6), pp. 1109–1120. Available from: https://doi.org/10.1111/0272-4332.216179 Accessed 19 May 2016.

Vianello, C., Maschio, G., 2014. Quantitative risk assessment of the Italian gas distribution network. J. Loss Prev. Process Ind. 32, 5–17. Available from: http://www.sciencedirect.com/science/article/pii/S0950423014001132.

Villa, V., Paltrinieri, N., Cozzani, V., 2015. Overview on dynamic approaches to risk management in process facilities. Chem. Eng. Trans. 43, 2497.

Watanabe, S., Tamura, Y., Suzuki, J., 2007. The new facility for hydrogen and fuel cell vehicle safety evaluation. Int. J. Hydrogen Energy 32 (13), 2154–2161. Available from: http://www.sciencedirect.com/science/article/B6V3F-4NXGS7Y-3/2/57777bfb84363d9ef38dc3bbc4ebdb71.

Wells, G., Wardman, M., Whetton, C., 1993. Preliminary safety analysis. J. Loss Prev. Process Ind. 6 (1), 47–60. Available from: http://linkinghub.elsevier.com/retrieve/pii/095042309380019I. (Accessed 28 March 2016).

Index

Note: Page numbers followed by *f* indicate figures, *t* indicate tables, and *s* indicate schemes.

E

F

T

U

V

W

Kolibri 360 GmbH
Ferdinand-Jühlke-Straße 7
99095 Erfurt, Deutschland
produktsicherheit@kolibri360.de